Recent Trends in Biotechnology

nova
science publishers

www.novapublishers.com

Recent Trends in Biotechnology

Exploring the Biotechnological Potential of C2 Alternative Substrates in Biocatalysis and Metabolic Engineering
Wenfa Ng
2022. ISBN: 978-1-68507-555-2 (Hardcover)
2022. ISBN: 978-1-68507-601-6 (eBook)

Fundamentals of Medical Biotechnology
Anita Mishra, M.Sc, B. Manivannan, PhD,
M. Kasturi, PhD and Rajneesh Prajapat, PhD (Editors)
2021. ISBN: 978-1-53619-328-2 (Hardcover)
2021. ISBN: 978-1-53619-393-0 (eBook)

Microbes, Environment and Human Welfare
Tilak Saha and Bipransh Kumar Tiwary (Editors)
2020. ISBN: 978-1-53617-945-3 (Hardcover)
2020. ISBN: 978-1-53617-946-0 (eBook)

Applied Algal Biotechnology
Muthu Arumugam, PhD, S. Kathiresan, PhD
and S. Nagaraj (Editors)
2020. ISBN: 978-1-53617-524-0 (Hardcover)
2020. ISBN: 978-1-53617-525-7 (eBook)

High Throughput Screening: Methods, Techniques and Applications
Mark Wigglesworth, David Murray,
Geoffrey Holdgate and Carolyn J. Blackett (Editors)
2020. ISBN: 978-1-53617-248-5 (Hardcover)
2020. ISBN: 978-1-53617-253-9 (eBook)

More information about this series can be found at https://novapublishers.com/product-category/series/recent-trends-in-biotechnology/

Shivaji H. Pawar
Editor

Silk Fibroin

Advances in Applications and Research

nova
science publishers

www.novapublishers.com

DOI: 10.52305/BOCB5641.

Library of Congress Cataloging-in-Publication Data

ISBN: 979-8-88697-402-7

Published by Nova Science Publishers, Inc. † New York

Contents

Preface

This book entitled, *Silk Fibroin: Advances in Applications and Research*, is an extensive collection of learned materials and new results focusing on advances in silk fibroin for their applications by the contributing authors who are experts working in the fields of nanoscience, material science, energy, agricultural, computer science and engineering, atmospheric nanoscience, medicine, and nanobiotechnology.

Silks from both spiders and silkworms have emerged as useful materials for addressing new challenges in material science. These protein-based materials are often superior to many other polymers from a mechanical point of view and provide the possibility to realize composites with remarkable biological properties. Thus, studying raw silks, as well as silk-based natural and artificial composites, while controlling and manipulating processing conditions, is important to achieve advances in silk utility for materials science and technology. In doing this, a multidisciplinary and multiscale approach is crucial for studying silk and its derivatives. This article collection focuses on some of the latest advances in silk sciences and technology and topics, including:

- Introduction to resources, physico-chemical properties and blends of silk fibroin
- Non-mulberry silk fibroin blends
- Silk fibroin thin films
- Silk fibroin sponge as a cryo device combined with transporter and scaffold functions in cellular transplantation
- Silk fibroin-based nanocomposite for futuristic electronics
- Silk fibroin-based systems for conversion and storage of energy
- Silk fibroin as a functional biomaterial for tissue engineering
- Silk fibroin: A diagnostic tool for targeted drug delivery system
- Silk fibroin: Advancement in wound healing technologies
- Engineering of silk fibroin bio-nanocomposites: A glimpse of advanced research on innovative formulations for potential applications

This book will be a precious piece and basic knowledge material for those looking for new opportunities in the field of progress and prospects of silk science and technology. I hope the readers will find them interesting and enlightening.

Prof. Dr. S.H. Pawar
Editor

Acknowledgments

The editor is thankful to Nova Science Publisher Inc., USA, for publishing this book. The editor and contributing authors of this book are especially indebted to the publishers of scientific journals and books and various authors for freely granting permission to reproduce their copyrighted works.

The editor would like to thank all authors for contributing their chapters as well as the excellent research work published by them. Dr. V. S. Mohite, my colleague in the Department of Physics, T.C. College, Baramati is gratefully acknowledged for fruitful discussions and technical support. The inspiration, fruitful discussions and strong support rendered by Principal Dr. Chandrashekhar V. Murumkar, Tuljaram Chaturchand College, Baramati and Prof. Dr. R. K. Mudgal, Vice-Chancellor, D. Y. Patil Education Society, Deemed to be University, Kolhapur are noteworthy to mention, without which this book could not have been brought into reality. The inspiration and strong support rendered by Dr. Swpnil Pawar, Dr. Miss. Aparna S. Pawar Ghalame and Dr. Santosh Ghalame, Baramati are noteworthy to mention, without which this book would not be possible.

Prof. Dr. S. H. Pawar
Editor

Chapter 1

Introduction to Resources, Physico-Chemical Properties and Blends of Silk Fibroin

Sunetra A. Deshmukh[1] and Shivaji H. Pawar[2,3,*]
[1] Department of Chemistry, School of Chemical Sciences,
Sanjay Ghodawat University, Atigre, Kolhapur, Maharashtra, India
[2] Center for Innovative and Applied Research (CIAR),
T. C. College, Baramati, Maharashtra, India
[3] Center for Interdisciplinary Research,
D. Y. Patil University, Kolhapur, Maharashtra, India

Abstract

This chapter views physical, chemical, mechanical, thermal and optical properties of silk fibroin of *Bombyx mori*. All these unique properties attract attention of scientists and researchers from all over the world. The chapter begins with the origin of silk, diseases of silk, types, and extraction of silk fibre. The primary and secondary structures of silk fibroin are discussed in detail under the section on physical properties. In addition, this chapter surveys the blends of silk fibroin with other biopolymers like chitosan, hyaluronic acid, collagen, alginate, synthetic polymers which are cross-linked with inorganic materials. Moreover, it discusses the summary of different types of blends and their usage in biomedical applications like tissue engineering, wound healing, regenerative medicine and drug delivery.

1. Introduction

Silk Fibroin is a fine product of the sericulture industries, one of the oldest industries in India and Asia. Sericulture is the agro-cottage, forest-based, labour intensive and commercially attractive economic activity falling under the cottage and small-scale sector. It particularly suits rural population in agriculture, entrepreneurs, and artisans that require low investment with potential for relatively higher returns. It provides income and employment to rural and poor farmers with small land-holding capacity or weaker sections of society. As a result, Indian government formulated the Central Board of Sericulture to nurture these industries

* Corresponding Author's Email: shpawar1946@gmail.com.

In: Silk Fibroin
Editor: Shivaji H. Pawar
ISBN: 979-8-88697-402-7

in 1948. Over the years, it has established nine research centres across the country to develop quality silk fibroins. However, these institutes and sericulture industries have not gone beyond major applications in silk fabrics. In recent years, however, different varieties of silk fibroin application have emerged in combination with nanoscience and nanotechnology which are leading to bionanotechnology of silk fibroin. Considering these developments, this chapter throws light on fundamental aspects of silk fibroin specifically on their resources, physicochemical properties, different types of blends and their applications.

Silk is made up of protein biopolymer of repetitive units of amino acids, which are hydrophilic and hydrophobic [1]. The chemical constituents of silk determine the nature, strength and properties of silk fibroin. The different anthropods such as silkworms, spiders, scorpions, mites and fleas spun the silk fibre [1]. However, the silk produced by mulberry silkworms is superior to the primary source of manufacturing silk in textile industries. Also, *Bombyx mori* silkworm is well suited to biomedical applications due to chemical and physical properties. Domestic silkworms are reared commercially before adult moths emerge. The pupae are killed by dipping into boiling water, which allows the whole cocoon that can be unraveled in one continuous thread. Silk fibroin of *Bombyx mori* possesses outstanding properties like mechanical, biocompatibility, and controlled biodegradation which make it highly valuable for innovative applications. Recent biomedical applications of silk-based materials are used for bone regeneration, gene therapy, wound healing, small molecule drug delivery and skin regeneration [1]. The different compositions of silk fibroin with natural/synthetic polymers are discussed from their perspective for future developments.

2. Resources of Silk Fibroin

2.1. History of Silk Fibroin

The word silk originates from ancient English: sioloc; from ancient Greek: σηρικός; Romanized: sērikós ("silken"); Mandarin: sī ("silk"). The biological scientific name of the silkworm is *Bombyx mori* (In Latin, mulberry Silkmoth). If we look at the history of silk in the Neolithic period, Republic of China is the first country in silk production from the last half of the first millennium BC. At that time, Republic of China was dominated by the luxury silk market. Later, silk cultivation was spread to Japan around 300AD. Then after to Arabian countries, Western Europe started production of silk which resulted in reducing the Chinese silk exports. In France, the epidemic diseases decreased silk production between 1865 to 1870. Later, Louis Pasteur discovered that disease was spread by two separate infections which were caused by parasites. After isolation and elimination of infected silkworms, disease was eliminated, and the French silk industry flourished again. Due to innovations in textile industries, it is found that spin cotton has become popular and much cheaper in production. Silk thread is a protein fibre while cotton and jute yarn is cellulose. Emergence of new fabrics like nylon, a synthetic fabric has reduced the prevalence of silk worldwide. In the twentieth century, Japan and China regained their former roles in silk production and at present time the Republic of China is the world's largest silk producer. India is the only country to produce all five known commercial varieties of silk (Mulberry, Tropical Tasar, Oak Tasar, Eri, and Coral). China is the largest silk producer, which produces 79.1% of the total whole worldwide silk production. Second largest producer is India after China in the distribution of silk production worldwide. India has 17.5% share in the whole raw silk production worldwide. Many other countries are also engaged in raw silk production including Japan, Brazil, Korean Republic, Uzbekistan, Thailand and Vietnam to name but a few [1].

2.2. Silk Fibroin Sources

Bombyx mori cocoons and *Nephilaclavipes* webs are the most common sources of silk fibroin. *Bombyx mori* (domestic silk moth) belongs to the moth family of *Bombycidae*. Cocoon Silk Mulberry silkworm (*Bombyx mori*) is well-known for production of silk throughout the world. However, the cultivation of silkworms from *Bombyx mori* is known as sericulture. It takes around 2,500 silkworms to produce one pound of raw silk. Due to well-established sericulture process, per annum about forty thousand tons of dry silkworm *Bombyx mori* cocoons are made available worldwide for different usages like textile industries and biomedical applications.

2.3. The Life Cycle of Silkworms

Holometabolism, also called complete metamorphosis, is a form of insect development, which includes the four life stages in the cycle of silkworm. These are: egg (ovum), larva, pupa, and imago (or adult/moth). To complete one cycle from egg to adult (moth) needs 50 to 60 days. Female silkworm after every sexual contact insulates 300-400 eggs on the mulberry leaves. In the first ten days, eggs produce small female insects, called larvae also known as caterpillars. This process of incubation is carried out in a controlled environment until they hatch into larvae. The *Bombyx mori* silkworms feed on mulberry leaves for six weeks to grow to their full potential. Once silkworms are fully grown, they stop eating and are ready to spin their cocoons. Insect spins continuously for the next three days his head round it and forms the silk solution of single long thread and spins around protective case or shell called cocoon or Koya. Salivary glands located in the head of caterpillar transform liquid silk from fully grown silkworm to the physical threads made up of proteins. The thread solidifies when it is exposed to air. The length of thread is 1000 meters long and after this, larvae are converted into pupae [2]. The worm stays protected inside the cocoon from harsh and unfriendly weather conditions without food and water throughout the winter. This stage of the silkworm considered as sleeping stage. After 12-15 days, pupa becomes silkworm, by alkaline secretion cuts the cocoon from one side and comes out (Figure 1). Due to this silk threads are broken into small pieces. To obtain good quality silk thread, the cocoons are boiled in water. As a result, the silkworm dies inside the shell and, then the threads are segregated from cocoon.

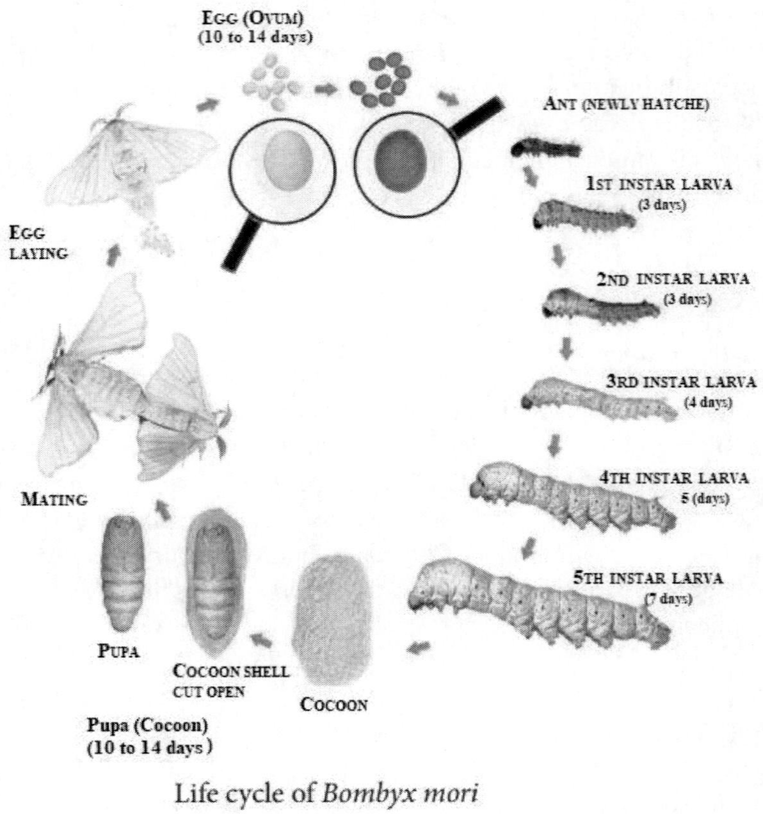

Figure 1. Life cycle of the silkworm under temperatures ranging from 25 to 27°C [2].

2.4. Silk Producing Insects

Silk produced by insects consists of fibrous protein; formed from water-insoluble fibre which exhibits high strength and toughness. The different glands of insects produce silk such as collateral, salivary, dermal glands and malpighian tubules. The larvae of the insects, which undergo complete metamorphosis, produce silk. However, the only silk produced by moth caterpillars is used for textile manufacturing. Apart from the pet *Bombyx mori* other insects such as web spinners and raspy cricket have lifelong silk [3].

Silk is also produced by Hymenoptera (bees, seeds, and ants), Silverfish, Mayflies, Thrips, Leafhoppers, Beetles, Lacewings, Flies and Midge. Other types of arthropods especially arachnids like spiders also produce silk. Variety of insect species, most of which are used commercially for silk are listed below:

- Honeybee and bumblebee larvae
- Bulldog ants spin cocoons
- Weaver ants
- Hornets
- Silverfish
- Mayflies

- Thrips
- Leafhoppers
- Beetles
- Lacewings
- Fleas
- Flies
- Midges
- Parasitic wasps

2.5. Production of Silk

The production of silk does not require harsh processing conditions [4]. Under the natural condition moth cut the outside layer, koya from one side and comes out from the other side. Hence, it is difficult to get continuous reeling of silk. In silk production, yarn is reeled by sequence of operations. There are different processes involved:

1. Stifling
2. Sorting
3. Deflossing
4. Reeling
5. Spinning

During "Stifling" process pupa is usually killed in the cocoon by using steam or hot air before it comes out of cocoon naturally. Then cocoons are sorted out according to their quality, colour, and percentage of silk fibre, (Figure 2). Inferior quality cocoons, reelable cocoons and non-reelable cocoons are separated. "Reel-able" quality cocoons undergo "cocoon-cooking" process. In this process, the excellent quality cocoons are treated with heat. As serine is water soluble, the cooking starts mild degumming process. In degumming process, some of the complex layers of sericin are dissolved. As a result, silk filament can be unwounded easily which makes reeling process easy. After "cocoon-cooking" the "Deflossing" of cocoons is followed to remove unreelable layers. Once deflossing process of cocoon is finished, the end of cocoon filaments is drawn together to make continuous silk yarn [2].

Figure 2. Fully developed cocoon.

2.6. Traditional Uses of Silk

Low conductivity of silk keeps warm air close to skin during cold weather. The absorption makes it comfortable to wear in warm weather and inactive conditions. Due to such soothing properties, silk clothing is preferred in all seasons. As silk has attractive lustre it suits to furnishing applications. Due to high tensile strength and mechanical properties, it has many industrial and commercial uses; they are used in parachutes, bicycle tires, comforter filling, artillery gunpowder bags etc. Being one of the best biomedical applications, it is suitable for non-absorbable surgical sutures and holograms.

Wild silks have been known and used in many countries for ages, although the amount of production is much lower than the cultivation of silkworms. Silk cocoons and nests are similar to paper or cloth and their use has appeared independently in many societies. Wild silks are more difficult to dye than silkworms from cultivated silkworms. A technique known as demineralizing allows removal of the mineral layer around the cocoons of wild silkworms that hinder sole variability of colour from creating a commercial silk industry based on wild silk in different parts of the world, such as Africa and South America. The word 'Vanya' is a Sanskrit source, meaning unfamiliar, wild, or forest-based. Muga, Tasar, and Eri silkworms are not completely hegemonic, and the world affectionately calls the silk they produce 'Wild Silks'.

2.7. Origins of Silk

2.7.1. Indian Silk

In archaeological discoveries at Harappa and Chanhu-daro evidence was found for sericulture of wild silkworms during the Indus Valley Civilization between 2450 BC and 2000 AD. But in China silk production evidence was located until 2570 BC. After China, India produced silk in large amounts in the world. Six states of India, Andhra Pradesh, Karnataka, Jammu and Kashmir, Tamil Nadu, Bihar and West Bengal contribute about 97% of raw mulberry silk.

There are many types of silk commercially known and produced around the world. Four types of silks are made in India. They are Mulberry, Tassar, Muga, and Eri.

A. *Tussar silk*

Tussar silk has different alternative names like tussah, tushar, tassar etc. It is also known as Kosa silk. Tussar silk due to its deep natural colour and rich texture is produced in many countries like China, Japan and Sri Lanka. Silkworm belonging to the moth genus Antheraea. These silkworms feed on leaves of the jamun and oak trees. The end use of Tussar silk is in weaving sarees, base material for handicrafts, furnishing fabrics and additives to soap.

B. *Mulberry Silk*

Mulberry silk is most commonly used silk about 90% of the world's supply. It is popular especially in China, Japan and Korea. Silk in Mulberry silkworm Bombyx is found in mori's larvae cocoon. Mulberry silkworms feed on Mulberry leaves (Figure 3). These silkworms are delicate and need extra care at the time of cultivation.

C. *Eri Silk*
It is also known as Endy or Errandi silk because castor is called Errand in Indian Hindi language. The Castor silkworms produce Eri silk is reared in the castor oil plant. Native species of silkworm are known as Samiaricini and Philosamia ricin. Mulberry silkworms and Castor silkworms both relay on human development (Figure 3).

D. *Muga Silk*
Muga silk is found in the state of Assam, India. It belongs to the Tassar silkworm family. Muga silk possesses extreme durability and has a natural yellowish-golden tint with shimmering, glossy texture. Muga silkworms feed on fragrant leaves of Somali trees and semi- domesticated animals. Muga silk is used to make traditional garments. This silkworm does not need delicate care.

E. *Pat Silk*
It is bright white or off-white. It is another variety of domestic silk in Assam. Larvae of this silkworm feed on Noni leaves. Such silks are mainly used for products like mekhelas, sheets and other textiles. Pat silk is produced from indigenous silkworms, so it needs reproduction with selective breeding. In Assam, sericulture is brought by the Tibetan-Burman Kachari tribes.

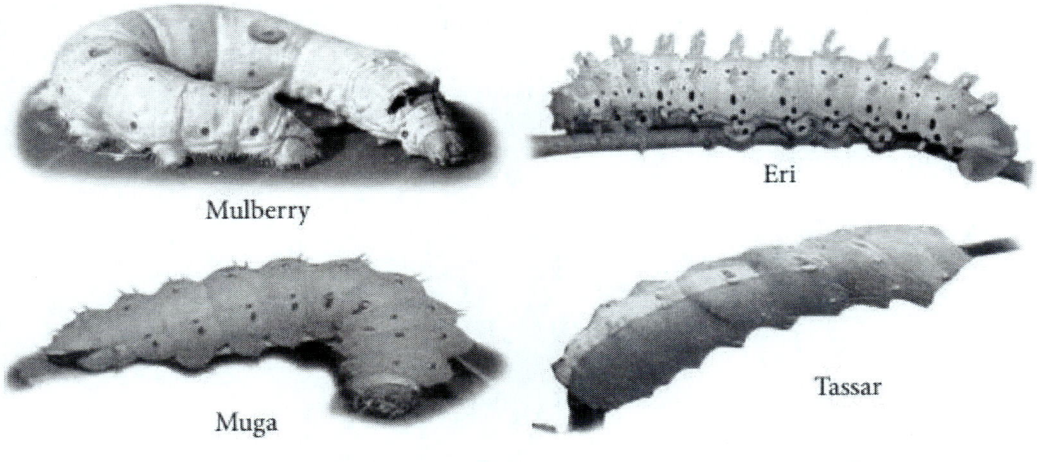

Figure 3. Types of silkworm.

2.7.2. China Silk

Sericulture was first developed in ancient China. Evidence is found at the Neolithic site Jiahu in Henan, in the soil samples of two tombs, dating back to 8500 years ago. Chinese empress, Leizu, developed sericulture in China. In ancient China, weaving and dying techniques were complicated. China tried their best to keep the secret knowledge of sericulture to maintain its monopoly in silk production. Nevertheless, sericulture techniques reached Korea in early 200BC and India in 140 AD.

2.8. Diseases of Silkworms

Despite tremendous developments in sericulture, techniques occurrence of diseases causes severe damage to sericulture industry. *Bombyx mori* life cycle affects at different stages due to deadly diseases. Signs and symptoms are specific to every disease. There are four types of major diseases [3]:

1. Pebrine (protozoan disease)
2. Flecherie (bacterial disease)
3. Grasserie (viral disease)
4. Muscardine (fungal disease)

2.8.1. Pebrine (Protozoan Disease)

Silkworm *Bombyx mori* a monophagous lepidopteran insect, is most important in the sericulture economic industry.

2.8.1.1. Causative Agents

Microsporidian, *Nosemabombycis Nosema*sp. NIK-2r, NIK-3h and NIK-4 m.

It can occur in any season in silkworms. Primary infection is transovarial transmission from infected mother to the offspring (Figure 4). In early 4th larval stage secondary infection occurs in flimsy cocoon. The disease is transmitted by secondary infection of larvae feeding from lower quality mulberry leaves, faecal matter, contaminated rearing house, egg surface, diseased silkworm etc. If the disease spreads through egg, larval death occurs in 1st and 2nd instar larva. During infection, the spore enters the healthy silkworm body through contaminated leaves is noticed during late larva stage, or such larva infects moths and lay diseased eggs. The wings of diseased moths are curled, and the scales fall off easily and appear wounded. Out of total crop losses in India, 2.32% loss occurs due to Pebrine (microsporidia) [5, 6].

Figure 4. Pebrine infected larva.

2.8.2. Flacherie (Bacterial Disease)

Bombyx mori infectious flacherie virus (BmIFV) is caused by seven types of silkworm bacteria as follows:

2.8.2.1. Causative Agents

1. Streptococcus bombycis
2. Staphylococcus sp.
3. Serratia marcescens,
4. Micrococcus sp.
5. Pseudomonas sp.
6. Bacillus sp.
7. Bacillus thuringiensis (Bacillus sotto Ishiwata).

Out of these five are more pathogenic causing considerable cocoon crop loss. Out of the four major diseases, Flacherie causes 57.22% signicant loss. In addition, Flacherie associated with various viruses e.g., IFV, CPV, DNV and Kenchu. This disease occurs at late age of silkworms due to fluctuation in temperature; poor quality of mulberry leaves and high humidity, resulting in death of silkworms within 10 to 15 days. Infected silkworms look weak and are coloured dark brown before they die. This disease is common in summer and rainy seasons due to humid and hot weather. The primary infection is caused by consumption of contaminated leaves. Bacteria enter the body through mouth. Some bacteria enter through wounds in the body and symptoms appear within 2 to 3 days. The disease symptoms are of general types. Larvae lose appetite, become lethargic and the dead larvae putrify and become black, rot and emit foul smell.

2.8.3. Grasserie (Viral Disease)

Grasserie viral disease to silkworm is caused by different causative agents such as:

1. Bm infectious flacherie virus (IFV)
2. Bmdensonucleus virus (DNV1, Bm DNV2)
3. Bm nuclear polyhedrosis virus (NPV)
4. Bm cytoplasmic polyhedrosis virus (CPV)

Amongst these, nucleopolyhedro virus (BmNPV) which is also called poly- hydrolysis a milky disease, or hanging disease, is one of the most destructive diseases. At the final stage of larval development silkworm shows disease symptoms. Larva dies without cocoon formation resulting in waste of expenses. Due to short life cycle of silkworm, this infection is difficult to cure. Symptoms of the disease are restlessness, shining and fragile skin, oozing of milky white haemolymp and swollen inter-segmental region. Out of total crop losses in India, 33.88% loss is due to grasserie. Hygiene maintenance is necessary to control infections. Good quality mulberry leaves, proper spacing and ventilation are required to prevent diseases.

2.8.4. Muscardine and Aspergillosis (Fungal Disease)

Muscardine is a fungal disease caused by causative agents like white muscardine and green muscardine. These are caused by fungus, *Beauveriabassiana and Metarrhiziumanisopliye respectively.*

The favourable conditions for Muscardine are the low temperature and high humidity. Black spots are noticed on the body surface, spiracles and legs in the early stage of infection (Figure 5A). This infection can occur at any stage of the silkworm including larva, pupa and moth. The symptoms of diseases are losing appetite, inactivity, accidity eventual death and body getting harder and finally mummied due to toxins released by the fungus. Mummified silkworm develops white or green colour (Figure 5B) because of an overgrowth of fungus which develops fruiting body and produces conidia on the body surface. In chawki stage in 2 to 3 days death/mummification is possible. Out of total loss 0.487% loss occurs due to muscardine (fungus).

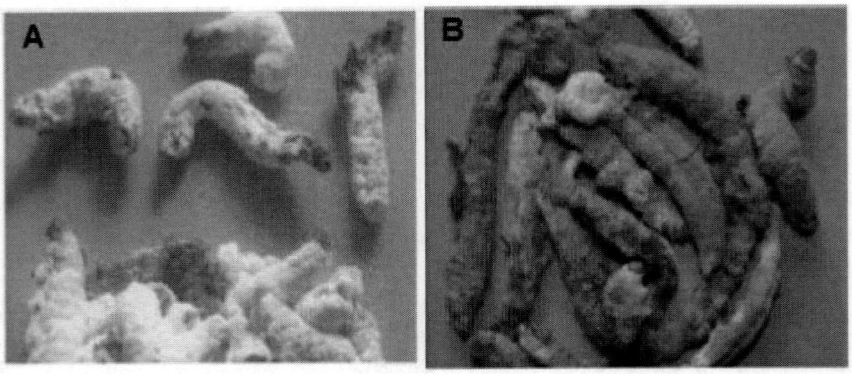

Figure 5. A. White Muscardine diseased silkworm; B. White Muscardine diseased silkworm [7].

2.8.5. Aspergillosis

This disease is caused by three causative agents such as *Aspergillus flavus*, *A. oryzae* and *A. tamari.*

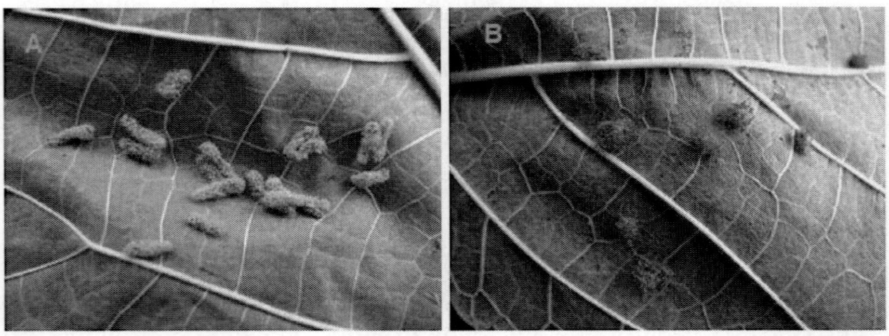

Figure 6. A. Silkworms infected with *A. flavus*; B. Silkworms infected with *A. tamarii* [7].

High temperature and high humidity are favourable conditions for Aspergillosis. Conidia is a sexual spore formed at the tip of specialized hypha (conidiophore). After infection in early stage, the larvae become inactive, hind region shrink and die without clear morphological symptoms. Dead larvae develop brownish-green colour due to mycelial growth. Due to this brownish colour it is difficult to identify *Aspergillus* that infect chawki silkworms in the rearing bed. The duration from time of infection to death/mummification is 3 to 4 days (Figure 6).

3. Physical Properties of Silk

3.1. Structural Properties

Silk is natural protein made up of repetitive units of naturally occurring amino acids. Overall, silk fibroin contains more than 5000 different amino acids and other minor components like wax, fat and pigments leading to different structures, giving rise to variety of innovative and exciting properties. These properties emerge a variety of applications. To understand the complex nature of silk fibroin, it is essential to elaborate structural and physical properties.

In primary structure of silk fibroin, different amino acids such as serine (12%), alanine (30%) and glycine (43%) are present (Figure 7). Silk is naturally occurring biopolymer and molecular weight of silk fibroin varies from low to high molecular weights. E.g., Yunusov L. et al. reported highest molecular weight of 370 000 Da for silk fibroin with its macro chain length of 150 nm and diameter of 0.45 nm [8].

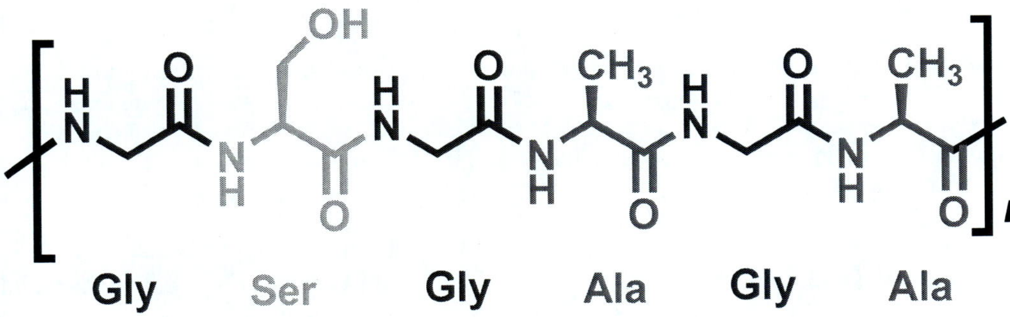

Figure 7. Silk fibroin primary structure: Gly-Ser-Gly-Ala-Gly-Ala.

Sericulture cocoons consist of mainly two proteins:

1. Fibroin (water-insoluble) and
2. Sericin (water-soluble)

Sericin is a glue-like protein coated on protein fibroin. The fibroin and sericin content vary according to cocoon strain [8]. In *Bombyx mori* 66.5-73.5 wt% fibroin and 26.5-33.5 wt% sericin are present. Sericin in the silk-fibroin joins two core brings together and acts as a fungicidal and bactericidal agent [9]. The Physicochemical properties of silk-fibroin are mainly dependent on the constituent and structure of silk fibroin. In the process of degumming, the water-soluble sericin is removed, leaving behind the silk-fibroin. Silk fibroin is mainly divided into two parts namely primary and secondary structures.

3.1.1. Primary Structure of Silk Fibroin

Shimura et al. demonstrated that the basic unit of silk-fibroin composed of three-protein chains consists of light chain (L-fibroin), heavy chains (H-fibroin) and glycoprotein P25 in a molar ratio of 6:6:1 [10, 11]. These three proteins make the primary structure of silk fibroin. Two protein subunits heavy chain (H-chain) and light chain (L-chain) are linked covalently by a disulfide bridge at the carboxy terminus of the two subunits and glycoprotein P25 attached by non-covalent interactions to covalently bonded heavy and light chain complex. The Molar ratio of silk-fibroin suggests that P25 is connected to a set of six (H-L fibroin) dimmers [$(H\text{-}L)_6$ (P25)] [12]. The Molecular weight of heavy chain (H-fibroin), light chain (L-fibroin) and glycoprotein P25 are approximately 390 kDa, 26kDa and 25kDa respectively [11, 13-15].

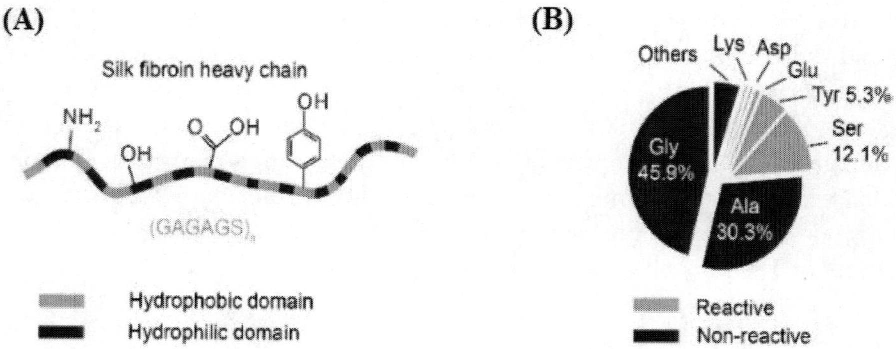

Figure 8. (A) The silk fibroin heavy chain consists of alternating hydrophobic and hydrophilic domains. The hydrophobic domain is rich in the hexapeptide motif, GAGAGS, where G, A, and S are glycine, alanine, and serine, respectively. Some amino acid residues provide polar reactive groups such as amine, hydroxyl, carboxyl, and phenol for chemical cross-linking.
(B) Amino acid composition of the silk fibroin heavy chain [19].

In a few species of silkworms (Saturniidae family) only the heavy chain is the main component of silk, light chain and glycoprotein P25 are missing [16]. The heavy chain component has two structures, such as primary and secondary. The primary structure is formed by a highly repetitive sequence of GAGAGS, GAGAGY, and GAGAGVGY [17, 18] which are mainly hydrophobic. Zhou et al. [19] reported the primary structure of heavy chain (H-fibroin) primarily consists of amino acid composition Gly (45.9%), Ala (30.3%), Ser (12.1%), Tyr (5.3%), Val (1.8%), and 4.7% of other 15 amino acid types (Figure 8) which occupies most of the silk fibroin.

Though the silk fibroin consists of more than 5000 amino acids, 16 different amino acids composition predominantly present in primary as well as secondary structures are listed in Table 1 [20]. From the table, it may be noted that as the total mol % composition of the amino acids decreases, their heavy areas decrease. However, there are no systematic variations in light areas of amino acids. Physical and chemical properties like solubility, swelling etc., are determined by the nature of polar or non-polar side chains of proteins (Figures 9 and 10). Various types of interactions such as hydrogen bonding, dipole interactions and electrostatic interactions are present between polar and non-polar side chains resulting in different properties of silk fibroin. The Asp, Gly, Lys, and Arg are the polar charged [21] species and will always be found at the protein surface and participate in hydrogen bonding with other charged species. The Non-repetitive sequences of amino [22] and carboxyl [23] terminal domains are hydrophilic [24]. Non-polar side chains are hydrocarbon residues that indicates hydrophobicity, and the polar residues are hydrophilic, as shown below.

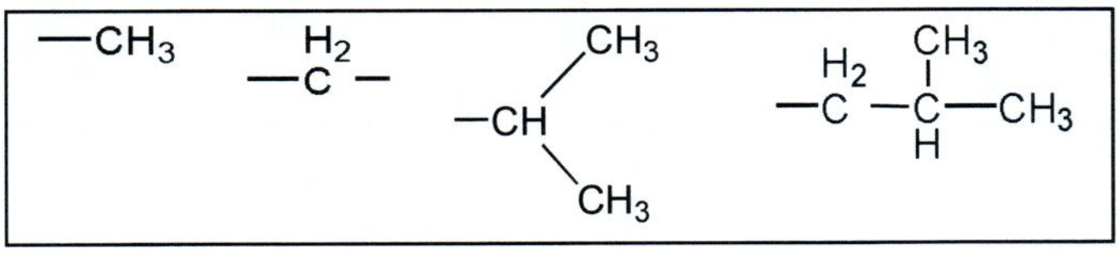

Figure 9. Non-polar side chains.

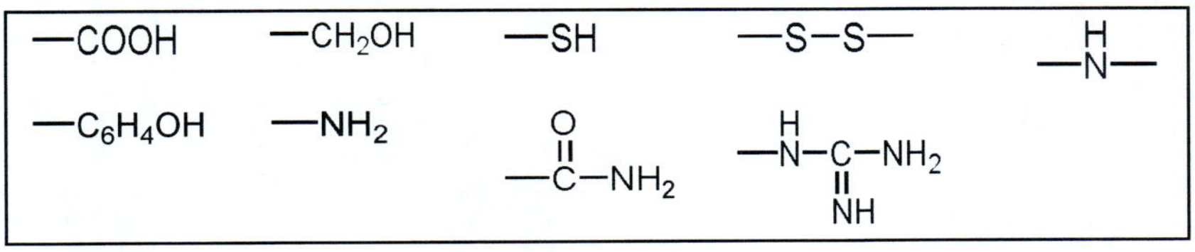

Figure 10. Polar side chains.

Table 1. Structures of different amino acids are present in silk fibroin [20]

Amino acid	Composition, mol %				
	Total	Heavy areas	Light areas	Structure of side chain	Properties (PKa values of side chains)
Glycine	42.9	49.4	10.0	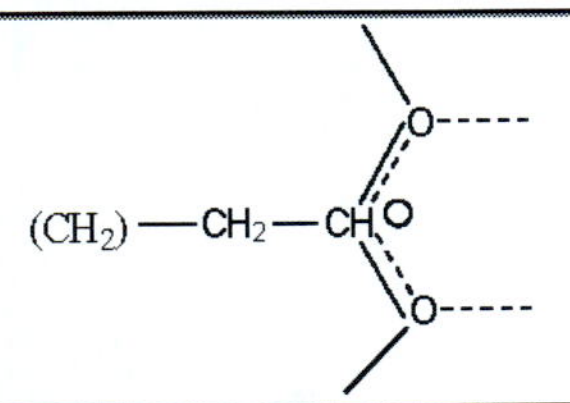	Highly flexible
Alanine	30.0	29.8	16.9	$-CH_3$	Hydrophobic
Serine	12.2	11.3	7.9	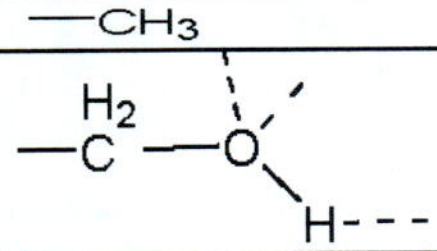	Polar, uncharged
Tyrosine	4.8	4.6	3.4	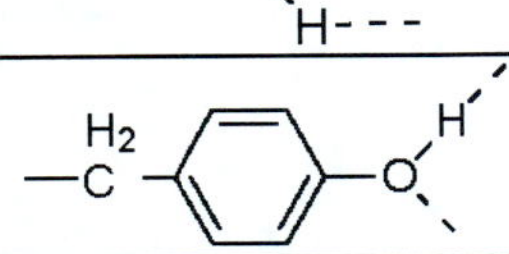	Polar, but almost as hydrophobic, as phenylalanine (10.46)
Valine	2.5	2.0	7.4	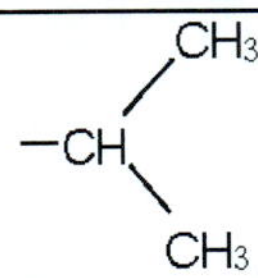	Hydrophobic, sterically hindered α-carbon
Aspartic acid	1.9	0.65	15.4	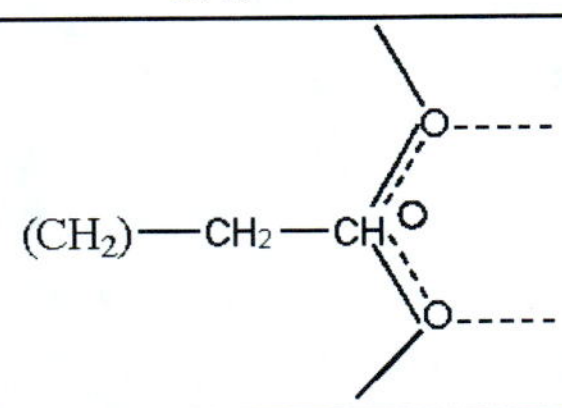	Polar charged (3.90)

Table 1. (Continued)

Amino acid	Composition, mol %				
	Total	Heavy areas	Light areas	Structure of side chain	Properties (PKa values of side chains)
Glutamic acid	1.4	0.70	8.4	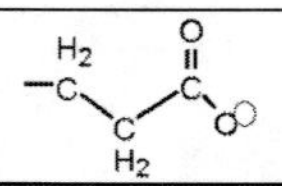	Polar, charged (4.07)
Threonine	0.92	0.45	2.8	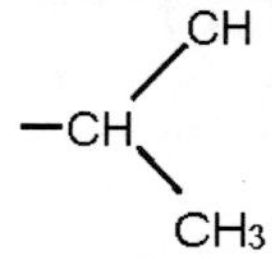	Polar, uncharged, relatively hydrophobic
Phenylalanine	0.67	0.39	2.7	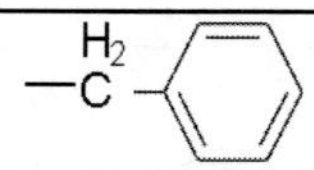	Hydrophobic
Methionine	0.37	-	0.37	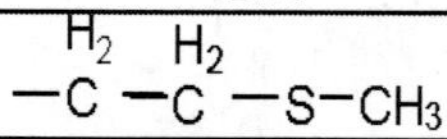	Hydrophobic
Isoleucine	0.64	0.14	7.3	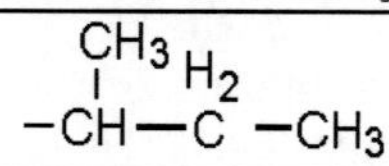	Hydrophobic, sterically hindered carbon
Leucine	0.55	0.09	7.2	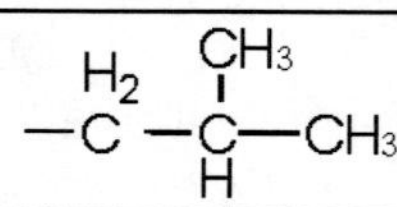	hydrophobic
Proline	0.45	0.31	3.0	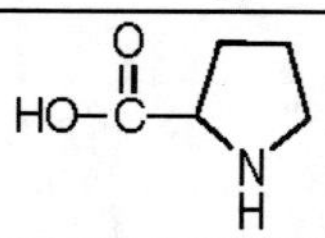	Amino acid, can form *cis*-peptide bonds
Arginine	0.51	0.18	3.8	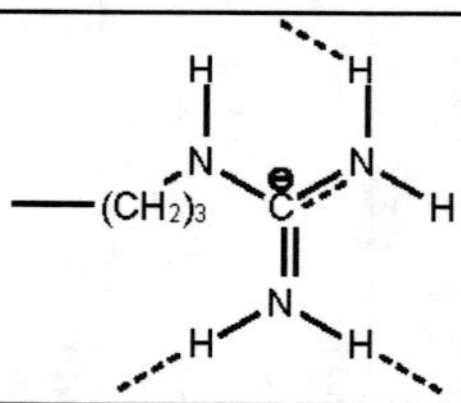	Polar, charged (12.48)
Histidine	0.19	0.09	1.6	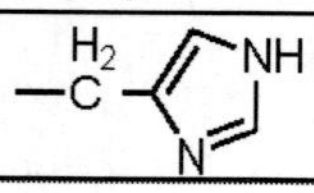	Hydrophobic (06.04)
Lysine	0.38	0.06	1.5	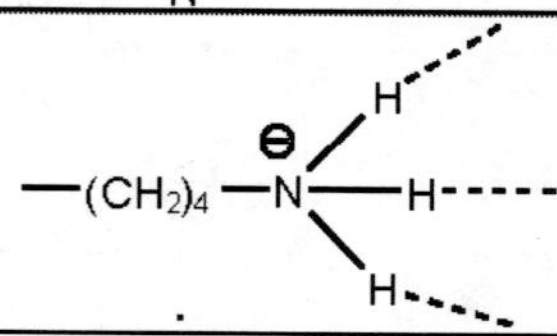	Polar, charged flexible side chain (10.54)

3.1.2. Secondary Structure

Due to specific interactions between polar and nonpolar amino acids side chains, various secondary structures such as random coil, α-helices and β-sheets are formed. β-sheets which include a flattened layer, whereas α-helices form a coiled or helical structure (Figure 11).

Silk fibroin shows two types, Silk I and Silk II. Silk I has an α-helices structure and consists of primarily amorphous random coils. Silk II mainly consists of a crystalline β-sheet structure. Silk fibroin is made up of crystalline regions and connected by amorphous regions or non- crystalline regions [25]. The chemical, conformational and oriental information of the secondary structure of silk fibroin is examined by NMR, circular dichroism (CD), IR and Raman spectroscopy [26-28]. Solid-state NMR (Nuclear Magnetic Resonance) spectroscopy can determine the Silk I structure. As silk I structure is not in single crystal form so X-ray diffraction technique is not suitable. α-Helices have a spiral structure of polypeptide chain. In α-helices, polypeptide chains are twisted and turn around clockwise as one moves along them. Silk I, the spiral chains are held together by two patterns of hydrogen bonding in between polar >C=O and >N-H groups. An alternating pattern of hydrogen bond formed within a molecule (intramolecular hydrogen bonds) and hydrogen bound formed between adjacent molecules (intermolecular hydrogen bonds). Intramolecular hydrogen bonding with a hydrophobic fragment is displaced to the periphery which leads to the formation of α-helical structure. After the spinning of silk by silkworms, intramolecular hydrogen bonds are broken by force which simultaneously forms new intermolecular hydrogen bonds with adjacent molecules. In this way, strong silk fibroin is included [29].

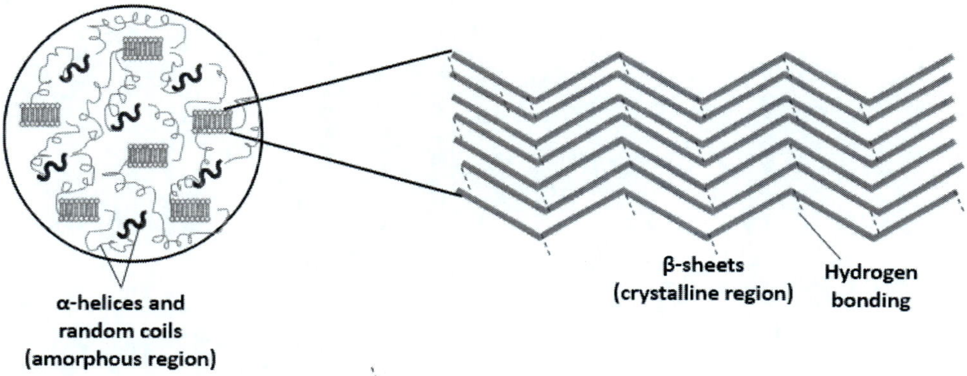

Figure 11. Schematics of the secondary structure of silk fibroin such as random coil, α-helices and β-sheets.

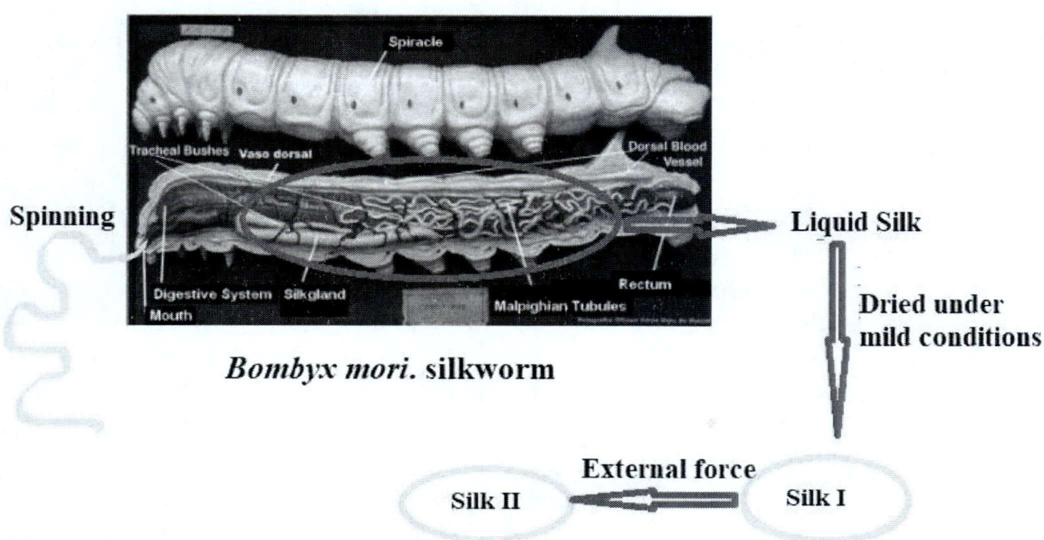

Figure 12. Silk I and Silk II.

Silk II is a product of the polymerization of silk I. Silk I is a water soluble protein, that changes quickly to Silk II structure by an external force like spinning, pressure at press part and dehydration by silkworm (Figure 12). Silk II consists of two crystalline β-sheet structures, parallel and anti-parallel. β-sheets are the dominant secondary structure in silk fibroin. Arrangement of macromolecules in parallel or anti-parallel modes forms β-folded sheets or layers. Similarity (Figure 13), as well as anti-parallel β-sheets (Figure 14) are hydrogen bonded through >C=O– and >N-H– groups between two neighbouring chains. The dominance of β-sheets is confirmed in the Raman spectrum of *Bombyx mori* silk fibroin [30].

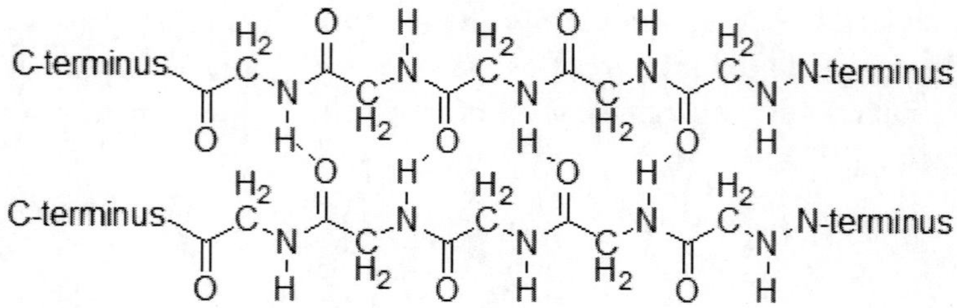

Figure 13. The structure of parallel β-sheet.

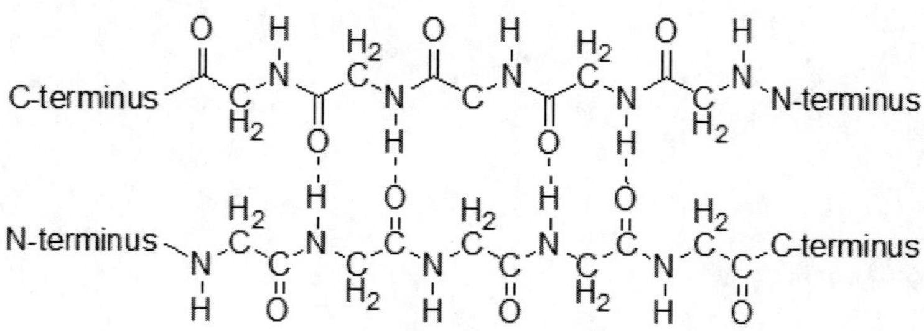

Figure 14. The structure of anti-parallel β-sheet.

Secondary structures of silk are responsible for the excellent mechanical properties of silk. Amorphous regions (α-helices and random coils) give elasticity and permeability to the silk fibroin, and crystallinity (*β*-sheets) imparts rigidity [31]. These structures show variations between different silk proteins produced by various species. To build up crystalline region amino acids sequence is very important. Specific interactions between polar and non-polar groups of amino acids are changing as the sequence alters. The Amino acids sequence is divided into four motifs. Motif 1, constitutes the crystalline regions of silk fibroin, which have a highly repetitive AGSGAG sequences. The Crystalline region plays an important role in mechanical properties. The total number of AGSGAG sequence repeats is 433 (2598 amino acid residue). Therefore, motif 1, occupies almost half part of the total number of amino acid residues 5263 in the H chain. Motif 2, contains hydrophobic and, or aromatic residues GAGAY, GAGAGV and GAGAGVGY. Motif 3, is very similar to the sequence of motif 1, except for the presence of the AAS motif. Motif 4, contains TGSSGFGPYVANGGYSGYEYAWSSESDFGT sequence of amino acid which is bulky hydrophobic, negatively charged, polar, and, or aromatic residue and makes an amorphous region of the silk fibroin as a heavy chain. However, Hu et al. reported that Silk I is neither an α-helical structure nor a β-sheet structure, but it is an orthorhombic crystal system.

Whereas Silk II belongs to the monoclinic crystal structure, which has a layered structure that is formed by β-antiparallel folding. In addition, silk III crystalline form is discovered by Valluzzi et al. at the interface between silk fibroin solution and air, a structure similar to a 3-fold helix with a hexagonal crystal system. These crystal structures of silk fibroin affect the performance of silk fibroin material [32]. Nevertheless, silk II dominated β-sheets with anti-parallel assembly in secondary structure imparts stability, crystallinity and mechanical properties in the spun fibre [33].

3.2. Mechanical Properties

Mechanical properties such as Young's modulus, ultimate tensile strength, breaking strain and toughness are structure-dependent.

3.2.1. Structure-Dependent Mechanical Properties of Silk

Due to inter- and intramolecular hydrogen bonding in-between the polar groups of amino acids present in parallel and anti-parallel β-sheets and van-der Waals interactions affect the mechanical properties. A combination of stiffness and toughness makes material of the outstanding mechanical properties with applications in several areas including biomedicine and textile manufacture. Rigid crystalline regions are connected by amorphous regions, which impart exceptional mechanical properties like high strength, elasticity, and toughness.

Raw silk fibres of *Bombyx mori* with sericin are referred to as "bave" [34], two individual silk fibrils referred as monofilament "brin" [35] and woven yarn [36]. Raw silk (bave), and yarn show tensile strength of 5-12, 15-17 and 16-17 gigapascals (Gpa), respectively. Breaking elongations for bave, brin, and yarn are 19%, 4-16% and 14-17% respectively. The mechanical properties of silkworm silk is studied after removing the sericin gum coating of the fibroin filaments from the 'bave' thread. These properties are studied by different researchers for naturally occurring *Bombyx mori* silk fibroin, spider silk (Nephilaclavipes) as well as for synthetic polymers like collagen, nylon and polypropylene and these are listed in Table 2.

Table 2. Comparison of mechanical data of *Bombyx mori* silkworm with other natural/synthetic material

Material	Young's modulus (GPa)	Ultimate Strength (MPa)	Breaking strain (%)	Toughness (MJm^{-3})	Ref.
Silkworm silk fibroin (*Bombyx mori.*)	10-17	300-740	4-26	70-78	[37, 38-41]
β-sheet crystallites (*Bombyx mori.*)	16-18	-	-	-	[39, 42]
Spider silk (Nephilaclavipes)	10.9	875	16.7	-	[39]
Collagen	0.0018-0.046	0.9-7.4	24-68	-	[43, 44]
Nylon	1.8-5	430-950	18	80	[41]
Polypropylene	4.6	490	23	-	[45]

From Table 2, it may be noted that β-sheet crystallites show higher Young's modulus this depicts the higher strength and stiffness of silk. The higher value of Young's modulus observed in β-sheet crystallites attribute to hydrogen bonding, hydrophobic and intersheet van der Waals interaction from β-sheet crystallites [46]. The semi-amorphous matrix and β-sheet crystallites of silk govern mechanical behaviour, respectively during the tensile loading. Semi-amorphous matrix gives extensibility and toughness to silk [40]. When the material undergoes homogenous stretching from tensile loading, there is a stretching until onset of yielding. When material comes to an end to behave elastically and begin to behave plastically, at this point semi- amorphous components start to undo and the tensile load is transferred to β-sheet crystallites.

There are two types of β-sheet crystallites:

1. Intermolecular β-sheet crystallites, and
2. Intramolecular β-sheet crystallites.

A high content of intramolecular β-sheet crystallites gives high mechanical strength and toughness to the silk. The 18% (of total β-sheet crystallite content) of intramolecular β-sheet crystallites are present in silkworm silk. The reason for lower content is heavy chains form interactions to form intermolecular β-sheet crystallites when the domains are well aligned during spinning [47]. Other factors also affect the mechanical strength of silk, e.g., parallel, and anti-parallel β-sheet crystallites, size of β-sheet crystallites, the orientation of β-sheet crystallites etc. [48].

Mechanical parameters are varied for different silks, obtained from various insects. For example, silk from *Bombyx mori* is strong silk, but less extensible than spider dragline silk. Naturally, occurring silk is spun by moving silkworms. The artificial reeling of silk forms superior silk than the natural spun one. Variability in mechanical measures results from spinning conditions, which are affected by the insects or animal body temperature and speed of drawing. Controlled conditions of reeling speed show changes in the properties of silk. Typical silk from *Bombyx mori* has a tensile strength of about 0.5 gigapascals (Gpa) at 4mms^{-1} [49]. At higher spinning, speed of 13 mms^{-1} extensibility of silk fibroin is lost and decreased its toughness. These variations in mechanical properties with spinning speeds at 4 mms^{-1} and 13 mms^{-1} are listed in (Table 3).

Table 3. The rate of spinning changes the mechanical properties of silk fibroin [49]

	Slow spinning (4mms^{-1})			Faster spinning (13 mms^{-1})		
Bombyx mori.	Tensile Strength (breaking strength)	Breaking elongation	Breaking Toughness (breaking energy)	Tensile Strength (breaking strength)	Breaking elongation	Breaking Toughness (breaking energy)
	0.5 GPa	15%	6x10^{4} JKg^{-1}	0.7 GPa	37%	12 x 10^{4} JKg^{-1}

3.3. Chemical Properties

Natural silk fibroins are coated with a water-soluble hydrophilic protein called sericin consisting of a high content of polar charged amino acids (Ser, Thr, Asp, Asn, Glu and Gln). Sericin can be removed because of its water solubility and this process is called degumming [50]. Silk II is composed of non-polar, hydrophobic glycine and alanine amino acids. But silk protein neither dissolves in water as a polar protic solvent nor in most organic solvents.

Instead of dissolving, silk protein swells by 30.0-40.0% and 66.7%. Due to polar and non-polar functional groups, large amounts of inter- and intramolecular hydrogen bonding present in β-sheet crystallites of fibroin makes silk fibroin soluble in a limited number of organic solvents. A solvent system that can break its hydrogen bonding and hydrophobic interactions only dissolve the silk fibroin. Solubility of fibroin depends upon the molecular structure of fibroin, which is made up of different amino acids and has polar and non-polar functional groups. Dissolution and regeneration of silk fibroin is an essential step for the application of silk protein-based cosmetics, biomedicine and biomaterials because the dissolution process directly affects the properties of the resulting silk fibroin protein. So, selecting solvent systems for the dissolution process is crucial because it directly affects the characteristics of silk fibroin and impacts on resulting silk biomaterials mechanical properties in that processing.

In a concentrated solution of acids, the groups like amino and amido are protonated. Protonated ions show interactions with each other which leads to solvation and solubility in acidic solution. Whereas in the salt systems, interaction occurs between solvent ions and functional groups of fibroin may lead to less solubility of fibroin. Some of the known solvent systems like Ajisawa's ternary solvent system ($CaCl_2$-Ethanol-H_2O) [51], LiBr-H_2O and $Ca(NO_3)_2$- Ethanol-H_2O [52] are the most common solvents for degummed SF fibres $CaCl_2$-Ethanol-H_2O (Ajisawa's ternary solvent system), $Ca(NO_3)_2$-Ethanol-H_2O system possesses high ionic strength [51, 52].

Silk-I (α-helices) can be soluble in water, while Silk-II (β-sheet) is soluble only in aqueous inorganic salts, e.g., LiCNS, LiBr, $CaCl_2$, $Ca(CNS)_2$, $ZnCl_2$, NH_4CNS, $CuSO_4$ + NH_4OH, $Ca(NO_3)_2$. The concentration of anions and cations affects solvency. The concentration of anions and cations in increasing order: for cations, $Ca^{2+} < Sr^{2+} < Ba^{2+} < Li^{+} < Zn^{2+}$, for anions, sulfate < citrate < tartrate < acetate < chloride < nitrite < bromide < iodide < thiocyanate < dichloroacetate. Solubility is higher due to the presence of polar groups like 1.4% carboxylic acid groups (aspartic and glutamic acids), 0.8% amine groups (lysine and arginine) and 18.1% hydroxyl groups (11.9% serine,1.0% threonine and 5.2% tyrosine) [53, 54].

Table 4. Common solvent systems for dissolution of silk fibroin

Sr. No.	Solvent systems (ratio)	Soluble type	Reference
1.	5.0 M CaCl2 solution	Degummed SF fibre	[56]
2.	CaCl2/ethanol/water (1:2:8 M)	Degummed SF fibre	[56, 57]
3.	CaCl2/methanol (6:94, weight ratio)	Degummed SF fibre	[58]
4.	Ca(NO3)2.4H2O/ methanol (75:25, weight ratio)	Degummed SF fibre	[59]
5.	9.3 M or 9.5 M LiBr solution	Degummed SF fibre	[60-62]
6.	LiBr/ethanol (40:60, weight ratio)	Degummed SF fibre	[63]
7.	A 45:44:11 (weight ratio) mixture of LiBr, ethanol and water	Degummed SF fibre	[64]
8.	Contain less than 15 wt% water (phenol/LiSCN/water)	Raw SF fibre	[65]
9.	1.0 M solution of LiCl in DMSO	Degummed SF fibre	[66]
10	7.0 wt% LiCl in DMAc	SF film or sponge	[67]

To prepare devices based on SF, one needs the purified SF solution. After dissolving degummed SF, SF-solvent mixture must be dialyzed to remove salts before casting SF films, sponges or nanoparticles. Lithium halide in organic solvents like Dimethyl acetamide (DMAA) and dimethyl formamide (DMF) show high solvency. 9.5-10 M solution of LiBr in DMAA or DMF at 50°C is reported for dissolution of 10.0% solution of silk fibroin [55]. Various solvent systems are used and reported in the literature by different groups of scientists listed in (Table 4).

From Table 4, it is noted that, these are complex solvent systems. In such types of complex solvent systems, anions act as nucleophiles. Anion attacks the polar electrophilic centre like hydroxyl, amide, and amino groups of fibroin macromolecules. In non-aqueous and aprotic solvents, anions are not solvated, so nucleophilicity is higher [68]. Thiocyanates (^{-}SCN) in organic solvents are less nucleophilic and Cl^{-} is more nucleophilic ($Cl^{-} > {}^{-}SCN > Br^{-}$). Whereas, in an aqueous solution, the ^{-}SCN is more nucleophilic, and it interacts with H_2O more strongly than the polymers. So, aqueous solution of LiSCN readily dissolves the polymer compared to thiocyanates in an organic solvent. An aqueous solution of $ZnCl_2$ is the most effective solvent, in the case of cations. The solvent system like $ZnCl_2$ in organic solvents like DMAA, DMF, and *N*-methyl-2-pyrrolidone, fibroin dissolves at 155°C to form dilute solutions [69]. Silk fibroin aqueous solution can be stored for less than 30 days at 4°C and only for 7 days at RT [70].

3.3.1. Inorganic Acids/Alkalis

Silk fibroin is not soluble in water as a polar protic solvent. To dissolve it, one needs to add inorganic acids or salts.

3.3.2. Concentrated Inorganic Alkalis/Acid

SF can be dissolved in conc. H_3PO_4, HCl and NaOH. For example, SF dissolves completely in 6.0 N HCl at 95-100°C [71]. Sometimes, extensive depolymerization of SF molecules occurs in strong inorganic acids or alkalis. So, SF fibre is degraded into peptides or amino acids in which 90% portion is glycine, alanine and serine.

3.3.3. Dilute Alkali-Containing Protease

Water-soluble silk peptide is obtained by treatment with dil. Alkali and then hydrolyzed with a protease [72].

3.3.4. HCl with Neutral Salt System

The mixture of conc. HCl or H_2SO_4 and neural salts (LiBr or $CaCl_2$) system dissolve degummed silk fibroin fibre. The resulting solution is hydrolyzed by heating at 80° to 100°C for 30-40 hrs [73].

3.3.5. NaOH with Strong Urea Solution

Niu et al. first reported the dissolution of degummed silk fibroin fibre in NaOH/urea solution at -12°C with sonication [74]. But NaOH/urea solvent cannot dissolve it entirely. NaOH/urea solution can destroy hydrogen bonds of SF and can exfoliate the fibres into nanoribbons, which resemble β-sheet layer.

3.3.6. Organic Acid Complex System

In formic acid, regenerated silk fibroin is easily soluble but not silk fibroin. Degummed silk fibroin hardly dissolves in formic acid, but at a ratio of 70/30-60/40, the maximum solubility reaches 20% (w/v) [75]. However, silk fibroin can be soluble in a formic acid complex solvent system containing $CaCl_2$, LiBr, phosphoric acid (PA) or hydroxyapatite (HAp). With a concentration of 1-12% (w/v) of $CaCl_2$ with formic acid, degummed silk fibroin was soluble at room temperature. This disintegrates silk fibroin into nanofibrils. Degummed silk fibroin in 0.05% (w/v) Na_2CO_3 or 0.005% (w/v) Na_2CO_3 will completely dissolve in the formic acid-LiBr approximately in 40 min and 300 min, respectively. In addition to this, formic acid- hydroxyapatite (with 1.0–5.0 wt%) solvent system dissolves silk fibroin fibre. The dissolved protein shows the diameter of approximately 103.6 ± 20.4 nm as nanofibrils. The degummed silk fibroin can depolymerize when treated with concentrated phosphoric acid. To retard depolymerization, this mixture of formic acid-phosphoric acid can be used [76-77].

3.3.7. Other One Components Organic Solvents

In the case of raw silk, it can initially swell and then dissolve in some polar aprotic solvents. These polar aprotic solvents can break the intermolecular hydrogen bonding by penetrating which results in swelling and then dissolution. Order of swelling of raw silk in organic solvents is as follows: dimethylsulfoxide (DMSO) > dimethylformamide (DMF) > dimethylacetamide (DMAc) > tetramethylurea (TMU) > dioxane. Solubility is not observed at room temperature, but under reflux conditions [78]. Regenerated silk fibroin can dissolve at room temperature. Organic solvent-silk fibroin is more stable and transparent than an aqueous-silk fibroin system. Aqueous 40 wt% tetrabutylammonium hydroxide (TBAOH) can dissolve degummed silk fibre at room temperature. Other one component organic solvents are reported, e.g., HFIP and hexafluoroacetone, for the dissolution of silk fibroin [79].

3.3.8. Ionic Liquids

Phillips et al. first reported the use of ionic liquids to dissolve silk fibroin in 2004 [80]. Three main types of ionic liquids can be used, e.g., 3-methyl imidazolium-based ionic liquids, protic ionic liquids, and choline salts [81]. The 3-Methylimidazolium-based ionic liquid which is most studied and needs constant stirring and heating at 90-100°C for 2hrs to dissolve silk fibroin.

3.3.9. Metal Complex System

Solubility of silk fibroin is possible at room temperature with a cuprammonium hydroxide solution (Cuoxam complex). In addition, other Cuoxam systems are copper hydroxide ammonia solution, copper hydroxide-alkali-ammonia solution, copper hydroxide-alkali-glycerol aqueous solution and the copper ethylene-diamine aqueous solution are good solvent systems for silk fibroin [82].

Overall, considering all solvent systems, the eco-friendly solvent is N-methyl morpholine N-oxide (NMMO) for silk fibroin. For the first time in 1999, Freddi et al. [83] found that NMMO was an excellent direct solvent for silk fibroin. The regenerated silk fibroin is easily dissolved in NMMO, concentration higher than 10 wt% of silk fibroin which was challenging to obtain because it undergoes an extensive degradation above 110°C. The 10 to 25 wt% concentration of RSF in NNMO can be obtained [84].

NMMO-water-organic solvent ternary systems with different concentrations are suitable for silk fibroin dissolution. When the water content is less than 10 wt% at this condition silk fibroins dissolve completely [85]. If the temperature is above 110°C, silk fibroin degrades extensively [86]. The low molecular weight organic solvents, e.g., protic (alcohol) and aprotic solvents (DMSO, DMF and DMAc) with NMMO-water reduce dissolution temperature. In the ternary system NMMO:H_2O:Methanol if the ratio is 1:0.3:**0.6**, the solubility is observed at 120°C. Whereas if the ratio is 1:0.3:**0.7**, the solubility of silk fibroin is kept at 90°C.

3.4. Thermal Properties: Stability

The thermal stability of silk fibroin is determined by thermogravimetric analysis (TGA). The mass loss of silk fibroin in the thermogravimetric study is observed during heating from room temperature to 450°C. Silk fibroin shows weight loss in three main stages:

- *Stage I:* Weight loss of random coiled material (SF-R), Silk I material (SF-I) and silk fibroin fibre is 10.80%, 9.08% and 7.02% respectively at 100°C. Mainly moisture present in the material is lost up to 100°C. At 150°C, loosening of inter- and intramolecular interactions occurs before the glass transition. At 177°C, mobility of the chain increases in the amorphous region. Silk fibroin has its glass transition temperature of about 178°C [87-89].
- *Stage II:* In stage II, the temperature rises from 200°C to 300°C, which is the slow decomposition stage. The initial decomposition temperature of (SF-R), (SF-I) and silk fibroin fibre is 266.70°C, 260.61°C and 286.37°C respectively. By observing given decomposition temperature, no silk fibroin degrade below 200°C, which indicates that the silk fibroin material has good thermal stability. β-sheet crystals of silk fibroin do not melt. Silk II, an anti-parallel secondary structure can be induced to exposure of an aqueous solution of silk fibroin to organic solvents, mechanical stress, and thermal treatment. Hydrogen bonding N-H···O present between the anti-parallel 'polyamide-like' chains in the β-sheet secondary structure plays a crucial role in different properties of silk fibroin [88, 89].
- *Stage III:* In the third stage, all structures show rapid weight loss between 300°C and 400°C. Breaking of peptide bonds and side groups causes weight loss. Thermal decomposition of Silk I crystals is slightly easier than that of silk II crystals. Heat induces conversion of Silk I to Silk II via β-sheet formation. The maximum degradation temperatures for (SF-R), (SF-I), and for (Silk fibre) are 285.43°C, 281.38°C and 313.73°C respectively.

To improve thermal stability of silk fibroin an addition of polymer like Nylon-6 is reported by Chen et al. Nylon-6 and silk fibroin have similarities in chemical structure, crystal structure and pattern of hydrogen bonding. Nylon-6 have the hydrogen bonding between anti-parallel polyamide chains in α-crystallographic phase and parallel polyamide chain in γ- crystallographic phase. Strong interaction between N6 and SF is observed due to strong hydrogen bonding, which helps to improve the thermal stability of the blend. Just above 210°C silk fibroin degrades, which is the melting endotherm of N6. In the thermogravimetric study of the blend, the measured mass of SF is slightly greater which indicates that the blending improved thermal stability and no β-sheet crystals of silk fibroin are observed in FTIR [90].

3.5. Optical Properties

Silkworm silk has been considered luxurious textile wearable for more than 5000 years due to its attractive colors. The transparency, mechanical robustness, easy surface patterning and degradability at a controlled rate make silk biocompatible and biofunctional optical material and interfaces. Functionalized silk material exhibits optical applications. This is because silk fibroin shows a variety of fluorescence due to the presence of more than 5000 amino acids. For example, tryptophan is present in sufficiently high concentration with absorption in UV region and excitation with 280nm UV light which emits light approximately at 350 nm.

The optical properties of materials depend on the reflection, absorption and transmission of light. For brighter visibility of object needs less absorption, transmission and hence more reflection. For brighter wearables based on silk fibroin, it needs more reflection of light. The triangular shape of silk fibroin reflects light, which imparts luster to the silk fibre. Regenerated silk fibroin shows colour changes upon UV irradiation. Presence of aromatic amino acids such as tyrosine (5.1%), phenylalanine (0.4%) and tryptophan (0.1%) in SF is responsible for excitation with a suitable wavelength of UV light [91]. The main peak around 280 nm is attributed to aromatic amino acids and emitted light at approximately 350nm. The refractive index for regenerated silk, n= 1.554 at 500nm, are increased to n = 1.559 at 500 nm upon crystallization [92]. The absorption coefficient is less than or equal to 10–4 for wavelengths longer than 500 nm. The residual water content and crystallinity in silk fibroin infuence the refractive index. The molecular weight of silk fibroin does not play a role in the refractive index. Bucciarelli et al. [93] reported that the refractive index of sericin is slightly higher than regenerated silk fibroin.

The native silk fibroin (extracted from purple gland line) showed a higher refractive index $\lambda = \pm 566 \pm 0.001$at 500nm compared to regenerated silk. The difference in refractive index is probably due to inter-molecular aggregates of the β-sheet. In contrast to native silk, regenerated silk is amorphous with low amount of β-sheet. The amount of β-sheet can be increased by water vapour and methanol annealing in the regenerated silk films, which leads to increase in mechanical stiffness and decrease in the solubility of regenerated silk in water. Functionalization of silk fibroin by incorporating an organic molecule or inorganic nanoparticles exhibit fluorescence. The uses of functionalized silk fibroin material exhibit different optical applications such as nonlinear optics, photochemical reactions, UV resistance, photocatalysis, light energy conversion, sensors, optical elements, bioimaging and biomedical engineering, photonic and nanophotonics etc.

4. Blends of Silk Fibroin and Their Applications

To increase the strength of individual silk fibroin, the addition of one or two components with different properties helps to improve the strength of silk. The naturally originated silk fibroin, due to its unique features and blending with another natural/synthetic polymer, makes it a versatile material (Table 5) and indicates that the silk fibroin has rigidity which also shows limitations like lack of proper elasticity. To get rid of this and increase the use in different areas there is a need to blend composites of SF. Blending shows properties like biodegradability, biocompatibility used in many crucial fields of biomedicine e.g., cartilage regeneration, skin tissue engineering, nerve regeneration, bone tissue engineering etc. (Figure 15).

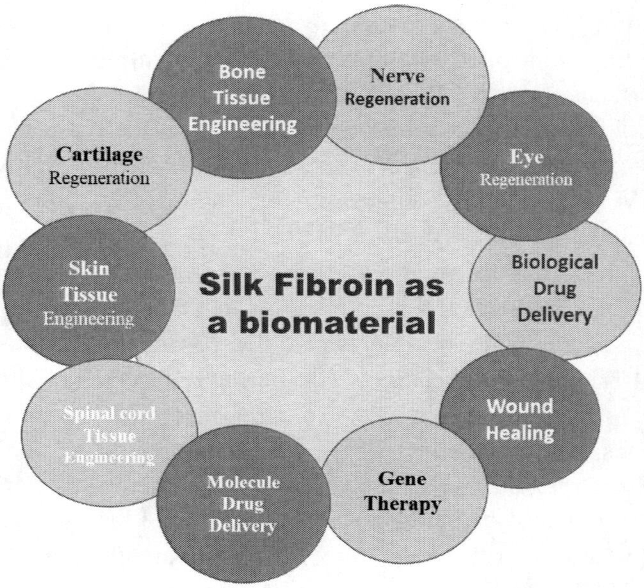

Figure 15. Biomedical applications of silk fibroin blending.

4.1. Physical Forms of Blends for Their Applications

For a variety of SF applications as shown in Figure 14, the SF and its blends need to be formulated in a variety of physical forms. These silk fibroin-based biomaterials can be divided into the following types: Fibres, 3D porous structures, Particles, Films, Hydrogels and Aerogels.

The 3D porous structures of silk fibroin are sponges, scaffolds or foams. The 3D scaffolds with controlled pore size and pore structure can be obtained by different methods e.g., electrospinning, lyophilisation, particulate leaching, and gas foaming [94].

Traditionally, Silk fibroin fibres are obtained by reeling natural cocoons. Moving silkworms spins silk naturally. The mechanical properties of silk crucially depend upon the spinning conditions. Reeling silk straight from the silkworm at steady and controlled conditions produces superior fibres compared to naturally spun ones. The silk fibre is strong or stretchable depending on the reeling speed [95].

Pore size and pore structure define the utility of biomaterial. Because specific pore size is required for cell migration and proliferation. Pore size can be controlled by different parameters like adjusting the pH of the solution, the amount of organic solvent in the lyophilization process and freezing temperature (Figure 16). Also, the addition of sucrose or salt particles of a desired pore-size diameter, as porogenes can be carried out to the polymeric solution to get the 3D scaffold. Ideal pore sized 3D scaffold can be obtained by 3D printing.

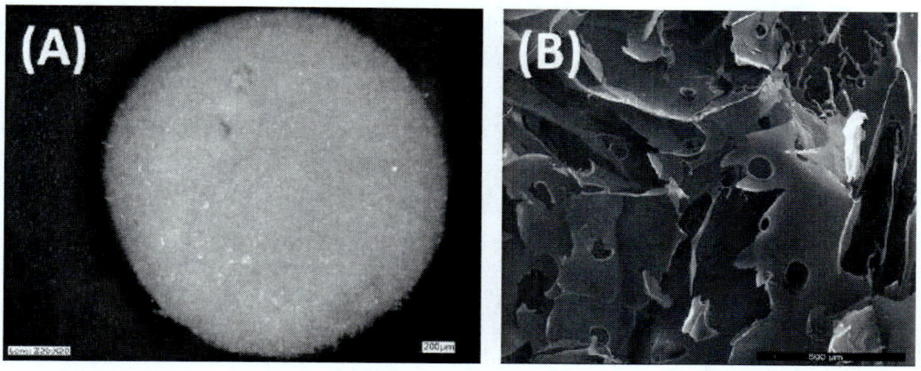

Figure 16. Silk fibroin scaffolds were obtained by lyophilization methods (A) optical microscope photograph, and (B) SEM image [95].

Microspheres, nanolayers, nanocoatings, nanoparticles, and microcapsules are silk fibroin-based biomaterials used in the controlled release of drugs. Different ways e.g., spray drying, freeze-drying, freeze-thawing, desolvation or PVA blending etc. can be prepared micro- and nanoparticles [96].

Micro- and nanoparticles can be prepared without the use of chemicals. Due to stability and non-toxicity, they can be used in drug delivery system in anti-cancer therapy. Silk-fibroin based films are obtained by casting method [97].

The degumming process of raw silk is necessary for which the sericin layer from fibroin is removed. After that washing with distilled water, drying, and dissolving in appropriate solvent steps are important processes. As fibroin is mainly composed of non-polar glycine and alanine amino acid makes it is highly hydrophobic and insoluble in water and most organic solvents. Aqueous-organic solvents can be used e.g., $Ca(NO_3)_2$ x $4H_2O$, 10M LiSCN, $CaCl_2/H_2O/C_2H_5OH$ (molar ratio: 1:8:2.5) [98]. By pouring aqueous solution on petri dishes the film is obtained (Figure 17).

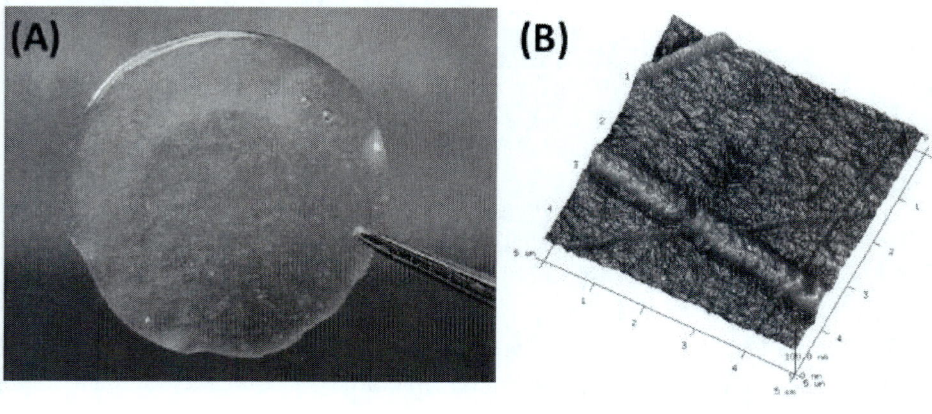

Figure 17. (A) Casting method applied to obtain Silk fibroin-film; (B) Atomic Force Microscopy (AFM) picture of silk fibroin-based film [95].

The Langmuir-Blodgett (LB) process involves the treatment of 50% methanol to modify properties, such as mechanical properties, degradability and solubility of silk fibroin films [99].

Hydrogels possess properties like, swelling in aqueous solutions but do not dissolve in these solutions, and they can be used in the biomedical field as three-dimensional polymer networks. Lyophilization or sonication technique can be used for the sol-gel transition of aqueous solution in the presence of an acid, ions, or other additives [100]. The rate of gelation is dependent on protein concentration, temperature, pH, higher Ca^{2+} amount. But the rate of water removal and extension determine the mechanical properties, morphology of silk fibroin hydrogel [101]. Calcium plays important role in silk fibroin.

Aerogels of silk fibroin are ultralight materials with large surface areas (500–1000 m^2/g), high porosity and low densities (0.003–0.15 kg/m^3). Aerogels are microporous solids characterized by an open pore structure. Aerogel can be prepared by ambient drying, freeze drying, direct supercritical drying and supercritical drying by CO_2 extraction. Ambient drying is done at ambient pressure, room or slightly elevated temperature. Vacuum pressure < 100 mbar and temperature $< -20^0C$ is required for freeze drying. High temperature is needed for direct supercritical drying and at temperature $> 31^0C$, $p > 74$ bar supercritical drying by CO_2 extraction drying technique is used for aerogels. [102-106].

4.2. Blends of Silk Fibroin

Silk fibroin as scaffold possesses several disadvantages like poor mechanical properties, weak cell attachment and stability in an aqueous solution. They are susceptible to ageing Mildew and promote microbial growth. To get rid of disadvantages and make silk broin available as a biomaterial with unique properties, blending of SF is essential. There are different types of blends of SF shown in (Figure18).

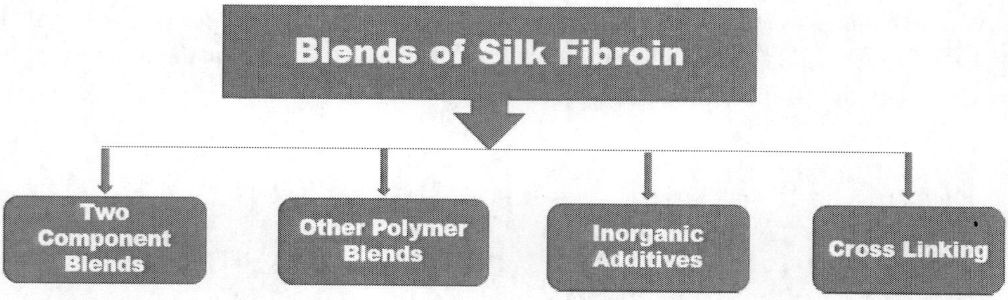

Figure 18. Blends of silk fibroin.

4.2.1. Two-Component Blends

The silk fibroin can be mixed with biopolymer/natural polymer/synthetic polymer imparts unique structural and mechanical properties to the composite and can be used in broadly understood biomedical applications. Different types of combinations have been proved in the medical field due to their unique characteristics of composites. Due to adequate mechanical strength, good biocompatibility and biodegradability, collagen is a suitable natural scaffold for cell proliferation (Figure 19).

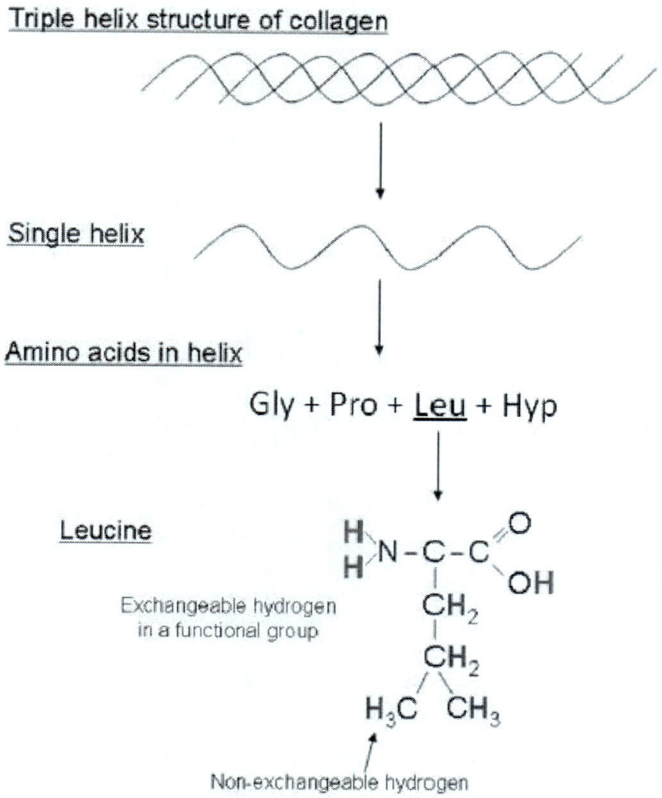

Figure 19. Structure of collagen [107].

SF blending with Collagen affects the mechanical properties like tensile strength and Young's modulus. The addition of collagen to silk fibroin led to the decrease of tensile strength and increase in Young's modulus with an increasing amount of collagen. This type of hybridization shows promising outcomes for cutaneous wound healing [107]. Miscibility of SF and Collagen defines the properties of blending and 3D scaffolds. Sionkowaska and group studied silk fibroin and collagen miscibility depending on the pH of the solution and ionic strength. Miscibility can be improved by using salt ions, which induce electrostatic interactions with the highest miscibility ratio between silk fibroin and collagen [108]. Collagen type II is the most used form in composites. SF70:Coll30 ratio shows good tensile mechanical behaviour and biocompatibility. Such compositions lead to a new idea of scaffolding for treatment of osteoarthritis because of better proliferation ability [109]. For human skin regeneration, the hybridization of silk fibroin and collagen provides excellent moisturizing property, excellent water absorption and retention capacity, interconnected structures, and synergistic effects on cell viability [110].

Chitosan is another primarily used natural polysaccharide for blending. It can be derived from chitin shells of Shrimp and other sea crustaceans. Chitosan is composed of β-(1→4)-(1-4)- linked D-glucosamine (deacetylated unit) and N-acetyl-D-glucosamine (acetylated unit) (Figure 20).

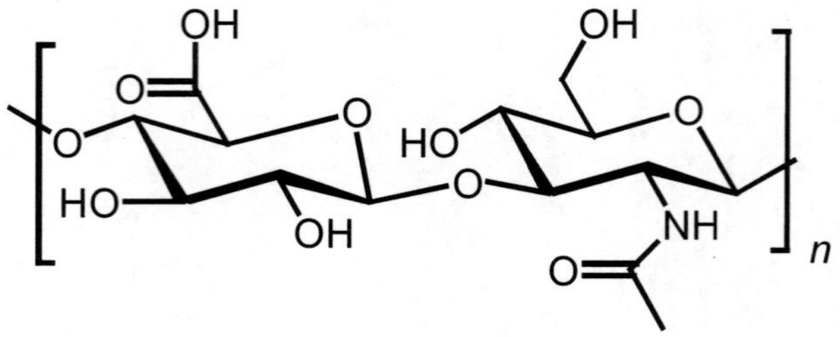

Figure 20. Structure of chitosan.

The amino group in chitosan has pka value of ~6.5, due to that it easily undergoes protonation. Protonation makes chitosan water-soluble, which can form hydrogel. This hydrogel can swell fully due to the protonation of amino group. The free amino group can form cross-linked polymeric network. All these properties make chitosan biocompatible, biodegradable and non-toxic. Silk fibroin can exist in two conformations, random coil and β-sheet forms in which, β-sheet structure is resistant to water and has good mechanical properties. Chitosan can induce the transition from random coil to β-sheet structure because the strong molecular hydrogen bonds between chitosan and fibroin. This film is used for polar drug delivery. Gobin et al. [111] reported that the 75:25 ratio of silk fibroin and chitosan 3D scaffold shows highest ultimate tensile strength and elastic modulus. Chitosan can form insoluble ionic complexes with a variety of anionic polymers.

Composites sponges of silk fibroin/Chitosan with ratio 80:20 show 75% porosity and homogeneous pore size from 20 to 150 μm appropriate for cell growth as it can mimic extra-cellular matrix. Pores of silk fibroin/chitosan scaffolds are interconnected with each other compared to each individual component. This makes it a unique sponge for tissue engineering. Such types of sponges can be used as a temporary scaffold for the formation of new tissue and organs [112].

Miscibility of both components affects the swelling property, mechanical properties and rate of degradation of films. The Young modulus value for Chitosan is less, but once mixed with silk fibroin value increases because its brittleness also increases. In addition, chitosan content upto 30% increases the mechanical properties of chitosan/silk fibroin films. The weight fraction of chitosan in the mixtures (W_{ch}) > 0.5, tensile strength is higher, less than < 0.5 reduction in tensile strength by 15% observed [113]. Ruan et al. [114] prepared 3D Fibroin/Chitosan blend porous scaffold for tissue engineering with regenerated silk fibroin and chitosan without crosslinker. Rate of water absorption enhanced with RSF/CS blend.

Also, porous 3D chitosan/SF scaffold can be used for osteoblasts through a freeze-drying method. Freeze –dried SF/CS scaffold three-dimensional material is prepared with different component ratio with 90% porosity. The 40% SF-60% CS is a good material. This combination has good cell affinity and pore which size is suitable for the growth of osteoblasts. The scaffold morphology of the 40% SF-60% CS showed porous network structure. The pore sizes range from 100 to 350 μm and clear connection between pores. Cell growth is highly dependent on pore size which, allows cell to adhere to the interior of the scaffold [115].

The SF/CS scaffold with tunable property exhibited the potential scaffold use for reconstructive surgical applications. Li et al. [116] prepared series of silk fibroin/chitosan

(SF/CS) scaffolds by using freeze-drying method, and the rat bone marrow-derived mesenchymal stem cells (BM-MSCs). BM-MSCs cultured in the SF/CS scaffold allowed to form new blood vessels to grow and excellent proliferation.

Sodium alginate is natural polymer (fibrous proteins and polysaccharides) (Figure 21), a combination with silk fibroin can be substituted for extracellular matrix proteins. Tissue regeneration needs natural extracellular matrix environment. Zhang et al. [117] fabricated nano-fibrous 1SF/1SA, 3SF/1SA and 5SF/1SA composite scaffolds by a thermally induced phase-separation method to mimic natural fibrous structure of the tissue extracellular matrix. The composites with 5SF/1SA and 3SF/1SA scaffolds possessed porosities higher than 90%, better than that of 1SF/1SA scaffold. In addition, composite scaffolds show good connectivity for tissue regeneration.

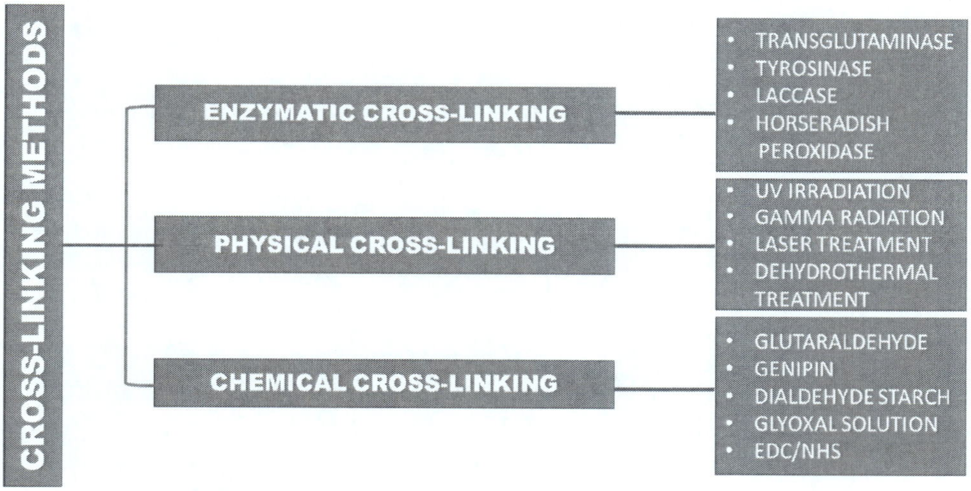

Figure 21. Structure of alginate.

Sodium alginate has higher hydrophilicity and degradability due to the presence of abundant carboxyl and hydroxyl groups. In addition, these groups can affect the hydrophilicity of silk fibroin/sodium alginate blend. Due to hydrogen bonding and Van-der Waals force of attraction between the amide groups of silk fibroin and carboxyl groups of sodium alginate lead to the highest solubility of the SF/SA blend scaffolds in aqueous solution. To get proper morphology of scaffold, solubility of SA is reduced by substituting sodium ion of alginate with calcium ion. Hydrophilicity and degradability of silk fibroin/sodium alginate blend scaffolds are reduced when blend solution treated with calcium ions. Such types of biomaterials are widely applicable in tissue engineering [118].

In contrast to sodium alginate, silk fibroin has amino groups shows hydrophobic property. Hydrophobic nature does not allow swelling and water plasticization, which explains low values of SF film thickness in the wet state and low swelling capacity. Sodium alginate film has two times higher water vapour permeability than silk fibroin film. De Moraes et al. [119] reported SF/SA blend film with ratio 25/75 exhibited tensile strength and elongation at the break of SF/SA film that was higher than the values observed for SF films. Such types of blends are found suitable for designing sophisticated high performance wound dressings.

Another natural linear polysaccharide is hyaluronic acid (HA) (Figure 22). Hyaluronic acid is a natural substance found in the fluids of eyes and joints. The structure consists of alternating disaccharide units of (1,4)-linked α-D-gluconic acid and (1,3)-linked β- N-acetyl- D-glucosamine. Hyaluronic acid is an anionic, non-sulphated glycosaminoglycan. It is one of the main components of the extracellular matrix. It has signicant role in tissue regeneration, cell proliferation and migration. Hyaluronic acid is an essential component of articular cartilage.

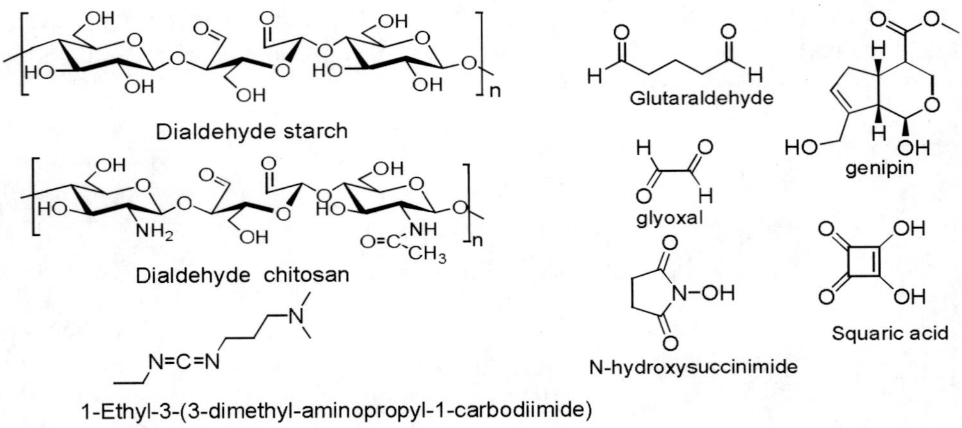

Figure 22. Structure of hyaluronic acid.

As hyaluronic acid is the main component of extracellular matrix, the blend of silk fibroin and hyaluronic acid nanofibrous scaffold can be suitable for biomimic. Zhang et al. prepared the SF/HA blend by electrospinning, to mimic the porous structure of natural extracellular matrix. Focus is on fabrication of fibrous scaffolds of SF/HA from aqueous solutions. Electrospinning in aqueous solution is difficult due to high viscosity and surface tension of an aqueous solution. As water is an ideal green solvent, Zhang and group [120] reported electrospinning of HA in aqueous phase. Electric force is used for electrospinning of nanoscale diameter fibre. The resulting nanofiber treated with 75% (v/v) ethanol vapour. The transformation of SF from random coil or α-helix to β-sheets is induced by water vapor. SF/HA nanofibrous scaffolds might benefit to cell growth for tissue-engineering applications [121, 122].

4.2.2. Other Polymer Blends

The two components blended with natural/synthetic polymers and SF are listed in Table 5.

Table 5. Silk fibroin and natural/synthetic polymers [98]

Scaffold	Composition Reference
Natural polymers	
Silk Fibroin + Gelatin	[123–125]
Silk Fibroin + Cellulose	[126, 127]
Silk Fibroin + Agarose	[128]
Silk Fibroin + Keratin	[129]
Silk Fibroin + Elastin	[130]
Silk Fibroin + Chitin	[131]

Scaffold	Composition Reference
Natural polymers	
Silk Fibroin + Heparin	[132]
Silk Fibroin + Carrageenan	[133]
Synthetic polymers	
Silk Fibroin + Poly(ethylene Glycol) (PEG)	[134]
Silk Fibroin + Poly(vinyl Alcohol) (PVA)	[135]
Silk Fibroin + Polyacrylamide(PAM)	[136]
Silk Fibroin + Polycaprolactone (PCL)	[137]
Silk Fibroin + Poly(Lactic-co-glycolic Acid) (PLGA)	[138]
Silk Fibroin + Polyurethane (PUR)	[139]
Silk Fibroin + Polylactide (PLA)	[140]

4.2.3. Three Component Blends

The three component blends of SF are also prepared and studied which are listed in Table 6.

Table 6. The different silk fibroin-based materials compositions with their morphology [98]

Composition	Morphology	Reference
Silk Fibroin + Chondroitin Sulphate +	3D scaffold	[141]
Hyaluronic Acid		
Silk Fibroin + Polylactide + Gelatin	3D scaffold	[142]
Silk Fibroin + Poly(lactic-co-glycolic Acid)	3D scaffold	[142]
+ Collagen 3D scaffold		
Silk Fibroin + Nanochitosan + Hyaluronic	3D scaffold	[143]
Acid 3D scaffold		
Silk Fibroin + Polyethylene Glycol +	3D scaffold	[144]
Keratin 3D scaffold		
Silk Fibroin + Hyaluronic Acid + Sodium	3D scaffold	[145]
Alginate 3D scaffold		
Silk Fibroin + Hyaluronic Acid + Heparin	3D scaffold	[146]
3D scaffold		
Silk Fibroin + Chitosan + Collagen 3D	3D scaffold	[146, 147]
scaffold		
Silk Fibroin + Chitosan + Polylactide 3D scaffold	3D scaffold	[142]
Silk Fibroin + Chitosan + Heparin 3D	3D scaffold	[142]
scaffold		
Silk Fibroin + Collagen + Hyaluronan	3D scaffold	[142]
Silk Fibroin + Collagen + Heparin	3D scaffold	[148]
Silk Fibroin + B-Cyclodextrin +	hydrogel	[141]
Polyethyleneimine		
Silk Fibroin + Calcium Alginate +	hydrogel	[141]
CarboxymethylCellulose		
Silk Fibroin + Chitosan + Alginate	film	[149]
Dialdehyde		
Silk Fibroin + Chitosan + Polylactide	film	[150]
Silk Fibroin + Cellulose + Chitin	fibers	[151]
Silk Fibroin + Poly(caprolactone) + Hyaluronic Acid	Nanofibrousmatrices	[141]
Silk Fibroin + Hyaluronic acid + Heparin	aerogel	[152]

4.2.4. Inorganic Additives

The addition of inorganic materials to the silk fibroin helps to increase the mechanical properties. Ceramics like titanium dioxide, hydroxyapatite or magnetic particles can add to make composites with silk fibroin. Longer biodegradation process is the major drawback of addition of ceramics. Inorganic material must have three important characteristics; it must be non-toxic, have excellent mechanical properties and have the proper degradation rate for tissue regeneration [153].

Natural bone contains both organic parts such as proteins, polysaccharides, lipids and inorganic nanohydroxyapatite crystals [Hap, $Ca_{10}(PO_4)_6(OH)_2$]. Addition of nanohydroxy-apatite improves the mechanical properties of sponge. Sionkowaska et al. reported synthesis of silk fibroin/collagen/nanohydroxyapatite blend porous 3D scaffold, according to actual consideration of organic/inorganic compositions in natural bones. Hydroxyapatite is non-toxic, nonimmunogenic and biocompatible. Hydroxyapatite is successfully added in silk fibroin-based matrices to use to orthopaedic and dental materials [154].

A freeze-drying technique is used to prepare the porous three-dimensional sponge, combination of silk fibroin/collagen type-I/nanohydroxyapatite 3D scaffold. A total 29 different types of collagens are reported in literature. Out of which, the most abundant in human organisms is collagen type-I. At room temperature and at dry conditions, compression strength and Young's modulus of 3D scaffolds are tested. Silk fibroin/collagen/nano-hydroxyapatite sponge shows smaller flexibility after addition of 50% HAp to biopolymer mixture in comparison to sponge made with fibroins only. Young modulus is smaller when sponge is made up of silk fibroin only. With 50% addition of nanohydroxyapatite for sponge SF90:10Coll the Young's modulus is 5.32 kPa. Therefore, the stiffness of 3D scaffold increases, and flexibility decreases as increase in addition to inorganic nanohydroxyapatite. Pure silk fibroin has bigger compressive strength, resistant to crushing and more flexible. After addition of inorganic particles to SF, compressive strength is decreased because of small volume of protein interacting with inorganic nanohydroxyapatite crystals. Compressive strength is observed 48.2 kPa for SF75:25Coll + 50% Hap blend [155].

Hydroxyapatite/chitosan composite resembles the natural bone composition. Crosslinking between HA clusters and SF fibrils give the three-dimensional porous composite. The SF increases the microhardness of HA ceramics. However, resulted HA-SF composite has insufficient formability and flexibility. Therefore, HA-SF composite cannot be used for bone substitution. Wang et al. reported novel hydroxyapatite/chitosan–silk fibroin (HA/CTS– SF) composite preparation for application of bone repair. The brittleness of HA is suppressed by addition of chitosan. Chitosan/HA composite reported by Yamaguchi et al. is flexible. By using this advantage, author used chitosan and SF together as a complex organic matrix and then used granules of HA. HA/CTS–SF as a bone scaffold material showed good osteoconductivity, enhanced mechanical strength and sufficient formability and flexibility [156].

Titanium dioxide (TiO_2) is lightweight in material and it is frequently used in the fabrication of bone implants. It is non-toxic, shows anti-bacterial properties, resistant to corrosion, inexpensive nanoparticles. In severer injuries, wound infections and their dressing play signicant role in wound healing process. Silk fibroin/Chitin/TiO_2 bio composites reported as wound dressing bandages with strong anti-bacterial properties. Chitin is derivative of chitosan. At neutral pH, chitin does not have antibacterial properties. To improve antibacterial properties titanium dioxide is used. Due to electrostatic interactions, results are in controlled release of TiO_2 nanoparticles. Released nanoparticles stick to the bacterial cell wall, leading to morphological changes and cell wall material transfer disturbance, which showed silent antibacterial and antifungal activities against all strains [157]. Zinc oxide has received growing interest in drug delivery because of non-toxicity, cost-effectiveness [158].

Metal oxide nanoparticles can interact with plasma proteins by electrostatic attraction. Therefore, silk fibroin modified disulfiram and ZnO nanocomposite developed for pH-triggered zinc ion can be released for anti-cancer activity. Dithiocarbamate family, which is able to strongly chelate metal ions, such as Zn^{2+}, and DSF with Zn^{2+} increases the cell-killing effect. SF/DSF@ZnO nanosystem shows synergistic antitumor effect not only a drug delivery carrier but also a metal ion reservoir for Zn^{2+} quick release [159]. Inorganic magnetic nanoparticles commonly used in biomedical applications like tissue repair and cellular therapy [160].

Niamsa et al. [161] reported chitosan/SF blend films with nanoparticles of methoxypoly(ethylene glycol)-b-poly(D,L-lactide) (MPEG-b-PDLL).The nanoparticle containing CS/SF blend films showed increased tensile strength and Young's modulus, with decrease in elongation. Added nanoparticles play important role in reinforcing filler of the blend films.

4.2.5. Cross Linking

The two-component and three-component blending of silk fibroin with natural/synthetic polymer shows different biomedical applications. Such modified composites can be treated with cross-linkers to improve different parameters like stability in water conditions and the improvement in degradation resistance. There are three types of cross- linkings: enzymatic cross-linking, physical cross-linking and chemical cross-linking (Figure 21) [162].

In continuation with addition of inorganic material, the use of chemical cross-linking agents to blend with Silk fibroin/collagen has increased to enhance different parameters (Figure 23). The structural formulas of some chemical cross-linkers are genipin [163], dialdehyde starch [164], glyoxal [165], EDC 1-Ethyl-3-(3-dimethyl-aminopropyl-1-carbodiimide), N-hydroxysuccinimide (NHS) and EDC/NHS mixture [166].

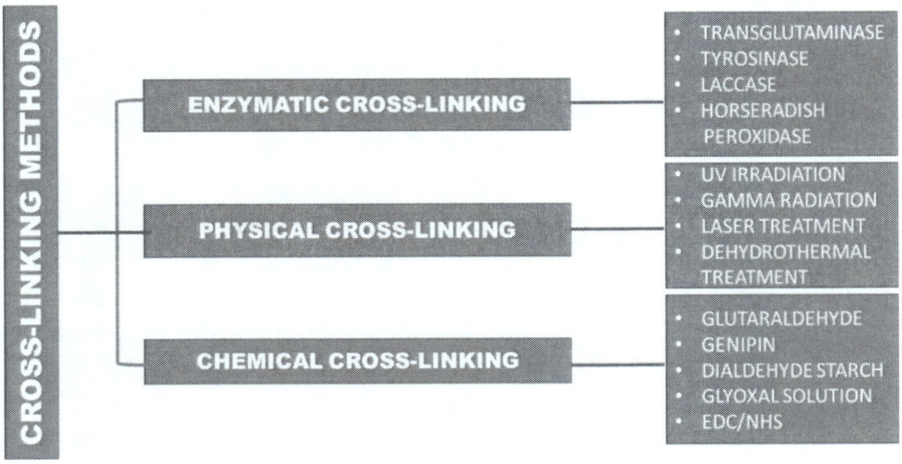

Figure 23. Types of cross-linking.

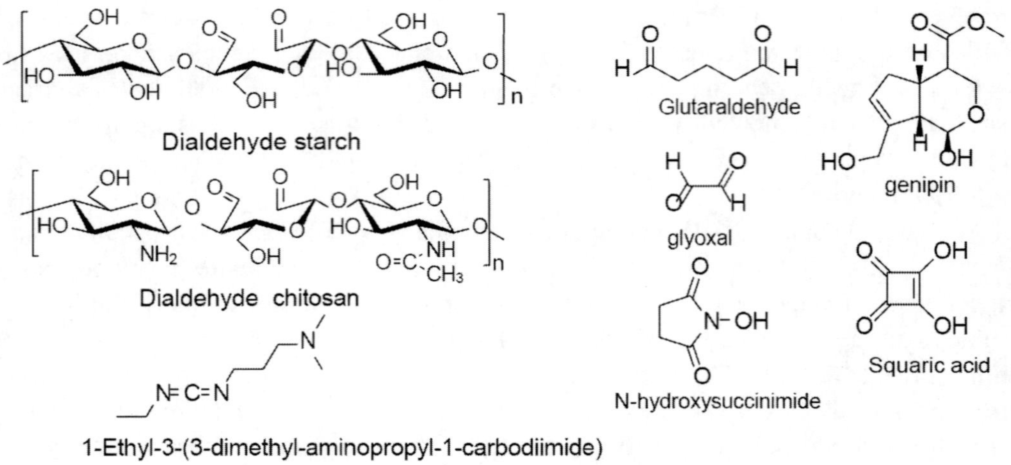

Figure 24.The structural formulas of some chemical cross-linkers.

The porous sponge 3D scaffold of silk fibroin/collagen is cross-linked with 5% solution of EDC/NHS in 98% ethanol solution. Young Modulus is a measurement of wet as well dry samples showing drastic differences. Dry sample with SF50:50Coll shows bigger value and indicates more flexibility. In contrast, wet cross-linked 3D scaffolds are more rigid and less elastic. Collagen content in the samples affects the Young Modulus value. Therefore, the dry sample and high content of collagen resulted in two to three times bigger Young Modulus values. Also cross-linking increases the compressive strength [166].

Glutaraldehyde as a crosslinking agent mixed with blend of chitosan and silk fibroin can form hydrogel with semi-interpenetrating polymer network. Silk fibroin/chitosan blend film is used as a drug delivery carrier for different drugs e.g., Theophylline, salicylic acid, diclofenac sodium, and amoxicillin. The blend film with 80% chitosan content could release maximum amount of drug and shows highest swelling ability of blend film. In acidic medium, protonation of amino group on chitosan is results in dissociation of hydrogen bond between silk fibroin and chitosan. The difference in the rate and amount of drug release is the result of solubility of drugs in blend solutions. The role of crosslinking agent glutaraldehyde is to retain gel structure of blend after swelling [167].

Conclusion

Silkworm silk is one of the natural proteins. Its primary structure is unique and forms crystalline region with β-sheet conformation. These β-sheet possess various molecular interactions like hydrogen bonds, van der Waals interaction and hydrophobic interactions. The crystalline region contributes to the strength of silk fibroin. In contrast to crystalline region, amorphous region composed of random/α-helix conformations, contribute to the elasticity of SF. Through different chemical processes, changes can be induced in structure of the silk fibroin to increase the mechanical, optical, thermal as well as chemical properties. The desired modifications in properties of silk fibroin enable it to use in different fields like biomedical, tissue engineering and optical devices.

References

[1] (a) Altman, G. H., Diaz, F., Jakuba, C., Calabro, T., Horan, R. L., Chen, J., Lu, H., Richmond, J. and Kaplan, D. L. (2003). Silk-based biomaterials. *Biomater.*, 24: 401–416; (b) Nguyen, T. P., Nguyen Q. V., Nguyen, V.-H., Le, T.-H., Huynh V. Q. N., Vo, D.-V. N., Trinh, Q. T., Kim, S. Y., Le, Q. V., (2019). Silk Fibroin-Based Biomaterials for Biomedical Applications: A Review. *Polymers*, 11: 1933; (c)"Textile Exhibition: Introduction." *Asian Art.* Retrieved 2007-08-02.

[2] Banno, Y., Shimada T., Kajiura, Z., and Sezutsu, H. (2010). The Silkworm—An Attractive Bio Resource Supplied by Japan. *Exp. Anim.* 59(2): 139–146.

[3] Tayal, M. K. and Chauhan, T. P. S. (2017). Silkworm Diseases and Pests. In: Omkar (eds.) *Industrial Entomology*. Springer.

[4] Reddy, R. M., (2009). Innovative and multidirectional applications of natural fiber, silk—*A review. Acad. J. Entomol.* 2 (2): 71–75.

[5] Rahul, K., Manjunatha, G. R. and Sivaprasad V. (2021). Pebrine monitoring methods in sericulture. *Methods in Microbiology*. 49: 79-96, ISSN 0580-9517.

[6] Näegeli, A. (1857). Ueber die neueKrankhiet der Seidenraupe und verwandteorganismen *[On the new disease of the silkworm and related organisms]. Botanische Zeitung,* 15: 760–761.

[7] http://silkwormmori.blogspot.com/2010/06/aspergillosis-in-silkworm.html.

[8] Yunusov, L., (1978). *Fiziko-khimicheskiesvoistvanatural'nogoshelka v protsesse pererabotki kokonov [Physicochemical Properties of Natural Silk in Cocoon Processing], Tashkent: Fan.*

[9] Padamwar, M. N. and Pawar, A. P. (2004). Silk sericin and its applications: A review. *J. Sci. Ind. Res.* 63(4): 323–329.

[10] Shimura, K., Kikuchi, A., Ohtomo, K., Katagata, Y. and Hyodo, A. J. (1976). Chemical composition and biosynthesis of silk protein. *Biochem.* 80: 693−702.

[11] Inoue, S., Tanaka, K., Arisaka, F., Kimura, S., Ohtomo, K. and Mizuno, S. J. (2000). Silk Fibroin of *Bombyx mori* is secreted, Assembling a High Molecular Mass Elementary Unit Consisting of H-chain, L-chain, and P25, with a 6:6:1 Molar Ratio. *J. Biol. Chem.* 275: 40517−40528.

[12] Tanaka, K., Inoue, S. and Mizuno, S. (1999). Hydrophobic interaction of P25, containing Asn-linked oligosaccharide chains, with the H-L complex of silk fibroin produced by *Bombyx mori. Insect. Biochem. Mol. Biol.* 29: 269−276.

[13] Takei F., Kikuchi Y., Kikuchi A., Mizuno S., Shimura K. (1987). Further evidence for importance of the subunit combination of silk fibroin in its efficient secretion from the posterior silk gland cells. *The Journal of Cell Biology*. 105(1): 175–180.

[14] Tanaka K., Kajiyama N., Ishikura K., Waga S., Kikuchi A., Ohtomo K., Takagi T., Mizuno S. (1999). Determination of the site of disulfide linkage between heavy and light chains of silk fibroin produced by *Bombyx mori. Biochim. Biophys. Acta, Protein Struct. Mol. Enzymol.* 1432(1): 92–103.

[15] Tanaka, K., Mori, K. and Mizuno, S. (1993). Immunological identification of the major disulfide-linked light component of silk fibroin. *J. Biochem.* 114: 1–4.

[16] Takei, F., Oyama, F., Kimura, K., Hyodo, A., Mizuno, S. and Shimura, K. (1984). Reduced level of secretion and absence of subunit combination for the fibroin synthesized by a mutant silkworm, Nd(2). *J. Cell Biol.* 99 (6): 2005−2010.

[17] Yamaguchi, K., Kikuchi, Y., Takagi, T., Kikuchi, A., Oyama, F., Shimura, K. and Mizuno, S. J. (1989). Primary structure of the silk fibroin light chain determined by cDNA sequencing and peptide analysis. *Mol. Biol.* 210: 127−139.

[18] Garel, J. (1982). The silkworm, a model for molecular and cellular biologists. *Trends Biochem. Sci.* 7: 105−108.

[19] (a) Zhou, C. Z., Confalonieri, F., Medina, N., Zivanovic, Y., Esnault, C., Yang, T., Jacquet, M., Janin, J., Duguet, M. and Perasso, R. and Li, Z. G. (2000). Fine organization of *Bombyx mori* fibroin heavy chain gene. *Nucleic Acids Res.* 28: 2413–2419; (b) Mu, X., Sahoo J. K., Cebe, P. and Kaplan, D. L. (2020), Photo- Cross-linked Silk Fibroin for 3D Printing. *Polymers* 12: 2936.

[20] Shimura, K., Kikuchi, A., Katagata, Y., and Ohotomo, K. (1982). *J. Seric. Sci. Jpn.*, 51, (1): 20 - 26.

[21] Sashina, E. S., Bochek, A. M., Novoselov, N. P. and Kirichenko, D. A. (2006). Structure and Solubility of Natural Silk Fibroin, *Russ. J. App. Chem.*, 79(6): 869- 876.

[22] Rising, A., Hjälm, G., Engström, W. and Johansson, J. (2006). N-terminal nonrepetitive domain common to dragline, flagelliform, and cylindriform spider silk proteins. *Biomacromolecules* 7(11): 3120–3124.

[23] Sponner, A., Vater W., Rommerskirch, W., Vollrath, F., Unger, E., Grosse, F. and Weisshart, K. (2005). The conserved C-termini contribute to the properties of spider silk fibroins. *Biochem. Biophys. Res. Commun.* 338(2): 897–902.

[24] Bini, E., Knight, D. P., Kaplan, D. L. (2004). Mapping domain structures in silks from insects and spiders related to protein assembly, *J. Mol. Biol.* Jan 2; 335(1):27-40.

[25] Termonia, Y. (1994). Molecular modeling of spider silk elasticity. *Macromolecules*. 27(25): 7378–7381.

[26] Simmons, A., Ray, E. and Jelinski, L. W. (1994). Solid-state ^{13}C NMR of Nephilaclavipes dragline silk establishes structure and identity of crystalline regions. *Macromolecules.* 27(18): 5235–5237.

[27] Asakura, T. and Yao, J. (2002). ^{13}C CP/MAS NMR study on structural heterogeneity in *Bombyx mori* silk fiber and their generation by stretching. *Protein Science* 11(11): 2706–2713.

[28] Boulet-Audet, M., Lefèvre, T., Buffeteau, T. and Pézolet, M. (2008). Attenuated total reflection infrared spectroscopy: An efficient technique to quantitatively determine the orientation and conformation of proteins in single silk fibers. *Applied Spectroscopy*, 62(9): 956–962.

[29] (a) Liang, C. X. and Hirabayashi, K. (1992), Improvements of the physical properties of fibroin membranes with sodium alginate. *J. Appl. Polym. Sci.* 45: 1937–1943; (b) Kratky, O., Schauenstein, E. and Sekora, A. (1950), An Unstable Lattice in Silk Fibroin. *Nature* 165: 319–320; (c) Hu, X., Shmelev, K., Sun, L., Gil, E. S., Park, S. H., Cebe, P. and Kaplan, D. L. (2011). Regulation of Silk Material Structure by Temperature-Controlled Water Vapor Annealing. *Biomacromolecules*, 12: 1686–1696; (d) Cebe, P., Partlow, B. P., Kaplan, D. L., Wurm, A., Zhuravlev, E. and Schick, C. (2017). Silk I and Silk II studied by fast scanning calorimetry. *Acta Biomater.* 55: 323–332; (e) Valluzzi, R., Gido, S. P., Muller, W. and Kaplan, D. L. (1999), Orientation of silk III at the air-water interface. *Int. J. Biol. Macromol.* 24: 237–242; (f) Asakura, T., (2021), Structure of Silk I (Bombyx mori. Silk Fibroin before Spinning) -Type II β-Turn, Not α- Helix. *Molecules*, 26: 3706; (g) Marsh, R. E., Corey, R. B. and Pauling, L. (1955).An investigation of the structure of silk fibroin. *Biochim. Biophys. Acta.* 16: 1–34.

[30] (a) Rousseau, M. E., Lefèvre, T., Beaulieu, L., Asakura, L. and Pézolet, M. (2004). Study of protein conformation and orientation in silkworm and spider silk fibers using Raman microspectroscopy. *Biomacromolecules*, 5(6): 2247– 2257; (b) Strassburg, S., Zainuddin, S., Scheibel, T. (2021). The Power of Silk Technology for Energy Applications, *Adv. Energy Mater.* 11, 2100519.

[31] Marsh, R. E., Corey, R. B. and Pauling, L. (1955). An investigation of the structure of silk fibroin. *Biochimica et Biophysica. Acta.* 16: 1–34.

[32] (a) Liang, C. X. and Hirabayashi, K. (1992), Improvements of the physical properties of fibroin membranes with sodium alginate. *J. Appl. Polym. Sci.* 45: 1937–1943; (b) Kratky, O., Schauenstein, E. and Sekora, A. (1950), An Unstable Lattice in Silk Fibroin. *Nature* 165: 319–320; (c) Hu, X., Shmelev, K., Sun, L., Gil, E. S., Park, S. H., Cebe, P. and Kaplan, D. L. (2011). Regulation of Silk Material Structure by Temperature-Controlled Water Vapor Annealing. *Biomacromolecules*, 12: 1686–1696; (d) Cebe, P., Partlow, B. P., Kaplan, D. L., Wurm, A., Zhuravlev, E. and Schick, C. (2017). Silk I and Silk II studied by fast scanning calorimetry. *Acta Biomater.* 55: 323–332; (e) Valluzzi, R., Gido, S. P., Muller, W. and Kaplan, D. L. (1999), Orientation of silk III at the air-water interface. *Int. J. Biol. Macromol.* 24: 237–242; (f) Asakura T., (2021), Structure of Silk I (*Bombyx mori*. Silk Fibroin before Spinning) -Type II β-Turn, Not α- Helix. *Molecules*, 26: 3706; (g) Marsh, R. E., Corey, R. B. and Pauling, L. (1955). An investigation of the structure of silk fibroin. *Biochim. Biophys. Acta.* 16: 1–34.

[33] Johari, N., Moroni, L. and Samadikuchaksaraei, A. (2020). Tuning the conformation and mechanical properties of silk fibroin hydrogels. *Eur. Polym. J.* 134: 109842.

[34] Perez-Rigueiro, J., Viney, C., Llorca, J. and Elices, M. (1998) Silkworm silk as an engineering material, *J. Appl. Poly. Sci.*, 70 (12): 2439–2447.

[35] Perez-Rigueiro, J., Viney, C., Llorca, J. and Elices, M. (2000b). Mechanical properties of single-brin silkworm silk. *J. Appl. Poly. Sci.*, 75 (10): 1270–1277.

[36] Aksakal, B., Tsobkallo, E. and Darvish, D. (2009). Mechanical properties of *Bombyx mori.* silk yarns studied with tensile testing method, *J Appl. Polym. Sci*, 113 (4):2514–2523.

[37] Vepari, C. and Kaplan, D. L. (2007). Silk as a biomaterial. *Prog. Polym. Sci.* 32: 991–1007.

[38] Sirichaisit, J., Brookes, V. L., Young, R. J. and Vollrath, F. (2003). Analysis of structure/ property relationships in silkworm (Bombyx mori.) and spider dragline (Nephilaedulis) silks using Raman spectroscopy. *Biomacromolecules*. 4: 387–94.

[39] Cunniff, P. M, Fossey, S. A., Auerbach, M. A., Song, J. W., Kaplan, D. L., Adams, W. W., Eby, R. K., Mahoney, D., Vezie, D. L. (1994). Mechanical and thermal properties of dragline silk from the spider Nephila clavipes. *Polym. Adv. Technol.*, 5:401–10.

[40] Shao, Z. and Vollrath, F. (2002). Surprising strength of silkworm silk. *Nature*. 418: 741.

[41] Gosline, J. M., Guerette, P. A., Ortlepp, C. S. and Savage, K. N. (1999). The mechanical design of spider silks: from fibroin sequence to mechanical function. *J Exp. Biol.* 202: 3295–303.

[42] Krasnov, I., Diddens, I., Hauptmann, N., Helms, G., Ogurreck, M., Seydel, T., Funari, S. S. and Muller, M. (2008). Mechanical properties of silk: interplay of deformation on macroscopic and molecular length scales. *Phys. Rev. Lett.* 100: 048104/1-4.

[43] Pins, G. D., Christiansen, D. L., Patel, R. and Silver, F. H. (1997). Self-assembly of collagen fibers Influence of fibrillar alignment and decorin on mechanical properties. *Biophys J.* 73: 2164–72.

[44] Yang L, van der Werf, K. O., Fitie, C. F., Bennink, M. L., Dijkstra, P. J. and Feijen, J. (2008). Mechanical properties of native and cross-linked type I collagen fibrils. *Biophys J.* 94: 2204–11.

[45] Kumar, S., Doshi, H., Srinivasara, M., Park, J. O. and Schiraldi, D. A. (2002). Fibers from polypropylene/nano carbon fiber composites. *Polymer*. 43: 1701–3.

[46] Keten, S., Xu, Z., Ihle, B. and Buehler, M. J. (2010). Nanoconfinement controls stiffness, strength and mechanical toughness of β-sheet crystals in silk. *Nat Mater.* 9: 359–67.

[47] Xu, G., Toh W. G., Du N. and Liu X. Y. (2012). Spider silk: the toughest natural polymer. *Natural polymers*, 1: 275–304.

[48] Koh, L.-D., Cheng, Y., Teng, C-P., Khin, Y.-W., Loh, X.-J., Tee, S.-Y., Low, M., Ye, E., Yu, H.-D., Zhang, Y.-W., and Han, M.-Y. (2015). Structures, mechanical properties and applications of silk fibroin materials. *Prog. Polym. Sci.*, 46: 86–110.

[49] Wilding, M. A. and Hearle, J. W. S. (1996). Textile Fibers: (Structure and Properties). *Polymeric Materials Encyclopaedia*. 11: 8307–8322. b) Shao, Z. and Vollrath F. (2002). Surprising strength of silkworm silk. *Nature*. 418: 741.

[50] Perez-Rigueiro, J., Elices, M., Llorca, J. and Viney, C. (2002) Effect of degumming on the tensile properties of silkworm (*Bombyx mori*) silk fiber, *J Appl Poly Sci*, 84(7): 1431–1437.

[51] Ajisawa A. (1997). Dissolution aqueous of silk fibroin with calcium chloride/ethanol solution. *J. Seric. Sci. Jpn.* 67:91–94.

[52] Hu, X., Cebe, P., Weiss A. S., Omenetto, F. and Kaplan, D. L. (2012). Protein- based composite materials. *Mater. Today*. 15(5):208–215.

[53] Zhou, C. Z., Confalonieri, F., Jacquet, M., Perasso, R., Li, Z. G. and Janin, J. (2001) Silk fibroin: structural implications of a remarkable amino acid sequence, *Proteins*, 44(2): 119–122.

[54] Shou, L. (1955). Degumming method with organic acid. *Jpn. Silk.*, 1:32–34.

[55] Furuhata, K.-I., Okada, A., Chen, Y., Xu, Y. Y. and Sakamoto, M. (1994). Dissolution of Silk Fibroin in Lithium halide/Organic Amide Solvent System. *J. Seric. Sci. Jpn.*, 63(4): 315-322.

[56] Miyaguchi, Y. and Hu, J. (2005). Physicochemical properties of silk fibroin after solubilization using calcium chloride with or without ethanol. *Food Sci. Technol. Res.*, 11:37–42.

[57] Liu, H., Yang, Q., Zhang, L., Zhuo, R. and Jiang, X. (2016). Synthesis of carboxymethyl chitin in aqueous solution and its thermo- and pH-sensitive behaviors. *Carbohydr. Polym.* 137:600–607.

[58] Shen, T., Wang, T., Cheng, G., Huang, L. Chen, L. and Wu, D. (2018). Dissolution behavior of silk fibroin in a low concentration $CaCl_2$-methanol solvent: From morphology to nanostructure. *Int. J. Biol. Macromol.* 113:458– 463.

[59] Ha, W., Park, Y. H. and Hudson, S. M. (2003). Dissolution of *Bombyx mori* silk fibroin in the calcium nitrate tetrahydrate-methanol system and aspects of wet spinning of fibroin solution. *Biomacromolecules*. 4(3): 488–496.

[60] Fang X., Shao, H. and Hu, X. (2006). Effect of storage time and concentration on structure of regenerated silk fibroin solution. *Int. J. Mod. Phys. B*. 20 (25-27): 3878–3883.

[61] Yin, J., Chen, E. Q., Porter, D. and Shao, Z. (2010). Enhancing the toughness of regenerated silk fibroin film through uniaxial extension. *Biomacromolecules*. 11(11): 2890–2895.

[62] Kook, G., Jeong, S., Kim, S. H., Kim, K. M., Lee, S. Cho, I.-J., Choi, N. and Lee, H. J. (2019). Wafer-scale multilayer fabrication for silk fibroin-based microelectronics. *ACS Appl. Mater. Interfaces*. 11(1):115–124.

[63] Chen, X., Knight, D. P, Shao, Z., and Vollrath, F. (2001). Regenerated *Bombyx silk* solutions studied with rheometry and FTIR. *Polymer*. 42(25):09969–09974.

[64] Wang, F., Cao, T.-T. and Zhang, Y.-Q. (2015). Effect of silk protein surfactant on silk degumming and its properties. *Mater. Sci. Eng. C*. 55:131–136.

[65] Wu, Z. W., Feng, X. X. and Zhu, H. L., et al. (2010). Effects of different solvent systems on molecular mass of silk fibroin and properties of the regenerated silk fibroin membranes. *Sci Sericulture*. 36:0707–0712.

[66] William C. U. (1992). Process for spinning polypeptide fibers from solutions of lithium thiocyanate and liquefied phenol. *US Patent* US5252277.

[67] Bucciarelli, A., Pal, R. K., Maniglio, A. D., Mulloni, V., Motta, A. and Yadavalli V. K. (2017). Fabrication of nanoscale patternable films of silk fibroin using benign solvents. *Macromol. Mater. Eng*. 302(7): 1700110–1700119.

[68] Parker, A. J., (1969). Protic-dipolar aprotic solvent effects on rates of bimolecular reactions. *Chem. Rev.,* 69(1):1-32.

[69] Wang, H. Y. and Zhang, Y. Q. (2013). Effect of regeneration of liquid silk fibroin on its structure and characterization. *Soft Matter*. 9:138–145.

[70] Sashina, E. S., Bochek, A. M., Novoselov, N. P. and Kirichenko, D. A. (2006). Structure and Solubility of Natural Silk Fibroin. *Russ. J. Appl. Chem.* 79(6): *869-876.*

[71] Ren, F., Zhu, Z., and Li, J., (1997). Nutritious foodstuff-silk extract made from natural silk and making method. *China Patent*. CN1152410A.

[72] Yamaguchi, T. and Mori, M. (1995). Production of water-soluble silk fibroin. *Jpn. Patent,* JP07-067687, Mar 14.

[73] Jin, M. G., (2003). Catalytic circular preparation of fibroin peptide. *China Patent*. CN1099424C, Jan 22.

[74] (a) Liu, H., Yang, Q., Zhang, L., Zhuo, R., & Jiang, X. (2016). Synthesis of carboxymethyl chitin in aqueous solution and its thermos and pH-sensitive behaviors. *Carbohydr. Polym.* 137:600–607; (b) Niu, Q., Peng, Q. F., Lu, L., Fan, S., Shao, H., Zhang, H., Wu, R., Hsiao, B. S. and Zhang, Y., (2018), Single molecular layer of silk nanoribbon as potential basic building block of silk materials. *ACS Nano,* 12 (12): 11860–1187.

[75] Ming, J., Liu, Z., Bie, S., Feng, Z. and Zuo, B. (2014). Novel silk fibroin films prepared by formic acid/hydroxyapatite dissolution method. *Mater. Sci. Eng. C*. 37:48–53.

[76] Zhang, F., Lu, Q., Yue, X., Zuo, B., Qin, M., Li, F., Kaplan, D. L. and Zhang, X. (2015). Regeneration of high quality silk fibroin fiber by wet spinning from $CaCl_2$-formic acid solvent. *Acta. Biomater*. 12:139–145.

[77] Liu, Q., Wang, F, Gu, Z., Ma, Q. and Hu, X. (2018). Exploring the structural transformation mechanism of Chinese and Thailand silk fibroin fibers and formic- acid fabricated silk films. *Int. J. Mol. Sci*. 19(11):3309.

[78] Kiyoshi, Y. (1990 Sep 25). Aqueous silk fibroin solution with excellent storage stability and preparation there of. *Japan Patent*. JP2240165A.

[79] Medronho, B., Filipe, A., Napso, S., Khalfin, R. L., Pereira, R. F. P., Bermudez, V. Z., Romano, A. and Cohen, Y. (2019). Silk fibroin dissolution in tetrabutylammonium hydroxide aqueous solution. *Biomacromolecules*. 20(11):4107–4116.

[80] Phillips, D. M., Drummy, L. F., Conrady, D. G., Fox, D. M., Naik, R. R., Stone, M. O., Trulove, P. C., de Long, H. C. and Mantz, R. A. (2004). Dissolution and regeneration of *Bombyx mori* silk fibroin using ionic liquids. *J. Am. Chem. Soc.* 126(44):14350–14531.

[81] Wang, Y., Wei, Z. G. and Zhang, Y. Q. (2020). Dissolution and regeneration of silk from silkworm *Bombyx mori* in ionic liquids and its application to medical biomaterials. *Int. J. Biol. Macromol.* 143:594–601.

[82] Kiyoshi, O. and Hisao, H. (1979 Jul 30). Preparation of fibroin dope. *Japan Patent*. JP54096126.

[83] Freddi, G., Pessina, G. and Tsukada, M. (1999). Swelling and dissolution of silk fibroin (*Bombyx mori*) in N-methyl morpholine N-oxide. *Int. J. Biol. Macromol.* 24(2-3): 251–263.

[84] Xu, Y., Zhang, Y., Shao, H., Hu, X., (2005). Solubility and rheological behavior of silk fibroin (Bombyx mori.) in N-methyl morpholine N-oxide. *Int. J. Biol. Macromol.* 35:155-61.

[85] Rockwood, D. N., Preda, R. C., Yücel, T., Wang, X., Lovett, M. L. and Kaplan, D. L. (2011). Materials fabrication from *Bombyx mori* silk fibroin. *Nat. Protoc.*, 6(10):1612–1631.

[86] Xu, Y., Zhang, Y., Shao, H., Hu, X. (2005). Solubility and rheological behavior of silk fibroin (*Bombyx mori*) in N-methyl morpholine N-oxide. *Int. J. Biol. Macromol.*, 35(3-4):155–161.

[87] Tsukada, M., Freddi, G., Nagura, M., Ishikawa H. and Kasai, N. (1992). Structural changes of silk fibers induced by heat treatment. *J. Appl. Polym. Sci.*, 46: 1945.

[88] Hu, X., Kaplan D. and Cebe, P. (2006). Determining Beta-Sheet Crystallinity in Fibrous Proteins by Thermal Analysis and Infrared Spectroscopy. *Macromolecules*, 39: 6161.

[89] Hu, X., Kaplan D. and Cebe, P. (2007). Effect of water on the thermal properties of silk fibroin. *Thermochim. Acta.*, 461 137.

[90] Chen, H., Hu X and Cebe P. (2008). Thermal Properties and Phase Transitions in Blends of Nylon-6 with Silk Fibroin. *Journal of Thermal Analysis and Calorimetry*, 93(1): 201–206.

[91] Hu, F., Lin, N. and Liu, X. Y. (2020), Interplay between Light and Functionalized Silk Fibroin and Applications. *iScience*, 23(4): 101035.

[92] Perotto, G., Zhang, Y., Naskar, D., Patel, N., Kaplan, D. L., Kundu, S. C. and Omenetto, F. G. (2017). The optical properties of regenerated silk fibroin films obtained from different sources. *Appl. Phys. Lett.*, 111: 103702.

[93] Bucciarelli, A., Mulloni, V., Maniglio, D., Pal, R. K., Yadavalli, V. K., Motta, A. and Quaranta, A. (2018). A comparative study of the refractive index of silk protein thin films towards biomaterial based optical devices. *Opt. Mater.*, 78: 407-414.

[94] Asakura, T., Tanaka, T. and Tanaka, R. (2019). Advanced Silk Fibroin Biomaterials and Application to Small-Diameter Silk Vascular Grafts. *ACS Biomater. Sci. Eng.* 5: 5561–5577.

[95] Grabska-Zielińska, S. and Sionkowska, A. (2021). How to Improve Physico-Chemical Properties of Silk Fibroin Materials for Biomedical Applications? — Blending and Cross-Linking of Silk Fibroin—A Review. *Materials, 14*(6): 1510.

[96] (a) Yoshimizu, H. and Asakura, T. (1990). Preparation and Characterization of Silk Fibroin Powder and Its Application to Enzyme Immobilization. *J. Appl. Polym. Sci.* 40: 127–134; (b) Yeo, J.-H., Lee, K.-G., Lee, Y.-W. and Kim, S. Y. (2003). Simple Preparation and Characteristics of Silk Fibroin Microsphere. *Eur. Polym. J.* 39:1195–1199; (c) Lammel, A. S., Hu, X., Park, S.-H., Kaplan, D. L. and Scheibel, T. R. (2010). Controlling Silk Fibroin Particle Features for Drug Delivery. *Biomaterials*, 31: 4583–4591; (d) Cao, Z., Chen, X., Yao, J., Huang, L. and Shao, Z. (2007). The Preparation of Regenerated Silk Fibroin Microspheres. *Soft Matter*, 3:910–915; (e) Wenk, E., Wandrey, A. J., Merkle, H. P. and Meinel, L. (2008). Silk Fibroin Spheres as a Platform for Controlled Drug Delivery. *J. Control. Release* 132: 26–34; (f) Wongpinyochit, T., Uhlmann, P., Urquhart, A. J. and Seib, F. P. (2015). PEGylated Silk Nanoparticles for Anticancer Drug Delivery. *Biomacromolecules*, 16: 3712–3722; (g) Wang, X., Yucel, T., Lu, Q., Hu, X. and Kaplan, D. L. (2010). Silk Nanospheres and Microspheres from Silk/Pva Blend Films for Drug Delivery. *Biomaterials*, 31: 1025–1035.

[97] Babu, M. K. (2019). Developments in the processing and applications of silk. In *Silk: Processing, Properties and Applications,* 2nd ed.; Elsevier Ltd.: Amsterdam, The Netherlands, 129–142. ISBN 978-0-08-102540-6.

[98] (a) Jin, H.-J., Park, J., Karageorgiou, V., Kim, U.-J., Valluzzi, R., Cebe, P. and Kaplan, D. L. (2005). Water-Stable Silk Films with Reduced β-Sheet Content. *Adv. Funct. Mater.*15:1241–1247; (b) Perry, H., Gopinath, A., Kaplan, D. L., Dal Negro, L. and Omenetto, F. G. (2008). Nano- and Micropatterning of Optically Transparent, Mechanically Robust, Biocompatible Silk Fibroin Films. *Adv. Mater.* 20: 3070–3072; (c) Rajkhowa, R., Levin, B., Redmond, S. L., Li, L. H., Wang, L., Kanwar, J. R., Atlas, M. D. and Wang, X. (2011). Structure and Properties of Biomedical Films Prepared from Aqueous and Acidic Silk Fibroin Solutions. *J. Biomed. Mater. Res. Part A* 97A: 37–45; (d) Kweon, H., Um, I. C. and Park, Y. H. (2001). Structural and Thermal Characteristics of Antheraea Pernyi Silk Fibroin/Chitosan Blend Film. *Polymer* 42: 6651–6656; (e) Sionkowska, A. and Planecka, A. (2011). The Influence of UV Radiation on Silk Fibroin. *Polym. Degrad. Stab.* 96: 523–528.

[99] Asakura, T., Yoshimizu, H. and Kakizaki, M. (1990). An ESR Study of Spin- Labeled Silk Fibroin Membranes and Spin-Labeled Glucose Oxidase Immobilized in Silk Fibroin Membranes. *Biotechnol. Bioeng.* 35: 511–517.

[100] Kundu, B., Rajkhowa, R., Kundu, S. C. and Wang, X. (2013). Silk Fibroin Biomaterials for Tissue Regenerations. *Adv. Drug Deliver. Rev.* 65: 457–470.

[101] Kim, U.-J., Park, J., Li, C., Jin, H.-J., Valluzzi, R. and Kaplan, D. L. (2004). Structure and Properties of Silk Hydrogels. *Biomacromolecules* 5: 786–792.

[102] Smirnova, I. and Gurikov, P. (2018). Aerogel Production: Current Status, Research Directions, and Future Opportunities. *J. Supercrit. Fluids.* 134: 228– 233.

[103] Baldino, L., Cardea, S., Scognamiglio, M. and Reverchon, E. (2019). A New Tool to Produce Alginate-Based Aerogels for Medical Applications, by Supercritical Gel Drying. *J. Supercrit. Fluids.*146: 152–158.

[104] Baldino, L., Cardea, S. and Reverchon, E. (2015). Natural Aerogels Production by Supercritical Gel Drying. *Chem. Eng. Trans.* 43:739–744.

[105] Li, Z. H., Wang, L., Dai, H. L., Wang, X. Y., Li, J. S. and Zhao, Z. (2019). Fabrication, characterization, and in vitro evaluation of biomimetic silk fibroin porous scaffolds via supercritical CO_2 technology. *J. Supercrit. Fluids.*150: 86– 93.

[106] Baldino, L., Concilio, S., Cardea, S. and Reverchon, E. (2016). Interpenetration of Natural Polymer Aerogels by Supercritical Drying. *Polymers*, 8: 106.

[107] Sionkowska, A., Lewandowska, K., Michalska, M. and Walczak, M. (2016). Characterization of Silk Fibroin 3D Composites Modified by Collagen. *J. Mol. Liq.* 215: 323–327.

[108] Ghaeli, I., de Moraes, M., Beppu, M., Lewandowska, K., Sionkowska, A., Ferreira-da-Silva, F., Ferraz, M. and Monteiro, F. (2017). Phase Behaviour and Miscibility Studies of Collagen/Silk Fibroin Macromolecular System in Dilute Solutions and Solid State. *Molecules* 22: 1368.

[109] Lin, X.-L., Gao, L.-L., Li, R., Cheng, W., Zhang, C.-Q. and Zhang, X. (2019). Mechanical Property and Biocompatibility of Silk Fibroin–Collagen Type II Composite Membrane. *Mat. Sci. Eng.* C105: 110018.

[110] Naomi, R., Ratanavaraporn, J. and Fauzi, M. B. (2020). Comprehensive Review of Hybrid Collagen and Silk Fibroin for Cutaneous Wound Healing. *Materials* 13: 3097.

[111] Gobin, A. S., Froude, V. E. and Mathur, A. B. (2005). Structural and Mechanical Characteristics of Silk Fibroin and Chitosan Blend Scaffolds for Tissue Regeneration. *J. Biomed. Mater. Res.* 74A: 465–473.

[112] Sionkowska, A. and Płanecka, A. (2013). Preparation and Characterization of Silk Fibroin/Chitosan Composite Sponges for Tissue Engineering. *J. Mol. Liq.* 178: 5– 14.

[113] Sionkowska, A., Lewandowska, K. and Płanecka, A. (2014). Miscibility and Physical Properties of Chitosan and Silk Fibroin Mixtures. *J. Mol. Liq.* 198: 354– 357.

[114] Ruan, Y., Lin, H., Yao, J., Chen, Z. and Shao, Z. (2011). Preparation of 3D Fibroin/Chitosan Blend Porous Scaffold for Tissue Engineering Via a Simplified Method. *Macromol. Biosci.* 11: 419–426.

[115] Zeng, S., Liu, L., Shi, Y., Qiu, J., Fang, W., Rong, M., Guo, Z. and Gao, W. (2015). Characterization of Silk Fibroin/Chitosan 3D Porous Scaffold and In Vitro Cytology. *PLOS ONE* 10: e0128658.

[116] Li, D.-W., Lei, X., He, F.-L., He, J., Liu, Y.-L., Ye, Y.-J., Deng, X., Duan, E. and Yin, D.-C. (2017). Silk Fibroin/Chitosan Scaffold with Tunable Properties and Low Inflammatory Response Assists

the Differentiation of Bone Marrow Mesenchymal Stem Cells. *Int. J. Biol. Macromol.*105: 584–597.

[117] Zhang, H., Liu, X., Yang, M. and Zhu, L. (2015). Silk Fibroin/Sodium Alginate Composite Nano-Fibrous Scaffold Prepared through Thermally Induced Phase- Separation (TIPS) Method for Biomedical Applications. *Mater. Sci. Eng. C*55: 8– 13.

[118] Ming, J. F., Zuo, B. Q. and Jiang, Y. X. (2012). Calcium Ion Treatment Behavior of Silk Fibroin/Sodium Alginate Scaffolds. *Adv. Mater. Res.* 528: 107–112.

[119] deMoraes, M. A., Silva, M. F., Weska, R. F. and Beppu, M. M. (2014). Silk Fibroin and Sodium Alginate Blend: Miscibility and Physical Characteristics. *Mater. Sci. Eng. C* 40: 85–91.

[120] Zhang, K., Fan, L., Yan, Z., Yu, Q. and Mo, X. (2012). Electrospun Biomimic Nanofibrous Scaffolds of Silk Fibroin/Hyaluronic Acid for Tissue Engineering. *J. Biomater. Sci.*23: 1185–1198.

[121] Foss, C., Merzari, E., Migliaresi, C. and Motta, A. (2013), Silk Fibroin/Hyaluronic Acid 3D Matrices for Cartilage Tissue Engineering. *Biomacromolecules*. 14:38−47.

[122] Guan, Y., You, H., Cai, J., Zhang, Q., Yan, S. and You, R. (2020). Physically crosslinked silk fibroin/hyaluronic acid scaffolds. *Carbohydr. Polym.* 239:116232.

[123] Luetchford, K. A., Chaudhuri, J. B. and De Bank, P. A. (2020). Silk Fibroin/Gelatin Microcarriers as Scaffolds for Bone Tissue Engineering. *Mater. Sci. Eng. C* 106: 110116.

[124] Du, Y., Gao, X., Wang, Z., Jin, D., Tong, S. and Wang, X. (2016). Construction and Characterization of Three-Dimensional Silk Fibroin-Gelatin Scaffolds. *J. Hard Tissue Biol.* 25: 269–276.

[125] Safdari, M., Shakiba, E., Kiaie, S. H. and Fattahi, A. (2016). Preparation and Characterization of Ceftazidime Loaded Electrospun Silk Fibroin/Gelatin Mat for Wound Dressing. *Fibers Polym.*, 17: 744–750.

[126] Burger, D., Beaumont, M., Rosenau, T. and Tamada, Y. (2020). Porous Silk Fibroin/Cellulose Hydrogels for Bone Tissue Engineering via a Novel Combined Process Based on Sequential Regeneration and Porogen Leaching. *Molecules* 25: 5097.

[127] Yang, J., Kwon, G.-J., Hwang, K., Byeon, J., Ghodake, G. S., Shinde, S. K., Hyun, S. and Kim, D.-Y. (2020). Influence of Silk Fibroin Content on Cellulose Blend Film Using LiBr Solution. *Bio Resources*. 15: 2459–2470.

[128] Yang, M., Wang, Y., Tao, G., Cai, R., Wang, P., Liu, L., Ai, L., Zuo, H., Zhao, P. and Umar, A. (2018). Fabrication of Sericin/Agrose Gel Loaded Lysozyme and Its Potential in Wound Dressing Application. *Nanomaterials*. 8, 235.

[129] Bhardwaj, N., Sow, W. T., Devi, D. Ng, K. W., Mandal, B. B. and Cho, N.-J. Silk Fibroin–Keratin Based 3D Scaffolds as a Dermal Substitute for Skin Tissue Engineering. *Integr. Biol.* (2015). 7: 53–63.

[130] Vasconcelos, A., Gomes, A. C. and Cavaco-Paulo, A. Novel Silk Fibroin/Elastin Wound Dressings. *Acta. Biomater.* 2012, 8, 3049–3060.

[131] Yoo, C. R., Yeo, I.-S., Park, K. E., Park, J. H., Lee, S. J., Park, W. H. and Min, B.-M. (2008). Effect of Chitin/Silk Fibroin Nanofibrous Component Structures on Interaction with Human Epidermal Keratinocytes. *Int. J. Biol. Macromol.*, 42: 324–334.

[132] Wang, S., Zhang, Y., Wang, H. and Dong, Z. (2011). Preparation, Characterization and Biocompatibility of Electrospinning Heparin-Modified Silk Fibroin Nanofibers. *Int. J. Biol. Macromol.*, 48: 345–353.

[133] Nourmohammadi, J., Roshanfar, F., Farokhi, M. and Nazarpak, M. H. (2017).Silk Fibroin/Kappa-Carrageenan Composite Scaffolds with Enhanced Biomimetic Mineralization for Bone Regeneration Applications. *Mater. Sci. Eng. C Mater. Biol. Appl.* 76, 951–958.

[134] Kweon, H., Yeo, J., Lee, K., Lee, H. C., Na, H. S., Won, Y. H. and Cho, C. S. (2008). Semi-Interpenetrating Polymer Networks Composed of Silk Fibroin and Poly (Ethylene Glycol) for Wound Dressing. *Biomed. Mater.* 3: 034115.

[135] Sayed, M. M., Mousa, H. M., El-Aassar, M. R., El-Deeb, N. M., Ghazaly, N. M., Dewidar, M. M. and Abdal-hay, A. (2019). Enhancing Mechanical and Biodegradation Properties of Polyvinyl Alcohol/Silk Fibroin Nanofibers Composite Patches for Cardiac Tissue Engineering. *Mater. Lett.* 255: 126510.

[136] Li, G., Kong, Y., Zhao, Y., Zhao, Y., Zhang, L. and Yang, Y. (2015). Fabrication and Characterization of Polyacrylamide/Silk Fibroin Hydrogels for Peripheral Nerve Regeneration. *J. Biomat. Sci. Polym. E* 26: 899–916.
[137] Lee, J. M., Chae, T., Sheikh, F. A., Ju, H. W., Moon, B. M., Park, H. J., Park, Y. R. and Park, C. H. (2016). Three Dimensional Poly ("-Caprolactone) and Silk Fibroin Nanocomposite Fibrous Matrix for Artificial Dermis. *Mater. Sci. Eng. C* 68: 758–767.
[138] Sahoo, S., Lok Toh, S. and Hong Goh, J. C. (2010), PLGA Nanofiber-Coated Silk Microfibrous Scaffold for Connective Tissue Engineering. *J. Biomed. Mater. Res. Part B Appl. Biomater.* 95: 19–28.
[139] Watcharajittanont, N., Putson, C., Pripatnanont, P. and Meesane, J. (2019). Layer- by-Layer Electrospun Membranes of Polyurethane/Silk Fibroin Based on Mimicking of Oral Soft Tissue for Guided Bone Regeneration. *Biomed. Mater.* 14: 055011.
[140] Hu, K., Lv, Q., Cui, F. Z., Xu, L., Jiao, Y. P., Wang, Y., Feng, Q. L., Wang, H. L., Huang, L. Y. (2007). A Novel Poly(L-Lactide) (PLLA)/Fibroin Hybrid Scaffold to Promote Hepatocyte Viability and Decrease Macrophage Responses. *J. Bioact. Compat. Polym.* 22: 395–410.
[141] Chouhan, D. and Mandal, B. B. (2020). Silk Biomaterials in Wound Healing and Skin Regeneration Therapeutics: From Bench to Bedside. *Acta Biomater.* 103, 24– 51.
[142] Kasoju, N.; Bora, U. (2012).Silk Fibroin in Tissue Engineering. *Adv. Health Mater.*, 1: 393–412.
[143] Gokila, S., Gomathi, T., Vijayalakshmi, K., Alsharani, F. A., Anil, S. and Sudha, P. N. (2018). Development of 3D Scaffolds Using Nanochitosan/Silk- Fibroin/Hyaluronic Acid Biomaterials for Tissue Engineering Applications. *Int. J. Biol. Macromol.* 120: 876–885.
[144] Kwak, H., Shin, S., Lee, H. and Hyun, J. (2019). Formation of a Keratin Layer with Silk Fibroin-Polyethylene Glycol Composite Hydrogel Fabricated by Digital Light Processing 3D Printing. *J. Ind. Eng. Chem.* 72: 232–240.
[145] Najberg, M., Haji Mansor, M., Taillé, T., Bouré, C., Molina-Peña, R., Boury, F., Cenis, J. L., Garcion, E. and Alvarez-Lorenzo, C. (2020). Aerogel Sponges of Silk Fibroin, Hyaluronic Acid and Heparin for Soft Tissue Engineering: Composition- Properties Relationship. *Carbohydr. Polym.*, 237: 116107.
[146] Grabska-Zielińska, S., Sionkowska, A., Coelho, C. C. and Monteiro, F. J. (2020). Silk Fibroin/Collagen/Chitosan Scaffolds Cross-Linked by a Glyoxal Solution as Biomaterials toward Bone Tissue Regeneration. *Materials*, 13: 3433.
[147] Grabska-Zielińska, S.; Sionkowska, A., Carvalho, Â, Monteiro, F. J. (2021). Biomaterials with Potential Use in Bone Tissue Regeneration— Collagen/Chitosan/Silk Fibroin Scaffolds Cross-Linked by EDC/NHS. *Materials*, 14, 1105.
[148] Lu, Q., Zhang, S., Hu, K., Feng, Q.; Cao, C.; Cui, F. (2007). Cyto compatibility and Blood Compatibility of Multifunctional Fibroin/Collagen/Heparin Scaffolds. *Biomaterials*, 28, 2306–2313.
[149] Gu, Z., Xie, H., Huang, C., Li, L. and Yu, X. (2013). Preparation of Chitosan/Silk Fibroin Blending Membrane Fixed with Alginate Dialdehyde For Wound Dressing. *Int. J. Biol. Macromol.*, 58, 121–126.
[150] Jones, G. L., Motta, A., Marshall, M. J., El Haj, A. J. and Cartmell, S. H. (2009). Osteoblast: Osteoclast Co-Cultures on Silk Fibroin, Chitosan and PLLA Films. *Biomaterials,* 30: 5376–5384.
[151] Hirano, S., Nakahira, T., Zhang, M., Nakagawa, M., Yoshikawa, M. and Midorikawa, T. (2002), Wet-Spun Blend Biofibers of Cellulose–Silk Fibroin and Cellulose–Chitin–Silk Fibroin. *Carbohydr. Polym.* 47: 121–124.
[152] Najberg, M., Haji Mansor, M., Taillé, T., Bouré, C., Molina-Peña, R., Boury, F., Cenis, J. L., Garcion, E. and Alvarez-Lorenzo, C. (2020). Aerogel Sponges of Silk Fibroin, Hyaluronic Acid and Heparin for Soft Tissue Engineering: Composition- Properties Relationship. *Carbohydr. Polym.*, 237: 116107.
[153] Kaczmarek, B. and Sionkowska, A. (2018). Chitosan/Collagen Blends with Inorganic and Organic Additive-A Review. *Adv. Polym. Technol.* 37: 2367–2376.
[154] (a) Lamlerd, T., Soontornvipart, K., Kesdangsakonwut, S., Kanokpanont, S. and Damrongsakkul, S. (2017). In Vivo Bone Regeneration of Thai Silk Fibroin Scaffolds with Gelatin, Hydroxyapatite

and Hyaluronic. *Thai J. Vet. Med.* 47: 165–172; (b) Mou, Z.-L., Duan, L.-M., Qi, X.-N. and Zhang, Z.-Q. (2013). Preparation of Silk Fibroin/Collagen/Hydroxyapatite Composite Scaffold by Particulate Leaching Method. *Mater. Lett.* 105: 189–191; (c) Zhou, H., Wang, Z., Cao, H., Hu, H., Luo, Z., Yang, X., Cui, M. and Zhou, L. (2019). Genipin- Crosslinked Polyvinyl Alcohol/Silk Fibroin/Nano-Hydroxyapatite Hydrogel for Fabrication of Artificial Cornea Scaffolds-a Novel Approach to Corneal Tissue Engineering. *J. Biomater. Sci. Polym. Ed.* 30: 1604–1619; (d) Sionkowska, A. and Tuwalska, A. (2020). Preparation and Characterization of New Materials Based on Silk Fibroin, Chitosan and Nanohydroxyapatite. *Int. J. Polym. Anal. Charact.* 25: 315–333.

[155] Sionkowska, A., Michalska, M. and Walczak, M. (2016). Preparation and Characterization of Silk Fibroin/Collagen Sponge with Nanohydroxyapatite. *Mol. Cryst. Liq. Cryst.* 640: 106–112.

[156] (a) Wang, L. and Li C. (2007). Preparation and physicochemical properties of a novelhydroxyapatite/chitosan–silk fibroin composite. *Carbohydr. Polym.* 68: 740–745; (b) Mehrabani, M. G., Karimian, R., Rakhshaei, R., Pakdel, F., Eslami, H., Fakhrzadeh, V., Rahimi, M., Salehi, R. and Kafil, H. S. (2018). Chitin/Silk Fibroin/TiO_2 Bio-Nanocomposite as a Biocompatible Wound Dressing Bandage with Strong Antimicrobial Activity. *Int. J. Biol. Macromol.* 116: 966–976; (c) Kim, J.-H., Sheikh, F. A., Ju, H. W., Park, H. J., Moon, B. M., Lee, O. J. and Park, C. H. (2014). 3D Silk Fibroin Scaffold Incorporating Titanium Dioxide (TiO_2) Nanoparticle (NPs) for Tissue Engineering. *Int. J. Biol. Macromol.* 68: 158–168.

[157] (a) Kim, J.-H., Kim, D.-K., Lee, O. J., Ju, H. W., Lee, J. M., Moon, B. M., Park, H. J., Kim, D. W., Lee, J. H. and Park, C. H. (2016). Osteoinductive Silk Fibroin/Titanium Dioxide/Hydroxyapatite Hybrid Scaffold for Bone Tissue Engineering. *Int. J. Biol. Macromol.* 82, 160–167; (b) Johari, N., Madaah Hosseini, H. R., Samadikuchaksaraei, A. (2020). Mechanical Modeling of Silk Fibroin/TiO2 and Silk Fibroin/ Fluoridated TiO_2 Nanocomposite Scaffolds for Bone Tissue Engineering. *Iran. Polym. J.* 29: 219–224; (c) Jao, W.-C., Yang, M.-C., Lin, C.-H. and Hsu, C.-C. (2012). Fabrication and Characterization of Electrospun Silk Fibroin/TiO_2 Nanofibrous Mats for Wound Dressings: Electrospun Silk Fibroin/TiO_2Nanofibrus Mats. *Polym. Adv. Technol.* 23: 1066–1076.

[158] (a) Salama, A. (2018). Chitosan/Silk Fibroin/Zinc Oxide Nanocomposite as a Sustainable and Antimicrobial Biomaterial. *Cell. Chem. Technol.* 52: 903–907; (b) Hadisi, Z., Farokhi, M., Bakhsheshi-Rad, H. R., Jahanshahi, M., Hasanpour, S., Pagan, E., Dolatshahi-Pirouz, A., Zhang, Y. S., Kundu, S. C. and Akbari, M. (2020). Hyaluronic Acid (HA)-Based Silk Fibroin/Zinc Oxide Core–Shell Electrospun Dressing for Burn Wound Management. *Macromol. Biosci.* 20: 1900328.

[159] Zhao, Y.-Z., Lin, M.-T., Lan, Q.-H., Zhai, Y.-Y., Xu, H.-L., Xiao, J., Kou, L. and Yao, Q. (2020). Silk Fibroin-Modified Disulfiram/Zinc Oxide Nanocomposites for PH Triggered Release of Zn^{2+} and Synergistic Antitumor Efficacy. *Mol. Pharm.* 17: 3857–3869.

[160] (a) Bunea, M. C., Vasile, E., Galateanu, B., Hudita, A., Serban, M. and Zaharia, C. (2017). Silk Fibroin Films Decorated with Magnetic Nanoparticles For Wound Healing Applications. *Mater. Plast.* 54, 83; (b) Aliramaji, S., Zamanian, A. and Mozafari, M. (2017), Super-Paramagnetic Responsive Silk Fibroin/Chitosan/Magnetite Scaffolds with Tunable Pore Structures for Bone Tissue Engineering Applications. *Mater. Sci. Eng. C*, 70: 736–744.

[161] Niamsa, N.; Srisuwan, Y.; Baimark, Y.; Phinyocheep, P. and Kittipoom, S. (2009). Preparation of Nanocomposite Chitosan/Silk Fibroin Blend Films Containing Nanopore Structures. *Carbohydr. Polym.* 78: 60–65.

[162] (a) Reddy, N., Reddy, R. and Jiang, Q. (2015). Crosslinking Biopolymers for Biomedical Applications. *Trends Biotechnol.* 33: 362–369; (b) Johari, N., Moroni, L. and Samadikuchaksaraei, A. (2020). Tuning the Conformation and Mechanical Properties of Silk Fibroin Hydrogels. *Eur. Polym. J.* 134: 109842.

[163] (a) Zheng, L., Lu, H. Q., Fan, H. S., and Zhang, X. D. (2013). Reinforcement and chemical cross-linking in collagen-based scaffolds in cartilage tissue engineering: a comparative study. *Iran. Polym. J.22*: 833–842; (b) Chang, S. T., Chen, L. C., Lin, S. B., & Chen, H. H. (2012). Nano biomaterials application: morphology and physical properties of bacterial cellulose/gelatin composites via crosslinking. *Food Hydrocoll, 27*, 137–144.

[164] Grabska-Zielińska, S., Sionkowska, A., Reczyńska, K and Pamuła, E. (2020). Physico-Chemical Characterization and Biological Tests of Collagen/Silk Fibroin/Chitosan Scaffolds Cross-Linked by Dialdehyde Starch. *Polymers*. 12: 372.

[165] Grabska-Zielińska, S., Sionkowska, A., Coelho, C. C. and Monteiro, F. J. (2020). Silk Fibroin/Collagen/Chitosan Scaffolds Cross-Linked by a Glyoxal Solution as Biomaterials toward Bone Tissue Regeneration. *Materials* 13: 3433.

[166] (a) Sionkowska, A., Michalska, M., Walczak, M., Śmiechowski, K. and Grabska, S. (2016). Preparation and Characterization of Silk Fibroin/Collagen Sponge Modified by Chemical Cross-Linking. *Mol. Cryst. Liq. Cryst.*, 640: 180–190; (b) Grabska-Zielińska, S., Sionkowska, A., Carvalho, Â. and Monteiro, F. J. (2021). Biomaterials with Potential Use in Bone Tissue Regeneration— Collagen/Chitosan/Silk Fibroin Scaffolds Cross-Linked by EDC/NHS. *Materials* 14: 1105; (c) Sionkowska, A. and Tuwalska, A. (2020). Preparation and Characterization of New Materials Based on Silk Fibroin, Chitosan and Nanohydroxyapatite. *Int. J. Polym. Anal. Charact.* 25: 315–333.

[167] Rujiravanit, R., Kruaykitanon, S., Jamieson, A. M. and Tokura, S. (2003). Preparation of Crosslinked Chitosan/Silk Fibroin Blend Films for Drug Delivery System. *Macromol. Biosci.* 3: 604–611.

Chapter 2

Non-Mulberry Silk Fibroin Blends: The Road Underexplored

Radhika Batra and Roli Purwar*
Department of Applied Chemistry, Delhi Technological University, Delhi, India

Abstract

Non- mulberry variety of silk fibroin, due to its fascinating properties especially, excellent biocompatibility and biodegradability, has attracted the attention of researchers worldwide. However, the area of non-mulberry silk fibroin-based blends is relatively underexplored because of the difficulties associated with the processing of silk fibroin itself and the discovery of a compatible solvent for the preparation of the mixture makes it further complicated. In the available limited literature, researchers have reported the preparation of non-mulberry silk fibroin-based blends with other varieties of silk fibroin or polymers like polyvinyl alcohol, agarose, gelatin, etc., in several different forms for a variety of applications, majorly in the biomedical field. This chapter discusses various constructions of non-mulberry silk fibroin-based blend materials and the impact of blending on their physio-chemical properties. The recent applications of such blends in the biomedical field have also been summarized, and the prospects for developing such materials are conferred.

Keywords: non-mulberry, silk fibroin, biomedical, blend, biopolymer

1. Introduction

Developments in the field of materials research over the past century have led to the progression of silk fibroin-based materials for numerous diverse applications from textile to wound healing to tissue engineering [1]. A variety of sources (e.g., spiders, silkworms, mites, scorpions, and flies) are available, from which silk can be extracted. Silk obtained from silkworms is considered a good source of development of material of biomedical interest [2]. There are different types of insects that produce silk fibre and are broadly classified into two varieties, mulberry, and non-mulberry (Figure 1). The silk produced by theBombycidae family (*Bombyx*

* Corresponding Author's Email: roli.purwar@dtu.ac.in.

In: Silk Fibroin
Editor: Shivaji H. Pawar
ISBN: 979-8-88697-402-7

mori) is known as mulberry, and the one made by the Saturniidae family (*Antheraea mylitta, Antheraea assamensis, Philosamia ricini,* etc.) is known as non-mulberry [3]. Natural silk fibre can either be extracted from silkworm glandsor cocoons. The non-mulberry silk fibroin shows a significant difference in its properties fromthe mulberry variety due to structural distinctions. It also possesses RGD sequence, which is known to be responsible for cell adherence and proliferation, making it a material of interest for researchers worldwide [4].

Silk fibroin (SF) has been used as medical sutures and in the textile field for a long time. In the past few decades, the utilization of SF in the biomedical field as a biomaterial has gotten wings due to its low cost, good mechanical properties, biocompatibility, exceptional biodegradability, non-toxicity, good oxygen and water vapour permeability, and easy availability.

Figure 1. Different varieties of silkworms with their cocoons.

Lately, great research studies have been reported using non-mulberry SF extracted from silkworm glands in various forms such as films [5], cryogels [6], nanofibers, and several scaffolds [7–9]. Kasoju et al. have prepared silk fibroin films by extraction of protein from glands of Antheraea Assama silkworm [10]. Kar et al. have reported the preparation of Muga gland extracted silk fibroin 2D films and have found them to be compatible with MG-63 human osteoblast cells [11]. It is pertinent to note that despite such fascinating features, the usage of non-mulberry SF is restricted due to its poor solubility in conventional solvents, which are used to dissolve mulberry SF, its less hydrophilicity, and poor rate of mass loss. Recently few research studies have been reported utilizing ionic liquids as a solvent for dissolution of cocoon extracted non-mulberry silk fibroin, which has introduced all-new prospective avenues of exploring this material for a wide range of applications. Goujan et al. have prepared Antheraea assamensis silkworm cocoon extracted silk fibroin films by dissolving in BMIMCl (1-butyl-3-methylimidazolium chloride), or BMIMAC (1-butyl-3- methylimidazolium acetate), ionic liquids. Electrospun nanofibrous mats of muga silk fibroin- ionic liquid (BMIMAC) solution have been prepared for skin tissue engineering applications [12]. Srivastava et al. have reported the preparation of flexible muga, tasar cocoon extracted silk fibroin films by dissolving in

BMIMAC ionic liquid and have them modified by adding dextrose for wound dressing application [13].

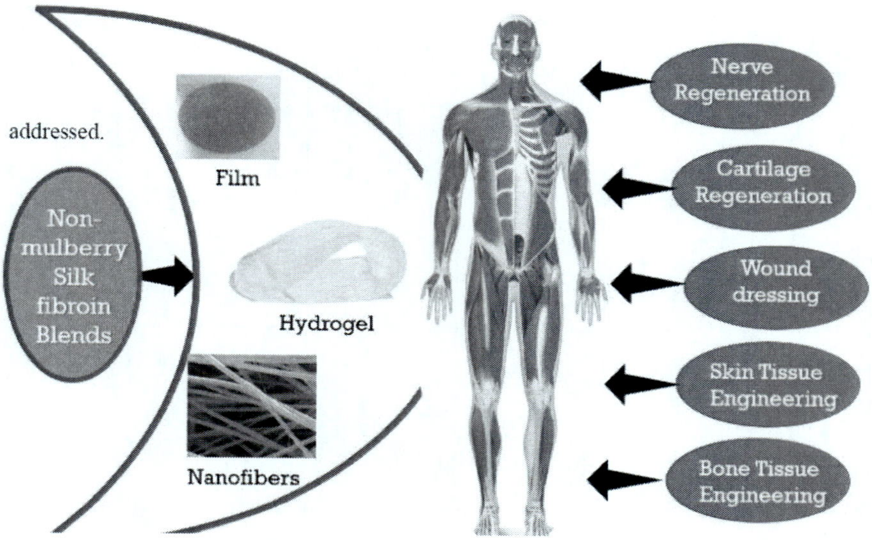

Figure 2. Non-mulberry silk fibroin blend-based constructs and their applications.

Polymer blending is a powerful tool that can generate a new improved class of materials with enhanced physicochemical properties by assembling the advantages of each component [14]. This tool has been utilized quite extensively to enhance the properties of mulberry silk fibroin by blending it with additional polymers such as gelatin [15–17], chitosan [18–22], PVA [23–26], polyethylene oxide [27–29], cellulose [30–34], etc. The research in non-mulberry silk fibroin blends is very limited to date. A few polymers like PVA, PCL, Chitosan, P(LLA-CL) agarose, and gelatin have been explored to make blends with regenerated non-mulberry silk fibroin in the form of films [35–37], scaffolds [6, 38], hydrogels [39] and nanofibers [40–47] for tissue engineering and other biomedical applications (Figure 2). This review article will discuss various forms of non-mulberry silk fibroin blends prepared along with the emphasis on the impact of blending on their structural, thermal, and mechanical properties and biodegradability. Moreover, the recent applications of non-mulberry silk fibroin-based blend for tissue engineering and wound healing applications are briefly addressed.

2. Regenerated Non-Mulberry Silk Fibroin Blends

Silk fibroin must be dissolved into a suitable solvent before blending with some other polymer. There is a wide range of protic and non-protic solvents available in which mulberry silk fibroin can be dissolved, but still these solvents fail or only partially dissolve cocoon- extracted non-mulberry silk fibroin in them. That is one of the primary reasons behind the under exploration of non-mulberry silk fibroin as one of the potential components for blend preparation. A few natural, as well as synthetic polymers have been considered to make blends with regenerated non-mulberry silk fibroin in the form of hydrogels, films, scaffolds, and nanofibers for a variety of applications, as shown in Table 1.

Table 1. Variety of non-mulberry silk fibroin-based blend materials with their respective applications

S. No.	Type of Silk	Blend Material	Construction form	Processing technique	Application	Reference
1	*A. Pernyi* (cocoon)	Poly (L-lactic acid-co-caprolactone)	Nanofibrous scaffolds	Electrospinning	Peripheral nerve tissue regeneration	[43]
2	Tussah (cocoon)	Poly (L-lactic-co-glycolic acid)	Nanofibers	Electrospinning	Bone tissue engineering	[44]
3	*Antheraea pernyi*	Chitosan	Scaffolds	Freeze drying	Bone tissue engineering	[38]
4	i) *B. Mori* (Cocoon) ii) *A. assama* (gland) iii) *P. ricini* (gland)	Poly(vinyl alcohol)	Nanofibrous Mats	Electrospinning	Accelerated wound healing	[40]
5	*B. Mori* (Cocoon)	i) *A. Mylitta* (Tasar) ii) *A. assama* (Muga) iii)*Philosamia ricini* (Eri) iv)Thai mulberry	Films	Casting	-	[36]
6	A. Mylitta (gland)	Poly(Ɛ-caprolactone)	Nanofibrous scaffolds	Electrospinning	Bone tissue engineering	[41, 46]
7	*Antheraea pernyi*	*Bombyx mori*	Films	Casting	-	[35]
8	*Antheraea pernyi*	Chitosan	Films	Casting	-	[37]
9	*Tussah* (cocoon)	*Bombyx mori*	Nanofibers	Electrospinning	Tissue engineering	[47]
11	*Philosamia ricini* (Eri) (cocoon)	*Antheraea* Mylitta (Tasar) (cocoon)	Nanofibrous scaffolds	Electrospinning	Bone tissue grafts	[42]
12	*B. Mori* (Cocoon) *Antheraea Mylitta* (gland)	Chitosan	Scaffolds	Freeze drying	Cartilage repair	[6]
13	*B. Mori* (Cocoon) *Antheraea assama* (gland)	Agarose	Hydrogels	Freeze drying	Cartilage repair	[39]
14	*Antheraea Mylitta* (gland)	Poly(vinyl alcohol)	Nanofibers	Electrospinning	Bone scaffolding material	[45]
15	*Antheraea Mylitta* (cocoon)	Gelatin	Films	Casting	Biomedical constructs	[50]
16	*Antheraea Mylitta* (cocoon)	Poly(vinyl alcohol)	Films	Casting	Biomedical constructs	[48]
17	*Antheraea Mylitta* (cocoon)	Poly(vinyl alcohol)	3D nanowebs	Electrospinning	Tissue engineering	[49]

In most of the studies, silk fibroin has been extracted from the gland of non-mulberry silkworm for the blend preparation [6, 39–41, 45, 46]. A. Mylitta SF extracted from silkworm glands has been blended with poly(e-caprolactone) [41, 46] and PVA [45] for bone tissue engineering applications.

PVA has been chosen as one of the blend materials for blending with non-mulberry SF due to its properties like hydrophilicity, semi crystallinity, non-toxicity, and biocompatibility which are suitable for biomedical applications. Moreover, an ample number of hydroxyl groups in PVA enables it to self-crosslink with fibroin. Our research group has recently developed biodegradable flexible films and 3D nanowebs by blending PVA with Tasar silk fibroin for biomedical applications [48, 49].

PCL is a hydrophobic polymer having suitable mechanical properties, ease in fabrication, biodegradability, and cytocompatibility but it lacks recognition sites, so it cannot provide an optimal environment for cell attachment and growth, so it is being blended with bioactive proteins chitosan, collagen, gelatin, and silk fibroin.

Tussah SF extracted from silkworm cocoons have been blended with poly (L-lactic-co-glycolic acid) [44] and B. Mori SF [47] to generate nanofibers by electrospinning for tissue engineering applications.

Poly(L-lactic-co-caprolactone) which is a copolymer of L-lactic acid and caprolactone, is a synthetic material with good biodegradability and exceptional mechanical properties but has limited use in tissue engineering scaffolds because of its poor hydrophilicity and unavailable cell recognition sites. It has been blended with A. pernyi SF, which possesses poor mechanical properties but has bioactive locations to generate improved tissue engineering materials with enhanced mechanical properties and biocompatibility [43].

Agarose is a transparent polysaccharide, has a neutral charge and thermo reversibility, and has been explored as a biomaterial for tissue engineering of cartilage. But, the applicability of agarose-based materials in animal models has been restricted due to their non- customizable mechanical properties and significantly low cell adhesion and proliferation. To overcome its valid drawbacks, it has been blended with other polymers like chitosan, gelatin, and silk. Non-mulberry silk is a superior choice for blending due to its excellent biocompatibility and tunable mechanical properties [39].

Gelatin is a water-soluble, biocompatible, biodegradable, non-immunogenic, and inexpensive natural protein that is usually obtained from collagen present in the animal tissues like bone, muscle, and skin. It also consists of RGD sequence, which provides it with the cell attachment sites [50]. Gelatin readily dissolves at physiological temperature because of its molecular transitions as a response to thermal change. This shortcoming can be addressed by blending it with non-mulberry silk fibroin which is stable at even higher temperatures. Recently, Antheraea mylitta silk fibroin was blended with gelatin to prepare flexible biodegradable blend films for biomedical applications [51].

3. Techniques for Developing Non-Mulberry Silk Fibroin Blends

To process non-mulberry silk fibroin blends in the form of films, hydrogels, fibres, or scaffolds, techniques like casting, freeze-drying, and electrospinning have been adopted by researchers.

3.1. Non-Mulberry Silk Fibroin Blends by Casting

Silk fibroin blend films can be prepared by casting the blend solution on polystyrene plates. After drying, the casted films are treated with organic solvents like methanol, ethanol, or water to improve the β-sheet crystallinity. Xue et al. studied the thermal and structural properties of cast films prepared by blending B. mori silk fibroin with Tussah, Muga, Eri, and Thai silks in different ratios [36]. The blending of non-mulberry silk fibroin into mulberry showed significant improvement in the thermal properties of the blend material. In another study B. mori/Antheraea pernyi SF blend films were prepared by casting, and their physical properties and thermal behaviour were investigated. The two protein components were suggested to show poor compatibility with clear phase separation. Kweon et al. have reported the preparation of cast blend films by mixing the aqueous Antheraea pernyi SF and chitosan acetic acid solutions. Structural and thermal analysis of these blend films was carried out and was reported to show phase separation but the presence of some intermolecular interactions between the two polymers [37]. Recently, films prepared by blending tasar silk fibroin with gelatin were reported by our research group, which were found to be biocompatible and more flexible than pure tasar silk fibroin films [51]. In another study, we reported successful preparation of flexible PVA/tasar blend films through casting technique without phase separation [48].

3.2. Silk Fibroin Blend Fibrous Scaffolds

Electrospinning is a fibre fabrication technique broadly utilized for forming polymer fibres in the nano to sub-micron range. The electrospun nanofibers have a very high surface- to-volume ratio, tunable porosity, malleability to adapt to a wide variation of shapes and sizes, and the ability to regulate the nanofiber composition to attain anticipated results from its functionality and properties, making them applicable for myriad applications like filtration, wound dressing materials [40], biological scaffolds [6, 43], tissue engineering scaffolds [38, 44, 47]. Recently, few researchers have reported the preparation of non- mulberry silk fibroin nanofibers through electrospinning [12, 36, 40, 42].

Nanofibrous scaffolds by blending Antheraea pernyi silk fibroin and P(LLA-CL) have been fabricated and they have been reported to show superior hydrophilicity and mechanical properties as compared to pure SF [43]. These nanofibrous scaffolds promoted cell growth quite significantly and are considered to be a potential peripheral nerve tissue regeneration material. In another study, electrospun nanofibrous scaffolds were prepared by blending tussah silk fibroin with poly (L-lactic-co-glycolic acid) for bone tissue regeneration [44]. These scaffolds were also bio mineralized with hydroxyapatite, and the resulting material showed excellent mechanical properties, osteoinductivity, and cytocompatibility. Bhattacharjee et al. have conducted a comparative study of gland-extracted A. mylitta SF/poly(e caprolactone) nanofibrous scaffolds with A. mylitta SF grafted aminolyzed electrospun nanofibrous PCL scaffolds [46]. They reported that the inclusion of A. mylitta SF into PCL in both cases improved the osteoinductivity and mechanical properties of scaffolds but grafted SF showed improved cell adhesion, growth, and formation of ECM as compared to blended SF/PCL scaffolds. These nanofibrous blend scaffolds were further investigated for in vivo studies by implanting the scaffolds into white New Zealand rabbits with bone defects and were reported to comprehend the early formation of bone, implant infiltration, and strong bonding at the bone-implant interface with slight immune reaction at initial stage but was normal within four weeks of implantation [41].

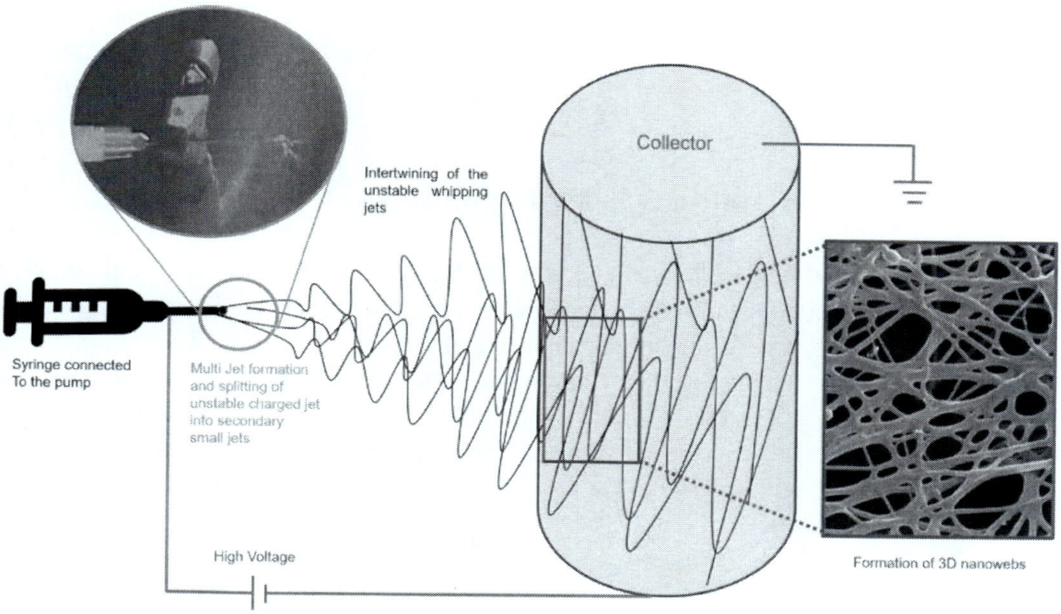

Figure 3. Electrospinning process for the development of tasar silk fibroin/PVA blends 3D nanowebs.

Zang et al. have prepared Bombyx mori SF/Tussah SF blend nanofibers using hexafluoroisopropanol (HFIP) as solvent by electrospinning and further treated them with ethanol for recrystallization [47]. The nanofibers obtained were found to have poor compatibility with each other but showed improvement in thermal degradation and hydrophilicity, the cell adhesion and biocompatibility were enhanced due to blending. In another study, Eri/Tasar SF blend nanofibers were prepared and were compared to B. mori and Gelatin scaffolds as control [42]. It was found that the eri-tasar blend scaffold had better hydrophilicity and mechanical strength than the control samples. Also, the blend scaffolds exhibited superior bioactivity with enhanced cell adhesion and higher metabolic activity of cells grown over them. Another blend was obtained from gland-extracted tasar SF and PVA. Electrospun nanofibrous matrices were fabricated by blending aqueous SF and PVA solutions.

The nanofibers exhibited mechanical strength similar to that of naïve bone tissue, enhanced cell attachment, growth, and formation of the extra cellular matrix compared to the pure PVA (control) samples showing osteoinductivity. In a research work reported by our group, porous 3D nanofibrous webs were developed using the electrospinning technique, as shown in Figure 3 [49].

3.3. Silk Fibroin Blends through Freeze Drying

Freeze drying is a low-temperature water removal technique in which the material is first frozen at low pressure, followed by the removal of ice by the sublimation process. This is also called lyophilization. A few non-mulberry silk fibroin blends have been prepared via this technique.

Bissoyi et al. have prepared freeze-dried A. pernyi SF/chitosan blended polyelectrolyte complex scaffolds for bone tissue engineering applications. In another study, silkworm gland extracted A. mylitta SF/chitosan blended porous 3D scaffolds have been prepared to utilize freeze-drying technique for cartilage tissue repair applications [6]. Singh and coworkers [39] have prepared A. assama SF/agarose hydrogels by freeze-drying the blend solution for cartilage repair applications.

4. Properties of Non-Mulberry Silk Fibroin Blends

Native silk fibroin proteins are known to possess β-sheet conformations, which on dissolution in solvents, get converted to the random coil. These regenerated random coil samples have poor mechanical stability and get easily dissolved in an aqueous atmosphere, so they are coagulated in some organic solvents like methanol, and ethanol, to induce β-sheet crystallinity. Various tools like FTIR, XRD, CD, TGA, DSC, DMA, and UTM are used to study the structural, thermal, and mechanical properties of blend materials. The biodegradation of the blend materials is analyzed in phosphate buffer saline (PBS) solution, and enzymatic degradation of materials is often studied.

4.1. Structural Properties

Protein materials are known to have specific conformations like α-helix, β-sheet, and random coil, and their conformations are determined in terms of structural properties. Any processing of these materials leads to a change in their conformation, as a result altering the properties. These changes in protein conformations can be analyzed by techniques like FTIR, CD, and XRD.

Fourier transform infrared (FTIR) spectroscopy is one of the most commonly used techniques to inspect the specific intermolecular interactions in polymeric blends. In the case of completely immiscible blend systems, where the two polymeric components are in different phases, it is assumed that since the components are not interacting in IR spectral terms, their IR spectra will be reflected as two individual components. If there is formation of new bands due to miscibility of the components, there will either be formation of new bands and, or some bands of the pure components might disappear, representing the formation of a miscible or partially miscible polymeric blend. Specific band shifts in the spectra can be associated with a switch between component- specific bonds to the formation of a bond among components [52]. In FTIR studies, the infrared region between 1700-1600, 1600-1500, and 1300-1200 cm-1 corresponds to the amide I region, which indicates the C = O stretching along the SF backbone, amide II, which represents the N–H deformation in the SF amide groups, and amide III indicates the vibration of O–C–O and N–H groups respectively, are used for the study of the secondary structure of fibroin proteins. The β-sheet and random coil conformation of silk fibroin proteins are characterized by absorption bands at 1650 and 1630 cm-1 (amide-I), 1540 and 1520 cm-1 (amide-II), 1230 and 1270 cm-1 (amide-III), respectively [13, 53].

Blending SF with other polymers has been shown to influence its structural properties. In the FTIR studies of tasar SF/ PVA blend nanofibers, Bhattacharya et al. have reported the presence of absorption peaks along with β-sheet corresponding to both the components without any signicant alterations, indicating structural stability of blends [45]. Moreover, the fibres on blending had slightly collapsed with a reduction in diameter (~50 nm), which as a result, enhanced cell migration. Poly(e-caprolactone) has been effectively blended with tasar gland SF as indicated by the smooth and uniform appearance of scaffolds without bead formation [46]. The FTIR spectra also indicated peaks of SF and PCL without deviation, showing the blending of both components. The reduction in fiber diameter upon blending from ~498 (pure PCL) to ~170 nm improved its bioactivity. Bissoyi et al. have shown that the amide groups of SF have been altered, and the C = O and NH2 functional groups peaks of Chitosan are missing in the FTIR spectra obtained for A. pernyi SF/ chitosan blend composite indicating their involvement in specified interactions. A. pernyi SF blended with 40% chitosan has been reported to show phase separation, as indicated by its SEM images [37]. Studies reported on A. pernyi SF/P(LLA-CL) blends have demonstrated the blending of two components without the involvement of any chemical bond reaction [43]. SF/PVA blends prepared by Chouhan and coworkers have demonstrated homogeneous blending with smooth, beadless, and randomly oriented, porous networks [40]. Electrospun B. mori/tasar SF blends have shown heterogeneous microstructures with the coexistence of α-helix and β-sheets (in the presence of 40% tasar SF content) with weak or no molecular interactions between the two components, indicated by FTIR [47]. A. pernyi/B. mori SF blend films were found to be phase separated [35]. Xue et al. have obtained completely miscible (in micrometer scale) blend films by mixing B. mori SF with tussah, muga, and eri SF, respectively [36]. The blending of muga SF with agarose resulted in transparent hydrogels (in a hydrated state) with no significant phase separation and intermolecular interaction between the two components indicated by shift in amide I and amide II peaks in the FTIR spectra [45]. Also, the pore size of blend material has increased quite significantly due to the incorporation of SF, resulting in higher extracellular matrix production.

4.2. Thermal Properties

The thermal properties of non-mulberry silk fibroin materials are affected by blending with other polymers and can be examined by techniques like TGA, DSC, DMTA. The effectiveness of blending, blend miscibility or phase separation, thermal degradation behaviour, percentage crystallinity and blend composition are certain properties of blend materials that can be analyzed through thermal analysis. The decomposition temperature of the chitosan component has been reported to be effectively enhanced by 14° in the A. pernyi silk and chitosan blend [37]. The DSC thermogram showed a mixed form of two components indicating incomplete miscibility of blends. Tasar silk fibroin blended with gelatin reported in one of our recent studies showed significant shifts in the DSC endotherms, indicating partial miscibility of the polymers [51]. Several studies are available in the literature on blending the B. mori SF with the non-mulberry varieties to generate different morphologies like films and nanofibers and the influence of blending on the thermal properties of blends has been studied [35, 36, 47]. In general, it was observed that the incorporation of non-mulberry SF fraction (Tasar, muga) had increased the blend's thermal decomposition temperature due to the more compact crystalline structure of non-mulberry SF, improving their thermal stability (Figure 4). The incorporation of tasar silk into gelatin has resulted in increased thermal decomposition temperature of blend films, as clearly visible in Figure 4. Also, the different blend ratios altered the thermal properties differently, showing varied functional performance in the case of A. pernyi/B. mori blends [35]. The films with higher A. pernyi showed more stability and stiffness while those with higher B. mori were found to be soft, as indicated by their thermal expansion and contraction properties in TMA.

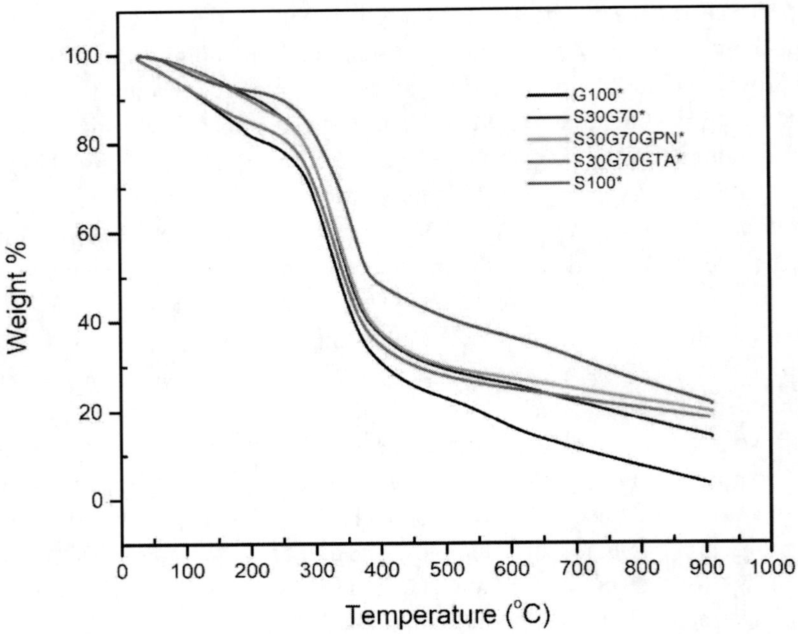

Figure 4. TGA of tasar silk fibroin/gelatin blend films in variable blend ratios. Here G100* is pure gelatin film, S30G70* is SF: Gelatin;30:70& S100 is pure silk film.

4.3. Mechanical Properties

Blending has been found to enhance the mechanical properties of blend materials, having one of the components as non-mulberry silk fibroin. Nanofibrous matrices of Tasar SF/PVA blend with 2%SF have shown higher tensile strength of value 4.87 ± 0.3 MPa as compared to naïve trabecular bones with the tensile strength of value 1.2 MPa, as well as pure PVA nanofibers due to the formation of inter-fibre bonds attributable to blending [45]. The tensile strength of PCL nanofibers was slightly improved from 2.3 ± 0.44 to 4.89 ± 0.68 MPa upon blending with tasar silk fibroin because of the formation of inter-fibre bonds [46]. Upon blending chitosan with A. pernyi SF, its compressive strength has been improved quite significantly [38]. Also, the equilibrium moduli of blended scaffolds were higher than the pure components, making them more suitable for bone tissue engineering. The stability and stiffness of Agarose/Muga SF were enhanced due to blending [39]. The enhanced mechanical properties upon blending PVA with regenerated tasar silk fibroin have been reported by our research group, as shown in Figure 5 [48].

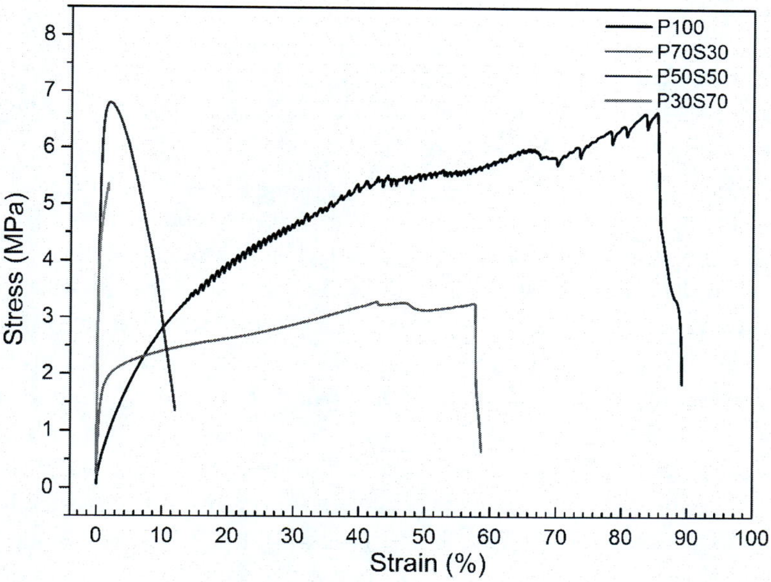

Figure 5. The stress-strain plot of tasar silk fibroin/PVA blend films. Here P100 is pure PVA, P70S30 is PVA:SF = 70:30, P50S50 is PVA:SF = 50:50 & P30S70 is PVA:SF = 30:70.

4.4. Biodegradability

Biodegradability is one of the vital features of biomaterials that are to be used for tissue engineering applications. The degradation should be optimum, neither too fast, resulting in the collapse of material, nor too slow. The degradability of materials is altered by blending. The addition of tasar SF into PVA in the blend films has increased its degradation rate making it more suitable as tissue engineering biomaterial [45]. Tasar/ agarose hydrogels have been reported to show an improved degradation process with a 35% reduction in mass due to the incorporation of SF into the blend as compared to no mass loss shown by pure agarose hydrogels after 28 days of incubation in enzymatic solution [39]. Blends of regenerated tasar silk fibroin with gelatin and PVA have shown variable biodegradability dependent upon the polymeric blend ratios [48, 49, 51]. The crystalline content amongst these blends playsa significant role in biodegradability. The blends with higher crystallinity showed slower biodegradation as compared to the ones with higher amorphous content.

4.5. Applications of Non-Mulberry Silk Fibroin Blends

Non-mulberry silk fibroin having very high toughness and load-bearing ability, along with exceptional biocompatibility, superior bioactivity, and controllable degradation has gained the interest of most researchers as an excellent candidate in tissue engineering and drug delivery applications. The utilization of non-mulberry silk fibroin as one of the components for constructing blend materials is relatively new. It is being focused upon for majorly tissue engineering and biomedical applications.

4.6. Tissue Engineering Applications

4.6.1. Bone Tissue Engineering

Bone scaffolding materials have been prepared by utilizing gland-extracted tasar silk fibroin and blending it with PCL and PVA, respectively [45, 46]. Incorporating of non-mulberry silk fibroin, having the cell recognition RGD sequence sites, into the blends results in significantly increased surface roughness, promoting cell adhesion which, as a result, improves the organization and production of extracellular matrix. The process of osteogenesis in scaffolds was followed by mineralization resulting in osteoinductive nanofibrous blended scaffolds for bone regeneration. Inclusion of tussah SF into P(LLA-GA) post mineralization has significantly improved the blend scaffold properties in terms of cell adhesion, proliferation, and mesenchymal stem cell (MSC) differentiation towards osteoblasts, making them versatile cytocompatible scaffolds for bone tissue engineering applications [44]. A. pernyi/chitosan blend polyelectrolyte complex has shown improved cell adhesion and osteogenic differentiation capability of MSC to be utilized for promising biomaterial for possible biomedical applications and bone tissue engineering [38]. Tasar SF blended with eri SF nanofibrous scaffolds also showed similar properties concerning cytocompatibility and bioactivity, so that can be used as potential bone tissue grafts [43].

4.6.2. Peripheral Nerve Tissue Regeneration

Blended nanofibrous scaffolds of A. pernyi/P(LLA-CL) have been prepared as a potential candidate for application in peripheral nerve tissue engineering [43]. These scaffolds have shown better growth and proliferation of Schwann cells (which play a vital role in the regeneration of peripheral nerve) seeded on them as compared to pure P(LLA-CL) scaffolds.

4.7. Cartilage Repair

Immunocompatible Muga SF/Agarose blended hydrogels showing significantly high levels of sulfated glycosaminoglycans and collagen have been prepared as an alternative biomaterial for cartilage tissue engineering [39]. In another study, gland extracted tasar SF/chitosan blended porous scaffolds with improved cell attachment, proliferation, and differentiation showing enhanced ECM deposition were developed for MSC-based cartilage repair [6].

5. Wound Healing Applications

Non-mulberry silk fibroin consists of inherent RGD motifs in its amino-acid sequence, due to which the binding to integrin receptors of the cell is facilitated. This provides the advantage of mediating the interaction between cells and biomaterial for wound healing. PVA has been explored as one of the non-protein blend materials which can provide stability to the dressing material in rich proteolytic surroundings of chronic wounds.

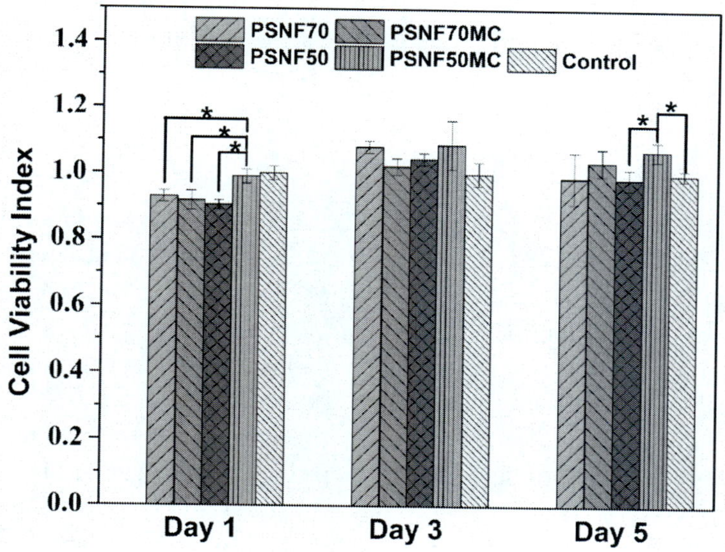

Figure 6a. Cytocompatibility of the nanoweb mats through Alamar Blue assay; error bars represent standard deviation for n = 3. ($p^* \leq 0.05$) where PSNF70 & PSNF50 are PVA: SF 70:30 and 50:50, respectively, MC represents samples incorporating modified clay.

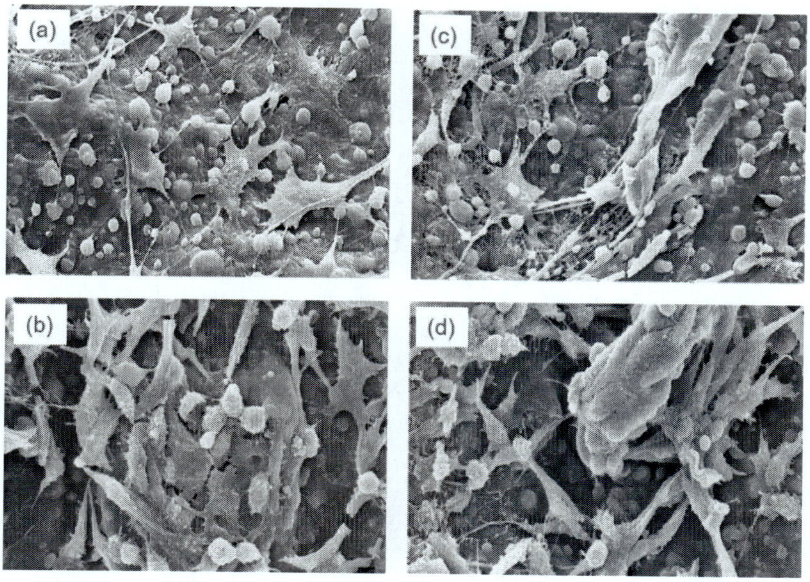

Figure 6b. SEM micrographs of nanoweb samples (a) PSNF50, (b) PSNF50MC, (c) PSNF70 and (d) PSNF70MC after 48 hours of seeding L929 fibroblast cells. Scale: 10 μm.

Bioactive nanofibrous wound dressing material has been prepared by blending B. mori, gland extracted A. assama, and P. ricini with PVA, respectively [40]. These blend materials functionalized with epidermal growth factor (EGF) or drug have shown scar less, accelerated wound healing with regulated ECM deposition. Recently developed cetyl trimethyl ammonium bromide modified nanoclay doped tasar silk fibroin and PVA-based 3D nanowebs (Figure 6a

& 6b) have shown excellent biocompatibility and antimicrobial activity for wound healing and tissue engineering applications [49].

Conclusion and Future Prospects

In this chapter, we have presented insights on non-mulberry silk fibroin blend materials for evolving biomedical applications. The polymeric components of these blend materials work together synergistically, enhancing their properties. Blending non-mulberry silk fibroin with suitable polymers can result in materials with tailor-made physio-chemical properties per the required end-use application. It is pertinent to note that despite the outstanding capability of non-mulberry silk fibroin blends, the transition from academic research to industrial settings still looks like a distant dream. As highlighted in this chapter, the source of extraction of non-mulberry silk fibroin in most of the research studies is from the gland of the silkworm, which is commercially non-feasible. So, attempts should be made to develop a manufacturing system where the processing is simple, eco-friendly, time-saving, and easy to scale up to an industrial level. Positively, with the appropriate alliance between academic innovators and industrial personnel, the non-mulberry silk fibroin-based blend materials can provide endless healthcare opportunities.

References

[1] Koh LD, Cheng Y, Teng CP, Khin YW, Loh XJ, Tee SY, Low M, Ye E, Yu HD, Zhang YW, Han MY (2015) Structures, mechanical properties and applications of silk fibroin materials. *Prog Polym Sci* 46:86–110. https://doi.org/http://dx.doi.org/10.1016/j.progpolymsci.2015.02.001.

[2] Kundu B, Rajkhowa R, Kundu SC, Wang X (2013) Silk fibroin biomaterials for tissue regenerations. *Adv Drug Deliv Rev* 65:457–470. https://doi.org/10.1016/j.addr.2012.09.043.

[3] Konwarh R, Bhunia BK, Mandal BB (2017) Opportunities and challenges in exploring Indian non-mulberry silk for biomedical applications. *Proc Indian Natl Sci Acad* 83:85–101. https://doi.org/10.16943/ptinsa/2017/41288.

[4] Rajkhowa R, Gupta VB, Kothari VK (2000) Tensile stress-strain and recovery behavior of Indian silk fibers and their structural dependence. J Appl Polym Sci 77:2418–2429. https://doi.org/10.1002/1097-4628(20000912)77:11<2418::AID-APP10>3.0.CO;2-Q.

[5] Li M, Lu S, Wu Z, Tan K, Minoura N, Kuga S (2002) Structure and properties of silk fibroin- poly(vinyl alcohol) gel. Int J Biol Macromol 30:89–94. https://doi.org/10.1016/S01418130(02)00007-7.

[6] Bhardwaj N, Kundu SC (2012) Chondrogenic differentiation of rat MSCs on porous scaffolds of silk fibroin/chitosan blends. *Biomaterials* 33:2848–2857. https://doi.org/10.1016/j.biomaterials.2011.12.028.

[7] Silva SS, Oliveira MB, Nayak S, Popa EG, Kundu SC, Gomes ME, Reis RL, Mano JF, Subia B (2013) Silk hydrogels from non-mulberry and mulberry silkworm cocoons processed with ionic liquids. *Acta Biomater* 9:8972–8982. https://doi.org/10.1016/j.actbio.2013.06.044.

[8] Hakimi O, Knight DP, Vollrath F, Vadgama P (2007) Spider and mulberry silkworm silks as compatible biomaterials. *Compos Part B Eng* 38:324–337. https://doi.org/10.1016/j.compositesb.2006.06.012.

[9] Mallepally RR, Marin MA, Surampudi V, Subia B, Rao RR, Kundu SC, McHugh MA (2015) Silk fibroin aerogels: Potential scaffolds for tissue engineering applications. *Biomed Mater* 10:1–10. https://doi.org/10.1088/1748-6041/10/3/035002.

[10] Wang G, Hu X, Lin W, Dong C, Wu H (2011) Electrospun PLGA–silk fibroin–collagen nanofibrous scaffolds for nerve tissue engineering. *Vitr Cell Dev Biol - Anim* 47:234–240. https://doi.org/10.1007/s11626-010-9381-4.

[11] Freddi G, Romano M, Massafra MR, Tsukada M (1995) Silk Fibroin/Cellulose Blend Films: Preparation, Structure, and Physical Properties. *J Appl Polym Sci* 56:1537–1545. https://doi.org/10.1002/app.1995.070561203.

[12] Srivastava CM, Purwar R (2016) Fabrication of robust Antheraea assama fibroin nanofibrous mat using ionic liquid for skin tissue engineering. *Mater Sci Eng C* 68:276–290. https://doi.org/10.1016/j.msec.2016.05.020.

[13] Srivastava CM, Purwar R, Gupta A, Sharma D (2017) Dextrose modified flexible tasar and muga fibroin films for wound healing applications. *Mater Sci Eng C* 75:104–114. https://doi.org/10.1016/j.msec.2017.02.021.

[14] Sionkowska A (2011) Current research on the blends of natural and synthetic polymers as new biomaterials: Review. *Prog. Polym. Sci.* 36:1254–1276.

[15] Gil ES, Frankowski DJ, Spontak RJ, Hudson SM (2005) Swelling behavior and morphological evolution of mixed gelatin/silk fibroin hydrogels. *Biomacromolecules* 6:3079–3087. https://doi.org/10.1021/bm050396c.

[16] Gil ES, Frankowski DJ, Bowman MK, Gozen AO, Hudson SM, Spontak RJ (2006) Mixed protein blends composed of gelatin and Bombyx mori silk fibroin: Effects of solvent-induced crystallization and composition. *Biomacromolecules* 7:728–735. https://doi.org/10.1021/bm050622i.

[17] Marcolin C, Draghi L, Tanzi MC, Faré S (2017) Electrospun silk fibroin–gelatin composite tubular matrices as scaffolds for small diameter blood vessel regeneration. *J Mater Sci Mater Med* 28:1–12. https://doi.org/10.1007/s10856-017-5884-9.

[18] He FL, Deng X, Ye Y, Liu YL, Li D, He J, Liu Y, Yin DC (2017) From 2D to 3D: The morphology, proliferation and differentiation of MC3T3-E1 on silk fibroin/chitosan matrices. *Carbohydr Polym* 178:69–77. https://doi.org/10.1016/j.carbpol.2017.09.035.

[19] Liu Y, Yin D, Ye Y, Lei X, He FL, Li D, Duan E, Deng X, He J (2017) Silk fibroin/chitosan scaffold with tunable properties and low inflammatory response assists the differentiation of bone marrow mesenchymal stem cells. *Int J Biol Macromol* 105:584–597. https://doi.org/10.1016/j.ijbiomac.2017.07.080.

[20] Campana-Filho SP, Ribeiro SJL, Delezuk JAM, Shimizu FM, Pavinatto A, Rodrigues VC, Oliveira ON, Moraes ML (2016) Silk fibroin organization induced by chitosan in layer-by-layer films: Application as a matrix in a biosensor. *Carbohydr Polym* 155:146–151. https://doi.org/10.1016/j.carbpol.2016.08.060.

[21] Rujiravanit R, Kruaykitanon S, Jamieson AM, Tokura S (2003) Preparation of Crosslinked Chitosan/Silk Fibroin Blend Films for Drug Delivery System. *Macromol Biosci* 3:604–611. https://doi.org/10.1002/mabi.200300027.

[22] Zhang Y, Cao CB, Lu Q, Ma XL, Zhu HS (2004) Conformation transition of silk fibroin induced by blending hydroxypropyl methylcellulose. *Acta Polym Sin* 556–560.

[23] Tsukada M, Freddi G, Crichton JS (1994) Structure and Compatibility of Poly(vinyl alcohol) -Silk Fibroin (PVA/SF) Blend Films. *J Polym Sci Part B Polym Phys* 32:243–248. https://doi.org/10.1002/polb.1994.090320205.

[24] Tanaka T, Kuranuki N, Tanigami T, Yamaura K (2000) Properties of Silk Fibroin / Poly (vinyl alcohol) Blend Solutions and Peculiar. *Polym Int* 42:107–111. https://doi.org/10.1002/(SICI)1097-0126(199701)42:1<107::AID-PI686>3.0.CO;2-9.

[25] Li M, Minoura N, Dai L, Zhang L (2001) Preparation of porous poly(vinyl alcohol)-silk fibroin (PVA/SF) blend membranes. *Macromol Mater Eng* 286:529–534. https://doi.org/10.1002/1439-2054(20010901)286:9<529::AID-MAME529>3.0.CO;2-S.

[26] Wang X, Yucel T, Lu Q, Hu X, Kaplan DL (2010) Silk nanospheres and microspheres from silk/pva blend films for drug delivery. *Biomaterials* 31:1025–1035. https://doi.org/10.1016/j.biomaterials.2009.11.002.

[27] Jin HJ, Park J, Valluzzi R, Cebe P, Kaplan DL (2004) Biomaterial films of Bombyx mori silk fibroin with poly(ethylene oxide). *Biomacromolecules* 5:711–717. https://doi.org/10.1021/bm0343287.

[28] Li C, Vepari C, Jin HJ, Kim HJ, Kaplan DL (2006) Electrospun silk-BMP-2 scaffolds for bone tissue engineering. *Biomaterials* 27:3115–3124. https://doi.org/10.1016/j.biomaterials.2006.01.022.

[29] Yang X, Yu F, Lou X, Chen M, He C, Wang H, Fan L, Jiang H, Lin S, Sun B (2015) Green electrospun grape seed extract-loaded silk fibroin nanofibrous mats with excellent cytocompatibility and antioxidant effect. *Colloids Surfaces B Biointerfaces* 139:156–163. https://doi.org/10.1016/j.colsurfb.2015.12.001.

[30] Marsano E, Corsini P, Canetti M, Freddi G (2008) Regenerated cellulose-silk fibroin blends fibers. *Int J Biol Macromol* 43:106–114. https://doi.org/10.1016/j.ijbiomac.2008.03.009.

[31] Shang S, Zhu L, Fan J (2011) Physical properties of silk fibroin/cellulose blend films regenerated from the hydrophilic ionic liquid. *Carbohydr Polym* 86:462–468. https://doi.org/10.1016/j.carbpol.2011.04.064.

[32] Zhou L, Wang Q, Wen J, Chen X, Shao Z (2013) Preparation and characterization of transparent silk fibroin/cellulose blend films. *Polymer* (Guildf) 54:5035–5042. https://doi.org/10.1016/j.polymer.2013.07.002.

[33] Marsano E, Canetti M, Conio G, Corsini P, Freddi G (2007) Fibers based on cellulose-silk fibroin blend. *J Appl Polym Sci* 104:2187–2196. https://doi.org/10.1002/app.24856.

[34] Yang G, Zhang L, Cao X, Liu Y (2002) Structure and microporous formation of cellulose/silk fibroin blend membranes. *J Memb Sci* 210:379–387. https://doi.org/10.1016/s0376-7388(02)00419-2.

[35] Tsukada M, Freddi G, Kasai N (1994) Physical properties and phase separation structure of Antheraea pernyi/Bombyx mori silk fibroin blend films. *J Polym Sci Part B Polym Phys* 32:1175–1182. https://doi.org/10.1002/polb.1994.090320705.

[36] Xue Y, Jao D, Hu W, Hu X (2017) Silk-silk blend materials: A comparative study of Mori-Tussah, Mori-Muga, Mori-Eri, and Mori-Thai silk films. *J Therm Anal Calorim* 127:915–921. https://doi.org/10.1007/s10973-016-5699-9.

[37] Kweon H, Um IC, Park YH (2001) Structural and thermal characteristics of Antheraea pernyi silk fibroin/chitosan blend film. *Polymer* (Guildf) 42:6651–6656. https://doi.org/10.1016/S0032-3861(01)00104-5.

[38] Bissoyi A, Kumar Singh A, Kumar Pattanayak S, Bit A, Kumar Sinha S, Patel A, Jain V, Kumar Patra P (2018) Understanding the molecular mechanism of improved proliferation and osteogenic potential of human mesenchymal stem cells grown on a polyelectrolyte complex derived from non-mulberry silk fibroin and chitosan. *Biomed Mater* 13:1–16. https://doi.org/10.1088/1748-605X/aa890c.

[39] Singh YP, Bhardwaj N, Mandal BB (2016) Potential of Agarose/Silk Fibroin Blended Hydrogel for in Vitro Cartilage Tissue Engineering. *ACS Appl Mater Interfaces* 8:21236–21249. https://doi.org/10.1021/acsami.6b08285.

[40] Chouhan D, Chakraborty B, Nandi SK, Mandal BB (2017) Role of non-mulberry silk fibroin in deposition and regulation of extracellular matrix towards accelerated wound healing. *Acta Biomater* 48:157–174. https://doi.org/10.1016/j.actbio.2016.10.019.

[41] Kundu SC, Nandi SK, Bhattacharya D, Maiti TK, Naskar D, Bhattacharjee P, Das P (2016) Potential of non-mulberry silk protein fibroin blended and grafted poly(Є-caprolactone) nanofibrous matrices for in vivo bone regeneration. *Colloids Surfaces B Biointerfaces* 143:431–439. https://doi.org/10.1016/j.colsurfb.2016.03.058.

[42] Panda N, Bissoyi A, Pramanik K, Biswas A (2015) Development of novel electrospun nanofibrous scaffold from P. ricini and A. mylitta silk fibroin blend with improved surface and biological properties. *Mater Sci Eng C* 48:521–532. https://doi.org/10.1016/j.msec.2014.12.010.

[43] Wang J, Sun B, Bhutto MA, Zhu T, Yu K, Bao J, Morsi Y, El-Hamshary H, El-Newehy M, Mo X (2017) Fabrication and characterization of Antheraea pernyi silk fibroin-blended P(LLA-CL) nanofibrous scaffolds for peripheral nerve tissue engineering. *Front Mater Sci* 11:22–32. https://doi.org/10.1007/s11706-017-0368-x.

[44] Gao Y, Shao W, Qian W, He J, Zhou Y, Qi K (2018) Biomineralized poly(L-lactic- co -glycolic acid) -tussah silk fi broin nano fi ber fabric with hierarchical architecture as a scaffold for bone tissue engineering. Mater Sci Eng C 84:195–207. https://doi.org/10.1016/j.msec.2017.11.047.

[45] Bhattacharjee P, Kundu B, Naskar D, Al E (2015) Nanofibrous nonmulberry silk/PVA scaffold for osteoinduction and osseointegration. *Biopolymers* 103:271–284. https://doi.org/10.1002/bip.

[46] Bhattacharjee P, Naskar D, Kim H, Maiti TK, Bhattacharya D, Kundu SC (2015) Non-mulberry silk fibroin grafted PCL nanofibrous scaffold: Promising ECM for bone tissue engineering. *Eur Polym J* 71:490–509. https://doi.org/10.1016/j.eurpolymj.2015.08.025.

[47] Zhang F, Zuo BQ, Zhang HX, Bai L (2009) Studies of electrospun regenerated SF/TSF nanofibers. *Polymer* (Guildf) 50:279–285. https://doi.org/10.1016/j.polymer.2008.10.053.

[48] Batra R, Bansal P, Yadav R, Purwar R, Kulanthaivel S, Mishra P (2022) Enhancement of functional properties by blending cocoon extracted Antheraea mylitta silk fibroin with polyvinyl alcohol for applications in biomedical field. *J. Appl. Polym. Sci.* 139: e51913 https://doi.org/10.1002/app.51913.

[49] Batra R, Purwar R, Kulanthaivel S, Mishra P (2021) Cetyl Trimethyl Ammonium Bromide Modified Montmorillonite-Doped Tasar Silk Fibroin/Polyvinyl Alcohol Blend 3D Nanowebs for Tissue Engineering Applications. *Macromol Mater Eng* 306:2100450. https://doi.org/10.1002/mame.202100450.

[50] Hoch E, Schuh C, Hirth T, Tovar GEM, Borchers K (2012) Stiff gelatin hydrogels can be photochemically synthesized from low viscous gelatin solutions using molecularly functionalized gelatin with a high degree of methacrylation. *J Mater Sci Mater Med* 23:2607–2617. https://doi.org/10.1007/s10856-012-4731-2.

[51] Batra R, Purwar R (2021) Deduction of a facile method to construct Antheraea mylitta silk fibroin/gelatin blend films for prospective biomedical applications. *Polym Int* 70:73–82. https://doi.org/10.1002/pi.6087.

[52] Kweon H, Ha HC, Um INC, Park YH (2001) Physical Properties of Silk Fibroin/Chitosan BlendFilms. *J Appl Polym Sci* 80:928–934.

[53] Mandal BB, Kundu SC (2008) Non-bioengineered silk gland fibroin protein: Characterization and evaluation of matrices for potential tissue engineering applications. *Biotechnol Bioeng* 100:1237– 1250. https://doi.org/10.1002/bit.21835.

Chapter 3

Production and Applications of Silk Fibroin Thin Films

Francisco Vieira dos Santos[1], Sergio Akinobu Yoshioka[2] and Marcia Cristina Branciforti[1,*]

[1] Department of Materials Engineering,
Sao Carlos School of Engineering,
University of Sao Paulo, Sao Carlos, Brazil
[2] Sao Carlos Institute of Chemistry,
University of Sao Paulo, Sao Carlos, Brazil

Abstract

According to recent research, silk fibroin (SF) has been driving the development of advanced materials in the areas of tissue engineering, biomaterials, sensors, biosensors, controlled drug delivery, and electronic devices. This review shows silk fibroin thin films of micro or nanometer scales have been used for several applications (e.g., wound healing, tissue engineering, disease models, implantable devices, and drug delivery carrier), since, in combination with natural and/or synthetic polymers, or organic and/or inorganic compounds and produced by casting, spin casting, layer-by-layer, electrospinning, lithography, spin, or dip coating techniques, they display excellent physical, chemical, biological, mechanical, thermal, and electrical properties capable of meeting the current demands of society. Such films are believed to continue driving scientific and technological advances in the production of new materials applied in the areas of biomaterials, sensors, biosensors, and controlled drug delivery.

Keywords: silk fibroin, thin films, tissue engineering, biomaterials, controlled drug delivery, electronic devices

1. Introduction

Over the past 15 years, several studies have promoted important advances in both fabrication and application of silk fibroin thin films in the areas of tissue engineering, biomaterials, optical devices, electronics, and controlled release of drugs.

* Corresponding Author's Email: marciacb@sc.usp.br.

In: Silk Fibroin
Editor: Shivaji H. Pawar
ISBN: 979-8-88697-402-7

According to Hu et al. [1], fibroin is one of the most promising materials of the 21st century, even after 5,000 years of its use in the area of textiles. Sun et al. [2] highlighted it draws special attention for the development of advanced materials, mainly as thin films applied to tissue engineering due to its excellent biocompatibility and biodegradability. Lima et al. [3] produced fibroin thin films incorporated with NS5A-1 antigenic peptide derived from hepatitis C (virus NS5A) protein and YVO4:Eu^{3+} luminescent nanoparticles (yttrium vanadate) for the detection of hepatitis C. Delezuk et al. [4] fabricated fibroin/chitosan thin films by layer-by-layer (LbL) for acting as matrices that immobilizephytase enzymes and detect phytic acid at up to 10^{-9} M concentrations through impedance spectroscopy, whereas Li et al. [5] produced the same films for tissue engineering (targeting applications such as skin, cornea, and adipose tissue) by the casting method. The *in vitro* results showed the films not only provided a favourable environment for the growth and proliferation of rat bone marrow- derived mesenchymal stem cells, but also enabled their osteogenic and adipogenic differentiation. Patil and Singh [6] fabricated fibroin thin films functionalized with phosphate and carboxylic groups by the casting method for the differentiation of human mesenchymal stem cells (hMSC) into osteocytes and chondrocytes, respectively. The *in vitro* results suggested the films can regenerate bone, cartilage, or the interface between them by hMSC differentiation. Srivastava et al. [7] fabricated thin fibroin films incorporated with dextrose (5-15% w/w) by casting for the development of wound dressings. According to the results of the mechanical properties, the flexibility of the films increased with an increasing dextrose content, showing a 3.2 to 40% elongation at break and a 16 to 18 MPa tensile strength increase. The glass transition temperature of the films decreased from 176 to 155ºC with an increasing dextrose content. Results of Fourier transform infrared spectroscopy (FTIR), X-ray diffraction (XRD), and atomic force microscopy (AFM) showed the dextrose content did not affect the crystalline structure of the films, but increased their surface roughness root-mean-square (Rq) from 0.9 to 2.16 nm, deviation, and arithmetic roughness average height (Ra) from 0.71 to 1.93 nm. Moreover, it increased their hydrophilicity ability to absorb water. *In vitro* assays revealed the films were non-toxic and promoted the adhesion and proliferation of L929 fibroblast cells. Mandal et al. [8] designed multilayer (3 to 5 layers) thin films composed of fibroin/gelatin by the layer-by-layer (LbL) method for the controlled release of drugs. The *in vitro* results suggested the release (initial burst and prolonged action of release) of the drug was dependent on both fibroin/gelatin multilayers and gelatin content in the films. Ritger-Peppas mathematical model, used in the analysis, showed the release behaviour was a combined mechanism of Fick's diffusion and polymer erosion/degradation.

Bucciarelli et al. [9] demonstrated fibroin thin films projected by the spin coating method had 50 μm thickness, a very low roughness over the nano and microscales, high transparency, and 1.5456 refractive index in water and 1.5429 in formic acid, whereas films of 500 nm thickness showed 1.5537 refractive index in water and 1.5541 in formic acid, enabling the fabrication of optical waveguides. Chakravartyet al. [10] produced thin fibroin films employing the soft lithography technique towards their micropatterning and detection of vitamin B12 via autofluorescence. The films showed 4-5 μm thickness and the results of detection based on photoluminescence, impedance, sensing, selectivity, selectivity interference, and real sample analysis suggested they can detect the presence of vitamin B12 in biological fluids. Haroone et al. [11] produced ultra-thin films of fibroin with cadmium tellurium (CdTe) quantum dots functionalized with 3-mercaptopropionic acid (3-MPA) by LbL for the development of biosensors capable of detecting immunoglobulin and IgG. Liu et al.[12] designed organic solar cells using fibroin thin films as a supporting substrate integrated into a mesh of silver nanowires (AgNWs) towards the development of wearable, biocompatible, and fully degradable electronic devices. The results showed the AgNW/fibroin thin film substrate exhibited ~11.0 Ω/sq electrical conductivity and~80% transmittance in the visible light range. The films maintained their conductivity even after being bent and unbent ~200 times and showed up to 6.62% maximum energy conversion efficiency. Stinson et al. [13] produced fibroin thin films by combining casting and freeze-drying techniques towards the stabilization and storage of inactivated polio vaccine (IPV) products. The results of long-term stability, immunogenicity after accelerated stability, tolerance, and safety testing showed the films maintained 70% potency of D antigen after approximately three-year storage at room temperature and potency higher than 50% for IPV-2 and IPV-3 serotypes at 45 °C for one year. Immunogenicity performed in Wistar rats after 2-week storage at 45 °C showed the films promoted neutralizing antibody responses equivalent to a commercial vaccine for IPV-1 and IPV-2. However, the lack of IPV-3 responses indicates the need for further investigations on intact IPV-3 D antigen. The results suggest the films can assist in the transport of IPV vaccine inputs.

Prakash et al. [14] claimed the scientific and technological development of biocompatible, environmentally friendly, and flexible sensors has been highlighted in recent decades. Among the polymers used, fibroin can be processed in solution form and then as thin films of controlled thickness (up to 125 nm). Moreover, it has stood out due to its excellent biocompatibility, adjustable biodegradability, ease of functionalization, and good mechanical properties (tenacity and mechanical strength), which are essential for the manufacture of promising sensors of high analytical performance (detection at low concentrations). The green electronics area has drawn the attention of both scientific and industrial communities regarding the development of self-sustainable and biodegradable polymeric materials capable of replacing the traditional materials used in inorganic electronics [15]. In this sense, fibroin thin films have adjustable biodegradability and flexibility (43 kPa to 3.9 GPa), as well as piezoelectric properties that make them promising candidates for the development of functional electronic devices such as piezoelectric sensors. Patil et al. [16] showed fibroin thin films are excellent bioabsorbable or resorbable platforms for the production of neural sensors, and Wen et al. [17] demonstrated such films are widely used as fundamental components for the construction of wearable, implantable, and biocompatible electronic devices, optical sensors, and energy collectors, filters, lenses, and biomemristors.

This review shows the scientific and technological advances made over the past twenty years in the production and applications of silk fibroin thin films in tissue engineering, biomaterials (applied in wound healing and cornea), electronic devices (such as electrodes, sensors, and biosensors), optics, and controlled drug release.

2. Silk Fibroin Thin Films

2.1. Physical, Chemical, Biological and Mechanical Properties

Silk fibroin thin films have driven the development of advanced materials capable of meeting the current demands of biomaterials, tissue engineering, sensors, biosensors, optical devices, and controlled drug delivery due to their excellent physical, chemical, biological, mechanical, thermal, and electrical properties. Such materials can be produced on micrometric and/or nanometric scales through the combination (or not) of production techniques such as casting, leaching, freeze-drying, spin coating, LbL, and electrospinning.

Uniform thin and ultra-thin polymeric films can be manufactured by casting, spin casting, LbL, dip, and spin coating techniques. Spin casting is the easiest fabrication method for the production of fibroin thin films of 100 to 1000 nm thickness, whereas ultrathin multilayer fibroin films of 1 to 100 nm thickness can be fabricated by LbL [18]. Nogueira et al. [19] manufactured 20 nm thick fibroin/chitosan thin films, formed by the deposition of six layers, via LbL for tissue engineering applications, and Moraes et al. [20] produced thin films by successive depositions of fibroin/chitosan/poly (allylamine hydrochloride) of 29 ± 6 nm thickness by LbL. The results showed the films were not toxic to NIH-3T3 fibroblast cells, evidencing they can be used as coatings for biomedical devices, controlled drug release, and biosensors. Also using the LbL technique, Cho et al. [21] produced thin films of fibroin/graphene oxide of 489 and 519 nm thickness with good mechanical properties (indentation modulus (GPa) and hardness (GPa)). Joseph et al. [22] fabricated ultrathin fibroin films of 900 ± 25 nm average thickness using spin coating and lithography techniques. The films showed excellent piezoelectric properties for the development of miniaturized, flexible, and wearable sensors. Chen et al. [23] developed fibroin thin films of 90, 130, and 230 μm thickness incorporated with Ag nanowires by casting. The films showed flexible, of good electrical conductivity and tensile strength, and excellent substrates for the fabrication of sensors and wearable devices. Yadav et al. [24] produced fibroin thin films of 30, 50, 70, 100, and 200 μm thickness via casting and incorporated withtitanium dioxide nanoparticles (TiO_2) in 0.1, 0.25, 0.50, 0.75, and 1.0% m/m proportions for the development of dielectric materials of high refractive indices for optoelectronic applications.

Kundu et al. [25] showed fibroin films are biocompatible and biodegradable, have good thermal and mechanical stability, and are compatible with soft and hard tissues, whereas Gil et al. [26] demonstrated the physical, chemical, morphological, and biological properties of fibroin thin films of 2μm thickness enable them to mimic the extracellular matrix (ECM) of muscle, neural, and corneal tissues. In another study, the authors [27] manufactured 2 μm thick thin films of fibroin/polyethylene oxide (PEO) functionalized with tripeptides Arg-Gly-Asp (RDG) by casting to mimic the structure of the stroma of the human cornea. The results showed the surface modification with RGD (arginine-glycine-aspartic acid) improved cell attachment, proliferation, alignment, and expression of corneal stromal markers, while the surface pattern increased the expression of corneal stromal markers. Ziemba et al. [28] demonstrated the coating of electrospunpoly(L-lactic acid) (PLLA) mats using 12-16 nm thick fibroin thin films by immersion in a coating solution. According to the results, the coatings formed by self-assembly uniformly coated PLLA nanofibers, thus providing a new strategy to increase the neuroregenerative capacity of neural cells such as neurites. Lin et al. [29] developed fibroin films incorporated with insulin-like growth factor I (IGF-I) for the treatment of chronic diabetic wounds. The *in vitro* results showed the films promoted an approximately 20% increase in the proliferation rate of BALB/3T3 fibroblasts (~10 μm) in comparison to the film with no IGF-I, thus indicating the films are promising candidates for the treatment of wounds in diabetic patients. Santos et al. [30] produced silk fibroin thin films (SF) with and without glycerol (G) of large macroscopic dimensions (1033 cm^2) and 36-51 μm average thickness by two different methods, namely M1 and M2, adapting the casting technique to thin layer chromatography equipment. SEM results revealed a smooth, homogeneous, transparent, and bubble-free surface of the films, and FTIR and XRD analyses showed a semi-crystalline structure, which provides the materials with good thermal stability (240°C) and storage modulus between 4-5 GPa. Furthermore, the contact angle showed their good hydrophilicity, with values ranging between 43-75°. The authors highlighted the new method enabled the production of promising fibroin films for the treatment of large-extension skin wounds. Figure 1(a-d) displays the large size (1033 cm^2), flexibility, and transparency of the fibroin films produced by casting.

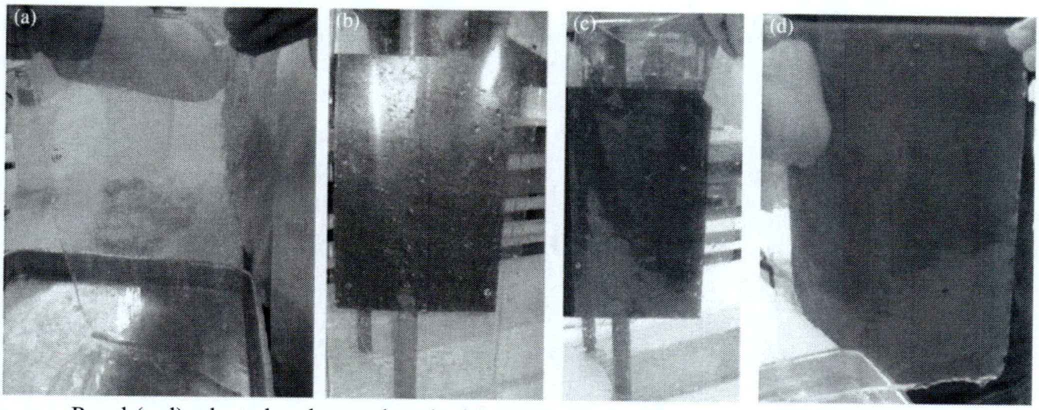

Source: Panel (a-d) adapted and reproduced with permission of ref. [30] Copyright 2021 Wiley – VCH GmbH.

Figure 1. (a) SF-M1, (b) SF-M2, (c) SF-M1G, and (d) SF-M2G silk fibroin films (59 cm width x 17.5 cm length).

Using the electrospinning technique, Safonova et al. [31] manufactured fibroin thin films incorporated with two recombinant spidroins, namely rS2/12 and rS2/12-RGDS, for the treatment of wounds in Wistar rats. According to the SEM and AFM results, the films showed 315 ± 26 nm average diameter fibres, 94.5 ± 1.4% porosity structure, 25.4 ± 4.2 μm^{-1} ratio between the surface and the volume of the fibres, 3.8 ± 0.6 nm roughness, and 328 ± 57 nm thickness. *In vivo* trials revealed the films accelerated skin healing by 19 days in comparison with the control group. Yang et al. [32] designed thin films of fibroin/poly(vinyl alcohol) (PVA) incorporated with Cys- QHREDGS peptide through the spin-coating technique towards wound healing in diabetic patients. Films produced in water and those produced in hexafluoro-2-propanol (HFIP) showed 95 and 98 nm thickness, respectively. Tensile strength, modulus of elasticity, and elongation at break of the film produced with the use of HFIP were 50.35 ± 0.25 kPa, 24.89 ± 0.46 MPa, and 50%, respectively, whereas the film produced in an aqueous medium showed 13.5 ± 0.3 kPa, 17.87 ± 0.07 MPa, and 40%, respectively. *In vitro* and *in vivo* assays revealed the presence of Cys-QHREDGS enabled the film to adhere to the wound surface during the healing process and promoted both adhesion and proliferation of human umbilical vein endothelial cells (HUVEC), hence formation of blood vessels, hair roots, and collagen matrix remodelling.

Yadav and Purwar [33] produced fibroin thin films incorporated with Zn and Cu nanoparticles by casting for the development of optoelectronic devices. The SEM images show the 30 and 70 μm thickness of the films. The addition of metallic oxide nanoparticles to the 30 μm films reduced the percentage of crystallinity and increased the electrical conductivity from 10^{-6} to 10^{-3} S.cm^{-1}. The conductivities of the 30 and 70 μm films were 35.86 and 32.65 x 10^{-6} S.cm^{-1}, respectively. Using casting technique, Reizabal et al. [34] developed 40-60 μm thick fibroin thin films incorporated in 5, 10, and 20% proportions of cobalt ferrite nanoparticles (35-

55 nm) for bone regeneration. AFM images revealed a surface of 100 nm average roughness (Ra) with dispersednanoparticles and no apparent fractures. *In vitro* assays showed the films promoted adhesion and proliferation of MC3T3-E1 pre-osteoblast cells after four-day testing. Luo et al. [35] designed 200 nm thick fibroin thin films with Ag nanoparticles through lithography and blade-coated techniques towards the development of highly sensitive wearable transparent moisture sensors for respiratory monitoring. According to the results, the dielectric constant of the films showed an excellent response to vapour penetration along with human breathing and they were able to distinguish different intensities and rates of human breathing in ultra-fast response (4 Hz). They are flexible and showed good mechanical stability in the bending test; therefore, they can be integrated directly into a commercial capacitance test module. Kopp et al. [36] prepared fibroin solutions in CH_2O_2 and $CaCl_2$ towards the fabrication of films by casting. The results of thickness, tensile strength limit, and strain at break were (145 μm, 1.16 MPa, and 14%), (164 μm, 1.35 MPa, and 29%), (208 μm, 1.40 MPa, and 27%), (252 μm, 1.35MPa, and 28%), and (289 μm, 1.45 MPa, and 23%) for fibroin films with 6, 8, 10, 12, and 14% (w/v) concentrations, respectively. The light throughput analysis revealed higher transparency of the 10% w/v fibroin films, which are more permeable to electromagnetic radiation, and the *in vitro* test evidenced the films are not cytotoxic. Tsao et al. [37] produced fibroin thin films incorporated with gold nanoparticles (AuNPs) for the development of linear photothermoelectronic devices. Silkworm cocoons were cut into pieces and degummed in a 0.2 M Na_2CO_3 solution for 30 min. The degummed silk fibres were washed with ultrapure water for 1 h with gentle agitation, dried under ambient conditions for 12 h, and dissolved in 9.3M LiBr at 10% (w/v) content and heating at 60°C for 4 h. They were then dialyzed for 48 h at 25°C with the use of cellulose membranes (MWCO = 3500). The fibroin solution wascentrifuged for 20 min at 4°C twice and the films were produced by casting, as shown in Figure 2.

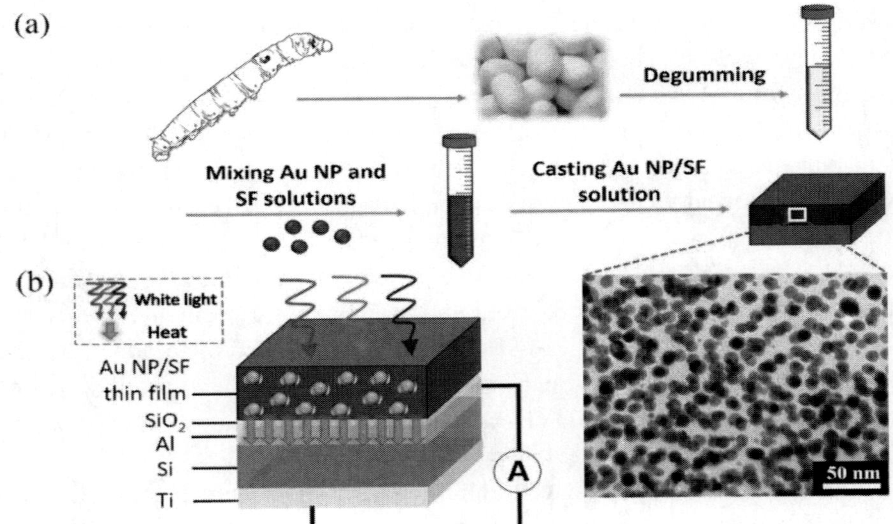

Source: Panel (a-b) adapted and reproduced with permission of ref. [37]. Copyright 2015 American Chemical Society.

Figure 2. Schematic representations of (a) fabrication of a hybrid Au NP/SF thin film upon a substrate and (b) white light irradiating an Au NP/SF thin film upon an Al/Si Schottky diode. Cross-sectional TEM image of a small gold nanoparticle/SF thin film (Inset).

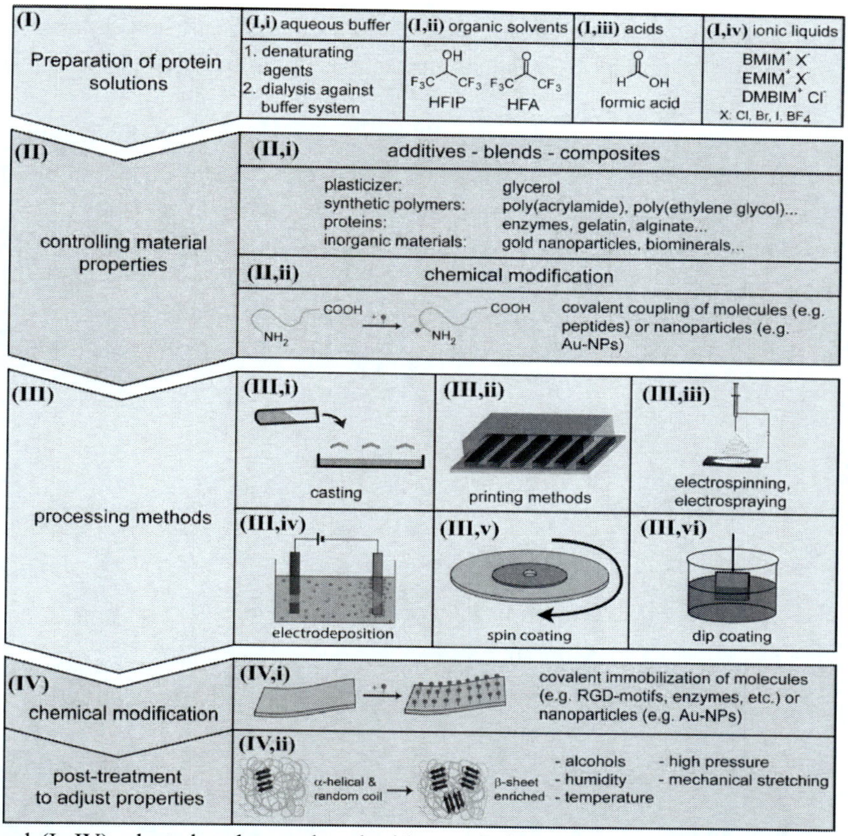

Source: Panel (I- IV) adapted and reproduced with permission of ref. [42]. Copyright 2014 American Chemical Society.

Figure 3. Schematic overview of silk processing. After (I) the preparation of the silk solutions, the properties of the material can be controlled (II) before and (IV) after (III) the processing steps.

Amsden et al. [38] highlighted the characteristics of silk fibroin films are appropriate for an ideal nanoimprinting, which, in combination with their optical properties and biocompatibility, make the material an interesting new technological platform for the production of self-sensitive optofluidic devices. Liang et al. [39] designed fibroin thin films to act as a dielectric layer on the gate of organic transistors (OTFTs), thus simplifying the manufacturing process and reducing its costs. The authors used poly(3-hexylthiophene) (P3HT) polymer dissolved in chlorobenzene or 1,2,4-trichlorobenzene. Tullii et al. [40] demonstrated the steps of the production of micro and nanostructured fibroin thin films by the soft lithography technique for the development of materials capable of preventing the formation of bacterial biofilms in biomedical implants. The results of SEM, AFM, contact angle, *in vitro* tests (HEK-293cells), and antibacterial assays (Escherichia coli) showed the materials are nanostructured, hydrophilic, biocompatible, and antibacterial. Kaushik et al. [41] developed fibroin/chitosan thin films to understand the physicochemical interfacial phenomena and the rapid growth of bacterial biofilms in polymeric films. The analysis of bacterial biofilm growth revealed the films strongly supported a rapid biofilm growth not only for *Synechococcus sp*. BDU 140432, but also for the other two bacteria of industrial importance, namely *E. coli* and *L. plantarum*. Borkner et al. [42] demonstrated the possibility of producing isotropic and anisotropic coated fibroin thin films for biomedical applications such as tissue engineering, cell adhesion, implant coatings, optics, and biosensors. Figure 3 (I), (II), (III) and (IV) illustrates the preparation of solutions and control of properties, production methods, and possible chemical changes in the fibroin-based material.

Rastogi and Kandasubramanian [43] demonstrated functionalization strategies significantly alter the properties of fibroin thin films in comparison to native fibroin fibres. In this sense, the tensile strength values of thin films of fibroin/chitosan irradiated with ultraviolet (UV) and of fibroin/polyurethane (PU) treated with plasma and native fibroin fibres treated with ozone were 70 ± 3.5 MPa (240 min), 1.11 ± 0.04 MPa (2 min), and 0.2660 MPa, respectively. Li et al. [44] claimed the production of sub-micron fibroin thin films by spin coating can potentially develop humidity sensors capable of detecting anti-counterfeit labels, as illustrated in Figure 4.

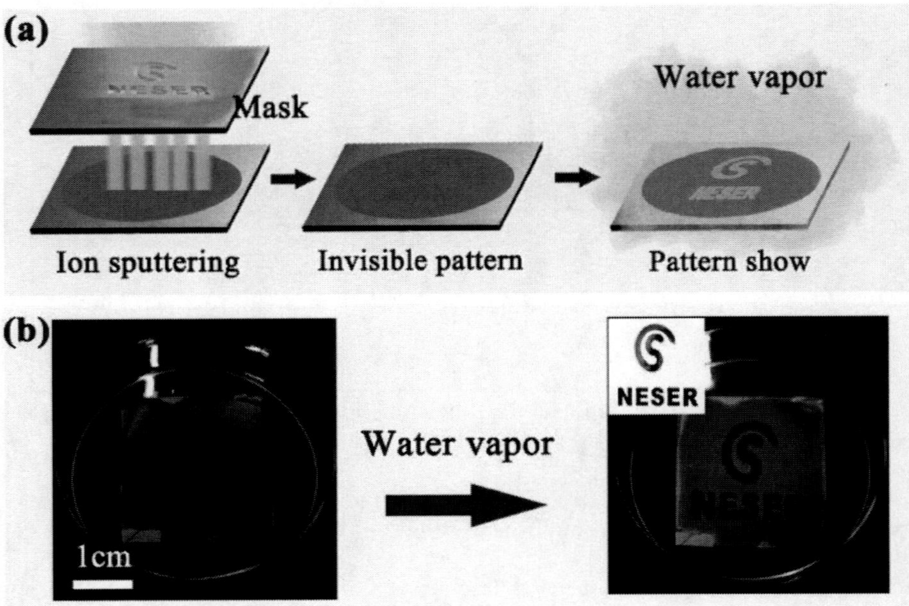

Source: Panel (a-b) adapted and reproduced with permission of ref. [44]. Copyright 2017. The Royal Society of Chemistry.

Figure 4. Anti-counterfeit labelling with the use of humidity-responsive fibroin thin films; (a) Schematic illustration of the anti-counterfeit label application on a fibroin film by ion sputtering; (b) The label of the SF film appears after exposure to water vapour.

Li et al. [45] designed superhydrophobic membranes incorporated with silver nanowires (AgNWs) and produced from fibroin thin films by the spraying technique for the development of flexible sensors that detected human movements under humid conditions. The results of TEM, SEM, AFM, FTIR, XRD, and contact angle (water, n- propanol, n-hexadecane, DMSO, and olive oil) revealed the materials are semicrystalline and superhydrophobic (>151° and <162°). Mechanical tensile testing showed the modulus of elasticity, ultimate tensile strength, and tenacity of the fibroin membrane were 8 ± 3 GPa, 112 ± 5 MPa, and 0.7 ± 0.2 $MJ.m^{-3}$, respectively, whereas the membrane composed of fibroin/AgNWs showed 30 ± 6 GPa, 152 ± 8 MPa, and 0.9 ± 0.1 $MJ.m^{-3}$, respectively. Transmittance tests showed a high transparency of the membranes (see Figure 5).

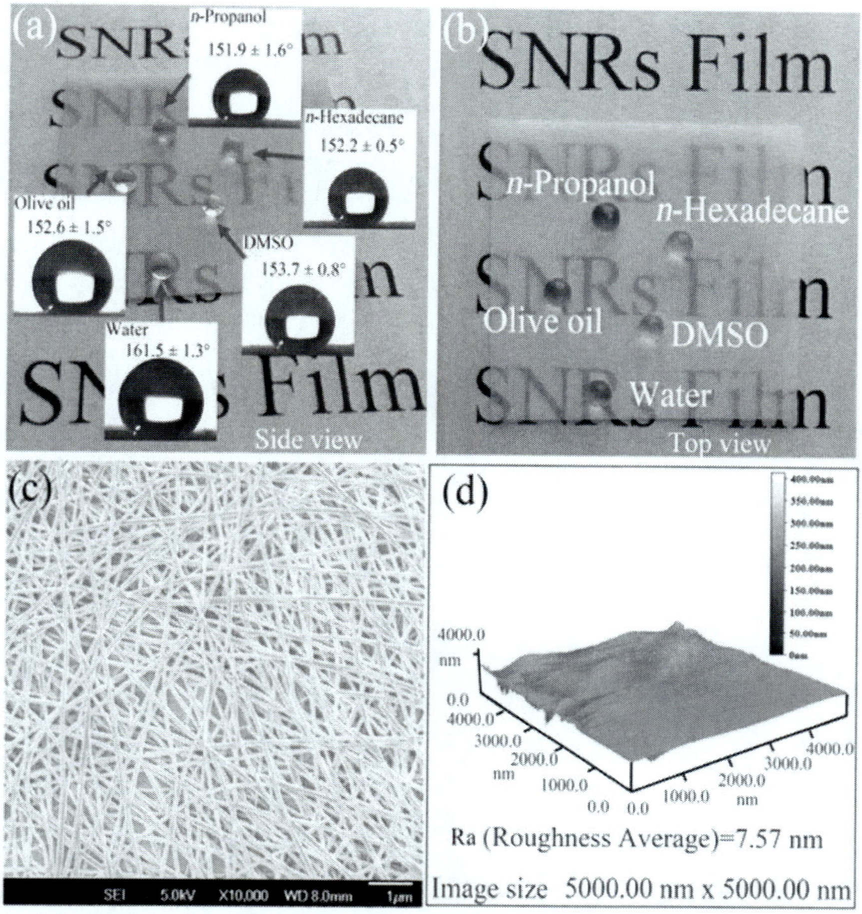

Source: Panel (a-d) adapted and reproduced with permission of ref. [45]. Copyright 2020 American Chemical Society.

Figure 5. Antiwetting behaviour of SF/AgNWs membrane. (a) Side and (b) top-viewed photographs of droplets with different surface tensions on the transparent SF/AgNWs membrane surface. (c) Typical three-dimensional hierarchical structures of AgNWs network array. (d) AFM image of the SF/AgNWs membrane prepared by spraying.

Hajiabbas et al. [46] produced nanostructured films containing fibroin/alginate/gelatine that mimic ECM. The materials were manufactured through the combination of electrospinning, leaching, and crosslinking *in situ* and freeze-drying techniques. The structures showed 44-55% crosslinking degree, 65-85% porosity, and 387-451 μm pore size. Furthermore, they are non-toxic and enable the migration and proliferation of adipose tissue-derived mesenchymal stem cells (AMSC). Qin et al. [47] introduced a method for the production of 100 μm thick fibroin films incorporated with platinum (Pt) and Ag via spin-coating combined with metallic sputtering. The filmsshowed very satisfactory transmittance and conductivity and an almost linear conductivity-temperature correlation. According to the results, they are good candidates for the development of conductive, wearable devices capable of acting as temperature sensors. Rocha et al. [48] showed fibroin thin films incorporated with trivalent Europium ion (Eu^{+3}) and produced by spin coating are highly transparent, enabling the obtaining of flat, flexible, and biodegradable films. The presence of europium ion makes Eu^{+3} act as a local probe, and the luminescence results confirmed the films produced can act as vapour sensors. Kim et al. [49] developed fibroin thin films of 20- 50 μm thickness via casting; they were implanted in animals towards neural mapping and fabrication of implantable devices.

3. Applications to Tissue Engineering

Materials developed for tissue engineering must have porosity and interconnected pores, mechanical strength, modulus of elasticity compatible with the tissue to be regenerated or replaced, adjustable biodegradability, excellent biocompatibility, and favourable surface chemistry. Such properties enable the mimicking of both structure and function of the natural extracellular matrix that provide a three-dimensional (3D) environment conducive to cell adhesion, proliferation, and differentiation. In this sense, fibroin has excelled in the development of thin films for tissue engineering, biomaterials, cornea, wound healing, electronic devices, controlled drug delivery, sensors and biosensors, as illustrated in Figure 6.

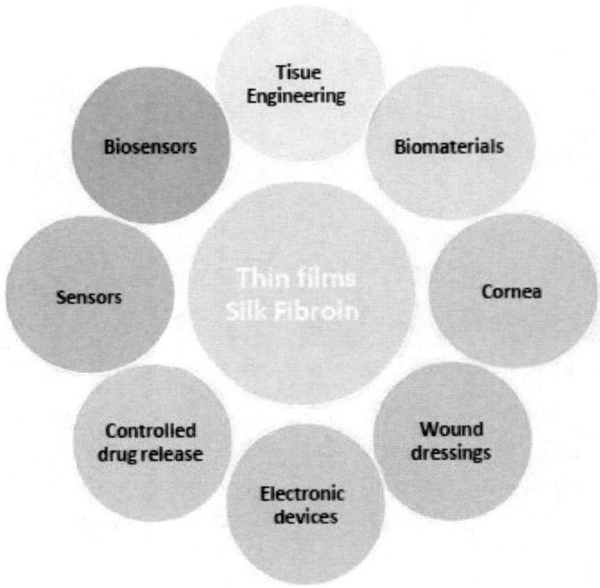

Figure 6. Versatility of silk fibroin thin films in the development of materials.

3.1. Wound Healing

Fibroin films can potentially be used for clinical applications in wound healing, due to their good performance of controlled drug release, flexibility, and permeability to gases (water vapour); moreover, they have been widely used as dressings for skin wounds [50].

Arthe et al. [51] manufactured films composed of fibroin and/or paramyl on carbohydrate by casting aiming at the healing of chronic wounds. According to the results, the fibroin/paramylon (25:75) films showed high thermal stability, good hydrophilicity (52°), and water absorption (70%) and are non-toxic and hemocompatible. Zhu et al. [52] developed dressings for skin wounds from fibroin thin films incorporated with nanoparticles of gold functionalized with 4,6-diamino-2- pyrimidinethiol (DAPT). The contact angle indicated the films are hydrophilic and controlled release assays showed the release of the gold nanoparticles was highly effective. The antibacterial activity was maintained regardless of the antimicrobial susceptibility, and *in vitro* tests revealed the skin coatings promoted wound healing in rats infected with multi-resistant bacteria (*Escherichia coli*). Wang et al. [53] produced fibroin films incorporated with polydopamine nanoparticles (PDA) by casting for the development of wound dressings. The results of AFM, contact angle, and X-ray photoelectron spectroscopy (XPS) showed the incorporation of PDA significantly increased both roughness and hydrophilicity of the films and *in vitro* wound healing tests showed the film/PDA promoted the adhesion and migration of mesenchymal stem cells (MSCs). *In vivo* testing confirmed the wound was healed and completely covered after 14 days. A histological examination indicated the fibroin/PDA film did not cause significant wound inflammation in comparison with the film with no PDA and gauze (control) and promoted higher epithelialization and well-organized deposition of collagen in the dermis. Xu and Yadavalli [54] developed a biosensor of 2–3 μm, thick fibroin thin films by the photolithography/poly(3,4-ethylenedioxythiophene) polystyrene sulfonate (PEDOT:PSS) technique for the monitoring of wound healing and the early diagnosis of diseases through the biorecognition of vascular endothelial growth factor (VEGF). The results of electrochemical impedance, mechanical stability under bending, and *in vitro* assays revealed the engineered biosensors are mechanically compatible and adaptable to soft tissue for the *in situ* impedimetric detection of protein biomarkers. Besides, a range of biologically relevant fluids (buffer, serum, and urine) were tested and the biosensors showed high selectivity and sensitivity (pg.mL^{-1}). Figure 7 displays SEM images of the electrodes and their use in the soft tissue for the *in situ* detection of biomarkers (such as VEGF) between injured tissue interfaces.

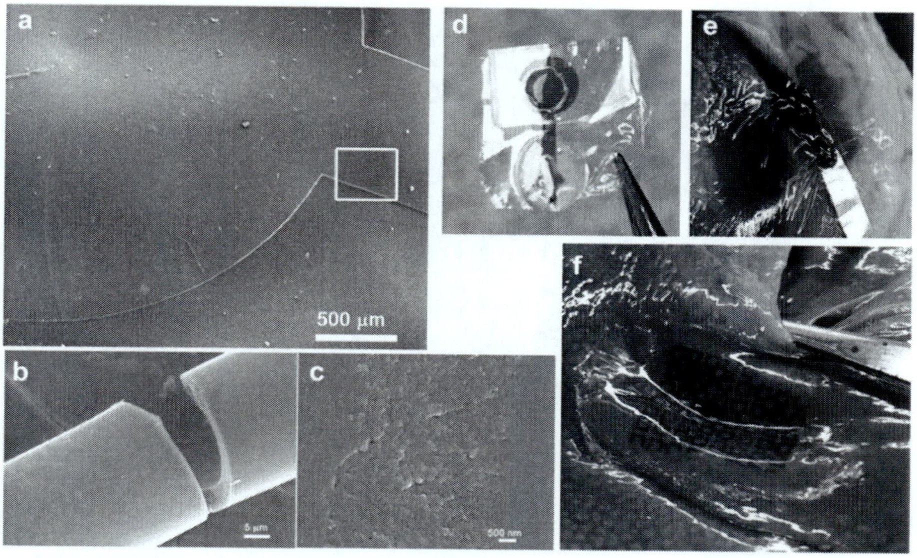

Source: Panel (a-f) adapted and reproduced with permission of ref. [54]. Copyright 2019 American Chemical Society.

Figure 7. SEM images of (a) the sensor electrode, (b) the sensor electrode in a rolled-up condition, and (c) close-up of the electrode surface marked by the white rectangle in the panel (a). The electrodes are 2–3 μm thick. (d) The free-standing biosensor is flexible and can be conformable to soft tissue (panels (e) and (f)) for a potentially *in situ* detection of biomarkers at tissue interfaces. Various electrode configurations can be easily achieved via photolithography.

Zhang et al. [55] produced thin 65 µm thick fibroin films for the development of materials capable of healing wounds. *In vitro* tests performed by ISO 10993 showed the films met the acceptance criteria, indicating the material is biocompatible for its intended use. *In vivo* tests with rabbits and pigs revealed the films effectively reduced the mean wound healing time when compared with commercial Suprathel® and Sidaiyi® dressings of 266 and 1789 µm thicknesses, respectively. Randomized double-blind controlled clinical trials with 71 patients showed the films significantly reduced both healing time and incidence of adverse events when compared with the commercial Sidaiyi® dressing, as illustrated in Figure 8.

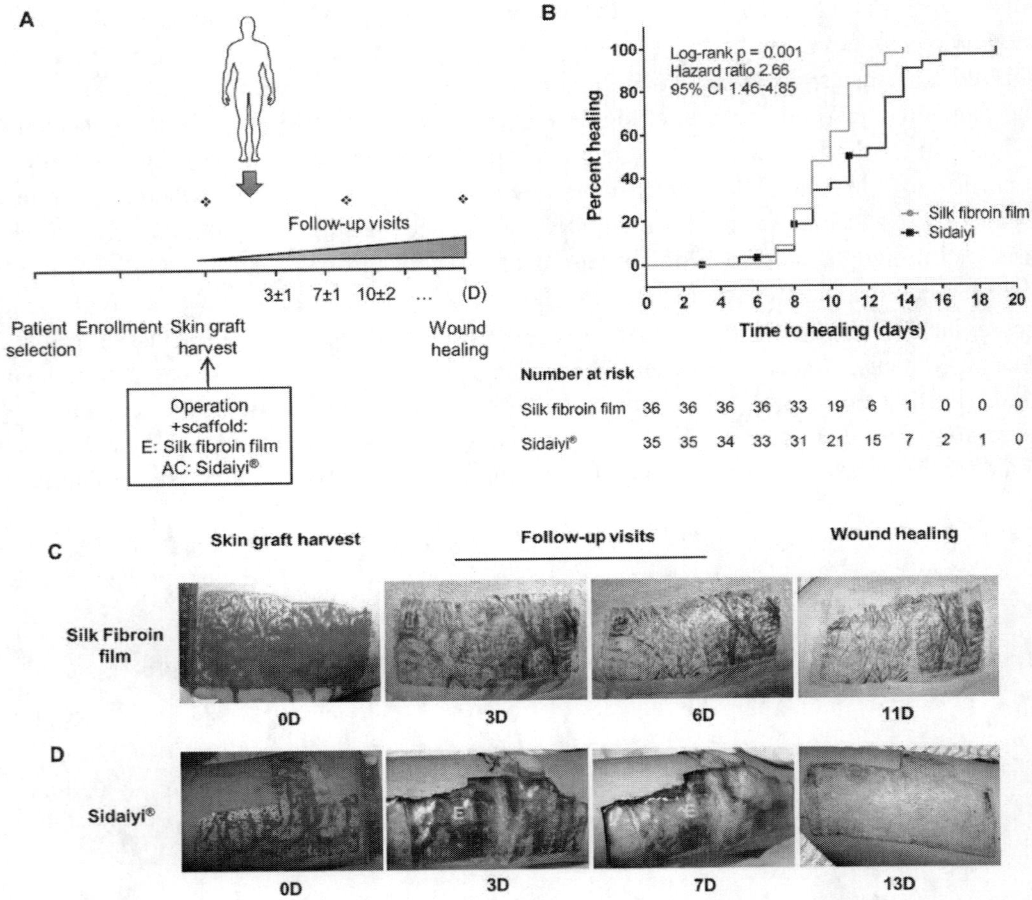

Source: Panel (A-D) adapted and reproduced with permission of ref. [55]. Copyright 2017 Wiley – VCH GmbH.

Figure 8. Clinical trial: Study design and follow-up evaluation. (A) patients were selected according to the inclusion/exclusion criteria (71 eligible patients were enrolled for the clinical trial). After the harvest of the skin graft, the donor site wounds were covered with either silk fibroin film or Sidaiyi. Follow-up visits were performed on days 3 ± 1, 7 ± 1, 10 ± 2 until the wound had been completely healed. A macroscopic evaluation was conducted after skin graft harvest, during the follow-ups and when the wound had been healed; (B) Kaplan–Meier curves comparing cumulative healing per treatment group; (C) Representative case from the silk fibroin film group. Macroscopic images were captured after skin graft harvest (day 0), at the 1st follow-up (day 3), at the 2nd follow-up (day 6), and when the wound had been healed (day 11); (D) Representative case from the Sidaiyi group. Macroscopic images were captured after skin graft harvest (day 0), at the 1st follow-up (day 3), at the 2nd follow-up (day 7), and when the wound had been healed (day 13). "E" stands for (exudate).

Liu et al. [56] developed a method for the production of porous fibroin thin films for wound dressings by autoclave and leaching techniques. FTIR and XRD results confirmed the films are semi-crystalline and SEM and permeability analysis revealed the pores are interconnected (34.85 ± 2.12%) with their size ranging from 50-300 nm. *In vitro* degradation tests of the films were assessed using phosphate buffer solution (PBS) and protease XIV solutions and the results showed the films degraded more slowly in the PBS than in protease XIV solution. *In vitro* cell culture tests (HS 865.SK human dermal fibroblasts) showed the films are promising candidates for the promotion of growth and migration of epithelial and fibroblast cells and *in vivo* tests revealed after 21 days, the dermal thickness promoted by the porous fibroin film was 48.65 ± 6.54 μm, that of non-porous film was 57.31 ± 3.62 μm, and the one of the control was 73.68 ± 3.62 μm. Wang et al. [57] demonstrated the production (casting) of fibroin/κ- carrageenan thin films in 1:1, 2:1, and 4:1 proportions incorporated with bioactive glass doped with zinc (6 and 10%) towards the development of wound dressings. The contact angle result showed angle values of 36, 6, and 70° for films (1:1), (2:1), and (4:1) and SEM micrographs displayed the porous structure of the materials. According to mechanical tests, film (1:1) (2:1) showed tensile strength, elongation at break, and Young's modulus of 5.7 MPa, 260%, and 16, 9 MPa; 7.8 MPa, 355%, and 18.1 MPa, respectively, and the swelling test revealed film (4:1) increased its volume from ~200% to 1200%, whereas that of film (1:1) increased 1700% in 4 to 20 min. *In vitro* tests (3T3 and L929 cells) indicated an excellent cell viability of the films, i.e., 96% and 97% for 3T3 and L929 cells, respectively, after 24 h culture. The antibacterial assay (*S. aureus* and *E. coli*) showed the incorporation of 6 and 10% of Zn^{+2} in the glass inhibited bacterial growth and *in vivo* testing suggested the films can accelerate collagen fibre production and promote wound closure. Safonova et al. [58] produced fibroin thin films by casting (SF) and electrospinning with no (SF-1) and with gelatin (SFG-30%) for the development of coverings for cutaneous wounds.

SEM and AFM results showed a 36.5 ± 10.6 nm roughness of SF, whereas the thickness of SF-1 and SFG-30% films ranged from 300–700 nm. A scanning probe nanotomography analysis revealed dense globule structures ranging from 10 to 30 nm of all films, and SF-1 and SFG-30% films showed 81.3 ± 12.6 and 86.4. No film showed cytotoxicity, thus enabling cell adhesion and proliferation (mice 3T3 fibroblast). According to *in vivo* tests performed in up to 40 days, the films promoted healing in 17 days, on average, compared with the control group, and no statistically significant differences were observed among SF, SF-1, and SFG-30% in the healing rate. Therefore, fibroin thin films are promising candidates to act as coverings for cutaneous wounds. Patil et al. [59] developed a fibroin/poly(vinyl alcohol) film incorporated with zinc nanoparticles to treat skin wounds. The TEM result showed a 13 nm average diameter of the zinc nanoparticles and XRD analysis revealed amorphous regions of the film and crystalline domains associated with the (100), (002), (101), (102), and (110) planes. A thermal analysis performed by DSC indicated the incorporation of nanoparticles increased the melting temperature from 429 to 452°C, and the film still showed 3.80 MPa tensile strength and 207% elongation to rupture, which are higher than the values of the film with no nanoparticles. Antibacterial tests showed an antibacterial activity of the film against *Staphylococcus aureus*, *Escherichia coli*, *Pseudomonas aeruginosa*, *Proteus mirabilis,* and *Streptococcus pyogenes*, and blood clotting and cytotoxicity studies revealed zinc nanoparticles denature anticoagulant proteins, thus reducing clotting time and causing no toxicity to L929 fibroblasts. Singh et al. [60] developed a porous fibroin/PEO film incorporated with Amplex™ reagent by electrospinning for the detection of hydrogen peroxide in the wound healing process.

SEM micrographs showed the SF/PEO film is formed by nanofibers of 50 nm average diameter and FTIR evidenced Amplex™ was successfully incorporated into the semi-crystalline structure of the film. *In vitro* testing showed the film was able to release Amplex™ and detect the presence of hydrogen peroxide (25-0.25 µg/mL) by changing the colour of the film from colourless to pink. Cell viability tests performed with human skin keratinocytes (HaCat) showed the film containing Amplex™ was not toxic to keratinocytes, and, according to an *in vivo* assay, the SF/PEO film with Amplex™ (1-5 mg/mL) detected the oxidative stress promoted by the high release of hydrogen peroxide in wounds in mice.

Özen et al. [61] produced fibroin/hyaluronic acid/gelatin films containing boric acid by casting for the treatment of wounds. Tensile tests showed the incorporation of 1% w/w boric acid improved the mechanical behaviour of the film. The film with 1% w/v boric acid showed 0.73 MPa tensile strength and approximately 86.4%. elongation at a break. Such values were 156.7% and 119.3% higher than those of the film with no boric acid, respectively. The absorption test carried out in PBS (pH ~7.4) at 37°C for 24 h showed the stronger the presence of boric acid, the greater the absorption of water from the film - the maximum absorption value was 160%. The cell viability indicated boric acid did not make the films cytotoxic, according to ISO 10993-5 standard, and an *in vitro* assay revealed films containing the acid promoted both adhesion and proliferation of L929 cells. Yerra and Mamatha [62] developed fibroin films incorporated with various antibiotics to treat bacterial infections in burn wounds. The antibacterial assay showed an antibacterial activity of films containing amoxicillin (15 µg), ciprofloxacin (4-6 µg), amikacin (30 µg), lincomycin (25 µg), tetracycline (20 µg), and gentamicin (20 µg) against *Staphylococcus aureus*, *Pseudomonas aeruginosa*, *Acinetobacter baumannii*, *Klebsiella pneumonia,* and *Escherichia coli*. According to the *in vitro* assay, after 1, 2, and 3 h seeding, L929 cells stably adhered to the films, and a larger number of cells adhered to the films containing antibiotics was observed in the film composed of only fibroin.

3.2. Cornea

Due to the worldwide lack of corneal donors, considerable progress has been achieved in the development of living corneal substitutes produced by tissue engineering. Corneal stromal tissue emulation is believed to be the most challenging step in the development of an artificial human cornea caused by the difficult reproduction of its highly ordered microstructure, its robust biomechanical properties, and optical transparency of the tissue. In this context, fibroin thin films are promising candidates for the fabrication of constructs for both regeneration and repair of corneal stromal tissue [63, 64].

Harkin and Chirila [65] demonstrated fibroin films have specific properties such as transparency, thickness, permeability, and biocompatibility, which enable their use in corneal reconstruction. Choi et al. [66] showed the good hydrophilicity, transparency, and biocompatibility (adhesion, proliferation and expression of genes and proteins specific to CEnCs) of such films, which enable them to regenerate diseased or damaged CEnCs cells. Kim et al. [67] produced fibroin films incorporated with β-carotene towards enhancing corneal endothelial cell regeneration. The results showed the films are transparent, hydrophilic, non-toxic, and biocompatible and promoted cell proliferation (rabbit corneal endothelial cells). Sun et al. [68] demonstrated fibroin films are excellent candidates for the development of biomaterials capable of mimicking native corneal tissue; moreover, they stimulate wound regeneration, due to their optical transparency, biocompatibility, and mechanical strength. Lawrence et al. [69] reported the transparency, mechanical strength, biocompatibility, and slow biodegradability of fibroin films, which are important requirements for the development of biomaterials for ocular regeneration. The films produced were 2 μm thick and patterned on the surface towards guiding cell alignment. Pores of 0.5 to 5.0 μm diameters were introduced in the silk films for increasing thetranslamellar diffusion of nutrients and promoting cell-cell interaction. *In vitro* assays with keratinocytes and fibroblasts (rabbits) showed fibroin films are biocompatible and enable cell proliferation.

According to the results, the films can mimic the dimensions of the collagen lamellae present in the cornea. Madden et al. [70] produced 5 μm thick fibroin films to act as membranes that promote the growth of human corneal endothelial cells. Light transmission through silk fibroin membrane hydrated in PBS was greater than 96% for the full visible spectrum. *In vitro* assays indicated the films enabled both adhesion and proliferation of B4G12 type human corneal endothelial cells and are promising candidates for corneal endothelial transplantation. Higa et al. [71] developed porous films from fibroin for corneal epithelium regeneration. The results of *in vitro* (rabbit corneal epithelial cells and 3T3 fibroblasts) and *in vivo* (rabbit) tests showed they are biocompatible and enable the expression of important markers of corneal epithelial tissue such as corneal specific cytokeratin's K3 and K12, and pax6. Basement membrane components such as type IV collagen and β1 integrin were expressed in the stratified epithelial lamina. Furthermore, the efficiency of colony formation of dissociated cells was similar to that of primary corneal epithelial cells, showing progenitor cells were preserved. Farasatkia et al. [72] produced hybrid films composed of fibroin (SF) and methacryloylgelatin (GelMA) by casting and obtained materials capable of emulating the mechanical properties and water permeability.

The AFM results showed the addition of 30% w/w GelMa reduced the mean pore size from 28 ± 8 to 22 ± 7 μm. On the other hand, the film containing 70% m/m GelMa showed high transparency with light transmittance (85%) in the wet state, good hydrophilicity (64°),36.2 ± 7 kPa modulus of elasticity, 3.8 ± 1 MPa tensile strength (values close to those of the natural stroma of the cornea), and capacity to absorb up to $138 \pm 27\%$ of water. *In vitro* studies showed the SF/GelMa (30:70) film promoted adhesion, dissemination, andproliferation of stromal cells, thus increasing the metabolic activity of human corneal stromal fibroblasts (T0578) by twice as compared with SF. Cubayachi et al. [73] developed silk fibroin thin films incorporated with insulin by casting for the treatment of corneal wounds. The films showed good transparency at 340-620 nm wavelength, 184 ± 15 μm average thickness, and 4.85 ± 0.885 $g.mm.m^2.day^{-1} \cdot kPa^{-1}$ water vapour permeability, as well as 22.6 ± 2.2 MPa tensile strength,$7.5 \pm 1.6\%$ elongation at break, and 3.3 ± 0.4 MPa Young's modulus. A circular dichroism analysis revealed the films are semi-crystalline and the degradation test indicated they lost 10 to 15% mass after 24 h of contact with PBS 0.01M at pH = 7.0. Films in contact with protease showed a significant increase in mass loss over 24 h and, according to the release profile, insulin (10%) was released quickly in the first few minutes, followed by a slower and almost linear release in 14 h. Higuchi model established 1.2 ± 0.4 $IU/cm^2/h^{1/2}$ as the release rate.

Kim et al. [74] produced silk fibroin ultrathin films incorporated with aloe vera (1- 3 wt.%) by casting towards corneal regeneration (corneal endothelial cells (CECs)). The films showed optical intensity between 0.06 and 0.17 in the 400-650 nm wavelength range. The optical intensity of human acellular corneal stroma is between 0.1 and 0.13 in the same wavelength range. This result indicates the transparency of the produced film has an acceptable value for the cornea substitute for transplantation. According to the contact angle, the incorporation of aloe vera made the films more hydrophilic and *in vitro* tests showed they promoted adhesion and proliferation of corneal endothelium cells and were biocompatible. A field emission scanning electron microscopic (FESEM) analysis revealed the surface morphology of the films resembles the ECM of the cornea. Histological expression showed the materials produced are promising carriers of CECs for the delivery of transplantable intraocular cells and *in vivo* tests (rabbits' eyes) indicated the films (AV/SF 3 wt.%) can self-regenerate CECs after transplantation, as shown in Figure 9.

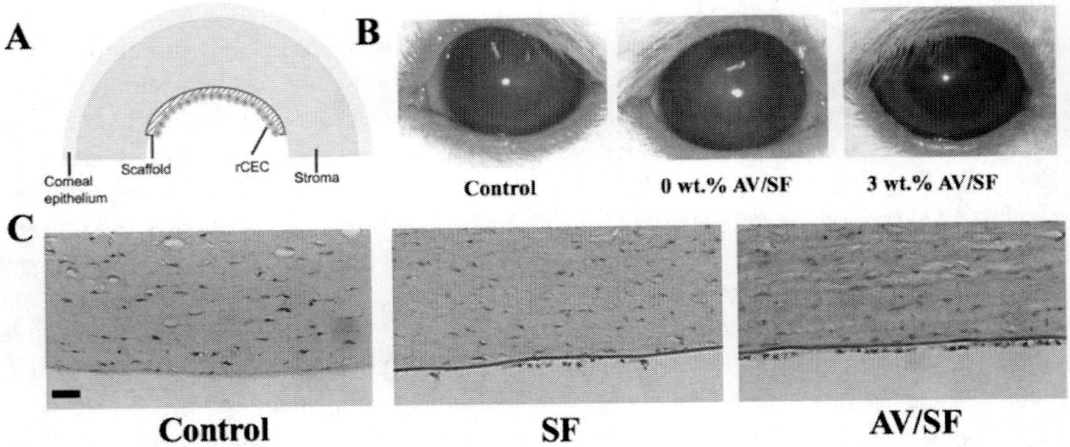

Source: Panel (A-C) adapted and reproduced with permission of ref. [74]. Copyright 2016 American Chemical Society.

Figure 9. (A) Schematic image of the DSEK surgical procedure; (B) Transplantation of SF-based film scaffolds. Descemet's membrane with rCECs was mechanically removed and scaffolds were implanted. Images of untransplanted scaffolds (control), transplanted scaffolds of 0 wt.% AV/SF (SF), and 3 wt.% AV/SF (AV/SF) film scaffolds with rCECs 4 weeks after surgery; (C) Hematoxylin and eosin staining of transplanted eyes shows adherent scaffolds on the surface of the corneal stroma with attached rCECs. Scale bar = 50 μm.

Aghaei-Ghareh-Bolagh et al. [75] evaluated the incorporation of tropoelastin in fibroin films produced by casting for the development of synthetic corneas. According to the tensile test results, the films composed of tropoelastin/SF in 100:0, 75:25, 50:50, 25:75, and 0:100 proportions showed 2.0, 0.6, 1.2, 1.4, and 1.8 MPa tensile strength, 150, 75, 60, 45, and 50% elongation to rupture, 0,7, 2, 5, 7, and 11 MPa modulus of elasticity, and, 70, 52, 35, 48, and 45% water absorption, respectively. According to *in vitro* assays performed on B4G12 immortalized corneal endothelial cells, the 75/25 film provided the cells with greater adhesion and proliferation in seven days and promoted the formation of two to three layers of stratified and differentiated epithelial cells of 44, 9 ± 13 μm average thickness. Vázquez et al. [76] developed casting fibroin films to act as endothelial grafts capable of correcting corneal deterioration or opacification.

The SEM result showed the films produced have 3.46 cm^2average area and 10 μm thickness and the mechanical tensile test revealed the dry films showed modulus of elasticity, elongation at break, and tensile strength of 346 ± 33 MPa, 8.3 ± 12%, and 21.5 ± 33 MPa, respectively. On the other hand, the wet films showed 22 ± 2.0 MPa, 23.5 ± 9.1%, and 2.1 ± 1.0 MPa. Optical tests indicated the films are optically transparent at all wavelengths of the electromagnetic spectrum of visible light with an 89.447 ± 0.84% average light transmission value. *In vitro* assays revealed they promoted both adhesion and migration of human endothelial cells and formation of a monolayer of compact cells of hexagonal endothelial morphology and 80,000 cells/cm^2 confluence after 30 days, as well as adhesion and proliferation of rabbit endothelial cells with 150,000 cells/cm^2 confluence. Figures 10(A-C) and (D) display, respectively, the film implantation in rabbits and the external appearance of the eyes during the follow-up period after surgery performed in six New Zealand white rabbits.

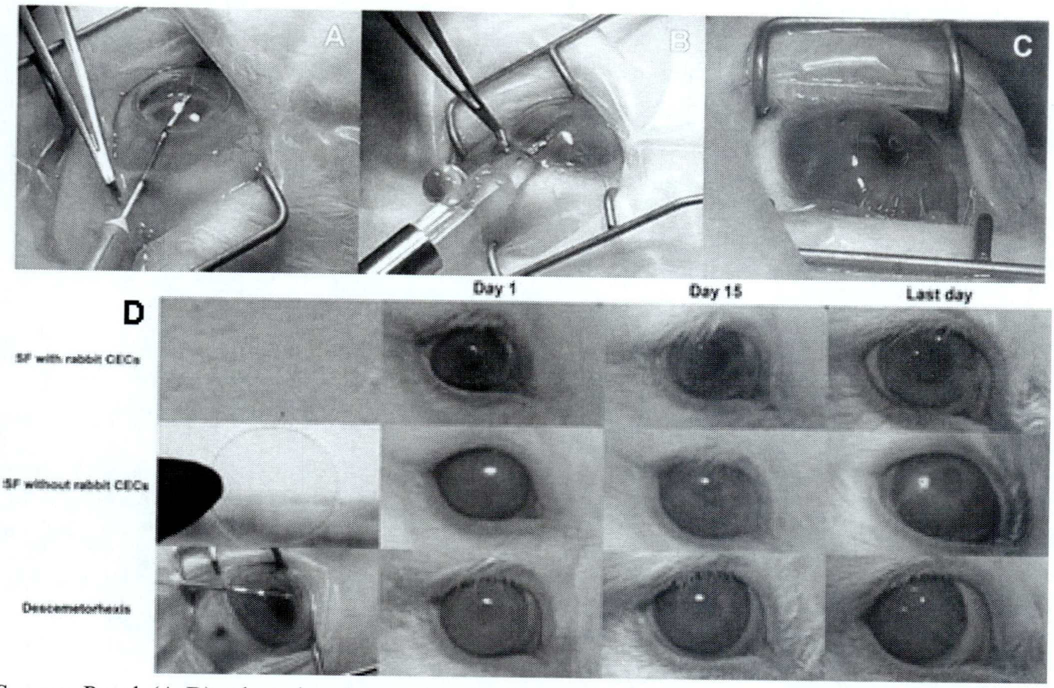

Source: Panel (A-D) adapted and reproduced with permission of ref. [76]. Copyright 2017 ARVO Journals.

Figure 10. DMEK surgical procedure for SF film transplantation. Descemetorhexis from the posterior stroma (A) through the injection of SF films with or without rabbit CECs into the anterior chamber (B), unfolded SF films inside the anterior chamber (C), and photographs of the exterior appearance of transplanted rabbit eyes during the follow-up period. *From top to bottom*: rabbits transplanted with or without rabbit CECs cultured on SF films and rabbits with only peeled Descemet's membrane (D).

Ghezzi et al. [77] created a high-fidelity tissue model of corneal stroma from multilayer fibroin/poly(ethylene oxide) thin films incorporated with human corneal stromal stem cells (hCSSCs) and P4 human corneal fibroblasts (hCFs). The SEM results revealed 4 μm thickness, 5 μm pores size, and 45 x 45 mm^2 area of the films and the FTIR analysis showed their semicrystalline structure formed by α-helix (amorphous phase) and β-sheet (crystalline domains). According to the mechanical tensile test, the film with hCSSCs showed the highest modulus of elasticity, i.e., 24.17 ± 5.83 MPa, whereas that containing hCFs showed the highest elongation at break, i.e., 48.34 ± 16.10%. *In vitro* assays indicated both films promoted the adhesion and proliferation of the cells after nine weeks and an immunohistochemistry test showed the film with hCSSCs led to a greater secretion of keratin sulphate.

The authors highlighted the materials produced can fill the gap between clinical models and the current 2D models used in eye regeneration. Combining lithography and casting techniques, Luo et al. [78] developed fibroin films coated with collagen I for the treatment of corneal wounds. SEM results revealed a standardized nanotopographic surface of different widths, namely 2000, 1000, and 800 nm, and 1 μm depth. *In vitro* assays performed in mouse corneal epithelial cells (MCEC) showed a spherical shape of MCECs on glass or flat silk film surfaces (control), whereas on the silk nanotopographien, the cells were elongated and aligned parallel to the direction of the patterned features. A spreading cell number on 800 nm pattern was twofold greater than that on 2000 nm ridge width films, 20-fold greater than that on flat silk films, and 50-fold greater than that on glass at the one-day time point. Figures 11(A) and (B), respectively, depict the surfaces of the films with linear topographic patterns of different sizes and the effects of the tested surfaces (glass, control, and nanotopographic films) in contact with MCECs. Figure 11(C) shows the cell proliferation was more pronounced as the cells were cultured for longer periods of time.

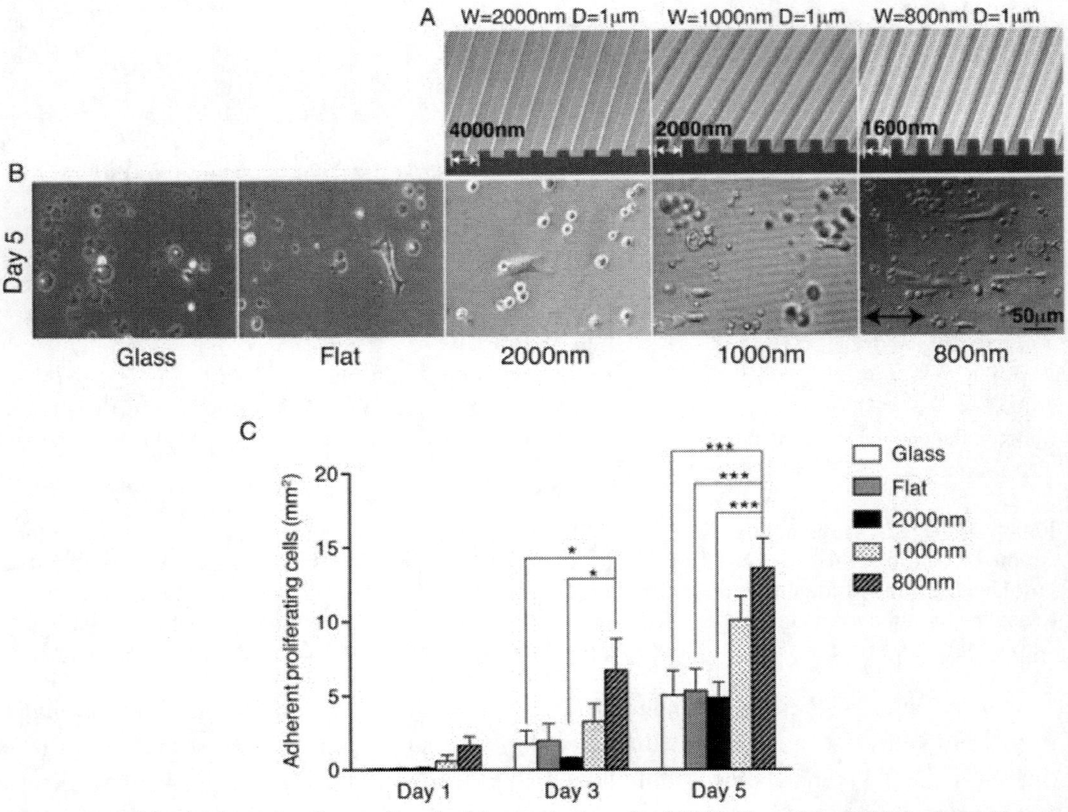

Source: Panel (a-c) adapted and reproduced with permission of ref. [78]. Copyright 2021 Springer Nature.

Figure 11. Nanotopography on silk films induced cell spreading and growth. (A) SEM images of silicon wafers with nanotopographic patterns. Ridge width sizes were 2000 nm, 1000 nm, 800 nm, and 1 μm deep. (B) MCECs were seeded on a glass surface and 4 different silk topographies, cellular spreading, and growth were followed up to 5 days. (C) The total cell number was analysed through the counting of phase-contrast images taken at 1-, 3- and 5-day time points. (* indicates $p < 0.05$, ** indicates $p < 0.005$. Values are means ± S.E.M. The arrow indicates the direction of the pattern axis).

Song et al. [79] produced fibroin films by casting as a potential substrate for corneal endothelial cell regeneration. Films with no glycerol (SF) and with 1 and 3% m/m glycerol (SFG1) and (SFG3) showed 10.39 ± 0.65, 7.25 ± 0.68, and 6.37 ± 0.86 μm average thickness, respectively. Transparency evaluated at 380 to 780 nm wavelength (visible light range) showed 0.097-0.052, 0.106-0.053, and 0.171-0.061 optical intensity of SF, SFG1, and SFG3 films seeded with rCEnCs cells, whereas those of films with no cells were 0.074–0.043, 0.086–0.043, and 0.159–0.059, respectively. The contact angle showed SF, SFG1, and SFG3 films are hydrophilic, and the values found were 65.3 ± 5.0°; 64.5 ± 3.5°, and 62.7 ± 2.9°, respectively. *In vitro* assays indicated the films promoted the adhesion of rCEnCs cells after 30 minutes. The number of cells adhered to SF, SFG1, and SFG3 films was 371 ± 634 cells/mm^2, 2666 ± 607 cells/mm^2, and 2610 ± 539 cells/mm2, respectively.

Kim et al. [80] designed fibroin films coated with type I collagen for corneal transplantation, and AFM results showed they displayed a regular uniform morphology and mean square roughness (Rrms = ~3.14 nm). Their contact angle was 58.1° after five minutes. The absorption test showed the water absorption capacity of the films was approximately 75% at 14 days; nevertheless, on the 20th day, it was 60%. *In vitro* assays (rCEnCs) performed at 1, 3, 5, and 7 days indicated the films promoted adhesion, proliferation, and differentiation of rCEnCs cells, which is associated with type I collagen coating on the fibroin film. Lee et al. [81] developed a casting method that uses the action of centrifugal force (4000 rpm) to produce fibroin films capable of regenerating corneal tissue.

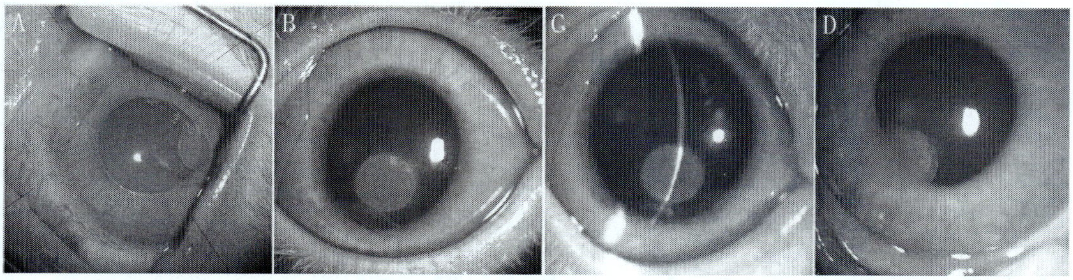

Source: Panel (A-D) adapted and reproduced with permission of ref. [83] Copyright 2015 Wiley Periodicals, Inc.

Figure 12. The same rabbit eye was photographed on successive occasions with a slit lamp. The position of the implants can be visualized in a postoperative observation. Lack of an immune or neovascular response: (A) immediately after the implant procedure, and after (B) 7 days, (C) 30 days, and (D) 90 days operative.

The results of SEM and AFM showed 50 μm thickness and2.42 ± 0.5 μm Rrms of the films. According to the transparency test, their transmittance (%) was 57.07 ± 66.66, 85.17 ± 0.41, and 86.44 ± 0.21 at 300, 500, and 700 nm wavelengths, respectively. The mechanical tensile test revealed 43 MPa tensile strength and 2.3% elongation at break, and *in vitro* tests showed the films produced by casting via centrifugal force enabled the proliferation of human corneal keratocytes. Nili et al. [82] developed a fibroin film incorporated with RGD by casting to investigate the growth of stromal and epithelial cells on the surface of the fibroin-RGD film. Themechanical tensile test revealed 11.1 ± 3.2 MPa modulus of elasticity and 2.5 ± 0.4 MPa tensile strength of the PEG/Horseradish Peroxidase (HRP) treated RGD-film, whereas the fibroin-RDG film showed 21.2 ± 5.8 MPa modulus of elasticity and 1.7 ± 0.2 MPa tensile strength. The elongation at break value of the films was similar to 40%. *In vitro* assays with stromal and limbal epithelial cells showed the fibroin-RDG film significantly reduced stromal cell agglomeration; however, the authors highlighted further studies are necessary for confirming whether the *in vitro* results were due to either changes in the integrin expression, or other transcriptional level events.

Wang et al. [83] developed 2 μm thick fibroin films functionalized with adhesion factor RDG that were used as biomaterials for the treatment of corneal wounds. *In vitro* and *in vivo* assays showed the fibroin-RDG films promoted cell adhesion and proliferation and no immunogenic, neovascular, or rabbit corneal stromal degradation responses. Moreover, type I collagen increased, whereas type III collagen and fibronectin initially increased and then gradually decreased. Figure 12(A-D) illustrates the surgery of fibroin-RDG film insertion.

3.3. Electronic and Optical Devices

Electronic devices, sensors, and biosensors are tools used in the biomedical area to diagnose and treat certain diseases (e.g., cancer, diabetes, COVID-19, etc.) in their initial phase.

Koh et al. [84] highlighted the latest technological advances from the versatility and the use of fibroin thin films for the manufacture of electronic, bioelectronic, and optoelectronic devices and addressed the use of fibroin in combination with other materials and standardization (nanometric scale) and functionalization of the devices. They also reported the development of fibroin-based bioelectronic materials such as e- sutures, e-skins, e-bandages, biosensors, wearable screens, implantable devices, artificial muscles, field-effect transducers, antennas to measure neurotransmitters, food sensors, light-emitting diodes (wound healing, corneal neovascularization), and memristors (bionic brains).

SuNa et al. [85] pointed to advances in the manufacturing methods for fibroin-derived materials (e.g., thin films for specific purposes such as cell detection and visualization) and wearable and smart devices. Li et al. [86] developed electrodes in the form of thin films based on fibroin, poly(3,4-ethylene dioxythiophene), and polystyrene sulfonate towards the fabrication of flexible sensors. The electrodes showed high elasticity (>250%), approximately 0.13°C.m^2.W^{-1} thermal insulation, and 23 Pa.m^2.W^{-1} evaporation resistance, and were 10 times lower than commercial gel electrodes, with a 117 g.m^{-2}.h^{-1} water vapour transmission rate under sweaty conditions. Figure 13(a-c) displays the electrodes on the skin, (d) the relation between impedance and frequency, (e) a comparison between the electrode produced and a commercial one, and (f-g) their ability to detect thermal transmission and temperature variation in the skin before and after physical activity.

Hu et al. [87] developed photonic amplifiers from silk fibroin thin films for subcutaneous energy transmission. The results showed the fluorescence achieved was four times higher than that of the control, thus ensuring greater efficiency of energy conversion in the near-infrared (NIR). Moreover, the thin films used as substrates showed reversible switching to NIR exposure, functioning as an on/off switch. Figure 14 depicts the composition of the fibroin-based optoelectronic device, as well as the electrical results after several tests that evidenced its functionality as an on/off switch when exposed to NIR.

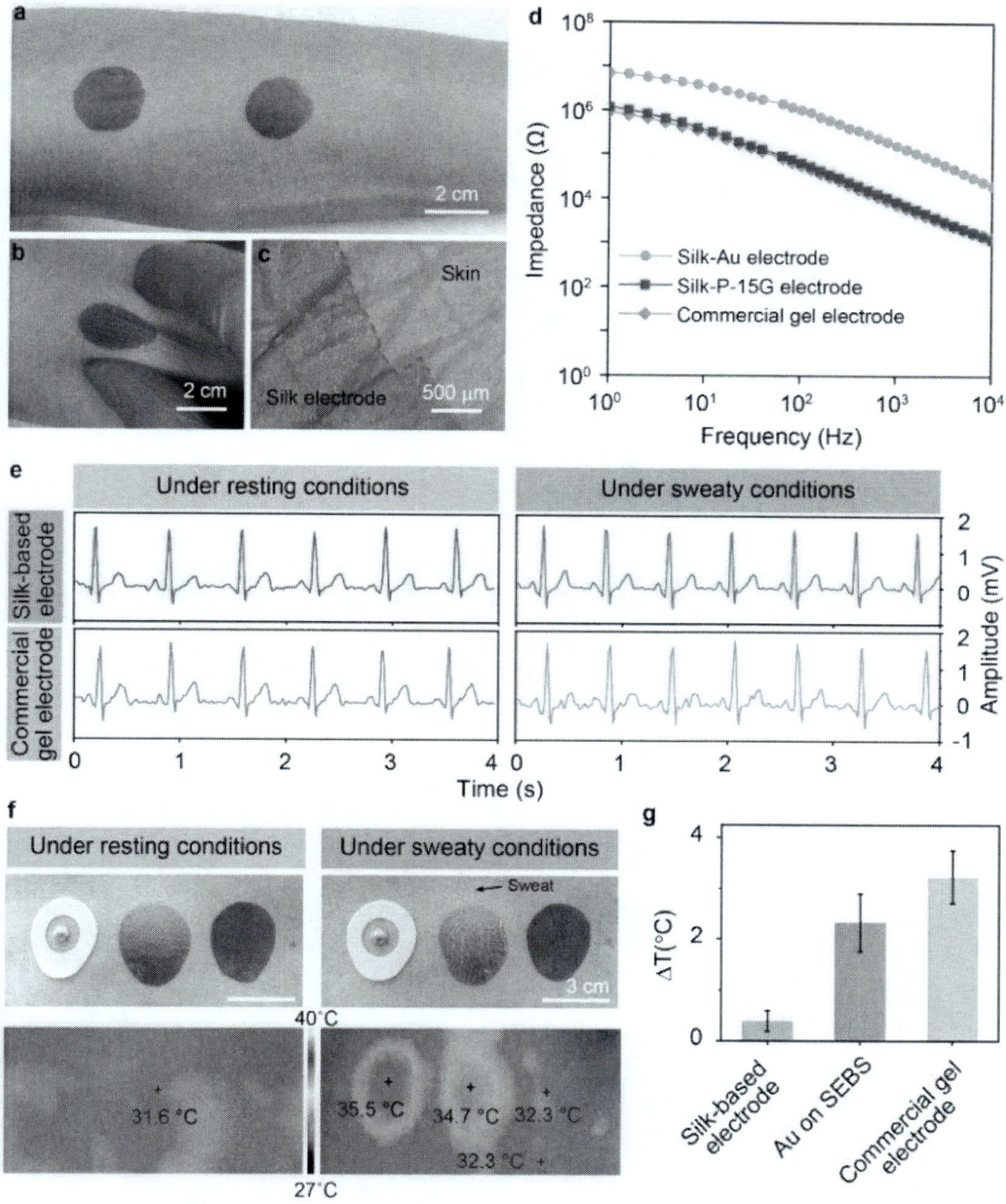

Source: Panel (a-g) adapted and reproduced with permission of ref. [86]. Copyright 2021 American Chemical Society.

Figure 13. On-skin electrophysiological monitoring with sweat tolerance. (a) Photographic image of silk- based electrode laminated on a human forearm at normal and (b) wrinkling states. (c) Optical microscopic image of the silk-based electrode conformability to the irregular structure of the human skin; (d) On-skin interfacial impedance of the silk-based electrodes in comparison with commercial gel electrodes and Silk-Au electrodes; (e) ECG signals recorded by silk-based and commercial electrodes before and after physical activity; (f) Optical and infrared photographs of several electrodes on skin before and after physical activity show the good thermal transmission of silk-based electrodes; (g) Central temperature variation of the skin with electrodes compared with the temperature of the surrounding naked skin after physical activity.

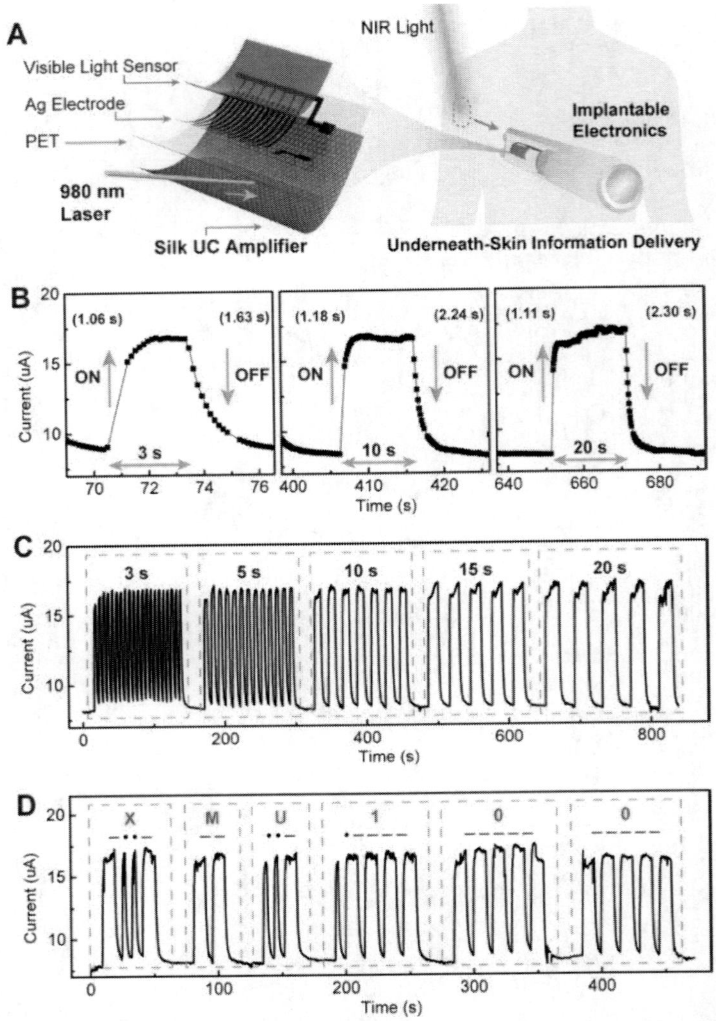

Source: Panel (a-d) adapted and reproduced with permission of ref. [87]. Copyright 2020 American Chemical Society.

Figure 14. (A) Schematic of the setup of photovoltaic measurements of NIR-light switches under the irradiation of a 980 nm laser. NIR laser is pointing at the SFUCPA (with PC structure); (B) Responses of the current for the NIR-light switches to an ON/OFF operation (3, 10, and 20 s). Response times (10−90%) shown in parentheses; (C) Response of the current for the NIR-light switches to a certain frequency of ON/OFF operation. The ON/OFF internals were approximately 3, 5, 10, 15, and 20 s; (D) Signal output for Morse code "XMU100" controlling an NIR-light system with the corresponding codes.

Lin et al. [88] designed ion exchange membranes from silk fibroin thin films for the development of reverse electrodialysis devices. The results of electrical measurements showed a 0.59 m.W.m^{-2} maximum output power density with 66 kΩ external load resistance. 10 membrane pairs produced a 1.58 V voltage, and the authors highlighted the fibroin-derived membranes provided insights for the construction of implanted medical devices. Chao et al. [89] produced a pressure sensor coated with titanium carbide (Ti3C2Tx - MXene, two-dimensional inorganic compounds) by combining electrospinning and serigraphy techniques, from membranes composed of silk fibroin thin films. The sensor showed a wide detection range (up to 39.3 kPa), high sensitivity, fast response (0.44 ms), good water vapour permeability, excellent stability (10,000 cycles), and good biocompatibility and degradability. It also promoted an excellent detection response in the monitoring of physiological signals in real-time, acting as artificial skin. It is therefore considered promising for the development of smart electronic skins, human motion detection, and disease diagnoses. Figure 15(a-f) displays pressure sensors made of fibroin thin films that detect human movement. Such MXene/fibroin-based sensors were mounted as artificial E-skin for the detection of various tactile signals and mapping of pressure distributions. They can also be coupled to a wireless transmitter for real-time pressure detection, as shown in Figure 15 (g), and when pressed by a finger, the sensing response can be timely and wirelessly obtained, Figure 15 (h-i).

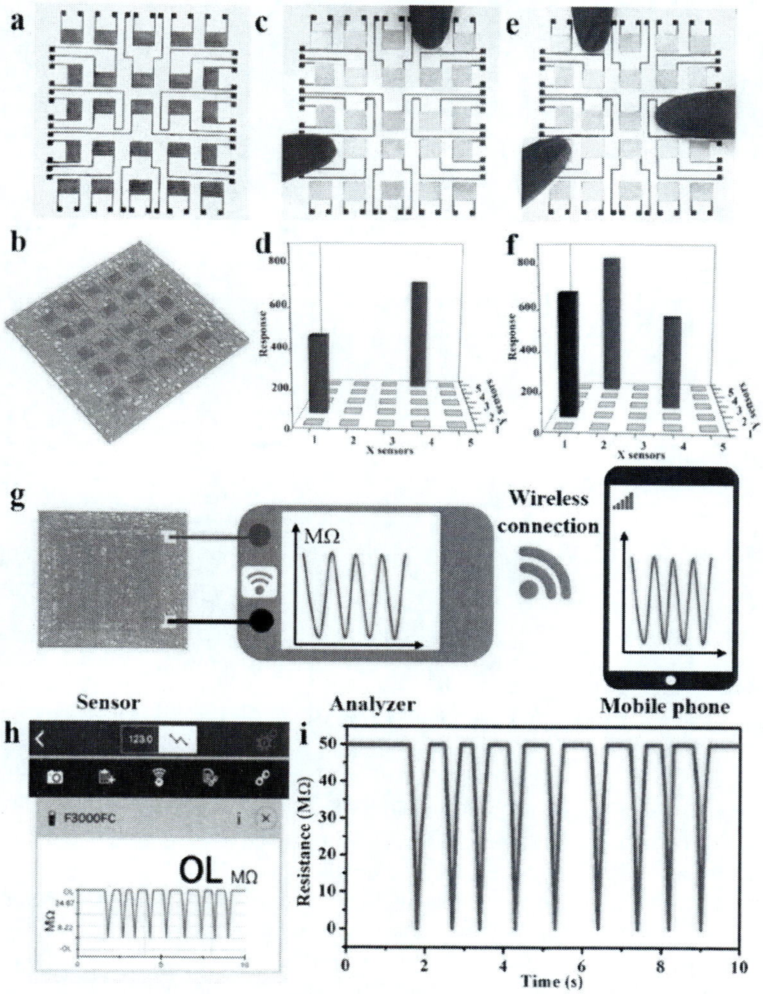

Source: Panel (a-i) adapted and reproduced with permission of ref. [89]. Copyright 2021 American Chemical Society.

Figure 15. Pressure mapping by the MXene/protein-based pressure sensing array and the wireless sensing performance. (a) Photograph and (b) schematic diagram of the assembled electronic skin with 5 × 5 pixels, (c-e) Optical images of different numbers of fingers pressing the E-skin and (d-f) relevant pressure distribution obtained from the current responses, (g) Concept of wireless transmission, (h) Real- time sensing responses obtained wirelessly on a mobile client during the detection of finger pressing, and (i) corresponding resistance changes.

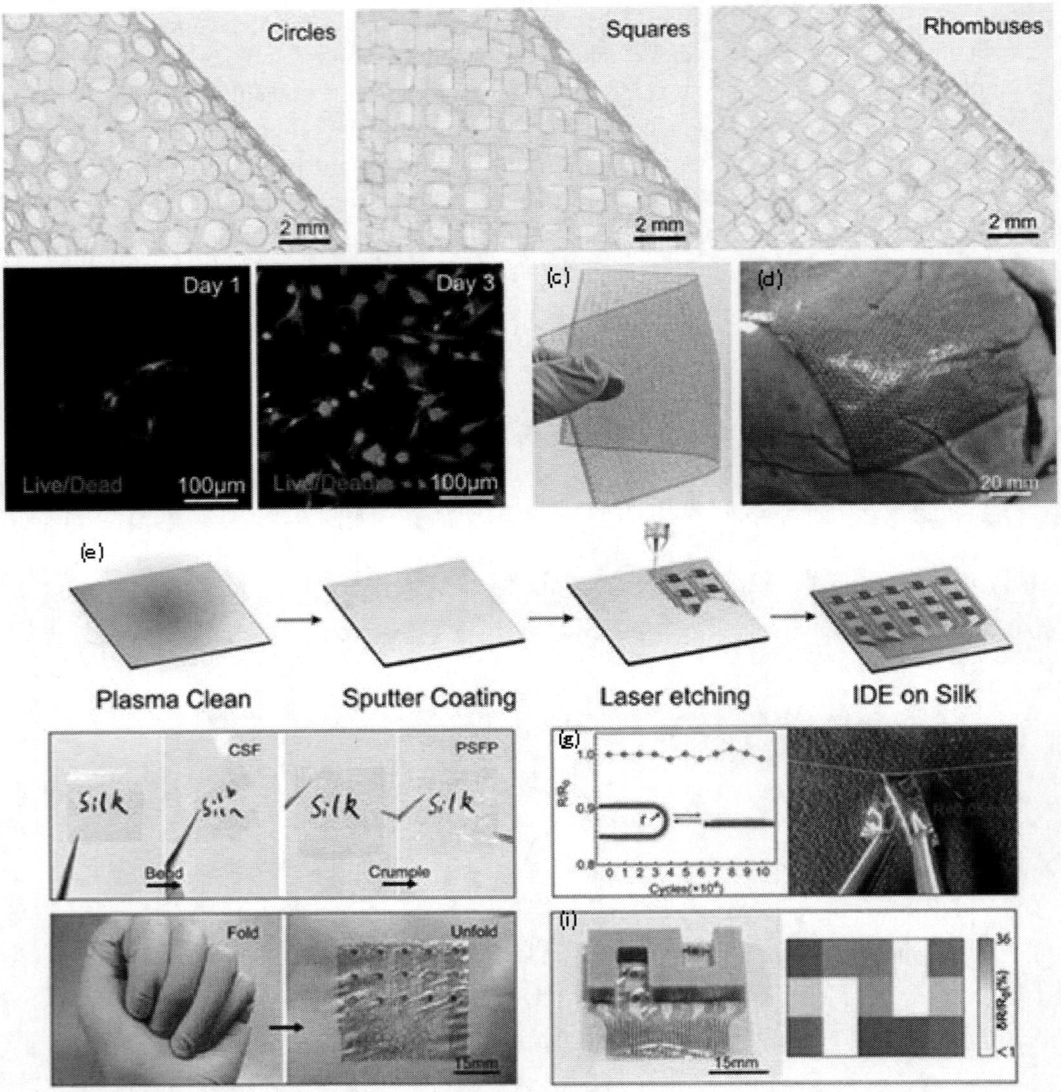

Source: Panel (a-i) adapted and reproduced with permission of ref. [92]. Copyright 2021 Wiley – VCH GmbH.

Figure 16. Engraving engineered PSFP biomeshes and their biocompatibility test: (a) Engraved holes with hexagonal circle, square, and rhombuse shapes fabricated; (b) Live/dead assay for rat BMSCs cultured on the surface of PSFP - images recorded on days 1 and 3; (c) Large area of the PSFP biomesh loaded with crystal violet (d) attached to an isolated porcine heart; (e) Steps for the fabrication of Au patterns (e.g., IDE array on PSFP); (f) Comparison of drop-casted silk film and PSFP in a bending and crumpling test; (g) (left) Conductive stability test of PSFP/Au electrode with 100 000 fold–unfold cycles r = 2 mm; (right) conductivity test of PSFP/Au electrode under 0.05 mm bending radius; (h) Tactile sensing array tightly attached to nitrile glove: (left) fold and (right) unfold; (i) Tactile sensing ability of the PSFP-based electronic device. Total weight of 3D printed "S" shape in (left) is 6.9 grams.

Wang et al. [90] developed a physicochemical strategy to fabricate composites of silk nanofibers and single-walled carbon nanotubes (CNTs) in an aqueous medium towards the development of electronic devices. The strategy showed a high yield (~ 87%), with films of 280 nm average diameter and 10-30 μm length, excellent dimensional stability, chemical durability, and 498.7 ± 64.4 MPa Young's modulus. They also exhibited excellent electrical properties, thus being promising candidates for bioelectronic applications. Zhang et al. [91] designed silk fibroin and silver (Ag) thin films by spin-coating for the development of bioelectronic devices (memristors). The 60 x 60 μm^2 area memristor showed excellent flexibility and transparency, operating at low power (0.7 μW) and current (1 μA). Such properties were attributed to the formation of microcrystals in the molecular structure of fibroin induced by the incorporation of Ag and the chemical treatment of the materials in ethanol. Liu et al. [92] developed an ion-trace assisted plasticization strategy for the fabrication of fibroin (PSFP) films in various shapes (periodic meshes, electronic prototypes, and kirigami-based devices).

AFM results showed, respectively, 1.83 and 2.31 nm roughness Ra and Rq of the films and TGA and DMA analyses revealed the glass transition and degradation temperatures of PSFPs were 210 and 255°C, respectively. In structural terms, the FTIR and XRD results evidenced the films have crystallites of 1-2 nm estimated grain size in the amorphous phase (α-helix). Moreover, PSFPs are highly transparent (~90%) in the visible region. Mechanical tests showed 23-47 MPa tensile strength and 13-19 MPa modulus of elasticity of the films, as well as their good toughness (14.4 ± 4 $kJ.m^{-2}$), which is higher than values of any plastic-electronic substrate. *In vitro* assays with rat bone marrow-derived mesenchymal cells (BMSCs) revealed the films are biocompatible and enable the proliferation adhesion of BMSCs cells. Figure 16 depicts (a) the shapes (hexagonal circle, square, and rhombus) of PSFPs, (b) biocompatibility, flexibility, and use in tissue engineering, (c) ability to be printed with large area, and (d) versatility, (e) Kirigami electrode architecture, (f) uniaxial and biaxial PSFPs/Au architecture, (g) comparison between biaxial PSFPs/Au and commercial electrode concerning frequency (Hz) and impedance (Ω), (h) their conformal adhesion ability, and (i) their tactile sensing ability.

Abadi et al. [93] highlighted the fabrication of porous and lightweight biomimetic (3D) materials is on the rise, since they can be applied as sensors, flexible energy storers, electromagnetic protection devices, and biomaterials. Despite recent progress, the current bottleneck for their fabrication is the combination of electrical conductivity, high flexibility, and shape memory. The authors first designed ultralight, flexible, electrically conductive and resistant thin, porous films by hybridising graphene oxide reinforced fibroin and Ti_3C_2Tx (MXene) nanosheets and combining freeze casting, freeze-drying, and gas-phase hydrophobic techniques. The films showed high porosity, low density (6-60 $mg.cm^{-3}$), hydrophobicity (~124°), high organic liquid absorption capacity (~120%), 30% reversible compressibility with no failure, 80% ultimate compression, and high sensitivity to deformation (50%) and pressure (~1 Pa to 2 kPa). The nanocomposites are promising candidates for the development of flexible pressure sensors and electronic films and for high-performance applications. Huang et al. [94] produced electrodes of silk fibroin thin films towards the development of devices that measure electromyographical signals. The methodology used enabled the production of electrodes of good rigidity (134 kPa - 1.9 GPa) and flexibility (~200%), and the epidermal electrodes did not cause irritation or inflammation in the skin after being fixed for more than 10 days. Figure 17 shows the flexible fibroin-based electrodes.

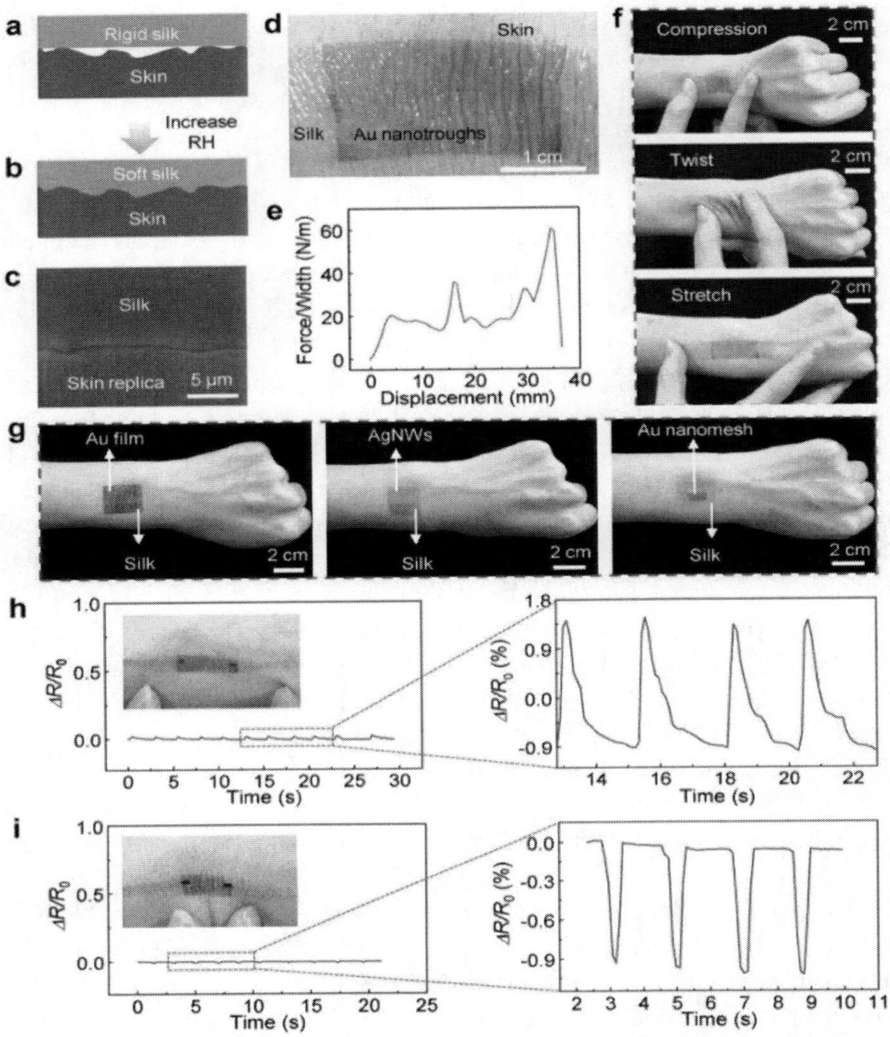

Source: Panel (a-i) adapted and reproduced with permission of ref. [94]. Copyright 2020 Wiley – VCH GmbH.

Figure 17. Mechanical and electrical stability of the silk epidermal electrode. (a-b) Schematic illustration of rigid silk/skin and soft silk/skin interfaces; (c) SEM image of an interface between the silk and a skin replica; (d) Optical image of a silk electrode on the skin; the electrode complies with the skin texture; (e) The adhesion strength between silk electrode and toughness was 60 $N.m^{-1}$; (f) Silk epidermal electrode in different mechanical states of squeezing, twisting, and stretching; (g) Photographs of silk electrode with Au film, Ag nanowire (AgNW) electrode, and Au nanomesh electrode laminated on the skin; (h-i) Normalized change in resistance of silk epidermal electrodes under repeated skin stretching and skin squeezing, respectively.
Insets show corresponding photographs.

Pradhan and Yadavalli [95] used flexible fibroin thin films as a substrate on which PEDOT:PSS conductive ink was micropatterned by spin-coating, followed by photolithography, for the development of customized temperature sensors capable of monitoring, detecting, and diagnosing diseases. According to the results of SEM, electrical conductivity, chronoamperometry, degradation, and stability in aqueous media, the sensors showed high sensitivity (−0.99% $°C^{-1}$) at 20−50°C temperature range, excellent stability at 10 to 90% humidity (relative humidity), high cyclic stability in several heating/cooling cycles, and thermal stability and biodegradability for 10 days under proteolytic conditions (1-3.5 U.ml^{-1} of protease solutions). Figure 18 shows (a) the size and shape of the electrode, (b) the SEM micrograph highlighting the morphology of the layers that compose the sensors, and (c) the device's degradability capacity under specific conditions.

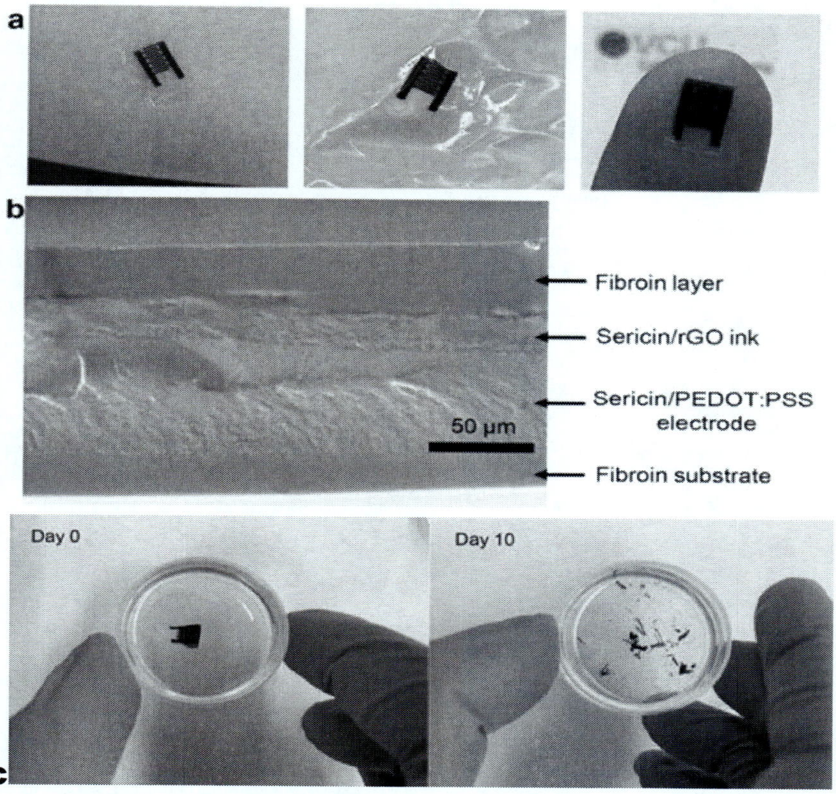

Source: Panel (a-c) adapted and reproduced with permission of ref. [95]. Copyright 2021 American Chemical Society.

Figure 18. Images of the flexible temperature sensor formed from silk proteins; (a) the sensor can be placed on the skin or integrated into a wrinkled surface (e.g., a textile) with a small form factor; (b) SEM imaging of the cross-section of the sensor showing the covalently integrated layers that prevent delamination and improve stability; (c) Degradation of the flexible temperature sensor in 10 days in enzymatic (protease) environment devices.

Kook et al. [96] reported a new process of microfabrication and micropatterning in multilayers of fibroin films intercalated with aluminium layers (AMoS) using the ultraviolet photolithography technique towards the fabrication of memristors and organic transistors for implantable biomedical applications. The SEM results showed the thin films are 10 µm wide and 1.2 µm thick and parameter Rrms revealed their surface roughness increased from 3.4 to 4.3 nm after the fabrication process. However, the increase was considered small and attributed mainly to thermal deposition and removal of aluminium on the fibroin surface. FTIR analysis showed the AMoS process did not cause significant disruption to the molecular structures in the fibroin films and, according to *in vitro* assays (primary neurons and NIH3T3 cells), the new manufacturing process is not cytotoxic and promoted cell proliferation.

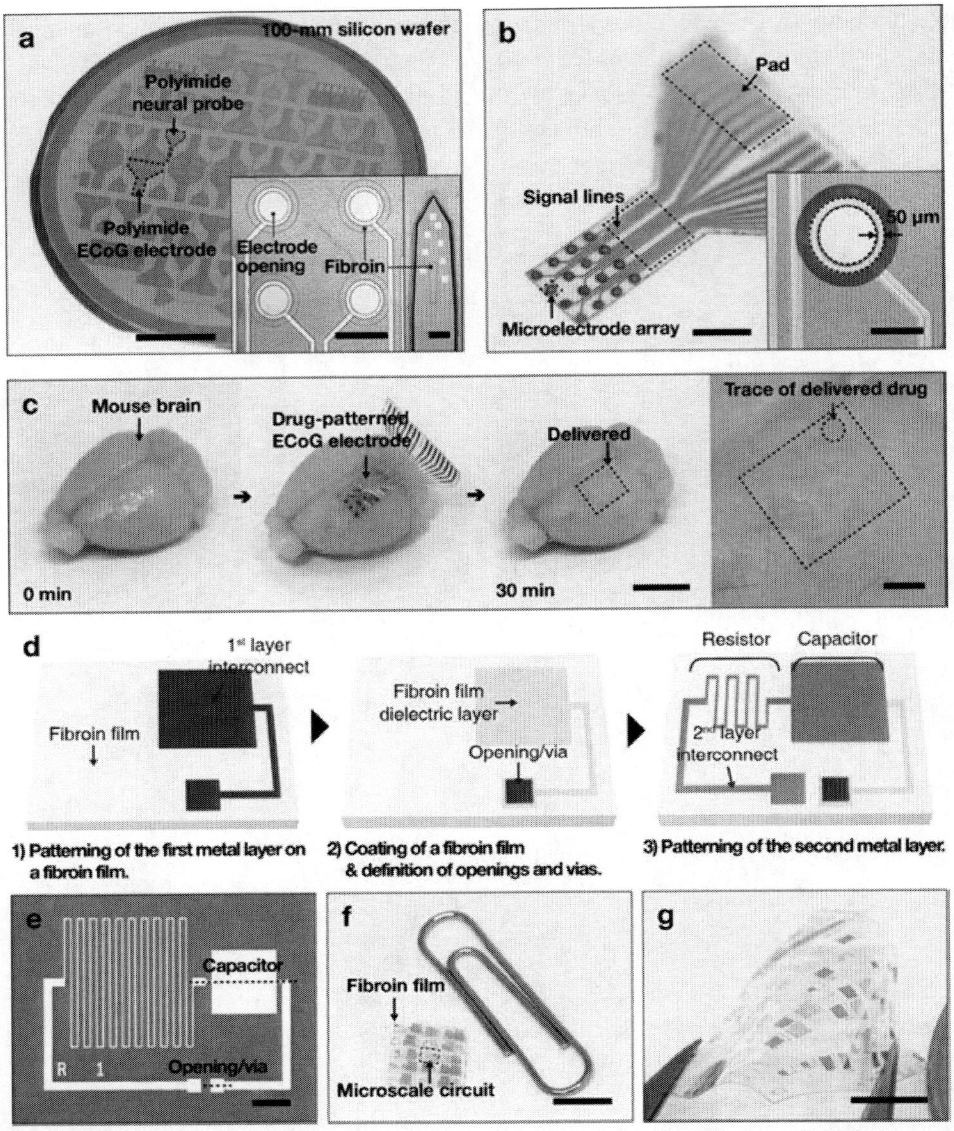

Source: Panel (a-g) adapted and reproduced with permission of ref. [96]. Copyright 2019 American Chemical Society.

Figure 19. (a) Polyimide-based flexible neural electrodes fabricated with fibroin patterns on a 100 mm silicon wafer (scale bar: 20 mm). Insets are microscopic images of the fibroin patterns on the electrodes. Left: ECoG electrodes (scale bar: 400 μm). Right: Neural probe (scale bar: 100 μm); (b) Photograph of the fabricated ECoG electrodes (scale bar: 2 mm). Rhodamine B loaded into fibroin for the visualization of fibroin patterns. The inset is a close-up view of one of the microelectrode arrays (scale bar: 200 μm): alignment margin between the electrode opening and the fibroin pattern was below 50 μm and easily achievable with a 1 μm alignment limit of the process; (c) Left: Photographs of the sequence of rhodamine B delivery onto a mouse brain from the ECoG electrode through micropatterned fibroin layers of high spatial resolution (scale bar: 5 mm). Right: Close-up view of the sites showing localized drug delivery (scale bar: 1 mm); (d) Wafer-scale fabrication process of μm-scale series RC components by the AMoS process and interleaving two fibroin layers and two metal layers; (e) Optical microscopic image of the fabricated device; (f) Photograph of fabricated microscale RLC circuits on a fibroin film (scale bar: 3 mm). Gold (yellow) and silver (white) patterned on different layers with a fibroin layer as a passivation layer in between; (g) Photograph of metal patterns on a flexible fibroin film (scale bar: 5 mm).

Figure 19(a-b) displays micropatterned fibroin obtained through high precision (20 μm) alignment to the microelectrode array in wafer-scale by AMoS process, (c) use of rhodamine B as a model drug to visualize the selective drug loading and delivery onto a mouse brain, (d) manufacture of passive fibroin-based electronic components such as resistors (R), capacitors (C), and inductors (L), where the fibroin layer is used as either a dielectric layer, or a substrate, (e) impedance of a resistor and a series-connected (RC) capacitor where fibroin was used as a dielectric and passivation layer, (f) a photography of microscale RLC circuits fabricated on a fibroin film, and (g) flexible, transparent, biodegradable fibroin film.

Malinowski et al. [97] designed optical silk fibroin thin films by the soft lithography method for the development of flexible and biocompatible optical and electronic devices. According to SEM, FTIR, optical transmittance, transmission optical haze, and photoelectronic results, the films showed >93% optical transparency and >65% optical transmission haze. The combination of silk fibroin thin film with the silicon photodiode can enhance the efficiency (high transmittance and high optical transmission haze effect) of optoelectronic devices. The efficiency increased by 6.96% for silk fibroin film with no patterns and 14.9% for nanopatterned silk fibroin film. Figure 20 displays transmittance and optical transmission haze - measured by Jasco V670 with integrated sphere - as a function of wavelength for glass (reference) of masters, polydimethylsiloxane (PDMS) molds, and silk fibroin films with no structures, and with linear and mesh structures.

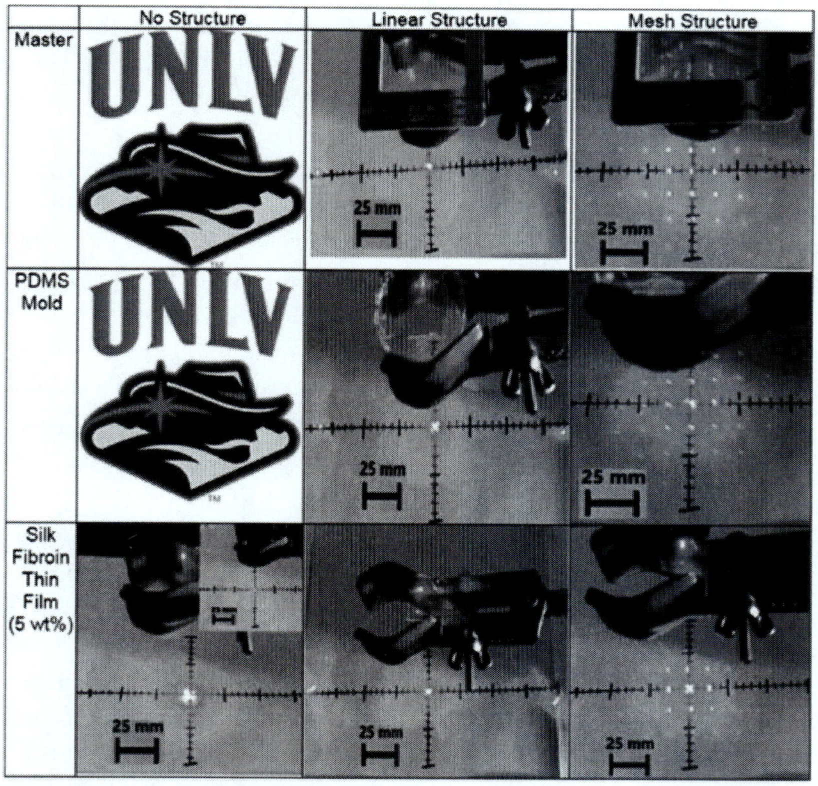

Source: Panel adapted and reproduced with permission of ref. [97]. Copyright 2019 The Royal Society of Chemistry.

Figure 20. Light scattering of the master, PDMS mould, and patterned silk fibroin film with the laser light test.

3.4. Controlled Drug Release

Silk fibroin thin films have been used as platforms for the delivery of different types of drugs (e.g., growth factors, anti-cancer agents, antibiotics, antioxidants, and ions) in a controlled manner for the areas of tissue engineering, regenerative medicine, and chronic diseases [98–100]. Choi et al. [101] produced fibroin/heparin and epirubicin (anti-cancer drug used in the treatment of breast cancer) thin films formed by 14.19 ± 6.27 nm thick bilayers via LbL. The controlled release results showed the release of epirubicin after 1h was 47.62 and 39.48% for glycerol and methanol treated films, respectively. Coburn et al. [102] demonstrated fibroin films enable the release control of cationic oncology drugs such as doxorubicin and vincristine, used in the treatment of bladder and lung tumours, respectively, through the insertion of $COOH^-$ (carboxylate) and SO_3H^- (sulfonate) ions. When functionalized, the films enable the controlled release of small-molecule drugs in immuno-oncology therapies.

Hines and Kaplan [103] showed the high potential of silk fibroin thin films as a polymeric biomaterial for systems of controlled release; they can also enable the control of a multiphase release by the mechanisms of release by diffusion and degradation of polymer chains. Huang et al. [104] claimed fibroin films display important characteristics for the controlled release of drugs and analyzed their controlled release behaviour regarding four types of hydrochlorides, namely ciprofloxacin, cisdiltiazem, tetracycline, and tetrahydroin), using permeability and diffusivity parameters. The results showed the behaviour is similar to those of other materials applied for controlled release and the films can potentially be used for the production of biomaterials applied to the controlled release of drugs. Tomoda et al. [105] produced fibroin films incorporated with diclofenac sodium by casting towards the obtaining of biomaterials for a controlled release of drugs in treatments of skin lesions. The release kinetic results showed the anti-inflammatory drug can be released in a controlled way and inhibits the growth of bacteria such as *Staphylococcus aureus, Pseudomonas aeruginosa, Escherichiacoli,* and *Candida albicans* without affecting the biocompatibility of the films. However, the authors highlighted the need for further studies towards improvements in the controlled release of the drug.

Srisuwan and Baimark [106] produced fibroin/alginate thin films by casting to evaluate the controlled release of tetracycline hydrochloride. The *in vitro* release assay showed their release profiles in a phosphate buffer solution (pH = 7.4) at 37°C for 24 h. Regarding alginate film, circa 95% of the loaded drug were released within 3 h, and the drug release level at 24 h was 64% for the fibroin film. The drug release levels in 24 h were 90%, 85%, 82%, and 71%, respectively, for films containing fibroin/alginate in proportions of 1:1, 2:1, 4:1, and 8:1 (w/w), showing the drug release decreased as the fibroin/alginate mixture ratio increased. Alam and Shubhra [107] projected a 200 μm thick thin film via casting; it is composed of fibroin/gelatin and coated with poly(ethylene glycol) for the release of antibiotic ciprofloxacin in a controlled way for healing a wound in a rat model.The *in vitro* release study showed the antibiotic (0.25, 0.50, and 1.0%) was incorporated into the films and the poly(ethylene glycol) coating improved ciprofloxacin release after 50 h. It also reported (at the first stage) a rapid antibiotic release associated with the non-trapped and free amount on the surface of the film. The second stage of release can be attributed to film degradation and antibiotic diffusion, which can occur simultaneously. Finally, the authors concluded that occurs the release of the drug molecules, which interacted with amino acids, as well as the drug molecules trapped inside the crystalline structure, and the film degradation occurs prior to the diffusion of ciprofloxacin. *In vivo* testing showed fibroin/gelatin thin film containing 0.5% ciprofloxacin promoted rapid wound healing in rats after 4 days. After seven days, an epidermis growth (from the margin to the centre) was observed at the wound sites.

Wang et al. [108] used silk fibroin thin films as a support to design microneedles by the vacuum compression technique. The microneedles were incorporated with proline for controlled insulin release in diabetic patients. FTIR and XRD results showed both fibroin and proline are compatible, and microneedles can act as drug carriers since they are non-toxic and degradable and release insulin slowly (*in vitro*). Mechanical tests revealed the needles can pierce the stratum corneum of the skin. Figure 21(A-E) shows the preparation of silk fibroin microneedles and (F-K) SEM images of the cross-sections and surface morphologies of different microneedles.

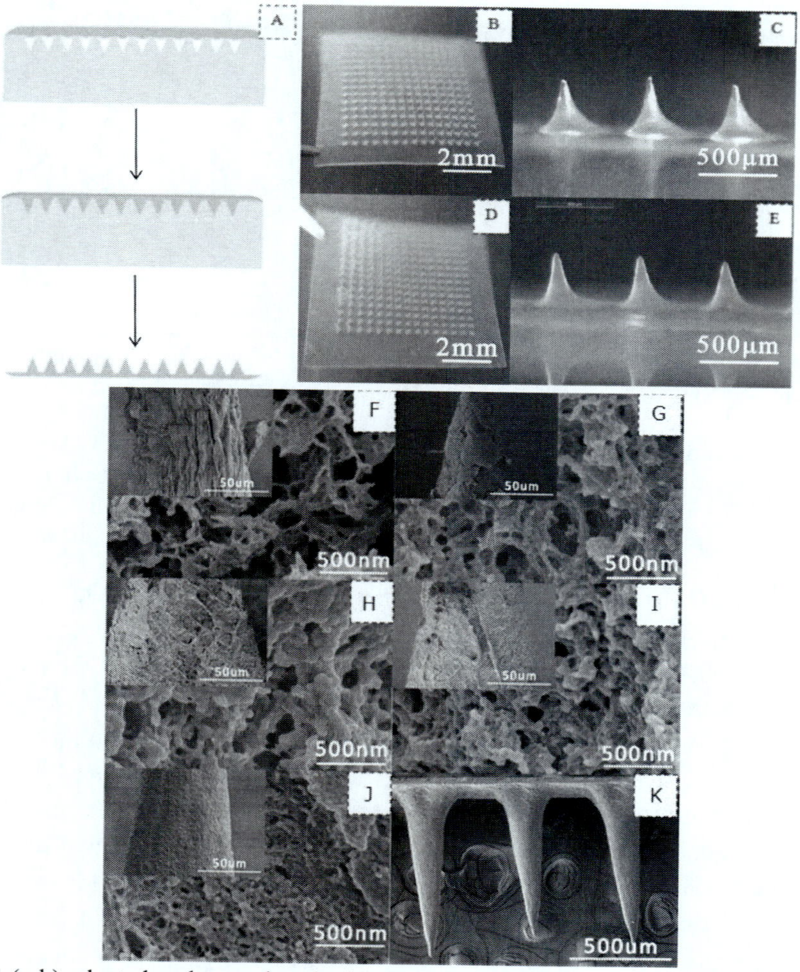

Figure 21. Schematic of the preparation of silk fibroin microneedles (needles); (A) preparation of needles; (B, C) Images of 15 × 15 arrays of untreated silk fibroin needles under the microscope, and (D, E) Images of silk fibroin needles after incubation with FITC-insulin under the microscope. SEM image of cross-sections and surface morphologies of different microneedles (needles); (A) pure fibroin; (F−I) proline/fibroin = 0.5/10, 1/10, 3/10; (J) methanol-treated; and (K) dry needles.

Zhu et al. [109] manufactured patches in the form of microneedles from silk fibroin thin films incorporated with insulin for the transdermal administration of drugs and reduction of pain caused by subcutaneous injections. The microneedles showed good dissolution properties and insulin was continuously released through the microporous channels. The microneedles also enabled a good storage stability (~90%) of insulin´s biological activity after 30 days. *In vitro* tests in diabetic rats suggested the patches exerted an apparent hypoglycaemic effect and controlled insulin release. Figure 22 displays (A) fibroin thin films in the *in vivo* assay (mouse), (B) and (C) action of the film on the wound, and (D) and (E) the microneedle morphology under a microscope before and after drug administration.

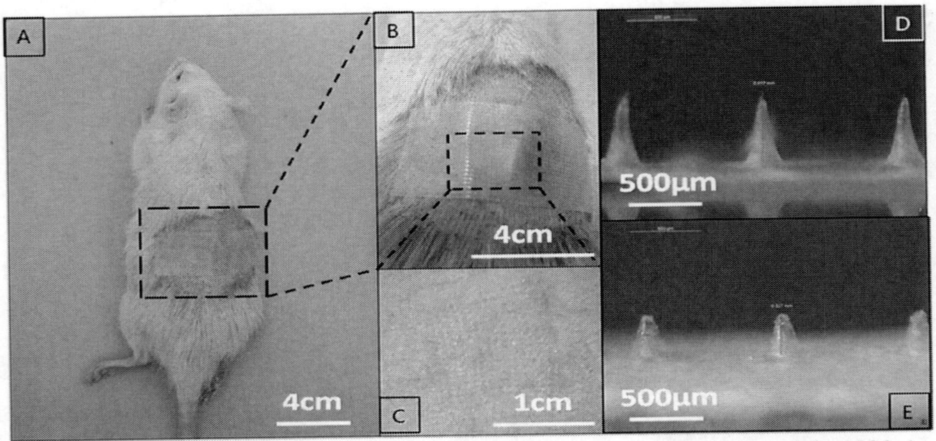

Source: Panel (a-e) adapted and reproduced with permission of ref. [109]. Copyright 2020 American Chemical Society.

Figure 22. *In vivo* experiments with insulin-loaded microneedles in rats: (A) and (B) photographs of rats treated with insulin-loaded microneedles, (C) photographs of the skin surface of rats after removal of microneedles, (D) microscopic morphology of the microneedles prior to administration, and (E) microscopic morphology of the microneedles after 4 h of administration in rats.

Pacheco et al. [110] developed multilayer membranes incorporated with sodium diclofenac composed of silk fibroin, chitosan, and alginate thin films deposited by the combination of casting and dip-coating techniques towards the production of high- performance wound dressing. SEM results showed the membrane´s microstructure with three layers (115, 178, and 227 µm) of which the thickest one is associated with chitosan. Solubility and TGA tests revealed a 73.96% ± 1.89% weight loss of the membrane in up to 7 days immersed in water and the crosslinking result by UV irradiation showed the average weight loss was reduced to 67.75% ± 2.72% in up to 7 days. According to the fluid handling capacity test, the membranes with and without drug incorporation can drain 2.39 ± 0.22 g/10 cm^2/24 h and 2.55 ± 0.29 g/10 cm^2/24 h of the fluid generated by the wound (exudate), respectively. Mechanical testing revealed 3.05 ± 0.41MPa tensile strength, 45.14 ± 6.23% elongation at failure, and 20.73 ± 4.64 MPa Young's modulus of the drug-free membrane, whereas the membrane with the drug showed 2.63 ± 0.35 MPa, 50.28 ± 8.44% and 17.84 ± 3.52 MPa, respectively. *In vitro* assay, performed according to ISO 10993-5, indicated no cytotoxicity of the membrane and controlled release tests showed 70% of diclofenac sodium were incorporated into the membrane. Peppas and Peppas-Sahlin models, which best describe the release of diclofenac sodium from the membrane, showed a 0.99 correlation coefficient. They used the value of 0.384 ± 0.065 mm as average membrane thickness to calculate the diffusion coefficient (D = 3.63×10^{-8} $cm^2.s^{-1}$) of the drug in the membrane.

Conclusion and Future Perspectives

Over the past few years, factors such as increased life expectancy, population growth, emergence of new diseases, automobile accidents, among others have led scientists to develop materials capable of meeting the needs of contemporary society. The materials are the result of scientific and technological advances in materials science, tissue engineering, biomaterials, characterization, sensors, and controlled drug delivery and among them, silk fibroin films have been driving the manufacture of biomaterials (skin and cornea), sensors, biosensors, optical devices, and controlled drug delivery systems capable of regenerating, treating, and/or replacing tissues and organs.

This review addressed the advances made over the past fifteen years in the production and application of silk fibroin thin films in biomaterials, sensors, biosensors, electronic devices, and controlled drug delivery. Fibroin thin films can be projected by casting, layer-by-layer, electrospinning, freeze-drying, lithography, dipping, and spin coating techniques and thin films with excellent physical, chemical, thermal, electrical, optical, mechanical, and biological properties can be produced by the combination of fibroin with natural polymers (e.g., chitosan, alginate, collagen, gelatin, and zein), synthetic polymers (e.g., PCL, PLLA, PVA, PEO, PEDOT, PSS, P3HT, and PU), organic compounds (e.g., antibiotics, antioxidants, anti-inflammatories, growth factors, stem cells, and drugs), and inorganic ones (e.g., silver, gold, zinc, titanium, calcium, sodium, phosphorus, potassium, iron, strontium, europium, tellurium, and cadmium) on micro or nanometer scales. All such versatility has led to the development of customized silk fibroin thin films for the aforementioned areas and their behaviours enable them to meet current and future demands of society.

Acknowledgments

The authors acknowledge the financial support of Sao Paulo Research Foundation (FAPESP), grant number: 2017/26194-7, and Coordination of Higher Level Staff Improvement (CAPES), finance code 001, and ACP Giampedro for the revision of the English language.

References

[1] Hu F, Lin N, Liu XY (2020) Interplay between Light and Functionalized Silk Fibroin and Applications. *iScience* 23:101035. https://doi.org/10.1016/j.isci.2020.101035.

[2] Sun W, Gregory DA, Tomeh MA, Zhao X (2021) Silk fibroin as a functional biomaterial for tissue engineering. *Int J Mol Sci* 22:1–28. https://doi.org/10.3390/pharmaceutics11100494.

[3] Lima, LR, Moraes, ML, Nigoghossian, K, Peres, MFS., & Ribeiro, SJL.(2016) Silk fibroin-antigenic peptides- YVO4:Eu3+ nanostructured thin films as sensors for hepatitis C. *J Lumin* 170:375–379. https://doi.org/10.1016/j.jlumin.2015.08.039.

[4] Delezuk, JAM, Pavinatto, A, Moraes, ML, Shimizu, FM, Rodrigues, VC, Campana-Filho, SP, Ribeiro, S.JL., & Oliveira, ON. (2017) Silk fibroin organization induced by chitosan in layer-by-layer films: Application as a matrix in a biosensor. *Carbohydr Polym* 155:146–151. https://doi.org/10.1016/j.carbpol.2016.08.060.

[5] Li, DW, He, J, He, FL, Liu, YL, Liu, YY, Ye, YJ, Deng, X, & Yin, DC. (2018) Silk fibroin/chitosan thin film promotes osteogenic and adipogenic differentiation of rat bone marrow-derived mesenchymal stem cells. *J Biomater Appl* 32:1164–1173. https://doi.org/10.1177/0885328218 757767.

[6] Patil S, Singh N (2018) Spatially controlled functional group grafting of silk films to induce osteogenic and chondrogenic differentiation of human mesenchymal stem cells. *Mater Sci Eng C* 91:796–805. https://doi.org/10.1016/j.msec.2018.06.008.

[7] Srivastava CM, Purwar R, Kannaujia R, Sharma D (2015) Flexible silk fibroin films for wound dressing. *Fibers Polym* 16:1020–1030. https://doi.org/10.1007/s12221-015-1020-y.

[8] Mandal BB, Mann JK, Kundu SC (2009) Silk fibroin/gelatin multilayered films as a model system for controlled drug release. *Eur J Pharm Sci* 37:160–171. https://doi.org/10.1016/j.ejps.2009.02.005.

[9] Bucciarelli, A, Mulloni, V, Maniglio, D, Pal, RK, Yadavalli, VK, Motta, A, & Quaranta, A.(2018) A comparative study of the refractive index of silk protein thin films towards biomaterial based optical devices. *Opt Mater* (Amst) 78:407–414. https://doi.org/10.1016/j.optmat.2018.02.058.

[10] Chakravarty, S, Gogoi, B, Mandal, BB, Bhardwaj, N, & Sarma, NS. (2018) Silk fibroin as a platform for dual sensing of vitamin B12 using photoluminescence and electrical techniques. *Biosens Bioelectron* 112:18–22. https://doi.org/10.1016/j.bios.2018.03.057.

[11] Sohail Haroone, M, Li, L, Ahmad, A, Huang, Y, Ma, R, Zhang, P, Hu, Y, Muhammad Kaleem, Q, & Lu, J. (2018) Luminous composite ultrathin films of CdTe quantum dots/silk fibroin co-assembled with layered doubled hydroxide: Enhanced photoluminescence and biosensor application. *J Mater* 4:165–171. https://doi.org/10.1016/j.jmat.2018.05.002.

[12] Liu, Y, Qi, N, Song, T, Jia, M, Xia, Z, Yuan, Z, Yuan, W, Zhang, KQ, & Sun, B. (2014) Highly flexible and lightweight organic solar cells on biocompatible silk fibroin. *ACS Appl Mater Interfaces* 6:20670–20675. https://doi.org/10.1021/am504163r.

[13] Stinson, JA, Palmer, CR, Miller, DP, Li, AB, Lightner, K, Jost, H, Weldon, WC, Oberste, MS, Kluge, JA, & Kosuda, KM.(2020) Thin silk fibroin films as a dried format for temperature stabilization of inactivated polio vaccine. *Vaccine* 38:1652– 1660. https://doi.org/10.1016/j.vaccine.2019.12.062.

[14] Prakash NJ, Mane PP, George SM, Kandasubramanian B (2021) Silk Fibroin As an Immobilization Matrix for Sensing Applications. *ACS Biomater Sci Eng* 7:2015–2042. https://doi.org/10.1021/acsbiomaterials.1c00080.

[15] Veronica A, Hsing I ming (2021) An Insight into Tunable Innate Piezoelectricity of Silk for Green Bioelectronics. *ChemPhysChem* 1–16. https://doi.org/10.1002/cphc.202100279.

[16] Patil AC, Xiong Z, Thakor N V. (2020) Toward Nontransient Silk Bioelectronics: Engineering Silk Fibroin for Bionic Links. *Small Methods* 4:1–25. https://doi.org/10.1002/smtd.202000274.

[17] Wen, DL, Sun, DH, Huang, P, Huang, W, Su, M, Wang, Y, Han, MD, Kim, B, Brugger, J, Zhang, HX, & Zhang, XS. (2021) Recent progress in silk fibroin-based flexible electronics. *Microsystems Nanoeng* 7:1–25. https://doi.org/10.1038/s41378-2.021-00261-2.

[18] Jiang, C, Wang, X, Gunawidjaja, R, Lin, YH, Gupta, MK, Kaplan, DL, Naik, RR, & Tsukruk, VV. (2007) Mechanical properties of robust ultrathin silk fibroin films. *Adv Funct Mater* 17:2229–2237. https://doi.org/10.1002/adfm.200601136.

[19] Nogueira GM, Swiston AJ, Beppu MM, Rubner MF (2010) Layer-by-layer deposited chitosan/silk fibroin thin films with anisotropic nanofiber alignment. *Langmuir* 26:8953–8958. https://doi.org/10.1021/la904741h.

[20] Moraes MA, Crouzier T, Rubner M, Beppu MM (2015) Factors controlling the deposition of silk fibroin nanofibrils during layer-by-layer assembly. *Biomacromolecules* 16:97–104. https://doi.org/10.1021/bm5012135.

[21] Cho, H, Lee, J, Hwang, H, Hwang, W, Kim, JG, & Kim, S. (2022) Mechanical properties of graphene oxide–silk fibroin bionanofilms via nanoindentation experiments and finite element analysis. *Friction* 10:282–295. https://doi.org/10.1007/s40544-021-0490-8.

[22] Joseph J, Singh SG, Vanjari SRK (2017) Leveraging Innate Piezoelectricity of Ultra-Smooth Silk Thin Films for Flexible and Wearable Sensor Applications. *IEEE Sens J* 17:8306–8313. https://doi.org/10.1109/JSEN.2017.2766163.

[23] Chen, Y, Duan, L, Ma, Y, Han, Q, Li, X, Li, J, Wang, A, Bai, S, & Yin, J. (2020) Preparation of transient electronic devices with silk fibroin film as a flexible substrate. *Colloids Surfaces A Physicochem Eng Asp* 600:1–7. https://doi.org/10.1016/j.colsurfa.2020.124896.

[24] Yadav R, Batra R, Bansal P, Purwar R (2022) N-type silk fibroin/TiO2 nanocomposite transparent films: electrical and optical properties. *Polym Int* 71:74–85. https://doi.org/10.1002/pi.6285.

[25] Kundu, B, Kurland, NE, Bano, S, Patra, C, Engel, FB, Yadavalli, VK, & Kundu, SC. (2014) Silk proteins for biomedical applications: Bioengineering perspectives. *Prog Polym Sci* 39:251–267. https://doi.org/10.1016/j.progpolymsci.2013.09.002.

[26] Gil, ES, Park, SH, Marchant, J, Omenetto, F, & Kaplan, DL. (2010) Response of human corneal fibroblasts on silk film surface patterns. *Macromol Biosci* 10:664–673. https://doi.org/10.1002/mabi.200900452.

[27] Gil, ES, Park, SH, Marchant, J, Omenetto, F, & Kaplan, DL. (2010) Helicoidal multi-lamellar features of RGD-functionalized silk biomaterials for corneal tissue engineering. *Biomaterials* 31:8953–8963. https://doi.org/10.1016/j.biomaterials.2010.08.017.

[28] Ziemba, AM, Fink, TD, Crochiere, MC, Puhl, DL, Sapkota, S, Gilbert, RJ, & Zha, RH. (2020) Coating Topologically Complex Electrospun Fibers with Nanothin Silk Fibroin Enhances Neurite Outgrowth in Vitro. *ACS Biomater Sci Eng* 6:1321–1332. https://doi.org/10.1021/acsbiomaterials.9b01487.

[29] Lin MJ, Lu MC, Chan YC, et al. (2021) An insulin-like growth factor-1 conjugated Bombyx mori silk fibroin film for diabeticwound healing: Fabrication, physicochemical property characterization, and dosage optimization in vitro and in vivo. *Pharmaceutics* 13:1–20. https://doi.org/10.3390/pharmaceutics13091459.

[30] Santos FV, Yoshioka SA, Branciforti MC (2021) Large-area thin films of silk fibroin prepared by two methods with formic acid as solvent and glycerol as plasticizer. *J Appl Polym Sci* 138:50759. https://doi.org/10.1002/app.50759.

[31] Safonova, L, Bobrova, M, Efimov, A, Davydova, L, Tenchurin, T, Bogush, V, Agapova, O, & Agapov, I. (2021) Silk fibroin/spidroin electrospun scaffolds for full-thickness skin wound healing in rats. *Pharmaceutics* 13:1–13. https://doi.org/10.3390/pharmaceutics13101704.

[32] Yang, H, Lai, C, Xuan, C, Chai, M, Liu, X, Chen, Y, & Shi, X. (2020) Integrin-binding pro-survival peptide engineered silk fibroin nanosheets for diabetic wound healing and skin regeneration. *Chem Eng J* 398:125617. https://doi.org/10.1016/j.cej.2020.125617.

[33] Yadav R, Purwar R (2020) Tailoring of electrical and optical properties of regenerated silk fibroin films with metal oxides. *J Mater Sci Mater Electron* 31:17784– 17797. https://doi.org/10.1007/s10854-02004332-4.

[34] Reizabal, A, Brito-Pereira, R, Fernandes, MM, Castro, N, Correia, V, Ribeiro, C, Costa, CM, Perez, L, Vilas, JL, & Lanceros-Méndez, S.(2020) Silk fibroin magnetoactive nanocomposite films and membranes for dynamic bone tissue engineering strategies. *Materialia* 12:100709. https://doi.org/10.1016/j.mtla.2020.100709.

[35] Luo, Y, Pei, Y, Feng, X, Zhang, H, Lu, B, & Wang, L. (2020) Silk fibroin based transparent and wearable humidity sensor for ultra-sensitive respiration monitoring. *Mater Lett* 260:126945. https://doi.org/10.1016/j.matlet.2019.126945.

[36] Kopp, A, Schunck, L, Gosau, M, Smeets, R, Burg, S, Fuest, S, Kröger, N, Zinser, M, Krohn, S, Behbahani, M, Köpf, M, Lauts, L, & Rutkowski, R. (2020) Influence of the casting concentration on the mechanical and optical properties of Fa/CaCl2-derived silk fibroin membranes. *Int J Mol Sci* 21:1–17. https://doi.org/10.3390/ijms21186704.

[37] Tsao, SH, Wan, D, Lai, YS, Chang, HM, Yu, CC, Lin, KT, & Chen, HL. (2015) White-Light-Induced Collective Heating of Gold Nanocomposite/Bombyx mori Silk Thin Films with Ultrahigh Broadband Absorbance. *ACS Nano* 9:12045–12059. https://doi.org/10.1021/acsnano.5b04913.

[38] Amsden, JJ, Domachuk, P, Gopinath, A, White, RD, Negro, LD, Kaplan, DL, & Omenetto, FG. (2010) Rapid nanoimprinting of silk fibroin films for biophotonic applications. *Adv Mater* 22:1746–1749. https://doi.org/10.1002/adma.200903166.

[39] Liang, FC, Huang, YH, Kuo, CC, Cho, CJ, Rwei, SP, Jia, Q, Ding, Y, Chen, Y, & Borsali, R. (2018) Thermally deposited silk fibroin as the gate dielectric layer in organic thin-film transistors based on conjugated polymer. *React Funct Polym* 131:368–377. https://doi.org/10.1016/j.reactfunctpolym.2018.08.010.

[40] Tullii, G, Donini, S, Bossio, C, Lodola, F, Pasini, M, Parisini, E, Galeotti, F, & Antognazza, MR. (2020) Micro- And Nanopatterned Silk Substrates for Antifouling Applications. *ACS Appl Mater Interfaces* 12:5437–5446. https://doi.org/10.1021/acsami.9b18187.

[41] Kaushik, S, Sarma, MK, Thungon, PD, Santhosh, M, & Goswami, P. (2016) Thin films of silk fibroin and its blend with chitosan strongly promote biofilm growth of Synechococcus sp. BDU 140432. *J Colloid Interface Sci* 479:251–259. https://doi.org/10.1016/j.jcis.2016.06.065.

[42] Borkner CB, Elsner MB, Scheibel T (2014) Coatings and films made of silk proteins. *ACS Appl Mater Interfaces* 6:15611–15625. https://doi.org/10.1021/am5008479.

[43] Rastogi S, Kandasubramanian B (2020) Processing trends of silk fibers: Silk degumming, regeneration and physical functionalization. *J Text Inst* 111:1794–1810. https://doi.org/10.1080/00405000.2020.1727269.

[44] Li, Q, Qi, N, Peng, Y, Zhang, Y, Shi, L, Zhang, X, Lai, Y, Wei, K, Kim, IS, & Zhang, KQ. (2017) Sub-micron silk fibroin film with high humidity sensibility through color changing. *RSC Adv* 7:17889–17897. https://doi.org/10.1039/c6ra28460d.

[45] Li D, Fan Y, Han G, Guo Z (2020) Superomniphobic Silk Fibroin/Ag Nanowires Membrane for Flexible and Transparent Electronic Sensor. *ACS Appl Mater Interfaces* 12:10039–10049. https://doi.org/10.1021/acsami.9b23378.

[46] Hajiabbas M, Alemzadeh I, Vossoughi M (2020) A porous hydrogel-electrospun composite scaffold made of oxidized alginate/gelatin/silk fibroin for tissue engineering application. Carbohydr Polym 245:116465. https://doi.org/10.1016/j.carbpol.2020.116465.

[47] Qin, X, Peng, Y, Li, P, Cheng, K, Wei, Z, Liu, P, Cao, N, Huang, J, Rao, J, Chen, J, Wang, T, Li, X, & Liu, M. (2019) Silk fibroin and ultra-long silver nanowire-based transparent, flexible and conductive composite film and its Temperature-Dependent resistance. *Int J Optomechatronics* 13:41–50. https://doi.org/10.1080/15599612.2019.1639002.

[48] Rocha EG, Pugina RS, Caiut JMA (2020) Luminescent sensor based on the lanthanide-fibroin composite. *Opt Mater* (Amst) 109. https://doi.org/10.1016/j.optmat.2020.110236.

[49] Kim, DH, Viventi, J, Amsden, JJ, Xiao, J, Vigeland, L, Kim, YS, Blanco, JA, Panilaitis, B, Frechette, ES, Contreras, D, Kaplan, DL, Omenetto, FG, Huang, Y, Hwang, KC, Zakin, MR, Litt, B, & Rogers, JA. (2010) Dissolvable films of silk fibroin for ultrathin conformal bio-integrated electronics. *Nat Mater* 9:511–517. https://doi.org/10.1038/nmat2745.

[50] Zhang, X, Chen, Z, Bao, H, Liang, J, Xu, S, Cheng, G, & Zhu, Y. (2019) Fabrication and Characterization of Silk Fibroin/Curcumin Sustained-Release Film. *Materials* (Basel) 12:1–16. https://doi.org/doi:10.3390/ma12203340.

[51] Arthe R, Arivuoli D, Ravi V (2020) Preparation and characterization of bioactive silk fibroin/paramylon blend films for chronic wound healing. *Int J Biol Macromol* 154:1324–1331. https://doi.org/10.1016/j.ijbiomac.2019.11.010.

[52] Zhu, G, Sun, Z, Hui, P, Chen, W, & Jiang, X. (2021) Composite film with antibacterial gold nanoparticles and silk fibroin for treating multidrug-resistant E. Coli-Infected Wounds. *ACS Biomater Sci Eng* 7:1827–1835. https://doi.org/10.1021/acsbiomaterials.0c01271.

[53] Wang, J, Chen, Y, Zhou, G, Chen, Y, Mao, C, & Yang, M.(2019) Polydopamine-Coated Antheraea pernyi (A. pernyi) Silk Fibroin Films Promote Cell Adhesion and Wound Healing in Skin Tissue Repair. *ACS Appl Mater Interfaces* 11:34736–34743. https://doi.org/10.1021/acsami.9b12643.

[54] Xu M, Yadavalli VK (2019) Flexible Biosensors for the Impedimetric Detection of Protein Targets Using Silk-Conductive Polymer Biocomposites. *ACS Sensors* 4:1040– 1047. https://doi.org/10.1021/acssensors.9b00230.

[55] Zhang, W, Chen, L, Chen, J, Wang, L, Gui, X, Ran, J, Xu, G, Zhao, H, Zeng, M, Ji, J, Qian, L, Zhou, J, Ouyang, H, & Zou, X. (2017) Silk Fibroin Biomaterial Shows Safe and Effective Wound Healing in Animal Models and a Randomized Controlled Clinical Trial. *Adv Healthc Mater* 6:1–16. https://doi.org/10.1002/adhm.201700121.

[56] Liu, J, Huang, R, Li, G, Kaplan, DL, Zheng, Z, & Wang, X.(2021) Generation of Nano-pores in Silk Fibroin Films Using Silk Nanoparticles for Full-Thickness Wound Healing. *Biomacromolecules* 22:546–556. https://doi.org/10.1021/acs.biomac.0c01411.

[57] Wang, R, Ruan, L, Li, P, Liu, T, & Jiang, G. (2021) Silk Fibroin and κ-Carrageenan Composite Films Containing Zinc-doped Bioactive Glass for Wound Closure. *J Bionic Eng* 18:1400–1412. https://doi.org/10.1007/s42235-021-00105-9.

[58] Safonova, L, Bobrova, M, Efimov, A, Lyundup, A, Agapova, O, & Agapov, I. (2021) A comparative analysis of the structure and biological properties of films and microfibrous scaffolds based on silk fibroin. *Pharmaceutics* 13:1561. https://doi.org/10.3390/pharmaceutics13101561.

[59] Patil, PP, Meshram, JV, Bohara, RA, Nanaware, SG, & Pawar, SH. (2018) ZnO nanoparticle-embedded silk fibroin-polyvinyl alcohol composite film: A potential dressing material for infected wounds. *New J Chem* 42:14620–14629. https://doi.org/10.1039/c8nj01675e.

[60] Singh, S, Cortes, G, Kumar, U, Sakthivel, TS, Niemiec, SM, Louiselle, A. E., Azeltine-Bannerman, M., Zgheib, C., Liechty, K. W., & Seal, S. (2020) Silk fibroin nanofibrous mats for visible sensing of oxidative stress in cutaneous wounds. *Biomater Sci* 8:5900–5910. https://doi.org/10.1039/d0bm01325k.

[61] Özen, N, Özbaş, Z, İzbudak, B, Emik, S, Özkahraman, B, & Bal-Öztürk, A. (2022) Boric acid-impregnated silk fibroin/gelatin/hyaluronic acid-based films for improving the wound healing process. *J Appl Polym Sci* 139:e51715. https://doi.org/10.1002/app.51715.

[62] Yerra A, Mamatha DM (2021) Antibiotic-based silk fibroin films for burn wound healing. *Polym Adv Technol* 32:861–871. https://doi.org/10.1002/pat.5137.

[63] Wu, J, Rnjak-Kovacina, J, Du, Y, Funderburgh, ML, Kaplan, DL, & Funderburgh, JL. (2014) Corneal stromal bioequivalents secreted on patterned silk substrates. *Biomaterials* 35:3744–3755. https://doi.org/10.1016/j.biomaterials.2013.12.078.

[64] Harkin, DG, George, KA, Madden, PW, Schwab, IR, Hutmacher, DW, & Chirila, TV. (2011) Silk fibroin in ocular tissue reconstruction. *Biomaterials* 32:2445–2458. https://doi.org/10.1016/j.biomaterials.2010.12.041.

[65] Harkin DG, Chirila T V. (2012) Silk fibroin in ocular surface reconstruction: What is its potential as a biomaterial in ophthalmics? *Future Med Chem* 4:2145–2147. https://doi.org/10.4155/fmc.12.155.

[66] Choi JH, Jeon H, Song JE, Oliveira, J, Reis, R, & Khang, G.(2018) Biofunctionalized lysophosphatidic acid/silk fibroin film for cornea endothelial cell regeneration. *Nanomaterials* 8:1–14. https://doi.org/10.3390/nano8050290.

[67] Kim DK, Sim BR, Kim JI, Khang G (2018) Functionalized silk fibroin film scaffold using β-Carotene for cornea endothelial cell regeneration. *Colloids Surfaces B Biointerfaces* 164:340–346. https://doi.org/10.1016/J.COLSURFB.2017.11.052.

[68] Sun, MG, Luo, Y, Teng, T, Guaiquil, V, Zhou, Q, McGinn, L, Nazzal, O, Walsh, M, Lee, J, & Rosenblatt, MI. (2021) Silk Film Stiffness Modulates Corneal Epithelial Cell Mechanosignaling. *Macromol Chem Phys* 222:1–12. https://doi.org/10.1002/macp.202100013.

[69] Lawrence, BD, Marchant, JK, Pindrus, MA, Omenetto, FG, & Kaplan, DL. (2009) Silk film biomaterials for cornea tissue engineering. *Biomaterials* 30:1299–1308. https://doi.org/10.1016/j.biomaterials.2008.11.018.

[70] Madden, PW, Lai, JNX, George, KA, Giovenco, T, Harkin, DG, & Chirila, TV. (2011) Human corneal endothelial cell growth on a silk fibroin membrane. *Biomaterials* 32:4076–4084. https://doi.org/10.1016/j.biomaterials.2010.12.034.

[71] Higa, K, Takeshima, N, Moro, F, Kawakita, T, Kawashima, M, Demura, M, Shimazaki, J, Asakura, T, Tsubota, K, & Shimmura, S. (2011) Porous silk fibroin film as a transparent carrier for cultivated corneal epithelial sheets. *J Biomater Sci Polym Ed* 22:2261–2276. https://doi.org/10.1163/092050610X538218.

[72] Farasatkia A, Kharaziha M, Ashrafizadeh F, Salehi S (2021) Transparent silk/gelatin methacrylate (GelMA) fibrillar film for corneal regeneration. *Mater Sci Eng C* 120:111744. https://doi.org/101016/j.msec.2020.111744.

[73] Cubayachi, C, Lemos, CN, Pereira, F, Dias, K, Herculano, RD, de Freitas, O, & Lopez, RFV. (2019) Silk fibroin films stabilizes and releases bioactive insulin for the treatment of corneal wounds. *Eur Polym J* 118:502– 513. https://doi.org/10.1016/J.EURPOLYMJ.2019.06.022.

[74] Kim DK, Sim BR, Khang G (2016) Nature-Derived Aloe Vera Gel Blended Silk Fibroin Film Scaffolds for Cornea Endothelial Cell Regeneration and Transplantation. *ACS Appl Mater Interfaces* 8:15160–15168. https://doi.org/10.1021/acsami.6b04901.

[75] Aghaei-Ghareh-Bolagh, B, Guan, J, Wang, Y, Martin, AD, Dawson, R, Mithieux, SM, & Weiss, AS. (2019) Optically robust, highly permeable and elastic protein films that support dual cornea cell types. *Biomaterials* 188:50–62. https://doi.org/10.1016/j.biomaterials.2018.10.006.

[76] Vázquez N, Rodríguez-Barrientos CA, Aznar-Cervantes SD, et al. (2017) Silk fibroin films for corneal endothelial regeneration: Transplant in a rabbit descemet membrane endothelial keratoplasty. *Investig Ophthalmol Vis Sci* 58:3357–3365. https://doi.org/10.1167/iovs.17-21797.

[77] Ghezzi, CE, Marelli, B, Omenetto, FG, Funderburgh, JL, & Kaplan, DL. (2017) 3D functional corneal stromal tissue-equivalent based on corneal stromal stem cells and multi-layered silk film architecture. *PLoS One* 12:1–18. https://doi.org/10.1371/journal.pone.0169504.

[78] Luo, Y, Kang, KB, Sartaj, R, Sun, MG, Zhou, Q, Guaiquil, VH, & Rosenblatt, MI. (2021) Silk films with nanotopography and extracellular proteins enhance corneal epithelial wound healing. *Sci Rep* 11:1–15. https://doi.org/10.1038/s41598-021-87658-1.

[79] Song, JE, Sim, BR, Jeon, YS, Kim, HS, Shin, EY, Carlomagno, C, & Khang, G. (2019) Characterization of surface-modified glycerol/silk fibroin film for application to corneal endothelial cell regeneration. *J Biomater Sci Polym Ed* 30:263–275. https://doi.org/10.1080/09205063.2018.1535819.

[80] Kim, EY, Tripathy, N, Cho, SA, Joo, CK, Lee, D, & Khang, G. (2015) Bioengineered neo-corneal endothelium using collagen type-I coated silk fibroin film. *Colloids Surfaces B Biointerfaces* 136:394–401. https://doi.org/10.1016/j.colsurfb.2015.09.041.

[81] Lee, MC, Kim, DK, Lee, OJ, Kim, JH, Ju, HW, Lee, JM, Moon, BM, Park, HJ, Kim, DW, Kim, SH, & Park, CH. (2016) Fabrication of silk fibroin film using centrifugal casting technique for corneal tissue engineering. *J Biomed Mater Res - Part B Appl Biomater* 104:508–514. https://doi.org/10.1002/jbm. b.33402.

[82] Nili, E, Harkin, DG, Dawson, RA, Richardson, NA, Suzuki, S, & Chirila, TV. (2021) Membranes prepared from recombinant rgd-silk fibroin as substrates for human corneal cells. *Molecules* 26:6810. https://doi.org/10.3390/molecules26226810.

[83] Wang, L, Ma, R, Du, G, Guo, H, & Huang, Y. (2015) Biocompatibility of helicoidal multilamellar arginine-glycine-aspartic acid-functionalized silk biomaterials in a rabbit corneal model. *J Biomed Mater Res - Part B Appl Biomater* 103:204–211. https://doi.org/10.1002/jbm.b.33192.

[84] Koh, LD, Yeo, J, Lee, YY, Ong, Q, Han, M, & Tee, BCK. (2018) Advancing the frontiers of silk fibroin protein-based materials for futuristic electronics and clinical wound-healing (Invited review). *Mater Sci Eng C* 86:151–172. https://doi.org/10.1016/j.msec.2018.01.007.

[85] Fan, S, Zhang, Y, Huang, X, Geng, L, Shao, H, Hu, X, & Zhang, Y. (2019) Special Topic: Smart and Functional Fiber Materials Silk materials for medical, electronic and optical applications. *Scinece China Technol Sci* 62:903–918. https://doi.org/https://doi.org/10.1007/s11431-018-9403-8.

[86] Li, Q, Chen, G, Cui, Y, Ji, S, Liu, Z, Wan, C, Liu, Y, Lu, Y, Wang, C, Zhang, N, Cheng, Y, Zhang, KQ, & Chen, X. (2021) Highly Thermal-Wet Comfortable and Conformal Silk-Based Electrodes for On-Skin Sensors with Sweat Tolerance. *ACS Nano* 15:9955–9966. https://doi.org/10.1021/acsnano.1c01431.

[87] Hu, F, Li, W., Zou, M, Li, Y, Chen, F, Lin, N, Guo, W, & Liu, XY. (2021) Subcutaneous Energy/Signal Transmission Based on Silk Fibroin Up-Conversion Photonic Amplification. *ACS Nano* 15:9559–9567. https://doi.org/10.1021/acsnano.0c09575.

[88] Lin, Z, Meng, Z, Miao, H, Wu, R, Qiu, W, Lin, N, & Liu, XY. (2021) Biomimetic Salinity Power Generation Based on Silk Fibroin Ion-Exchange Membranes. *ACS Nano* 15:5649–5660. https://doi.org/10.1021/acsnano.1c00820.

[89] Chao, M, He, L, Gong, M, Li, N, Li, X, Peng, L, Shi, F, Zhang, L, & Wan, P. (2021) Breathable Ti3C2TxMXene/Protein Nanocomposites for Ultrasensitive Medical Pressure Sensor with Degradability in Solvents. *ACS Nano* 15:9746–9758. https://doi.org/10.1021/acsnano.1c00472.

[90] Wang, Q, Yan, S, Han, G, Li, X, You, R, Zhang, Q, Li, M, & Kaplan, DL. (2020) Facile production of natural silk nanofibers for electronic device applications. *Compos Sci Technol* 187:107950. https://doi.org/10.1016/j.compscitech.2019.107950.

[91] Zhang Y, Han F, Fan S, Zhang Y (2021) Low-power and tunable-performance biomemristor based on silk fibroin. *ACS Biomater Sci Eng* 7:3459–3468. https://doi.org/10.1021/acsbiomaterials.1c00513.

[92] Liu, H, Wei, W, Zhang, L, Xiao, J, Pan, J, Wu, Q, Ma, S, Dong, H, Yu, L, Yang, W, Wei, D, Ouyang, H, & Liu, Y. (2021) Shape-Engineerable Silk Fibroin Papers for Ideal Substrate Alternatives of Plastic Electronics. *Adv Funct Mater* 2104088:1–10. https://doi.org/10.1002/adfm.202104088.

[93] Bandar Abadi, M, Weissing, R, Wilhelm, M, Demidov, Y, Auer, J, Ghazanfari, S, Anasori, B, Mathur, S, & Maleki, H. (2021) Nacre-Mimetic, Mechanically Flexible, and Electrically Conductive Silk Fibroin-MXene Composite Foams as Piezoresistive Pressure Sensors. *ACS Appl Mater Interfaces* 13:34996–35007. https://doi.org/10.1021/acsami.1c09675.

[94] Huang, J, Wang, L, Jin, Y, Lu, P, Wang, L, Bai, N, Li, G, Zhu, P, Wang, Y, Zhang, J, Wu, Z, & Guo, CF.(2020) Tuning the Rigidity of Silk Fibroin for the Transfer of Highly Stretchable Electronics. *Adv Funct Mater* 30:1–8. https://doi.org/10.1002/adfm.202001518.

[95] Pradhan S, Yadavalli VK (2021) Photolithographically Printed Flexible Silk/PEDOT:PSS Temperature Sensors. *ACS Appl Electron Mater* 3:21–29. https://doi.org/10.1021/acsaelm.0c01017.

[96] Kook, G, Jeong, S, Kim, SH, Kim, MK, Lee, S, Cho, IJ, Choi, N, & Lee, HJ. (2019) Wafer-Scale Multilayer Fabrication for Silk Fibroin-Based Microelectronics. *ACS Appl Mater Interfaces* 11:115–124. https://doi.org/10.1021/acsami.8b13170.

[97] Malinowski, C, He, F, Zhao, Y, Chang, I, Hatchett, DW, Zhai, S, & Zhao, H. (2019) Nanopatterned silk fibroin films with high transparency and high haze for optical applications. *RSC Adv* 9:40792–40799. https://doi.org/10.1039/c9ra07391d.

[98] Yucel T, Lovett ML, Kaplan DL (2014) Silk-based biomaterials for sustained drug delivery. *JControl Release* 190:381–397. https://doi.org/10.1016/j.jconrel.2014.05.059.

[99] Farokhi, M, Mottaghitalab, F, Reis, RL, Ramakrishna, S, & Kundu, SC. (2020) Functionalized silk fibroin nanofibers as drug carriers: Advantages and challenges. *J Control Release* 321:324–347. https://doi.org/10.1016/j.jconrel.2020.02.022.

[100] Hofmann, S, Wong Po Foo, CT, Rossetti, F, Textor, M, Vunjak-Novakovic, G, Kaplan, DL, Merkle, HP, & Meinel, L.(2006) Silk fibroin as an organic polymer for controlled drug delivery. *J Control Release* 111:219–227. https://doi.org/10.1016/j.jconrel.2005.12.009.

[101] Choi M, Choi D, Hong J (2018) Multilayered Controlled Drug Release Silk Fibroin Nanofilm by Manipulating Secondary Structure. *Biomacromolecules* 19:3096– 3103. https://doi.org/10.1021/acs.biomac.8b00687.

[102] Coburn JM, Na E, Kaplan DL (2015) Modulation of vincristine and doxorubicin binding and release from silk films. *J Control Release* 220:229–238. https://doi.org/10.1016/j.jconrel.2015.10.035.

[103] Hines DJ, Kaplan DL (2011) Mechanisms of controlled release from silk fibroin films. *Biomacromolecules* 12:804–812. https://doi.org/10.1021/bm101421r.

[104] Huang Y, Bailey K, Wang S, Feng X (2017) Silk fibroin films for potential applications in controlled release. *React Funct Polym* 116:57–68. https://doi.org/10.1016/j.reactfunctpolym.2017.05.007.

[105] Tomoda, BT, Corazza, FG, Beppu, MM, Lopes, PS, & Moraes, MA. (2020) Silk fibroin membranes with self-assembled globular structures for controlled drug release. *J Appl Polym Sci* 137:1– 9. https://doi.org/10.1002/app.48763.

[106] Srisuwan Y, Baimark Y (2013) Preparation of Biodegradable Silk Fibroin / Alginate Blend Films for Controlled Release of Antimicrobial Drugs. *Adv Mater Sci Eng* 2013:1–6. https://doi.org/http://dx.doi.org/10.1155/2013/412458.

[107] Alam AKMM, Shubhra QTH (2015) Surface modified thin film from silk and gelatin for sustained drug release to heal wound. *J Mater Chem B* 3:6473–6479. https://doi.org/10.1039/c5tb00920k.

[108] Wang, S, Zhu, M, Zhao, L, Kuang, D, Kundu, SC, & Lu, S.(2019) Insulin-Loaded Silk Fibroin Microneedles as Sustained Release System. *ACS Biomater Sci Eng* 5:1887–1894. https://doi.org/10.1021/acsbiomaterials.9b00229.
[109] Zhu, M, Liu, Y, Jiang, F, Cao, J, Kundu, SC, & Lu, S. (2020) Combined Silk Fibroin Microneedles for Insulin Delivery. *ACSBiomater Sci Eng* 6:3422–3429. https://doi.org/10.1021/acsbiomaterials.0c00273.
[110] Pacheco, MS, Kano, GE, Paulo, L de A, Lopes, PS, & de Moraes, MA. (2020) Silk fibroin/chitosan/alginate multilayer membranes as a system for controlled drug release in wound healing. *Int J Biol Macromol* 152:803–811. https://doi.org/10.1016/j.ijbiomac.2020.02.140.

Chapter 4

Silk Fibroin Sponge as a Cryodevice Combined with Transporter and Scaffold Functions in Cellular Transplantation

Shinichi Hochi*, **PhD and Yasushi Tamada, PhD**
Faculty of Textile Science and Technology,
Shinshu University, Ueda, Nagano, Japan

Abstract

Silk fibroin (SF) is a biocompatible structural protein extracted from the cocoon of *Bombyx mori* silkworm and is mechanically robust because of its unique β-sheet structure. The high biocompatibility of SF is favorable for preparing surgical sutures or scaffold materials in biomedical research. SF can be easily processed into hydrogels, films, or porous sponges. SF-derived porous sponges have attractive hydrophilic properties that endow these sponges with favorable water absorption and retention features. Thus, the SF sponge scaffold may be suitable as a transporter of pharmaceutical aqueous solutions. Furthermore, the SF sponge is a promising tool for cryopreservation by the minimum volume cooling procedure because successful post-warm cell retrieval requires prompt minimization of the surrounding cryoprotective solution before rapid cooling for vitrification. This chapter deals with the preparation and characterization of the SF sponge, the potential use of the SF sponge as a cryodevice for vitrification of pancreatic islets, and the consecutive application of the SF sponge as a cytokine transporter and scaffold in the subrenal transplantation of post-warm islets.

Keywords: cryodevice, DDS, MVC, pancreatic islets, SF sponge disc, SEM, subrenal transplantation, SSV, VEGF

1. Introduction

Bombyx mori silk comprises two proteins, the highly biocompatible fibroin and sericin. Silk fibroin (SF) is mechanically robust because of its unique β-sheet structure, which is attributed

* Corresponding Author's Email: shochi@shinshu-u.ac.jp.

In: Silk Fibroin
Editor: Shivaji H. Pawar
ISBN: 979-8-88697-402-7

to subdomains that are rich in glycine, alanine, serine, and tyrosine, and can be processed easily into hydrogels, films, or sponges (Altman et al., 2003; Vepari and Kaplan, 2007). The high biocompatibility of SF is favorable for preparing surgical sutures or scaffold materials in both basic and clinical biomedical studies (Meinel et al., 2004; Stenhahl et al., 2009; Mao et al., 2017). The functionality of the SF sponge can be reinforced by fusing the peptide sequence with biological activity through the production of transgenic silkworms (Kambe et al., 2010; Kambe et al., 2016). In tissue transplantation, the transplanted cell masses (e.g., pancreatic islets) should receive nutritional and physical support from the host by forming new blood vessels around and within the engraft (Emamaullee and Shapiro, 2007). To enhance vascularization in transplanted islets, vascular endothelial growth factor (VEGF), responsible for regulating blood vessel growth (Ferrara et al., 2003), has been co-delivered with islets in a scaffold or hydrogel with only a few safety concerns. The SF sponge is an attractive drug delivery system (DDS) because this sponge displays high hydrophilicity, and the interconnectivity of the pores is likely to be suitable for absorption and retention (and probably gradual release) of a cytokine solution in (from) the sponge.

Furthermore, the SF sponge is a promising device for cell/tissue storage by cryopreservation. Although the conventional slow freezing protocol is used for cryopreservation of single-cell suspensions or cultured cell lines, the ice-free vitrification protocol is more suitable for cryopreservation of a large cell type (e.g., unfertilized oocytes: Hwang and Hochi, 2014; Hochi, 2022) or cellular complexes composed from different cell types (e.g., pancreatic islets: Sasamoto et al., 2012; Nagaya et al., 2016; Yamanaka et al., 2016). A highly concentrated and toxic cryoprotective solution, termed vitrification solution (VS), is exposed to cell samples for a short period before rapid cooling in liquid nitrogen (LN_2). Minimizing the VS surrounding the cell samples during a limited exposure period facilitates the formation of an amorphous glass state by driving an extremely rapid cooling rate, which is called the minimum volume cooling (MVC) procedure. A novel SF sponge cryodevice with high water absorption should reduce the time cells are exposed to VS, not require special skills to operate and overcome the limitation of sample size in a single operation.

This chapter includes the preparation and characterization of the SF sponge, the potential use of the SF sponge as a cryodevice for islet vitrification, and the consecutive application of the SF sponge as a VEGF transporter and scaffold in subrenal transplantation of post-warm islets.

2. Preparation of the SF Sponge

Several reports describing the fabrication process to prepare SF sponges have been published. An aqueous fibroin solution is freeze-dried with an appropriate chemical crosslinker to make a fibroin porous material (Li et al., 2003). Repeated freeze-thawing of the fibroin aqueous solution at –42 to –12°C and subsequent freeze-drying leads to the formation of a spongy structure with different pore sizes depending on the thawing temperature (Li et al., 2002). The SF sponge has been prepared by the salt leaching method from a fibroin hexafluoroiso-propanol (HFIP) solution (Nazarovet al., 2004) and an SF aqueous solution (Kim et al., 2005) in the presence of NaCl particles. The SF sponge formed by the salt leaching process showed comparatively large pore sizes of more than 200 μm from HFIP and 500 μm from the aqueous solution. The gas-forming process by ammonium bicarbonate in a fibroin HFIP solution can also produce a fibroin porous material (Nazarov et al., 2004). These methods require complex processes and freeze-drying equipment, and a harmful solvent or chemical crosslinker must be used in some processes. As an alternative, a simple process was developed to fabricate SF sponges from SF aqueous solutions (Freeze-Thaw method) (Tamada, 2005). In this process, a small amount of water-miscible organic solvent is mixed with an SF aqueous solution prepared by a standard procedure from degummed silk fiber. The mixed solution is poured into a mold and frozen at –20°C for > 6 h (Kameda et al., 2011) and then thawed. The SF sponge structure forms during this process. This process is straightforward to perform, requires no freeze-drying or no repeated freeze-thaw treatments, and no cross-linking chemicals, harmful solvents, and other materials are needed. There are no restrictions on the shape of the SF sponge produced, and the shape is defined by the mold used and required for subsequent applications.

The SF sponge for the cryodevice is fabricated as follows. *Bombyx mori* silk cocoons (Shunrei × Shogetsu) are degummed by 0.02 M Na_2CO_3 at 95°C for 30 min, and the degummed silk fiber is dissolved in 9 M lithium bromide at ambient temperature for 12–18 h. This solution is dialyzed against pure water for 2 days, with the exchange of dialysis water at 8-h intervals. The SF aqueous solution is adjusted to a concentration of 1–4% (w/v). Diluted DMSO as the water-miscible organic solvent is mixed to a 1% (v/v) final concentration in the SF aqueous solution. The mixed solution is poured into an aluminum mold (50 × 50 × 1 mm), frozen in an ethanol brine bath to –20°C for 14 h under programmed temperature control, and then air-thawed at ambient temperature. The SF sponge is immersed in pure water to remove the remaining DMSO.

3. Characterization of the SF Sponge

The pore structure and size can be controlled over the range of 30–300 μm simply by changing the temperature or selecting the materials for the molds. For example, slower cooling during the freezing process leads to increased pore sizes. An acrylate mold forms larger pores in the SF sponge than an aluminum mold under the same freezing treatment conditions. The different pore shapes, such as round and lamellar types, can be formed by using different freezing methods. In addition, the SF concentration in the initial aqueous solution influences the pore size; increasing the concentration yields smaller pores in the SF sponge (Figure 1). The SF sponge fabricated by the Freeze-Thaw method has mechanical properties with appropriate softness and strength for use in cryopreservation. The mechanical properties of the SF sponge can be modified simply by changing the SF concentration in the aqueous solution. For example, the SF sponge formed from a 2% (w/v) aqueous solution shows 0.04 MPa and 0.8 kPa of tensile strength and Young's modulus, respectively, whereas increasing the concentration to 10% (w/v) increases the tensile strength and Young's modulus to 0.26 MPa and 7.8 kPa, respectively (our unpublished data).

The safety of the SF sponge is an important issue when used as a cryodevice. The biological safety of the SF sponge has been evaluated according to standard guidelines "Basic principles of biological safety evaluation required for approval to manufacture medical devices" (Tamada et al., 2015). As summarized in Table 1, sensitization test using guinea pigs, skin irritation and inflammation test using rabbits, genotoxicity Ames test, and the allergy reaction test with human skin have been performed. The biological safety of the SF sponge was upheld because the results from these evaluations were negative.

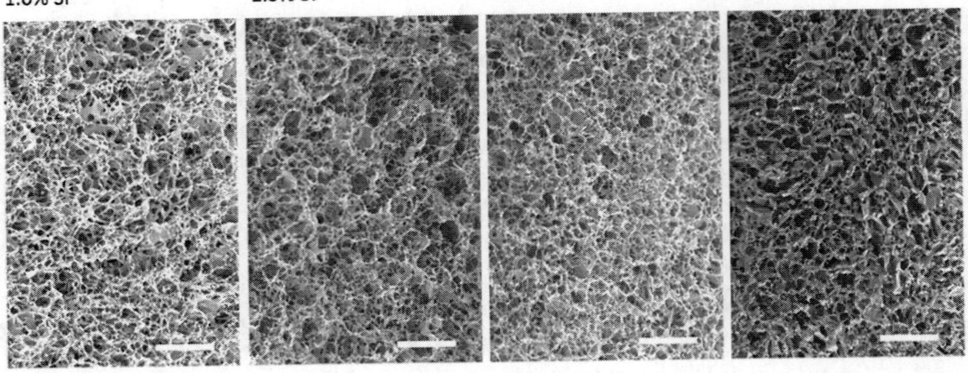

Figure 1. SEM images of SF sponges prepared from a 1.0, 1.5, 2.0, or 4.0% (w/v) SF solution. The pore size and porosity of the resultant sponge are likely to correlate with the initial SF concentration. Scale bar = 200 μm.

Table 1. Safety evaluation of the SF sponge

Evaluation	Method/Species	Result	Judgment
Sensitization	Maximization test/Guinea pig	Score [1] : 0.0	Negative
Skin imitation	Skin irritancy test/Rabbit	P.I.I. [2] : 0.1 to 0.3	Negative
Tissue damage Inflammation	Intradermal test/Rabbit	P.I.I. [2] :–0.1 to 0.0	Negative
Genotoxicity	Ames test/Bacteria	RR [3] : 0.6 to 4.9	Negative
Allergy reaction	Patch test/Human	Score [4] : 1.0 to 1.9	Negative

[1] Mean of Erythema and Edema score. 0 = No, 1 = Very slight, 2 = Well defined, 3 = Moderate, 4 = Severe.

[2] Primary Irritation Index. 0 to 0.4 = Negligible, 0.5 to 1.9 = Mild, 2.0 to 4.9 = Moderate, 5.0 to 8.0 = Severe. Expressed as (Sample index– Solvent index).

[3] Revertant Ratio. [(Colony on sample – Colony on negative)/Colony on negative] ×100.

[4] 0 to 106.07 = No experimental irritation, 106.07 to 427.5 = Probably mild in normal use, 427.5 to 961.07 = Possibly in normal use, 961.07 to 1243.93 = Experimental cumulative irritant, 1243.93 to 1350 = Experimental primary irritant.

Data are reproduced from Tamada et al. (2015), with approval from the Japanese Society of Silk Science and Technology.

4. SF Application to Cryopreservation

Various cell types, including pluripotent stem cells (ES cells/iPS cells), can be cryopreserved by conventional freezing regimens or ice-free vitrification procedures in the field of tissue engineering. The size of single-cell masses or multiple-cell type complexes affects cryosurvival. Thus, cryopreservation protocols must be optimized for each cell type. Vitrified chondrocyte sheets have been applied successfully to repeated transplantation therapies for a single patient (Maehara et al., 2013). MVC vitrification was originally developed to prevent harmful ice crystal formation and improve oocyte/embryo cryosurvival (Martino et al., 1996; Vajta et al., 1998; Kuwayama et al., 2005). This approach has been used to recover functional pancreatic islets for transplantation therapy in an insulin-dependent type-1 diabetes rodent model using several cryodevices such as Cryotop® (Yamanaka et al., 2016), open pulled straw (Sasamoto et al., 2012), and hollow fiber (Nagaya et al., 2016). However, MVC vitrification protocols require skillful capillary pipetting to load the samples onto a cryodevice with a minimal volume of the toxic VS under a tight schedule (Table 2), limiting the sample number per single operation. In the standard protocol of Cryotop® vitrification, cells are first exposed to an equilibration solution containing low concentrations of membrane-permeable ethylene glycol and DMSO for 3–15 min and then transferred to a VS containing high concentrations of the permeable cryoprotectants plus non-permeable sucrose for 1 min. Within this 1 min, cells must be loaded onto a strip of the Cryotop® device with a minimal volume of VS (< 0.1 μL) and be immersed into LN_2. After rapid warming, the cryoprotectants are removed from the cells in a stepwise manner using sucrose solutions.

Table 2. Typical solution composition and exposure time in pre-/post-vitrification steps

	Cryoprotectants	Concentration	Exposure time
(1) Equilibration solution	Ethylene glycol DMSO	7.5% (v/v) 7.5% (v/v)	3–15 min
(2) Vitrification solution	Ethylene glycol DMSO Sucrose	15% (v/v) 15% (v/v) 0.5 M	1 min*
(3) Warming solution	Sucrose	1.0 M	1 min
(4) Dilution solution	Sucrose	0.5 M 0.25 M 0 M	3 min 5 min 5 min

Basal medium for islets; RPMI-1640. Each solution contains 20% (v/v) fetal bovine serum.
*Within this 1 min, cell samples must be loaded onto the cryodevice with a minimal volume of VS and then immersed into LN_2.

Pancreatic islets responsible for blood glucose homeostasis are composed of five different endocrine cell types, α–, β–, δ–, γ–, and ε–cells that produce glucagon, insulin, somatostatin, pancreatic polypeptide, and ghrelin, respectively (Xavier, 2018). The size of isolated islets varies, ranging from 50 to 500 μm in diameter (Buchwald et al., 2009), and conversion of the individual islets to islet equivalent (IEQ; an islet with a diameter of 150 μm is defined as 1 IEQ [Ramachandran et al., 2015]) is set for clinical transplantation. The β-cell replacement therapy through islet transplantation has been a promising option for patients with type-1 diabetes since the establishment of the Edmonton protocol (Shapiro et al., 2000). Cryopreservation of islets without functional loss can overcome the inevitable shortage of donors and allow time to investigate immunological tissue matching between donors and recipients. Conventional freeze-thawing regimens are still attractive to cryopreserve a large quantity of islets in a single operation. However, earlier attempts at freeze-thawing pancreatic islets often resulted in impaired glucose responsiveness (Mendola et al., 1996; Corominola et al., 1998), whereas vitrification protocols have been successfully applied to recover insulin-secreting functional islets after warming (Taylor and Baicu, 2009; Kojayan et al., 2018).

Yamanaka et al. (2016) reported that MVC vitrification of rat islets using Cryotop® as a cryodevice was superior to conventional freezing with respect to cryosurvival, insulin secretion, and the expression of the β-cell function regulatory gene. In the glucose-stimulated insulin secretion (GSIS) assay, the stimulation index (SI) is determined as the stimulated insulin level in response to 20 mM glucose divided by the basal level in response to 3 mM glucose. There are no distinct SI criteria to distinguish functional and dysfunctional islets, but islets with a SI value of > 3 have been empirically regarded as those maintaining functional insulin secretion and are recommended for transplantation (Sakata et al., 2008). The SI values have been shown to differ significantly between fresh control and freezing groups, and the SI in a Cryotop® vitrification group gave an intermediate value (Table 3). Because the practical requirement of islet transplantation necessitates that a large quantity of islets is cryopreserved by MVC vitrification, Yamanaka et al. (2017) investigated the capacity of a novel nylon mesh (NM) device for MVC vitrification of large quantities of rat islets (n = 50–100) and obtained SI values of 2.8–3.8 in the GSIS assay. A 10-mm triangle NM sheet (a square opening of 37 μm on a side length, nylon 66A-fiber diameter of 35 μm) was processed to form a developed figure of a triangular pyramid by folding into quarters. Once islets are placed onto the center triangle of the NM device, samples can be handled using sterilized tweezers rather than glass capillary pipets. The VS volume surrounding the oocytes can be minimized by placing the device on a sterilized paper towel (medium removal through the mesh pores).

Table 3. Glucose-stimulated insulin secretion assay of rat pancreatic islets after cryopreservation by freezing or vitrification

	3 mM glucose	20 mM glucose	Stimulation index
Fresh control	0.12 ± 0.03 [a]	0.85 ± 0.29	6.7 ± 0.9 [a]
Bicell® freezing	0.74 ± 0.06 [b]	1.44 ± 0.43	1.9 ± 0.5 [b]
Cryotop® vitrification	0.27 ± 0.03 [a]	1.03 ± 0.36	3.9 ± 1.0 [ab]
Fresh control	0.13 ± 0.02 [a]	0.82 ± 0.04 [a]	6.5 ± 0.6
NM vitrification	0.26 ± 0.05 [b]	1.06 ± 0.06 [ab]	4.7 ± 0.7
SF sponge vitrification	0.30 ± 0.02 [b]	1.23 ± 0.17 [b]	4.2 ± 0.7

Insulin secretion level: Mean ± Standard Error of the Mean (ng/islet/h), determined by an ELISA kit for rat insulin.
[a,b] Different superscripts denote significantly different groups ($P < 0.05$, ANOVA-PLSD test).
Data are from Yamanaka et al. (2016) and Nakayama-Iwatsuki et al. (2020).

Nakayama-Iwatsuki et al. (2020) first applied an SF sponge as cryodevice material for vitrification of a large quantity of rat islets. The porous sponge was prepared from 4% (w/v) SF solution, and discs (8 mm in diameter, 1 mm in height) were cut out from the SF sponge (Figure 2A). Since the average pore size of the SF sponge disc is approximately 100 μm, loaded rat islets (standard size used for transplantation, 100–250 μm) can be deposited or moderately gripped on the top surface of the disc. The high two-dimensional porosity of the SF sponge disc serves to accelerate the absorption of excess VS around the islets (Figure 2B). Importantly, the rat islets/SF sponge disc must be cooled by solid surface vitrification (SSV) protocol rather than direct immersion into LN_2 to avoid mechanical damage to the SF sponge disc and "Leidenfrost effect" caused by LN_2 boiling ("Leidenfrost effect": air pocket generated by LN_2 boiling acts as an insulator, resulting in decreased thermal conductivity). Sasamoto et al. (2012) reported effective use of the device-free original SSV protocol in cryopreservation of rat islets, whereas Nagaya et al. (2016) reported a contradictory result in mouse islet SSV. In a direct comparison between the NM device and SF sponge disc (Table 3), post-warm islet functionality assessed by the GSIS assay was comparable. Streptozotocin-induced diabetic rats recovered from hyperglycemia after syngeneic subrenal transplantation of 800 IEQ islets vitrified using the NM device and the SF sponge disc (Nakayama-Iwatsuki et al., 2020). Among the cryodevices employed for islet MVC vitrification, the hollow fiber device (Nagaya et al., 2016), the NM device (Yamanaka et al., 2017; Nakayama-Iwatsuki et al., 2020), and the SF sponge disc (Nakayama-Iwatsuki et al., 2020; Yamanaka et al., 2021) are promising tools because they enable handling of a large quantity of islets by skipping the time-consuming and labor-intensive capillary operations. Another cryodevice with the high absorption potential of excess VS was prepared by the five-fold winding of a wet 1.5% (w/v) SF sheet around the strip of a Cryotop® device, and successfully applied to MVC vitrification of bovine mature oocytes (Nakayama et al., 2020).

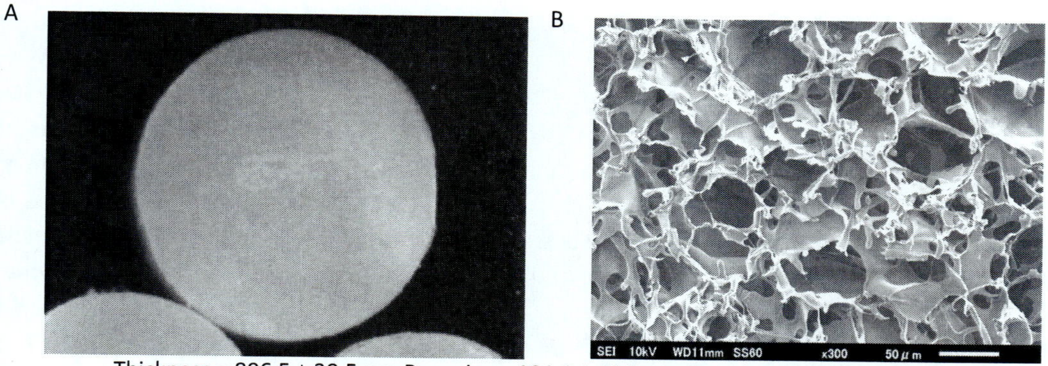

Thickness = 896.5 ± 39.5 μm;Pore size = 101.1 ± 12.9 μm;Porosity = 94.4 ± 1.0%

Figure 2. Structural characterization of the SF sponge-derived cryodevice. (A) SF sponge disc is 8 mm in diameter and 1mm in height. Rat pancreatic islets (100–200 μm in diameter; $n = 50$) have been loaded onto the center surface; (B) SEM image of the SF sponge surface. Measurement data is from Nakayama-Iwatsuki et al. (2020).

5. SF Application to Islet Transplantation

SF has been used as an excellent scaffold biomaterial in tissue engineering and applied to bone, cartilage, ligament, tendon, and blood vessels (Altman et al., 2003; Rockwood et al., 2011; Kasoju and Bora, 2012). In addition, the production of transgenic silkworms has facilitated modifications to SF characteristics, including fusion to an Arg-Gly-Asp (RGD) sequence of the cell adhesion domain (Kambe et al., 2010) or basic fibroblast growth factor (bFGF; Kambe et al., 2016). The high biocompatibility and hydrophilic properties of SF may contribute to the future development of a transplantable cryodevice/scaffold and serve as an extracellular matrix (ECM)-like structure by leaving SF fragments on the surface of the islets. A decellularized pancreas has been used as a native ECM scaffold for seeding and culturing mouse islets (Guruswamy Damodaran and Vermette, 2018). Recently, an electrospun hybrid nano-fiber scaffold with SF and ECM components derived from decellularized porcine pancreas improved islet survival and functionality (Zhu et al., 2021). Decellularized rat liver is also a possible source of functional ECMs, promoting islet functionality before and after vitrification (Nakayama-Iwatsuki et al., 2021).

5.1. Cytokine Transporter

Pancreatic islets have higher vessel density than surrounding exocrine tissues and receive 15–20% of the pancreatic blood supply despite making up 1–2% of the pancreas mass. A dense network of capillaries in islets and ECM is damaged by enzymatic digestion during the isolation process (Ricordi and Strom, 2004; Pepper et al., 2015). For successful engraftment, isolated islets should receive nutritional and physical support from the host by forming new blood vessels around and within the engraft (Emamaullee and Shapiro, 2007). To promote in vivo angiogenesis, several cytokines have been investigated, including bFGF and VEGF, which stimulate the recruitment of endothelial progenitors (Ferrara et al., 2003), as well as platelet-derived growth factor and transforming growth factor-β, which stabilize the newly formed blood vessels (Rouwkena et al., 2008).

Conjugated VEGF on an alginate hydrogel was found to enhance islet in vivo viability and functionality in mice (Yin et al., 2016). VEGF also increased functional β-cell mass by improving the angiogenesis of isolated human and murine pancreatic islets (Lai et al., 2005). Moreover, a heparin-releasing SF scaffold has been shown to enhance islet angiogenesis by stabilizing VEGF and inhibiting the instant blood-mediated inflammatory reaction in islet transplantation into the epididymal fat pad of diabetic mice (Mao et al., 2017). We have recently applied an SF sponge disc for transporting a VEGF solution and proven its effectiveness for glycemic control in the transplantation of vitrified-warmed islets beneath the kidney capsule of diabetic rats (Yamanaka et al., 2021).

5.2. Scaffold

Macroporous scaffolds have been used extensively as temporal artificial ECMs for cellular transplantation by offering an optimal site for cell accommodation (housing), proliferation, and differentiation, and/or facilitating the exchange of nutrients and gasses, cell infiltration, and neovascularization. Clinical islet transplantation has been performed by percutaneous infusion into the hepatic portal system (Shapiro et al., 2000), but alternative transplantation sites (e.g., subrenal site, intraperitoneal, or subcutaneous space) that are less invasive and more effective are desirable with the combined use of an optimal DDS and scaffold system (Cantarelli and Piemonti, 2011).

Co-encapsulation of islets with mesenchymal stem cells (MSCs), laminin, and collagen IV in an SF-based hydrogel resulted in prolonged viability and enhanced functionality of the islets in vitro (Davis et al., 2012). Co-transplanting a porous structure of an SF nonwoven mat with an adipose tissue-derived stromal vascular fraction (including MSCs, endothelial cells, fibroblasts, and growth factors) in the subcutaneous tissue of mice was shown to promote angiogenesis prior to islet transplantation (Perteghella et al., 2017). Hollow channels in SF scaffolds promoted pre-vascularization by facilitating endothelial cell growth, VEGF secretion, and capillary-like tube formation (Zhang et al., 2015). Co-delivery of MSCs and endothelial cells in SF hydrogels was also effective in the secretion of trophic factors, which contributes to blood vessel stabilization (Sun et al., 2016). An SF-based encapsulation platform has been developed for MSC delivery to promote mouse islet functionality (Hamilton et al., 2017). Interestingly, nanofibrous electrospun SF scaffolds enhance the differentiation of conjunctiva MSCs into insulin-producing β-cells (Barati et al., 2019). Direct interaction of SF scaffolds with β-cells has also been reported (Ashari et al., 2018; Parks et al., 2020).

We hereby restate the consecutive availability of an SF sponge disc as an SSV cryodevice, VEGF transporter, and islet scaffold for subrenal transplantation in a rat model. In a previous report (Nakayama-Iwatsuki et al., 2020), one hundred IEQ rat islets have been loaded onto an

SF sponge disc and then vitrified by the SSV protocol (Figure 3A). At least eight SF sponge discs were required to collect the post-warm islets (800 IEQ) for single subrenal transplantation. The SF sponge disc (ø = 8-mm) has a sufficient surface area to allow loading of > 500 IEQ islets into a single operation (Figure 2A). In an experiment designed by Yamanaka et al. (2021), a marginal rat model for subrenal transplantation was built to clarify the effect of VEGF supplementation into the SF sponge disc on the engraftment of the post-warm islets. When 550 IEQ post-warm islets were transplanted beneath the kidney capsule of diabetic rats (Figure 3B), only the co-transplantation of the VEGF-embedded SF sponge disc resulted in the successful achievement of the normoglycemia (Figure 3C), with a well-developed blood vessel network into the islet engrafts. The diameter/height and pore size of the SF sponge disc, the loading site and the number of the islets, and the co-delivered materials (support cells and/or growth factors) should be optimized by further investigations.

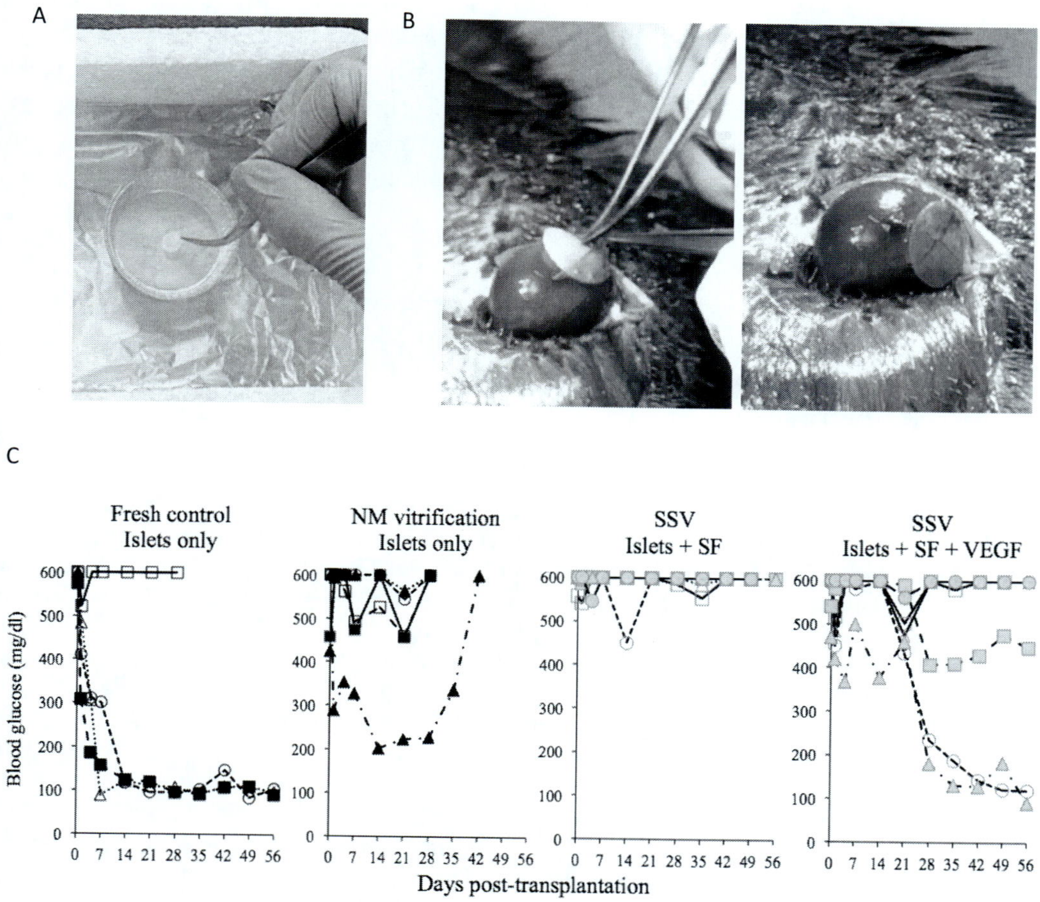

Figure 3. Consecutive triple use of an SF sponge disc throughout islet MVC vitrification to subrenal transplantation. (A) Solid surface vitrification (SSV). The islet-loaded disc is cooled on an aluminum boat without direct contact to LN_2; (B) Islet transplantation. Post-warm islets on the disc are co-transplanted beneath the kidney capsule of diabetic rats (blood glucose level > 350 mg/dL);
(C) Euglycemic control of diabetic rats by co-transplantation with or without angiogenesis-facilitating VEGF. Under the 550-IEQ (marginal mass) syngeneic transplantation, cured recipients (blood glucose

level < 200 mg/dL) were obtained only from the fresh control group (3/5) and SSV co-transplantation with VEGF group (2/6). Transplantation data is from Yamanaka et al. (2021).

Conclusion

The high biocompatibility and easy processing capability of SF can facilitate successful scaffold design in biomedical research and practical applications, including cell transplantation therapy. The porous SF sponge with high hydrophilicity also serves as a transporter in DDS because the SF sponge is suitable for absorption, retention, and gradual release of pharmaceutical aqueous solutions. Furthermore, the cryodevice derived from the SF sponge can be optimized for MVC vitrification of large cell complexes or tissues composed of different cell types. This system requires prompt minimization of exposure to the highly toxic VS around the cell masses immediately before cooling in LN_2 to guarantee the extremely rapid cooling rate and high post-warm cell viability. In fact, a single SF sponge disc (ø = 8-mm; prepared from 4.0% [w/v] SF solution) facilitated loading and MVC vitrification of rat 550 IEQ pancreatic islets using the SSV protocol, the holding of angiogenesis-facilitating cytokine VEGF, and the subrenal deposition with post-warm islets in diabetic recipient rats. Some of the recipients receiving the post-warm islets on the VEGF-embedded SF scaffold have been shown to reach normoglycemia within 8 weeks after transplantation.

References

Altman GH, Diaz F, Jakuba C, Calabro T, Horan RL, Chen J, Lu H, Richmond J, and Kaplan DL. (2003). Silk-based biomaterials. *Biomaterials*; 24: 401–416.

Ashari N, Pang HW, Simon T, Xiong Y, Cobum JM, Bromberg JS, Kaplan DL, McLenithan J, and Fontaine MJ. (2018). Silk fibroin preserves beta cell function under inflammatory stress while stimulating islet cell surface GLUT2 expression. *Cellular Immunology*; 329: 10–16.

Barati G, Rahmani A, and Nadri S. (2019). In vitro differentiation of conjunctiva mesenchymal stem cells into insulin producing cells on natural and synthetic electrospun scaffolds. *Biologicals*; 62: 33–38.

Buchwald P, Wang X, Khan A, Bernal A, Fraker C, Inverardi L, and Ricordi C. (2009). Quantitative assessment of islet cell products: estimating the accuracy of the existing protocol and accounting for islet size distribution. *Cell Transplantation*; 18: 1223–1235.

Cantarelli E, and Piemonti L. (2011). Alternative transplantation sites for pancreatic islet grafts. *Current Diabetes Reports*; 11: 364–374.

Corominola H, Mendola J, Esmatjes E, Sáenz A, Fernández-Cruz L, and Gomis R. (1998). Cryopreservation of pancreatic islets prior to transplantation: A comparison between UW solution and RPMI culture medium. *Cryobiology*; 37: 110–118.

Davis NE, Beenken-Rothkopf LN, Mirsoian A, Kojic N, Kaplan DL, Barron AE, and Fontaine MJ. (2012). Enhanced function of pancreatic islets coencapsulated with ECM proteins and mesenchymal stromal cells in a silk hydrogel. *Biomaterials*; 33: 6691–6697.

Desai T, and Shea LD. (2017). Advances in islet encapsulation technologies. *Nature Reviews. Drug Discovery*; 16: 338–350.

Emamaullee JA, and Shapiro AMJ. (2007). Factors influencing the loss of β-cell mass in islet transplantation. *Cell Transplantation*; 16: 1–8.

Ferrara N, Gerber HP, and LeCouter J. (2003). The biology of VEGF and its receptors. *Nature Medicine*; 9: 669–676.

Guruswamy Damodaran R, and Vermette P. (2018). Decellularized pancreas as a native extracellular matrix scaffold for pancreatic islet seeding and culture. *Journal of Tissue Engineering and Regenerative Medicine*; 12: 1230–1237.

Hamilton DC, Shih HH, Schubert RA, Michie SA, Staats PN, Kaplan DL, and Fontaine MJ. (2017). A silk-based encapsulation platform for pancreatic islet transplantation improves islet function in vivo. *Journal of Tissue Engineering and Regenerative Medicine*; 11: 887–895.

Hochi S. (2022). Cryodevices developed for minimum volume cooling vitrification of bovine oocytes. *Animal Science Journal*; 93: e13683.

Hwang I-S, and Hochi S. (2014). Recent progress in cryopreservation of bovine oocytes. *BioMedicine Research International*; 2014: 570647.

Kambe Y, Kojima K, Tamada Y, Tomita N, and Kameda T. (2016). Silk fibroin sponges with cell growth-promoting activity induced by genetically fused basic fibroblast growth factor. *Journal of Biomedical Materials Research*; 104: 82–93.

Kambe Y, Yamamoto K, Kojima K, Tamada Y, and Tomita N. (2010). Effects of RGDS sequence genetically interfused in the silk fibroin light chain protein on chondrocyte adhesion and cartilage synthesis. *Biomaterials*; 31: 7503–7511.

Kameda T. Hashimoto T., and Tamada Y. (2011). Effects of supercooling and organic solvent on the formation of a silk sponge with porous 3-D structure, and its dynamical and structural characterization using solid-state NMR. *Journal of Materials Science*; 46: 7923–7930.

Kasoju N, and Bora U. (2012). Silk fibroin in tissue engineering. *Advanced Healthcare Materials*; 1: 393–412.

Kim U-J, Park J, Kim HJ, Wada M, and Kaplan DL. (2005). Three-dimensional aqueous-derived biomaterial scaffolds from silk fibroin. *Biomaterials*; 26: 2775–2785.

Kojayan GG, Alexander M, ImagawaDK, and Lakey JRT.(2018). Systematic review of islet cryopreservation. *Islets*; 10: 40–49.

Kuwayama M, Vajta G, Kato O, and Leibo SP. (2005). Highly efficient vitrification method for cryopreservation of human oocytes. *Reproductive Biomedicine Online*; 11: 300–308.

Lai Y, Schneider D, Kidszun A, Hauck-Schmalenberger I, Breier G, Brandhorst D, Brandhorst H, Iken M, Brendel MD, Bretzel RG, and Linn T. (2005). Vascular endothelial growth factor increases functional β-cell mass by improvement of angiogenesis of isolated human and murine pancreatic islets. *Transplantation*; 79: 1530–1536.

Li M, Ogiso M, and Minoura N. (2003). Enzymatic degradation behavior of porous silk fibroin sheets. *Biomaterials*; 24: 357–365.

Li M, Zhang C, Lu S, Wu Z, and Yan H. (2002). Study on porous silk fibroin materials: 3. Influence of repeated freeze-thawing on the structure and properties of porous silk fibroin materials. *Polymers for Advanced Technologies*; **13**: 605–610.

Maehara M, Sato M, Watanabe M, Matsunari H, Kokubo M, Kanai T, Sato M, Matsumura K, Hyon S-H, Yokoyama M, Mochida J, and Nagashima H. (2013). Development of a novel vitrification method for chondrocyte sheets. *BMC Biotechnology*; 13: 58.

Mao D, Zhu M, Zhang X, Ma R, Yang X, Ke T, Wang L, Li Z, Kong D, and Li C. (2017). A macroporous heparin-releasing silk fibroin scaffold improves islet transplantation outcome by promoting islet revascularisation and survival. *Acta Biomaterialia*; 59: 210–220.

Martino A, Songsasen N, and Leibo SP. (1996). Development into blastocysts of bovine oocytes cryopreserved by ultra-rapid cooling. *Biology of Reproduction*; 54: 1059–1069.

Meinel L, Karageorgiou V, Hofmann S, Fajardo R, Snyder B, Li C, Zichner L, Langer R, Vunjak-Novakovic G, and Kaplan DL. (2004). Engineering bone-like tissue in vitro using human bone marrow stem cells and silk scaffolds. *Journal of BiomedicalMaterials Research. Part A*; 71: 25–34.

Mendola J, Corominola H, Gonzalez-Clemente JM, Esmatjes E, Saenz A, Fernandez-Cruz L, and Gomis R. (1996). Follow-up study of the revascularization process of cryopreserved islets of Langerhans. *Cryobiology*; 33: 530–543.

Nagaya M, Matsunari H, Kanai T, Maehara M, Nakano K, Umeki I, Katsumata Y, Kasai Y, Sakai R, Kobayashi M, Honda M, Abe N, Watanabe M, Umeyama K, and Nagashima H. (2016). An effective new cryopreservation procedure for pancreatic islets using hollow fiber vitrification. *Hormone and Metabolic Research*; 48: 540–549.

Nakayama K, Chinen S, Teshima J, Tamada Y, Hirabayashi M, and Hochi S. (2020). Slik fibroin sheet multilayer suitable for vitrification of in vitro-matured bovine oocytes. *Theriogenology*; 145: 109–114.

Nakayama-Iwatsuki K, Yamanaka T, Negishi J, Teshima J, Tamada Y, Hirabayashi M, and Hochi S. (2020). Transplantation of rat pancreatic islets vitrified-warmed on the nylon mesh device and the silk fibroin sponge disc. *Islets*; 12: 145–155.

Nakayama-Iwatsuki K, Yanagisawa K, Tanaka D, Hirabayashi M, Negishi J, and Hochi S. (2021). Acellular matrix derived from rat liver improves the functionality of rat pancreatic islets before or after vitrification. *Cryobiology*; 100: 90–95.

Nazarov R, Jin H-J, and Kaplan DL. (2004). Porous 3-D scaffolds from regenerated silk fibroin. *Biomacromolecules*; 5: 718–726.

Park SY, Kim B, Lee YK, Lee S, Chun JM, Suh JG, and Park JH. (2020). Silk fibroin promotes the regeneration of pancreatic β-cells in the C57BL/KsJ-*Lepr*$^{db/db}$ mouse. *Molecules*; 25: 3259.

Pepper AR, Gala-Lopez B, Pawlick R, Merani S, KinT, and Shapiro AMJ. (2015). A prevascularized subcutaneous device-less site for islet and cellular transplantation. *Nature Biotechnology*; 33: 518–523.

Perteghella S, Vigani B, Mastracci L, Grillo F, Antonioli B, Galuzzi M, Tosca MC, Crivelli B, Preda S, Tripodo G, Marazzi M, Chlapanidas T, and Torre ML. (2017). Stromal vascular fraction loaded silk fibroin mats effectively support the survival of diabetic mice after pancreatis islet transplantation. *Macromolecular Bioscience*; 17: 1700131.

Pileggi A, Molano RD, Ricordi C, Zahr E, Collins J, Valdes R, and Inverardi L. (2006). Reversal of diabetes by pancreatic islet transplantation into a subcutaneous, neovascularized device. *Transplantation*; 81: 1318–1324.

Ramachandran K, Huang HH, and Stehno-Bittel L. (2015). A simple method to replace islet equivalents for volume quantification of human islets. *Cell Transplantation*; 24: 1183–1194.

Ricordi C, and Strom TB. (2004). Clinical islet transplantation: advances and immunological challenges. *Nature Review. Immunology*; 4: 259–268.

Rockwood DN, Preda RC, Yucel T, Wang X, Lovett ML, and Kaplan DL. (2011). Materials fabrication from *Bombyx mori* silk fibroin. *Nature Protocols*; 6: 1612–1631.

Rouwkema J, Rivron NC, and van Blitterswijk CA. (2008). Vascularization in tissue engineering. *Trends in Biotechnology*; 26: 434–441.

Sakata N, Egawa S, Sumi S, and Unno M. (2008). Optimization of glucose level to determine the stimulation index of isolated rat islets. *Pancreas*; 36: 417–423.

Sasamoto H, Futami M, Ando Y, and Nakaji S. (2012). Cryopreservation of rat islets of Langerhans by vitrification. *Journal of Artificial Organs*; 15: 283–289.

Shapiro AM, Lakey JRT, Ryan EA, Korbutt GS, Toth E, Warnock GL, Kneteman NM, and Rajotte RV. (2000). Islet transplantation in seven patients with type 1 diabetes mellitus using a glucocorticoid-free immunosuppressive regimen. *New England Journal of Medicine*; 343: 230–238.

Stendahl JC, Kaufman DB, and Stupp SI. (2009). Extracellular matrix in pancreatic islets: relevance to scaffold design and transplantation. *Cell Transplantation*; 18: 1–12.

Sun W, Motta A, Shi Y, Seekamp A, Schmidt H, Gorb SN, Migliaresi C, and Fuchs S. (2016). Co-culture of outgrowth endothelial cells with human mesenchymal stem cells in silk fibroin hydrogels promotes angiogenesis. *Biomedical Materials*; 11: 035009.

Tamada Y. (2005) New process to form a silk fibroin porous 3-D structure. *Biomacromalecules*; 6: 3100–3106.

Tamada Y, Kobayashi K, and Sumi N. (2015). Fibroin sponge safety evaluations (In Japanese). *Journal of Silk Science and Technology of Japan*; 23: 71–73.

Taylor MJ, and Baicu S. (2009). Review of vitreous islet cryopreservation: Some practical issues and their resolution. *Organogenesis*; 5: 155–166.

Vajta G, Holm P, Kuwayama M, Booth PJ, Jacobsen H, Greve T, and Callesen H. (1998). Open pulled straw (OPS) vitrification: A new way to reduce cryoinjuries of bovine ova and embryos. *Molecular Reproduction and Development*; 51: 53–58.

Vepari C, and Kaplan DL. (2007). Silk as a biomaterial. *Progress in Polymer Science*; 32: 991–1007.

Xavier GDS. (2018). The cells of the islets of Langerhans. *Journal of Clinical Med*icine; 7: 54.

Yamanaka T, Goto T, Hirabayashi M, and Hochi S. (2017). Nylon mesh device for vitrification of large quantities of rat pancreatic islets. *Biopreservation and Biobanking*; 15: 457–462.

Yamanaka T, Nakayama-Iwatsuki K, Fujimoto S, Hirono N, Negishi J, Tamada Y, Hirabayashi M, and Hochi S. (2021). All-in-one silk fibroin sponge as the vitrification cryodevice of rat pancreatic islets and the VEGF-embedded scaffold for subrenal transplantation. *Transplantation Proceedings*; 53: 1744–1750.

Yamanaka T, Tashima K, Takahashi R, Takashima S, Goto T, Hirabayashi M, and Hochi S. (2016). Direct comparison of Cryotop® vitrification and Bicell® freezing on recovery of functional rat pancreatic islets. *Cryobiology*; 73: 376–382.

Yin N, Han Y, Xu H, Gao Y, Yi T, Yao J, Dong L, Cheng D, and Chen Z. (2016). VEGF-conjugated alginate hydrogel prompt angiogenesis and improve pancreatic islet engraftment and function in type 1 diabetes. *Material Science and Engineering. C, Materials for Biological Applications*; 59: 958–964.

Zhang W, Wray LS, Rnjak-Kovacina J, Xu L, Zou D, Wang S, Zhang M, Dong J, Li G, Kaplan DL, and Jiang X. (2015). Vascularization of hollow channel-modified porous silk scaffolds with endothelial cells for tissue regeneration. *Biomaterials*; 56: 68–77.

Zhu Y, Wang D, Yao X, Wang M, Zhao Y, Lu Y, Wang Z, and Guo Y. (2021). Biomimetic hybrid scaffold of electrospun silk fibroin and pancreatic decellularized extracellular matrix for islet survival. *Journal of Biometerials Science. Polymer Edition*; 32: 151–165.

Chapter 5

Silk Fibroin-Based Nanocomposite for Futuristic Electronics

Reetu Yadav and Roli Purwar*
Department of Applied Chemistry,
Delhi Technological University, New Delhi, India

Abstract

This chapter is focused on the contribution and future potential of regenerated silk fibroin-based nanocomposites in the electronics field. Here, the need of biomaterials for optoelectronics, microelectronics, and bioelectronics devices has been thoroughly discussed. Amongst the variable current and prospective materials, silk fibroin-based biomaterials incorporating nanofillers has shown immense potential in the aforementioned areas. This chapter explores the role of various inorganic fillers in silk fibroin and their applicability in electronics applications. Recently, the use of semiconductor-based bio-implantable optoelectronic devices has emerged for brain imaging devices and as optical stimulators for optogenetics or retinal prosthesis. Such research in this dimension has paved the pathway for biodegradable materials like e-skin, artificial muscles, bionic brains, and health monitoring devices with electronic and optical properties for biomedical applications. The substantial work detailed in this chapter stems from a decade of constant research on making silk fibroin as an active layer component in semiconductor devices. Silk fibroin-based nanocomposites which have shown their potential as a flexible substrate, dielectric layer, conductive layer, resistive layer, ion-transport layer, encapsulation layer in many electronic devices are discussed in detail.

Keywords: biomaterials, regenerated silk fibroin, nanocomposite, electronics, nanofillers

1. Introduction

Progressions in various semiconductor technologies have resulted in a massive fabrication of silicon-based electronic devices for use in high-speed computers, high-efficiency photovoltaic cells, and portable consumer electronics, dramatically altering our traditional modes of communication, work, and entertainment. Unfortunately, this would have resulted in the

* Corresponding Author's Email: roli.purwar@dtu.ac.in.

In: Silk Fibroin
Editor: Shivaji H. Pawar
ISBN: 979-8-88697-402-7

depletion of the non-renewable sources and the generation of undesirable large amounts of electronic waste which became a global problem. It's becoming increasingly difficult to boost the performance of traditional silicon technologies more organically because they've reached their fundamental limits. Organic materials have shown great promise in addressing the scientific challenge by constructing complementary electronics, which not only improve electronic and mechanical properties in comparison to their inorganic counterparts but also enable the manufacture of flexible and thin-film devices [1-8].

The aspiration to fabricate electronic devices using organic materials has opened new avenues in the field of biomaterial research [2,3,9,10]. Optoelectronic devices such as organic thin-film transistors, organic displays, light-emitting devices, and organic photovoltaics have all benefited from biomaterials as attractive building blocks for constructing next-generation bendable and bio-sustainable electronics [3,5, 11-15]. Using biomaterials in electronic devices has contributed to the solution for the electronic waste problem as well as showed the potential for advanced futuristic flexible electronic devices [3, 16-19].Electronic devices made from biomaterials like proteins, peptides, collagen, gelatin, DNA, keratin, reflectin, cellulose, water-soluble polymers are not only limited to electronics or microelectronic applications but also extending their utility in biomedical applications due to their attractive properties such as biocompatibility, biodegradability, bioresorbability, natural abundance, and light-weight [14,20-28].

Bombyxmori derived silk fibroin, one of the most abundant proteins, due to its unique structure and properties, has outperformed the other options and attracted more research interest [29-31]. Predominantly, silk fibroin is superior to other biodegradable materials in terms of mechanical toughness. Silk fibroin is a solution-processable material that can be easily fabricated into a variety of materials. Also, silk fibroin's physical properties and degradation behavior can be altered with its processing conditions. Multifunctionalpropertiesof silk fibroin like thermal stability, transparency, high refractive index, optical properties, biodegradability, and biocompatibility open its axis for applications in different fields including packaging, optics, electronics biomedicine, catalysis, tissue engineering, and drug delivery [16, 29, 39-48, 31, 49-52, 32-38]. In the last decade, silk fibroin-based materials are rooting their position in the field of optoelectronics which is considered beneficial for both electronics and biomedical fields.

2. Biomaterials in Electronics Applications

A biomaterial is no wonder taking the place of conventional polymers in electronics devices due to its sustainability [3]. The versatility involved with the fabrication of biomaterial in different structures or physical states along with properties like bioactivity, biocompatible, tunable biodegradation, mechanical stability provides substantial characteristics for its use in multifunctional applications [4]. Table 1 has been used to summarize the various biomaterials with their fabrication method embedded with electronic and optical properties to be used as electronic devices for different applications.

Because of the excellent mechanical properties, appealing electrochemical properties, low cost, and simple fabrication, cellulose and cellulose-based materials (i.e., cellulose fiber and paper) have been widely implemented to construct flexible electronic devices [53]. It has served as fundamental supporting components and functional components, such as organic light-emitting diodes (OLEDs), organic field-effect transistors (OFETs), and solar cells, but their biocompatibility has been a concern [54].

Table 1. Biomaterials used in electronic devices for various applications

S. no.	Name of biomaterial	Structure	Technique	Application	Ref.
1.	Tryptophan and tyrosine	Thin film	Patterning	Field effect transistor	4
2.	Bacteriorhodpsin	Thin film	Patterning	Memresistors	58
3.	Melanins	Film	Spin coating, Electrodeposition	Biosensor	26
4.	Carotenoids	Film	Spin coating	Active layer in transistor	4
5.	Indigo	Film	Spin coating	Organic thin film transistor	19
6.	Albumin	Film	Spin coating	Gate dielectric in thin film transistor	4
7.	DNA	Film	Spin coating	Gate dielectric in optical devices	59
8.	Silk	Film	Spin coating	Transistor, actuator	12
9.	Collagen	Thin film	Lithography patterning	Substrate for antennas and resistors	20
10.	Keratin	Film	Solvent casting	Optical elements and substrate in photonics	24
11.	Reflectin	Thin film	Lithography patterning	Photonic transistor	28
12.	Cellulose	Film	Casting	Printed circuit	60
13.	PLGA	Film	Spin coating	Substrate in transistor	20
14.	PVA	Film	Spin coating	Gate dielectric transistor	61

Pectin is another natural biomaterial derived from plant cell walls that have been used to construct flexible electrical devices [55]. Pectin is ideal for memory switching devices and sensing devices because of its good shape memory and great permeability to water molecules, as well as quick metal ion transport; nevertheless, pectin-based flexible materials often have poor mechanical qualities [56]. Chitin and its derivative chitosan have molecular structures comparable to cellulose, but they have greater biological activities, such as biocompatibility and biodegradability. Furthermore, chitosan and melanin films have great mechanical strength and outstanding processability, making them an excellent candidate for the fabrication of flexible natural biomaterials [57].

3. Potential of Silk Fibroin in Electronic Applications

Among so many biomaterials, silk fibroin is attracting researchers' attention for use in electronics applications due to its multifunctional properties, which include remarkable biocompatibility, adjustable biodegradability, and water solubility, excellent optical transmittance, and good mechanical strength; additionally, it can be developed into a variety of forms, including fibers, films, sponges, and hydrogels [16]. Figure 1 depicts the many structural forms of silk fibroin employed for various purposes due to the wide-ranging characteristics of silk. It has emerged as a promising choice and is extensively employed as a fundamental component in the construction of biocompatible wearable/implantable electronic devices and functional components in the formation of energy harvesters, sensors, filters, lenses, biomemristors, and other electronic devices [1,15]. Enhancement in functionality or improvement in the chemo-physical properties with tunable optical properties from transparent to highly reflective, dielectric properties as well as tailored mechanical properties have become the key characteristics of silk fibroin for electronic devices [62].

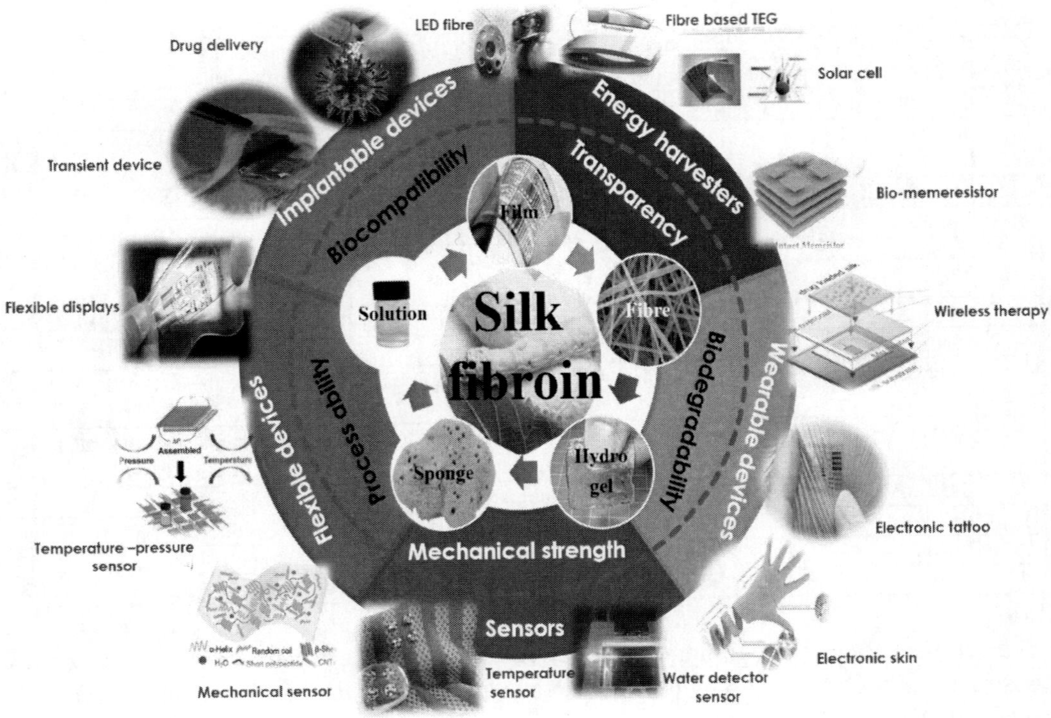

Figure 1. Potential of silk fibroin in various electronics applications in different structural forms.

Few research groups have employed silk fibroin as an active functional component for temperature sensing, which was linked with a double-chain thermoelectric generator to achieve sensing and energy harvesting integration. Temperature sensing works on two levels: a rise in temperature enhances the molecular motion of silk fibroin, and the intensity of molecular motion influences the dielectric constant of silk fibroin, which defines the capacitance of silk fibroin-based temperature sensors. The capacitance of the proposed silk fibroin-based temperature sensor established a strong linear relationship with the temperature in this study, showing an appealing possibility for temperature sensing [63].

In the last two decades, efforts have been made by several researchers to dissolve the silk fibroin and regenerate it in the form of films, hydrogel, electrospunfibers, and wet spun fibers with tailor-made properties for various applications [16]. However, the regenerated silk protein materials exhibit poorer mechanical properties and more brittleness compared with their origin form. Regenerated silk fibroin films are reported to be highly insulating with electrical conductivity in the range of 10^{-9} to 10^{-15} S/cm [3,64]. Efforts to enhance the conductivity of the regenerated silk fibroin films to be used as an active layer in biosensors, flexible organic light-emitting diodes, and wearable electronic textiles [62, 65-69] were made recently. Regenerated silk has been used as a flexible substrate [70], dielectric layer [71], resistive layer [10], conductive layer [64], coating layer [72] in manyelectronic devices. Regenerated silk fibroin has been also used as a resistive-switching layer and supportive substrates in memristors [73]. Wang et al., and Haoet et al., proposed regenerated silk fibroin as an active layer in logic or synaptic network for realizing e-skins or bionic brains [2,36]. Zhuanget et al., have used regenerated silk fibroin as an active layer while blending it with poly (vinyl alcohol) (PVA) at a blend ratio 7:5 which is used as a dielectric layer in organic field-effect transistor (OFET) [40]. Films show enhanced charge carrier mobility and current on/off ratio as compared to pure silk films. Free-standing films of silk fibroin were grafted with polypyrrole (PPy) for use in actuators as conductive layers [74]. Romero et al., have shown in-situ polymerization of cationic PPy on silk films having anionic sulfonate groups to form the interpenetrating network of polypyrrole. Using this strategy mechanically robust polymer electrodes with stable electrochemical performance and sheet resistivities in the order of $1 \times 102\ \Omega$/sq (conductivity ~1 S/cm) were produced [7]. Bradshaw et al., demonstrated that silk/polypyrrole composite can bend under an applied voltage using a simple bilayer device that can be used for application as artificial muscles [12]. Yadav & Purwaret et al., have fabricated the transparent regenerated silk fibroin films by incorporating various metal oxide nanoparticles as conductive layer as well as dielectric layer for optoelectronics applications and shown the conductivity in the range of 10-3S/cm [75].

4. Role of Inorganic Nanofiller-Based Nanocomposites in Electronic Applications

Regenerated silk fibroin nanocomposites using conductive nanostructure materials such as single-wall carbon nanotube, multiwall carbon nanotubes, graphene, silver & gold nanoparticles, silver nanowires, and metal oxide nanoparticleshave been used to fabricate various biomedical as well as electronic devices [34, 62, 66-68,76-80]. Free-standing silk fibroin films are often manufactured and employed in luminescent solar concentrators, optical devices, electrical conduction and dielectric relaxation, and a variety of other applications [81]. Use of regenerated silk fibroin nanocomposites in various application has been summarized in Table 2. The evenly dispersed reduced graphene oxide nanosheets significantly increase the conductivity of the silk fibroin nanocomposites, most likely by forming conductive networks across the host polymer matrix. This creates a conducting channel in the silk fibroin, resulting in higher conductivity [82]. It is also discovered that the resistance provided by the nanocomposite reduces as the graphene oxide level of the silk fibroin varies. This might be attributed to the shorter distance between the nanosheets. In the reduced-distance case, the generated electric field is sufficient to allow electrons to tunnel from one nanosheet to another [80]. A detailed knowledge of these nanocomposite's conduction characteristics can make them particularly valuable for biological applications such as biosensors and tissue engineering.

Inorganic nanofiller as a reinforcing agent introduced in silk fibroin to improve the properties of bio-nanocomposite for different applications. Nanostructured materials improve the mechanical, rheological, thermal, electrical, and optical properties [91]. These nanomaterials have a wide range of applications such as biomedical, sensing applications such as antibacterial coating, thin-film sensors, and plastic memory devices [92]. The addition of these nanomaterial structures has enhanced the conductivity of regenerated silk fibroin films in the range of 10-3 to 10-9 S/cm i.e., semiconducting range. These nanofillers also affect the type of semiconducting behavior of the silk fibroin material [75]. Zinc oxide (p-type) and copper oxide (n-type) nanoparticles influenced the p-type semiconducting behavior of silk fibroin reported by Yadav & Purwaret al., [75]. Metal oxide nanoparticles have outstanding electrical, magnetic, thermal and optical properties. They have high density, are abundant in size, and have different morphologies such as spherical, triangular, star, nanowires, nanotubes, and nanorods [5]. The presence of metal oxide nanoparticles can improve the thermal stability, toughness, glass transition temperature, optical and tensile strength properties of the polymer nanocomposites[93]. Nanoparticles such as Zinc oxide, Copper oxide, and Titanium dioxide have been used to tune the conducting properties of silk fibroin nanocomposite due to good intermolecular interaction. Zinc oxide has a high isoelectric point (IEP) of about 9.5, which makes it suitable for the adsorption of silk proteins with low IEPs via electrostatic interactions [94].

Table 2. Silk fibroin is incorporated with the different nanofillers and their applications

S.no.	Types of silk	Reinforcing agent	Technique	Structure/form of nanocomposite	Application	Ref.
1.	B. Mori fibroin	Graphene oxide	Spin LBL	Membrane	Electronics & optoelectronics	78
2.	B. Mori fibroin	Polyaniline	Electrospinning	Fiber	Optoelectronics	74
3.	B. Mori fibroin	Single wall CNT	Electrospinning	Fiber	Energy storage, electronics, vaccine delivery	77
4.	B. Mori fibroin	Gold nanoparticles	Electrospinning	Fiber	Tissue engineering	64
5.	Tasarfibroin	Carbon nanofibre	Casting	Films	Tissue enginering	76
6.	B. Mori fibroin	Na-MMT	Solvent Intercalation	Film	Bone regeneration	83
7.	B. Mori fibroin	Cellulose	Casting	Film	Tissue engineering	60
8.	B. Mori fibroin	Spiropyran	Electrospinning	Fiber	Optical sensing	84
9.	B. Mori fibroin	Multi wall CNT	Casting	Film	Optoelectronics	60
10.	B. Mori fibroin	Magnetic NP	Casting	film	Tissue engineering	50
11.	B. Mori fibroin	Na-hec clay	Casting	Film	Packaging	85
12.	B. Mori fibroin	Gold nanoparticles	Casting	Film	Light harvesting devices	69
13.	B. Mori fibroin	Graphene oxide	Electrospinning	Fiber	Tissue engineering, bioelectrincs, energy storage	86
14.	B. Mori fibroin	Laponitenanoclay	Chemical synthesis (Sol-gel analysis)	Hydrogel	Bone regeneration	87
15.	B. Mori fibroin	Closite MMT	Spin coating	Film	Bone formation	88
16.	B. Mori fibroin	Chitosan	Electrospinning	Fiber	Wound dressing	89
17.	B. Mori fibroin	Hydrotalcite	Spin coating	Film	Optoelectronics	62
18.	B. Mori fibroin	MMT Nanoclay	Spin coating	Film	Tissue scaffold	85
19.	B. Mori fibroin	Titanium dioxide nanoparticles	Casting	Film	Optoelectronics	90
20.	B. Mori fibroin	Silver nanowires	Casting	Film	Optoelectronics	81
21.	B. Mori fibroin	Polypyrole	Casting	Film	Optoelectronics	74
22.	B. Mori fibroin	Polyvinyl alcohol	Blend	Film	Optoelectronics	40
23.	B. Mori fibroin	Copper nanowires	Casting	Film	Optoelectronics	2
24.	B. Mori fibroin	Gold	Patterning	Film	Optoelectronics	64
25.	B. Mori fibroin	Silver nanoparticles	Casting	Film	Optoelectronics	67

S.no.	Types of silk	Reinforcing agent	Technique	Structure/form of nanocomposite	Application	Ref.
26.	B. Mori fibroin	Zinc oxide nanoparticles	Casting	Film	Optoelectronics	75
27.	B. Mori fibroin	Copper oxide nanoparticles	Casting	Film	Optoelectronics	75

Silk nanocomposite-based electronics are within the realm of possibility for advanced or futuristic bio-applications including e-skins, e-bandages, biosensors, wearable displays, implantable devices, artificial muscles, etc. [2]. The electrical, optical, magnetic, and catalytic properties of inorganic nanofillers have bought a new axis to silk fibroin bio-nanocomposites for electronic devices.

5. Structure-Property Relationship of Silk Fibroin-Based Bio-Nanocomposites

Silk fibroins are composed of relatively large hydrophilic chain end blocks (N and C-termini) with smaller hydrophilic internal blocks and large internal hydrophobic blocks. Therefore, the amphiphilic block copolymer-like properties of silk fibroin make it a good dispersing agent for nanofiller [95]. Amphoteric polyelectrolyte silk fibroin has an isoelectric point (pI) of 3.8-3.9. Positive charges are found below the pI in the silk fibroin chain [46]. The interaction between nanofiller and silk fibroin is primarily influenced by the balance of surface charges of these two moieties, which varies depending on the charge of the nanofiller. So, the pH of the silk solution is crucial to control the interaction between the silk fibroin and nanofillers which affect the final structure of the silk fibroin-based bio-nanocomposites.

The addition of nanofillers either affects the random coil conformation i.e., hydrophilic chain (amorphous region) of the silk fibroin or the β-sheet conformation i.e., hydrophobic chain (crystalline structure) dependent upon the pH of the silk solution. Different inorganic nanofiller have different ways to interact with silk fibroin chains to form a stable structure.Forzinc oxide nanoparticles, if the silk solution has a pH less than 4 gives the complex mainly joined with a carbonyl group (COO^-) of glycine in protein (i.e., hydrophobic chain) while for pH greater than 4 then tetragonal complexes with amide group (i.e., hydrophilic chain) can be formed [94]. For copper oxide nanoparticles, Cu2+ ion forms the tetragonal complexes with an amide group (i.e., hydrophilic chain) to form a stable structure irrespective of pH [96]. In the case of titanium dioxide nanoparticles (TiO2), S. Ranjan et al., have given the docking results for the interaction of silk fibroin with titanium dioxide nanoparticles to explain the mechanism. TiO2 has good interaction with the positive side chain group of amino acids. TiO2 has the least affinity with aromatic R-group-containing amino acids, which may be attributed to the fact that titanium dioxide nanoparticles do not form a stable hydrogen bond because of the resonance energy stabilization of the amino acid [97]. In this way, it can be analyzed that all the metal oxide nanoparticles have a distinguished way to interact with silk fibroin chains which in turn affect the properties of silk fibroin differently.These altered structures tend to affect the various properties of silk fibroin like rheology, morphology, optical, electrical, mechanical, and thermal properties. Many research groups have shown the effect of graphene oxide, zinc oxide, copper oxide, and titanium dioxide on silk fibroin conformation that loading of nanofiller leads to decrement in crystallinity of silk silk fibroin [75,78].

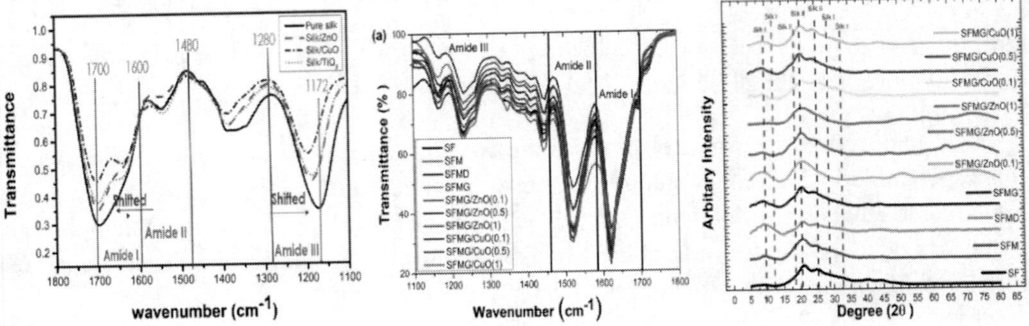

FTIR spectra of silk fibroin/metal oxide bionanocomposite solutions. FTIR spectra of silk fibroin/metal oxide bionanocomposite films. XRD spectra of silk fibroin/metal oxide bionanocomposite films.

Samples (solotins)	Ionic conductivity (x 10^{-3} S cm^{-1})	Crystallinity %
Formic acid (FA)	0.19	-
Pure ZnO/ FA	1.38	-
Pure CuO/ FA	1.89	-
Pure TiO_2 /FA	0.56	-
Pure Silk	4.81	44.23
SF/ZnO	6.46	22.77
SF/CuO	8.56	14.51
SF/TiO_2	5.98	30.66

Sample (Films)	β Sheet (%) (FTIR)	Crystallinity (%) (XRD)
SF	28.3	24.8
SFM	32.8	27.3
SFMD	39.7	34.9
SFMG	48.4	44.7
SFMG/ZnO(0.1)	48.1	43.1
SFMG/ZnO(0.5)	34.6	31.8
SFMG/ZnO(1)	29.5	26.4
SFMG/CuO(0.1)	47.3	44.2
SFMG/CuO(0.5)	45.8	42.9
SFMG/CuO(1)	34.7	29.3

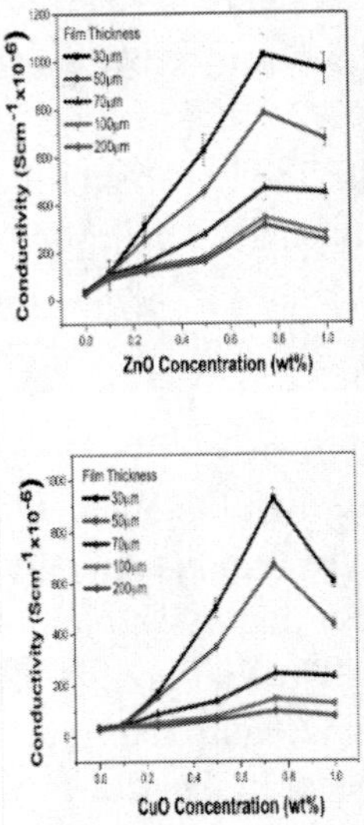

Figure 2. Effect of metal oxide nanoparticles on the silk fibroin conformation and its influence on the conductivity of the solution and film of silk fibroin.

The recent work of Yadav & Purwar et al., has shown in Figure 2 which depicts the effect of metal oxide nanoparticles on silk fibroin conformation in solution as well as in films and their influence on ionic conductivity and electrical conductivity respectively [75,90,94]. It was found that the ionic conductivity of pure silk fibroin solution was found to be higher than pure metal oxide nanoparticles solution. The incorporation of nanoparticles in silk fibroin enhances the ionic conductivity of the nanocomposite solutions. Suppression of β sheet content and enhancement in random coil conformation in nanocomposite solutions contributes toward higher ionic conductivity. Metal oxide nanoparticles may form a conducting pathway in the random coiled silk solution which leads to an increase in the ionic conductivity of the solution by increasing the ionic diffusivity. The observed conductivity of the nanocomposite solutions was found to be consistent with their random coil/β sheet structure content as observed from FTIR and XRD analysis. Among nanocomposite solutions, SF/CuO showed the highest conductivity of 8.56 x 10^{-3} S cm^{-1} may be due to higher random coil conformation content. A similar observation has been found for PVA/CuOnanocomposite by A. Manjunathet et al., who stated that with increment in amorphous nature in the nanocomposite, ionic diffusivity increases hence increasing the ionic conductivity [98].

The electrical conductivity of silk fibroin film depends upon the ionic impurities present during the preparation of the film, β sheet content, and method of measurement. In this case, ionic impurities such as calcium chloride and glycerol were added during the formation of the film. Therefore, pure silk films also show higher conductivity as compared to values reported in the literature. As the concentration of metal oxide nanoparticles increases, the conductivity of nanocomposite films increases up to 0.75% w/w after that there is no improvement observed. The nanocomposite films also show higher conductivity at lower thickness. The thickness of the films has an enormous impact on electrical conductivity. Similar results have been reported by Liang et al., for the conduction current of ultra-thin films of polyimides due to better orientation of polymer chain at lower film thickness [99]. The addition of metal oxide nanoparticles enhanced the conductivity of nanocomposite films from 10^{-6} $S.cm^{-1}$ to 10^{-3} $S.cm^{-1}$ with 30 μm thickness. The addition of metal nanoparticles in the silk fibroin matrix may affect the carrier (electron or hole) mobility in the matrix and change the conductivity of silk fibroin/metal oxides nanocomposite films. Dispersion of metal oxide nanoparticles, the type of carrier concentration, and β sheet content are some of the factors which may also play a major role to induce a different degree of conductivity in incorporating two different metal oxides in silk fibroin films.

6. Opto-Electronic Properties of Silk Fibroin-Based Bio-Nanocomposites

Silk fibroin has been gaining popularity as an optical material due to its unique features for bio-functional optical interfaces. The refractive index of silk fibroin determines how well it performs as an optical material [31]. The refractive index contrast between silk and the surroundings, in particular, is critical in determining the behavior of optical components such as optical waveguides, retro-reflectors, and opals [31]. It's crucial to know the value of the refractive index over a wide range of wavelengths, as well as the Cauchy parameters for refractive index dispersion, bandgap energy, and Urbach energy when designing optical systems [31,100]. Silk fibroin films have shown transparency in visible light more than 75 percent earlier [101]. Silk fibroin refractive index, absorption coefficient, transmittance, reflectance, and absorbance have all been found to be affected by residual water content, and crystallinity [102]. In comparison to the regenerated silk, the native silk fibroin isolated from the gland had a greater refractive index (n = 1.566 0.001 at 500 nm) and absorption coefficient [31].

The possibility of more dense packing of secondary structures for the variation in the refractive index has been discussed by a few researchers [102]. The structure of silk fibroin plays a vital role in the optical properties of silk fibroin.Its outstanding optical transmittance endows it with appealing potential in the realm of optical-electronic devices. Melikov and colleagues presented a silk hydrogel lens for light-emitting diodes (LEDs). The scattering phenomenon generated by the varied local refractive indices between proteins and water inside the silk hydrogel lens expanded the viewing angle [103]. It can be deduced from the experimental comparison that both the concentration and degree of crosslinking impact the optical characteristics of silk fibroin and the authors also demonstrated the excellent light extraction efficiency of silk hydrogel lenses, particularly for warm white LEDs. Furthermore, a wax coating method was examined in order to improve the durability of silk hydrogel lenses by more than three orders of magnitude [104].

Table 3. Exhibitions of silk fibroin in electronic devices as a different layer

Name of electronic device	Functions of layers of silk fibroin in devices	Properties of silk fibroin involved	Ref.
Wearable device	Encapsulate and substrate	Mechanical, optical, Biocompatibility, biodegradability	112-116
Implantable device	Encapsulate and substrate	Mechanical, Biocompatibility, biodegradability, solubility	117,118
Energy devices	Active and functional	Mechanical, conductivity, biocompatibility	119-122
Sensors application	Active and functional	Mechanical, environment sensitivity, pH sensitivity, biocompatibility	123
Transistor	Substrate, Active and functional	Mechanical, electric, dielectric biocompatibility, biodegradability	124,125
Membranes	Substrate, Active and functional	Mechanical, fiber-forming ability, water-solubility	126,127
Diode lense	Encapsulate, active and functional,	Optical, chemical crosslinking	103, 128
Mechanical actuators	Substrate, Active and functional	Mechanical, electric, dielectric, doping capability	129, 130

Silk fibroin plays a significant role in the manufacture of bio-based medical devices such as diffractive lenses, optical fibers, and gratings [1]. Silk fibroin has 18 amino acid residues with chemical activity and has unique self-assembly behavior and substrate recognition properties [105]. Silk fibroin is an insulating material. It has conductivity in the range of 10^{-9} to 10^{-15} $S.cm^{-1}$. The conductivity and dielectric properties of silk fibroin have been found to be co-dependent on the structural properties of silk fibroin by many research groups [90,106]. The presence of random coil conformation in silk fibroin provides the charge hopping mechanism for the conduction of electrons or holes in the silk fibroin film [67]. On the contrary, surface charge accumulation mechanisms have been proposed by researchers for the electron transfer in silk film with high crystallinity [90]. Whether silk fibroin has been used as a passive/substrate layer or active layer in electronic components, structure properties affect predominantly.

Flexible and wearable devices based on silk fibroin have been reported in the previous ten years for applications like flexible electrodes, flexible sensors, flexible power supply devices, electronics mentioned in Table [3, 10,17,18,107]. In most applications, silk fibroin has been used as a substrate layer. Biocompatibility or biodegradability qualities of silk fibroin have even been employed as a functional layer in devices. Other flexible functional devices that use silk fibroin as an active functional material, such as resistive switching memory devices, OFETs, filters, OLEDs, and actuators, have garnered significant interest in addition to energy harvesters and sensors based on silk fibroin [108]. Resistive switching memory devices, which can be switched between a high resistance state (HRS, OFF state) and a low resistance state (LRS, ON state) through a variable electric field, are viewed as attractive prospects for next-generation high-performance data memory [109]. Wang et al., used silk fibroin's adjustable water solubility, great biocompatibility, and remarkable dielectric characteristics to create transient resistive switching memory [110]. Very few research groups are making the use of conductivity and dielectric properties of silk fibroin in electronic devices. Because silk fibroin is a good precursor for chemical doping, it has also been employed as a chemically dopable functional material to build flexible electrochemical sensors [111]. A flexible sensor based on silk-derived carbon/metal composite material for the chemical detection of rutin has been developed. The gold nanoparticle/nitrogen-doped silk carbon composite material-based sensor developed by Wen et al., was shown to be effective and accurate, demonstrating the potential to expand the electrochemical applications of silk fibroin to the flexible sensing area [111].

Conclusion

Many superior properties of silk fibroin have been demonstrated, including remarkable biocompatibility, adjustable biodegradability, water-solubility, excellent optical transmittance, good mechanical robustness, lightweight, and ease of processing, making it ideal for the construction of next-generation biocompatible flexible electronic devices.Because of its desirable biological qualities, silk fibroin is widely employed as essential components, such as substrates and encapsulating materials, as well as scaffolds, of flexible wearable and implantable electronic devices such as electronic skins, bioabsorbable electronics, and therapeutic electronics. Furthermore, in recent years, there has been a surge in interest in flexible electronic devices that use silk fibroin as a functional material. In this chapter, the role of silk fibroin has been discussed as a substrate layer as well as a functional layerin electronic devices.

Nonetheless, several obstacles remain in the way of the actual development of silk fibroin-based flexible electronic devices. First, while silk fibroin-based electronic devices are superior in terms of flexibility and biocompatibility, their performance remains inferior to that oftraditional silicon-based electronics; thus, more efforts are needed to improve the performance of silk fibroin based flexible electronic devices so that they can replace rigid silicon-based electronics. Second, silk fibroin lacks mechanical strength at high water content and is brittle at low water content; so, the best equilibrium state must be researched, and novel technologies are expected to be created to keep silk fibroin's high mechanical strength and great flexibility at the same time.Third, metal electrodes or polymer conductors must be inserted into silk fibroin-based flexible electronic devices to improve conductivity, but this reduces the device's biocompatibility and biodegradability.The next step in research might be to incorporate biocompatible and even bioabsorbable conductive materials into silk fibroin-based flexible electronic devices.

In conclusion, silk fibroin has paved a new path for the flexible electronics industry, and while many hurdles remain, we are confident that further efforts will be made by researchers to improve the practical development of silk fibroin-based electronics in flexible wearable and implantable microsystems.

References

[1] Fan, S. N., Zhang, Y., Huang, X. Y., Geng, L. H., Shao, H. L., Hu, X. C., Zhang, Y. P. *Sci. China Technol. Sci.*, 2019, 62, 903–918.

[2] Koh, L. D., Yeo, J., Lee, Y. Y., Ong, Q., Han, M., Tee, B. C. K. *Mater. Sci. Eng. C*, 2018, 86, 151–172.

[3] Sun, Q., Qian, B., Uto, K., Chen, J., Liu, X., Minari, T. *Biosens. Bioelectron.*, 2018, 119, 237–251.

[4] Muskovich, M., Bettinger, C. J. *Adv. Healthc. Mater.*, 2012, 1, 248–266.

[5] Zilberberg, K., Meyer, J., Riedl, T. *J. Mater. Chem. C*, 2013, 1, 4796–4815.

[6] Vidhya, K., Saravanan, M., Bhoopathi, G., Devarajan, V. P., Subanya, S. *Appl. Nanosci.*, 2015, 5, 235–243.

[7] Romero, I. S., Bradshaw, N. P., Larson, J. D., Severt, S. Y., Roberts, S. J., Schiller, M. L., Leger, J. M., Murphy, A. R. *Adv. Funct. Mater.*, 2014, 24, 3866–3873.

[8] Shchipunov, Y. *Pure Appl. Chem.*, 2012, 84, 2579–2607.

[9] Liu, Y., Xie, Y., Liu, Y., Song, T., Zhang, K. Q., Liao, L., Sun, B. *Semicond. Sci. Technol.*, 2015, 30, 104004.

[10] Wang, H., Zhu, B., Ma, X., Hao, Y., Chen, X. *Small*, 2016, 12, 2715–2719.

[11] Lu, Z., Zhang, H., Mao, C., Li, C. M. *Appl. Energy*, 2016, 164, 57–63.

[12] Bradshaw, N. P., Severt, S. Y., Wang, Z., Fengel, C. V., Larson, J. D., Zhu, Z., Murphy, A. R., Leger, J. M. *Synth. Met.*, 2015, 209, 490–495.

[13] Luo, J., Yang, J., Zheng, X., Ke, X., Chen, Y., Tan, H., Li, J. *Adv. Healthc. Mater.*, 2020, 9, 1–9.

[14] Yang, N., Qi, P., Ren, J., Yu, H., Liu, S., Li, J., Chen, W., Kaplan, D. L., Ling, S. *ACS Appl. Mater. Interfaces*, 2019, 11, 23632–23638.

[15] Wang, C., Xia, K., Zhang, Y., Kaplan, D. L. *Acc. Chem. Res.*, 2019, 52, 2916–2927.

[16] Wang, C. H., Hsieh, C. Y., Hwang, J. C. *Adv. Mater.*, 2011, 23, 1630–1634.

[17] Zhu, B., Wang, H., Leow, W. R., Cai, Y., Loh, X. J., Han, M. Y., Chen, X. *Adv. Mater.*, 2016, 28, 4250–4265.

[18] Lee, J. H., Kwak, H. W., Park, M. H., Hwang, J., Kim, J. W., Jang, H. W., Jin, H. J., Lee, W. H. *Org. Electron.*, 2018, 59, 213–219.

[19] Tan, M. J., Owh, C., Chee, P. L., Kyaw, A. K. K., Kai, D., Loh, X. J. *J. Mater. Chem. C*, 2016, 4, 5531–5558.

[20] Moreno, S., Baniasadi, M., Mohammed, S., Mejia, I., Chen, Y., Quevedo-Lopez, M. A., Kumar, N., Dimitrijevich, S., Minary-Jolandan, M. *Adv. Electron. Mater.*, 2015, 1, 1–8.

[21] Sastry, T.*Asian J. Biomed. Pharm. Sci.*, 2014, 4, 11–14.

[22] Liu, Y., Xie, Y., Liu, Y., Song, T., Zhang, K. Q., Liao, L., Sun, B. *Semicond. Sci. Technol.*, 2015, 30, 104004.

[23] Paquin, F., Rivnay, J., Salleo, A., Stingelin, N., Silva, C. *J. Mater. Chem. C*, 2015, 3, 10715–10722.

[24] Haq, B. S., Khan, H. U., Dou, Y., Alam, K., Attaullah, S., Zari, I. *Appl. Phys. A Mater. Sci. Process.*, 2015, 120, 1415–1425.

[25] Tovar, J. D. *Acc. Chem. Res.*, 2013, 46, 1527–1537.

[26] Torculas, M., Medina, J., Xue, W., Hu, X. *ACS Biomater. Sci. Eng.*, 2016, 2, 1211–1223.

[27] Ordinario, D. D., Phan, L., Walkup IV, W. G., Jocson, J. M., Karshalev, E., Hüsken, N., Gorodetsky, A. A. *Nat. Chem.*, 2014, 6, 596–602.

[28] Qin, G., Dennis, P. B., Zhang, Y., Hu, X., Bressner, J. E., Sun, Z., Crookes-Goodson, W. J., Naik, R. R., Omenetto, F. G., Kaplan, D. L. *J. Polym. Sci. Part B Polym. Phys.*, 2013, 51, 254–264.

[29] Amsden, J. J., Domachuk, P., Gopinath, A., White, R. D., Negro, L. D., Kaplan, D. L., Omenetto, F. G. *Adv. Mater.*, 2010, 22, 1746–1749.

[30] Srisawasdi, T., Petcharoen, K., Sirivat, A., Jamieson, A. M. *Mater. Sci. Eng. C*, 2015, 56, 1–8.

[31] Bucciarelli, A., Mulloni, V., Maniglio, D., Pal, R. K., Yadavalli, V. K., Motta, A., Quaranta, A. *Opt. Mater. (Amst).*, 2018, 78, 407–414.

[32] Samal, S. K., Dash, M., Shelyakova, T., Declercq, H. A., Uhlarz, M., Bañobre-López, M., Dubruel, P., Cornelissen, M., Herrmannsdörfer, T., Rivas, J., Padeletti, G., De Smedt, S., Braeckmans, K., Kaplan, D. L., Dediu, V. A. *ACS Appl. Mater. Interfaces*, 2015, 7, 6282–6292.

[33] Zheng, H., Zuo, B. *J. Mater. Chem. B*, 2021, 9, 1238–1258.

[34] Kang, M., Jin, H. J. *Colloid Polym. Sci.*, 2007, 285, 1163–1167.

[35] Malinowski, C., He, F., Zhao, Y., Chang, I., Hatchett, D. W., Zhai, S., Zhao, H. *RSC Adv.*, 2019, 9, 40792–40799.

[36] He, F., You, X., Gong, H., Yang, Y., Bai, T., Wang, W., Guo, W., Liu, X., Ye, M. *ACS Appl. Mater. Interfaces*, 2020, 12, 6442–6450.
[37] Reizabal, A., Gonçalves, R., Fidalgo-Marijuan, A., Costa, C. M., Pérez, L., Vilas, J. L., Lanceros-Mendez, S. *J. Memb. Sci.*, 2020, 598, 117678.
[38] Johari, N., Madaah Hosseini, H. R., Samadikuchaksaraei, A. *Mater. Sci. Eng. C*, 2017, 79, 783–792.
[39] Yao, M., Su, D., Wang, W., Chen, X., Shao, Z. *ACS Appl. Mater. Interfaces*, 2018, 10, 38466–38475.
[40] Zhuang, X., Huang, W., Yang, X., Han, S., Li, L., Yu, J. *Nanoscale Res. Lett.*, 2016, 11.
[41] Luo, Y., Pei, Y., Feng, X., Zhang, H., Lu, B., Wang, L. *Mater. Lett.*, 2020, 260, 126945.
[42] Jia, X., Wang, C., Ranganathan, V., Napier, B., Yu, C., Chao, Y., Forsyth, M., Omenetto, F. G., Macfarlane, D. R., Wallace, G. G. *ACS Energy Lett.*, 2017, 2, 831–836.
[43] Min, K., Umar, M., Seo, H., Yim, J. H., Kam, D. G., Jeon, H., Lee, S., Kim, S. *RSC Adv.*, 2017, 7, 574–580.
[44] Lima, R. M. A. P., de Oliveira, M. C. A., de Oliveira, H. P. *Electrochem*, 2020, 1, 344–357.
[45] Andiappan, M., Sundaramoorthy, S., Panda, N., Meiyazhaban, G., Winfred, S. B., Venkataraman, G., Krishna, P. *Prog. Biomater.*, 2013, 2, 6.
[46] Kapoor, S., Kundu, S. C. *Acta Biomater.*, 2016, 31, 17–32.
[47] Garrido, I., Aznar-Cervantes, S., Aliste, M., Yáñez-Gascón, M. J., Vela, N., Cenis, J. L., Navarro, S., Fenoll, J. *Catalysts*, 2020, 10, 1–15.
[48] Srivastava, C. M., Purwar, R., Kannaujia, R., Sharma, D. *Fibers Polym.*, 2015, 16, 1020–1030.
[49] Zhu, B., Wang, H., Leow, W. R., Cai, Y., Loh, X. J., Han, M. Y., Chen, X. *Adv. Mater.*, 2016, 28, 4250–4265.
[50] Gianak, O., Pavlidou, E., Sarafidis, C., Karageorgiou, V., Deliyanni, E. *Separations*, 2018, 5.
[51] Janani, G., Kumar, M., Chouhan, D., Moses, J. C., Gangrade, A., Bhattacharjee, S., Mandal, B. B. *ACS Appl. Bio Mater.*, 2019, 2, 5460–5491.
[52] Ribeiro, M., Ferraz, M. P., Monteiro, F. J., Fernandes, M. H., Beppu, M. M., Mantione, D., Sardon, H. *Nanomedicine Nanotechnology, Biol. Med.*, 2017, 13, 231–239.
[53] Chen, C., Hu, L. *Acc. Chem. Res.*, 2018, 51, 3154–3165.
[54] Hoeng, F., Denneulin, A., Bras, J. *Nanoscale.*,8, 2016, 13131–13154.
[55] Xu, J., Zhao, X., Wang, Z., Xu, H., Hu, J., Ma, J., Liu, Y. *Small.*, 2019, 15, 1803970.
[56] Costanza, V., Bonanomi, L., Moscato, G., Wang, L., Suk Choi, Y., Daraio, C. *Appl. Phys. Lett.* 2019, 115, 193702.
[57] Jin, J., Lee, D., Im, H. G., Cheol Han, Y., Gyo Jeong, E., Rolandi, M., Cheol Choi, K., Bae, B. S. *Adv. Mater.*, 2016, 28, 5169–5175.
[58] Birge, R. R., Gillespie, N. B., Izaguirre, E. W., Kusnetzow, A., Lawrence, A. F., Singh, D., Song, Q. W., Schmidt, E., Stuart, J. A., Seetharaman, S., Wise, K. J. *J. Phys. Chem. B*, 1999, 103, 10746–10766.
[59] Shi, L., Xu, X., Ma, M., Li, L. *Appl. Phys. Lett.*, 2014, 104, 1–5.
[60] Feng, Y., Li, X., Li, M., Ye, D., Zhang, Q., You, R., Xu, W. *ACS Sustain. Chem. Eng.*, 2017, 5, 6227–6236.
[61] Aziz, S. B., Marf, A. S., Dannoun, E. M. A., Brza, M. A., Abdullah, R. M. *Polymers (Basel).*, 2020, 12, 1–17.
[62] Posati, T., Benfenati, V., Sagnella, A., Pistone, A., Nocchetti, M., Donnadio, A., Ruani, G., Zamboni, R., Muccini, M. *Biomacromolecules*, 2014, 15, 158–168.
[63] Wang, C., Xia, K., Zhang, M., Jian, M., Zhang, Y. *ACS Appl. Mater. Interfaces*, 2017, 9, 39484–39492.
[64] Ranjana, R., Parushuram, N., Harisha, K. S., Asha, S., Narayana, B., Mahendra, M., Sangappa, Y. *J. Mater. Sci. Mater. Electron.*, 2020, 31, 249–264.
[65] Park, M. H., Kim, J., Lee, S. C., Cho, S. Y., Kim, N. R., Kang, B., Song, E., Cho, K., Jin, H. J., Lee, W. H. *RSC Adv.*, 2016, 6, 5904–5914.
[66] Jung, R., Kim, H. S., Kim, Y., Kwon, S. M., Lee, H. S., Jin, H. J. *J. Polym. Sci. Part B Polym. Phys.*, 2008, 46, 1235–1242.
[67] Shivananda, C. S., Lakshmeesha Rao, B., Sangappa *J. Mater. Sci. Mater. Electron.*, 2020, 31, 41–51.
[68] Zhang, H., Zhao, J., Xing, T., Lu, S., Chen, G. *Polymers (Basel).*, 2019, 11.
[69] Tsao, S. H., Wan, D., Lai, Y. S., Chang, H. M., Yu, C. C., Lin, K. Te, Chen, H. L. *ACS Nano*, 2015, 9, 12045–12059.

[70] Pal, R. K., Kurland, N. E., Wang, C., Kundu, S. C., Yadavalli, V. K. *ACS Appl. Mater. Interfaces*, 2015, 7, 8809–8816.
[71] Urn, I. C., Kim, T. H., Kweon, H. Y., Ki, C. S., Park, Y. H. *Macromol. Res.*, 2009, 17, 785–790.
[72] Kaewpirom, S., Boonsang, S. *RSC Adv.*, 2020, 10, 15913–15923.
[73] Wang, H., Zhu, B., Wang, H., Ma, X., Hao, Y., Chen, X. *Small*, 2016, 12, 3360–3365.
[74] Xia, Y., Yun Lu *Compos. Sci. Technol.*, 2008, 68, 1471–1479.
[75] Yadav, R., Purwar, R. *J. Mater. Sci. Mater. Electron.*, 2020.
[76] Naskar, D., Bhattacharjee, P., Ghosh, A. K., Mandal, M., Kundu, S. C. *ACS Appl. Mater. Interfaces*, 2017, 9, 19356–19370.
[77] Ayutsede, J., Gandhi, M., Sukiraga, S., Ye, H., Hsu, C. M., Gogotsi, Y., Ko, F. *Biomacromolecules*, 2006, 7, 208–214.
[78] Wang, Y., Ma, R., Hu, K., Kim, S., Fang, G., Shao, Z., Tsukruk, V. V. *ACS Appl. Mater. Interfaces*, 2016, 8, 24962–24973.
[79] Cohen-Karni, T., Jeong, K. J., Tsui, J. H., Reznor, G., Mustata, M., Wanunu, M., Graham, A., Marks, C., Bell, D. C., Langer, R., Kohane, D. S. *Nano Lett.*, 2012, 12, 5403–5406.
[80] Jia, M., Qi, N., Dong, Q., Wang, H., Zhang, K. Q., Qi, M. L., Dong, N., Wang, Q., Zhang, H.*IJNT*, 2014,11,1-234
[81] Min, K., Umar, M., Seo, H., Yim, J. H., Kam, D. G., Jeon, H., Lee, S., Kim, S. *RSC Adv.*, 2017, 7, 574–580.
[82] Guo, X., Huang, L., Zhou, X., Chang, Q., Cao, C., Xiao, G., Shi, W. *Adv. Funct. Mater.*, 2020, 30, 1–7.
[83] Dang, Q., Lu, S., Yu, S., Sun, P., Yuan, Z. *Biomacromolecules*, 2010, 11, 1796–1801.
[84] Genovese, M. E., Caputo, G., Nanni, G., Setti, C., Bustreo, M., Perotto, G., Athanassiou, A., Fragouli, D. *ACS Appl. Mater. Interfaces*, 2017, 9, 40707–40715.
[85] Doblhofer, E., Schmid, J., Rieß, M., Daab, M., Suntinger, M., Habel, C., Bargel, H., Hugenschmidt, C., Rosenfeldt, S., Breu, J., Scheibel, T. *ACS Appl. Mater. Interfaces*, 2016, 8, 25535–25543.
[86] Zhang, C., Zhang, Y., Shao, H., Hu, X. *ACS Appl. Mater. Interfaces*, 2016, 8, 3349–3358.
[87] Su, D., Jiang, L., Chen, X., Dong, J., Shao, Z. *ACS Appl. Mater. Interfaces*, 2016, 8, 9619–9628.
[88] Mieszawska, A. J., Llamas, J. G., Vaiana, C. A., Kadakia, M. P., Naik, R. R., Kaplan, D. L. *Acta Biomater.*, 2011, 7, 3036–3041.
[89] Cai, Z. X., Mo, X. M., Zhang, K. H., Fan, L. P., Yin, A. L., He, C. L., Wang, H. S. *Int. J. Mol. Sci.*, 2010, 11, 3529–3539.
[90] Yadav, R., Batra, R., Bansal, P., Purwar, R.*Polym. Int.*, 2022, 71, 74-85.
[91] Seentrakoon, B., Junhasavasdikul, B., Chavasiri, W. *Polym. Degrad. Stab.*, 2013, 98, 566–578.
[92] Kharlampieva, E., Zimnitsky, D., Gupta, M., Bergman, K. N., Kaplan, D. L., Naik, R. R., Tsukruk, V. V. *Chem. Mater.*, 2009, 21, 2696–2704.
[93] Haldorai, Y., Shim, J. J. *Adv. Polym. Sci.*, 2014, 267, 249–282.
[94] Yadav, R., Purwar, R. *Polym. Test.*, 2021, 93, 106916.
[95] Dang, Q., Lu, S., Yu, S., Sun, P., Yuan, Z. *Biomacromolecules*, 2010, 11, 1796–1801.
[96] Sigel, H., Martin, B. Coordinating Properties of the Amide Bond. Stability and Structure of Metal Ion Complexes of Peptides and Related Ligands. *Chem. Rev.*, 1982, 82, 4, 385–426
[97] Ranjan, S., Dasgupta, N., Sudandiradoss, C., Ramalingam, C., Kumar, A. *Int. J. Nanomedicine*, 2018, 13, 47–50.
[98] Manjunath, A., Irfan, M., Anushree, K. P., Vinutha, K. M., Yamunarani, N. *Adv. Mater. Phys. Chem.*, 2016, 06, 263–273.
[99] Liang, T., Makita, Y., Kimura, S. *Polymer (Guildf).*, 2001, 42, 4867–4872.
[100] Parker, S. T., Domachuk, P., Amsden, J., Bressner, J., Lewis, J. A., Kaplan, D. L., Omenetto, F. C. *Adv. Mater.*, 2009, 21, 2411–2415.
[101] Zheng, K., Hu, Y., Zhang, W., Yu, J., Ling, S., Fan, Y. *Nanoscale Res. Lett.*, 2019, 14.
[102] Perotto, G., Zhang, Y., Naskar, D., Patel, N., Kaplan, D. L., Kundu, S. C., Omenetto, F. G. *Appl. Phys. Lett.*, 2017, 111.
[103] Melikov, R. Daniel Aaron Press, Baskaran Ganesh Kumar, Itir Bakis Dogru, Sadra Sadeghi, Mariana Chirea, İskender Yılgör &Sedat Nizamoglu. *Sci. Rep.*, 2017, 7, 7258.

[104] Min, K., Kim, S., Kim, S. *Proc. Natl Acad. Sci.*2017, 114, 6185–6190.

[105] Nguyen, T. P., Nguyen, Q. V., Nguyen, V., Le, T., Le, Q. Van 2019, 1–25.

[106] Dickerson, M. B., Fillery, S. P., Koerner, H., Singh, K. M., Martinick, K., Drummy, L. F., Durstock, M. F., Vaia, R. A., Omenetto, F. G., Kaplan, D. L., Naik, R. R. *Biomacromolecules*, 2013, 14, 3509–3514.

[107] Tao, H., Brenckle, M. A., Yang, M., Zhang, J., Liu, M., Siebert, S. M., Averitt, R. D., Mannoor, M. S., McAlpine, M. C., Rogers, J. A., Kaplan, D. L., Omenetto, F. G. *Adv. Mater.*, 2012, 24, 1067–1072.

[108] Wang, H., Zhu, B., Ma, X., Hao, Y., Chen, X. *Small*, 2016, 12, 2715–2719.

[109] Wang, H., Meng, F., Zhu, B., Ru Leow, W., Liu, Y., Chen, X. *Adv. Mater.* 2015, 27, 7670–7676.

[110] Gao, S., Yi, X., Shang, J., Liu, G., Li, R. W. *Chem. Soc. Rev.*, 2019,48, 1531–1565.

[111] Patil, A. B., Huang, Y., Ma, L., Wu, R., Meng, Z., Kong, L., Zhang, Y., Zhang, W., Liu, Q., Liu, X.Y. *Biosens. Bioelectron.*, 2019, 142, 111595.

[112] Zhang, S., Zhou, Z., Zhong, J., Shi, Z., Mao, Y., Tao, T.H. *Adv. Sci.*, 2020, 7, 1903802.

[113] Wang, Q., Ling, S., Liang, X., Wang, H., Lu, H., Zhang, Y. *Adv. Funct. Mater.* 2019, 29, 1808695.

[114] Chen, F., Lu, S., Zhu, L., Tang, Z., Wang, Q., Qin, G., Yang, J., Sun, G., Zhang, Q., Chen, Q.*J. Mater. Chem. B.* 2019, 7, 1708–1715.

[115] Huang, J., Wang, L., Jin, Y., Lu, P., Wang, L. L., Bai, N., Li, G., Zhu, P., Wang, Y., Zhang, J., Wu, Z., Guo, C. F. *Adv. Funct. Mater.* 2020, 30, 2001518.

[116] Tao, H., Hwang, S. W., Marelli, B., An, B., Moreau, J.E., Yang, M., Brenckle, M.A., Kim, S., Kaplan, D.L., Rogers, J.A., Omenetto, F.G. *Proc. Natl Acad. Sci.* 2014, 111, 17385–17389.

[117] Wang, Y., Boero, G., Zhang, X. S. & Brugger, J. *Adv. Mater. Interfaces,* 2020,7, 2000733.

[118] Zhang, X. S., Brugger, J. & Kim, B. *Nano Energy*, 2016, 20, 37–47.

[119] Guo, Y., Zhang, X., Wang, Y., Gong, W., Zhang, Q., Wang, H., Guo, Y., Zhang, X. S., Wang, Y., Gong, W., Brugger, J. *Nano Energy* 2018, 48, 152–160.

[120] Seo, J. W., Kim, H., Kim, K. H., Choi, S. Q. & Lee, H. J. *Adv. Funct. Mater.* 2018, 28, 1800802.

[121] Konstantaki, M., Skiani, D., Vurro, D., Cucinotta, A., Selleri, S., Secchi, A., Iannotta, S., Pissadakis, S. *IEEE PhotonicsTechnol. Lett.* 2020, 32, 514–517.

[122] Burke, K. A., Brenckle, M. A., Kaplan, D. L. & Omenetto, F. G. *ACS Appl. Mater. Interfaces* 2016, 8, 16218–16226.

[123] Wang, Q., Jian, M., Wang, C. & Zhang, Y. *Adv. Funct. Mater.* 2017, 27, 1605657.

[124] Mukherjee, C., Hota, M. K., Naskar, D., Kundu, S. C. & Maiti, C. K. *Phys. StatusSolidi A,*2013, 210, 1797–1805.

[125] Xing, X., Chen, M., Gong, Y., Lv, Z., Han, S. T., Zhou, Y. *Sci. Technol. Adv. Mater.* 2020, 21, 100–121.

[126] Gao, X., Gou, J., Zhang, L., Duan, S., Li, C.*RSC Adv*. 2018, 8, 8181–8189.

[127] Moon, B. M., Choi, M. J., Sultan Md. T., Yang, J. W., Ju, H.W., Lee, J. M., Park, H. J., Park, Y.R., Kim, S. H., Kim, D. W., Lee, M. C., Jeong, J. Y., Lee, O. J., Sung, G. Y., Park, C.H. *J. Biomed. Mater. Res. Part B Appl. Biomater.* 2017, 105, 2136–2144.

[128] Fernández-Luna, V. Juan P. Fernández-Blázquez, Miguel A. Monclús, O Francisco Javier Rojo, Rafael Daza, Daniel Sanchez-de Alcazar, Aitziber L. CortajarenaandRubén D. Costa. *Mater. Horiz.* 2020, 7, 1790–1800.

[129] Srisawasdi, T., Petcharoen, K., Sirivat, A., Jamieson, A. M. *Mater. Sci. Eng. C.* 2015, 56, 1–8.

[130] Fengel, C. V., Bradshaw, N. P., Severt, S. Y., Murphy, A. R. Leger, J. M. *Smart Mater. Struct.* 2017, 26, 055004.

[131] Reizabal, A., Correia, D. M., Costa, C. M., Perez-Alvarez, L., Vilas-Vilela, J.L., Lanceros-Méndez, S.*ACS Appl. Mater. Interfaces,* 2019,11, 30197–30206.

Chapter 6

Silk Fibroin-Based Systems for Conversion and Storage of Energy

V. S. Mohite[1], M. M. Darade[2],
P.A. Patil[1], S. B. Kulkarni[1],
R. D. Kale[1], P. C. Pingale[1] and S. H. Pawar[2,3,*]
[1] Department of Physics,
T. C. College, Baramati, Maharashtra, India
[2] Center for Interdisciplinary Research,
D. Y. Patil University, Kolhapur, Maharashtra, India
[3] Centre for Innovative and Applied Research (CIAR),
T. C. College, Baramati, Maharashtra, India

Abstract

This book chapter gives an overview of the research that has been done on how to use smart textiles, especially those made from silk fibroin, to harvest energy from renewable sources in the human body and in its environment. We start with a short explanation of how important smart textiles are in the coming energy crisis, pollution of the environment and the need for small-scale energy devices to be made for public health. We examine smart textiles in terms of their ability to produce energy in various forms like biomechanical, body heat, biochemical and solar energy, along with the hybrid forms. We discuss the variety of forms of energy generation technologies like triboelectric generators, piezoelectric generators and thermoelectric energy conversion gazettes. Since the storage of energy has equal importance as that of energy generation, we have discussed in depth the application of silk fibroin in the supercapacitors. Finally, we offer a critical examination of smart fabrics, as well as insights into future difficulties and prospects. The improvements in chemistry and materials mentioned in this book chapter will push the boundaries of smart textiles, which will soon alter our lives in the Internet of Things age, with global efforts.

Keywords: triboelectric generator, piezoelectric generator, thermoelectric generator, supercapacitor, energy storage

* Corresponding Author's Email: shpawar1946@gmail.com.

In: Silk Fibroin
Editor: Shivaji H. Pawar
ISBN: 979-8-88697-402-7

1. Introduction

The early development of industrial revolution was fueled by traditional energy sources such as coal, wood and biomass. Then came the era of natural gas and oil. However, the technological advancement and rate of development soon dwarfed the existing supply exerting tremendous pressure on the limited energy source. The world population in this period also increased exponentially, thereby compounding the problem. Pollution from these sources is another factor, which should also be considered. Not only is pollution hazardous to human and animal life, but it also has a severe effect on the ecological balance of the planet itself. The CO_2 emission from fossil fuels has a substantial role in global warming, which consequently is responsible for worldwide climate change. The current CO_2 levels are dangerously close to effect a rise in global temperature by 2°C. The average rise in 2°C will read catastrophe to the planet's climatological balance. Clean energy options such as wind, solar, geothermal, tidal, etc. should be considered seriously.

Silks can be used as components in flexible electronic devices, as described in the previous chapters of this book. Surprisingly, new studies reveal that silk can convert energy into the energy needed for these devices. The possibilities of converting mechanical, thermal, chemical and solar energy into electrical energy, as well as the significance of silk materials in energy storage devices, are discussed in this chapter.

2. Overview of the Energy Crisis and the Need for Smart Textiles to Capture Energy

Energy conversion and storage have become more important in the recent years, among many other things. White's law says that, if everything else stays the same, "culture changes as the amount of energy captured per person per year goes up or as the efficiency of the tools used to put the energy to work goes up."

The term "culture" refers to the whole of all human cultural activity worldwide, including its technological, sociological, and ideological facets. Current state of affairs indicates that more energy is being produced and consumed than in the past, and this tendency is likely to continue in the years to come. This means that technology will have to develop solutions to the energy crisis and society, will have to evolve to gather more energy and use it wisely.

Traditional energy sources such as coal, wood, and biomass were employed to power the early stages of the industrial revolution. However, the rate of technological growth and development quickly outpaced the available supply, putting enormous strain on the limited energy sources. During this time, the world's population grew at an exponential rate. Another thing is to consider the pollution generated by above sources. Pollution not only endangers human and animal life, but also harms the planet's ecological balance. CO_2 emissions from fossil fuels contribute significantly to global warming and thus to global climate change. CO_2 levels are terrifyingly close to causing a 2°C rise in global temperature. The climatological balance of the world will be ruined if the average rise of 2°C continues.

As the world enters the second year of the Covid-19 epidemic, the annual Global Energy Review assesses the path of energy demand and carbon dioxide emissions in 2021. The most recent statistical data and in-the-moment analysis confirm earlier projections for 2020 energy use and CO_2 emissions while demonstrating how both are rising. As the world transitions to the Internet of Things (IoT), wireless sensor networks, big data, robots, and artificial intelligence, among other factors, are pushing people to use a new kind of pervasive energy to sustainably power billions of scattered devices. [1-4]. The need for a universal energy solution exceeds the capability of current centralized power supply systems, which presents a formidable obstacle for people to overcome [5]. The obvious choice for a widespread energy solution is a battery [6–8]. For customized healthcare, information technology, and communication, wearable electronics, particularly skin electronics, are crucial elements of the Internet of Things (IoT) [9–11]. On the other hand, the following factors will hinder the widespread use of battery systems for wearable technology. First off, skin interfaces shouldn't be used with devices that have heavy, rigid battery systems [12–15]. Second, the bulk of battery systemizes are made up of harmful compounds, limiting their ability to be used in bio-integrated applications and posing environmental dangers [16-18]. Finally, because batteries have a limited life span, they are not suitable for use in widely spread device systems [19-21]. As a result, a ubiquitous, ecologically friendly, and long-term energy solution to power body electronics is desperately in demanding need. The human body and its environment, which includes biomechanical motions, body heat, bodily fluids, and sunlight, are abundant renewable energy sources. Figure 1 shows schematic representation of the human body and its surroundings, including biochemical activity, body heat, and the potential for harvesting biofuels from the body through smart textiles composed of silk fibroins. [22-24].

Body heat and biomechanical motions, for example, have energy values of up to 4.8 W and 67 W [25, 26], respectively, whereas body fluids have energy values of up to 100 W [27]. Aside from the body's energy, the environment is abundant with renewable energy sources, such as solar irradiance, which has an energy density of upto 100 mW/cm^2 [28, 29]. Modern on-body electronics have lower energy requirements of 200 W^{-1} W [30], which can be met by harvesting energy from the human body. To gather this type of energy, various working mechanisms and wearable device designs have emerged in recent years, including triboelectric nanogenerators (TENGs), [31, 32] piezoelectric nanogenerators (PENGs), [33, 34] thermoelectric generators (TEGs), [35, 36] biofuel cells (BFCs), [37, 38] solar cells (SCs) [39, 40], and hybrid generators (HGs) [41, 42].

3. Silk Fibroin-Based Conversion of Energy

As mentioned in the earlier chapters of this book, silks can be employed as parts of flexible electronic devices. Incredibly recent investigations have also proven that silk; can transform the renewable energy sources represented in Figure 1 into the energy required for the devices used in applications related to healthcare. This chapter discusses the potential for turning mechanical, thermal, chemical, and solar energy into electrical energy, as well as the role of silk materials in energy storage systems.

3.1. Mechanical Energy Conversions with SF

Mechanical energy from wind and water motion has been harnessed for millennia to power ships, windmills, and watermills. After the industrial revolution, generators were mounted to wind or water-driven turbines to convert kinetic energy to electrical energy. Hydroelectric dams and windmill farms have since made significant contributed to the global energy mix. [43] While these primary mechanical energy sources have been extensively exploited, less attention has been dedicated to alternative kinds of exploitable mechanical energy such as triboelectric and piezoelectric generators. Silk fibroin and its bio-nanocomposites exhibit both of these properties and have lately been utilized in the developing these technologies, notably for wearable electronics used in health care applications.

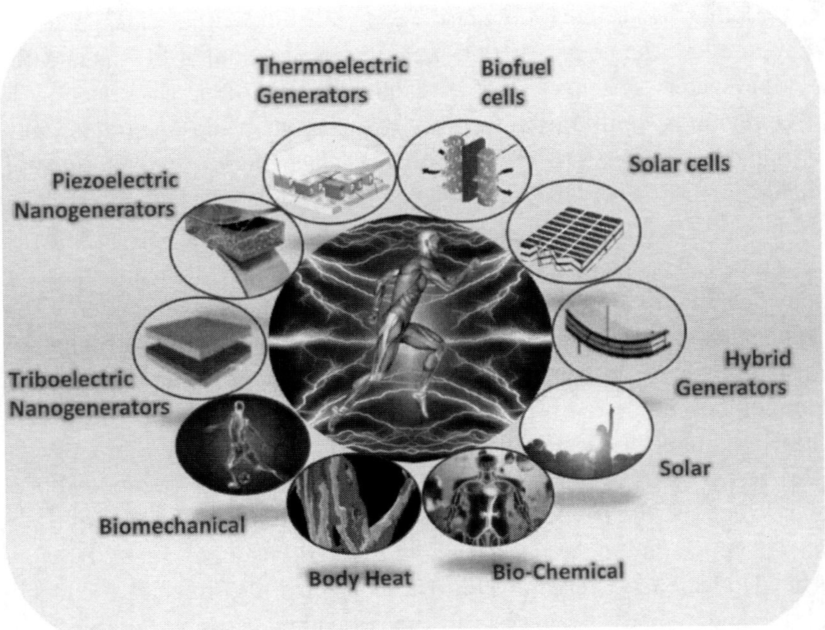

Figure 1. Schematic representation of the human body and its surroundings, including biochemical activity, body heat, and the potential for harvesting biofuels from the body through smart textiles composed of silk fibroins.

3.1.1. Triboelectric Generators with SF

The material with the higher surface electron affinity pulls electrons to its surface, making it negatively charged, in contrast the other material becomes positively charged when two dissimilar materials come into contact. This event is brought on by the triboelectric effect. A high enough voltage can be produced when two sufficiently triboelectrically distinct materials are rubbed together to electrical power equipment [44]. A different group of researchers have attempted to design, fabricate and test the performance of triboelectric generators. Some of these are summarized below.

In their investigation, Fan et al. [45] taped together a rectangle of PET (220 m) film measuring 4.5 cm by 1.2 cm and a Kapton (125 m thick) film on the 1.2 cm edges to create a triboelectric nanogenerator. The upper and bottom sections of the electrode layers were then sputtered-coated with 100 nm gold. Charge separation resulted from the inner surface of the films rubbing against one another as they were bent. They measured output voltages of upto 3.3 V with a 10.4 $mWcm^3$ power density, enough to light an LED. In addition, 10 capacitors (22F) were charged to 0.2 V in 450 deformation cycles using the nanogenerator. By maintaining a steady voltage output even after 105 cycles of deformation, the nanogenerators' durability was demonstrated (spanning about 30 days).

Another example of a triboelectric generator quoted here is based on silk fibroin. Silk fibroin is a good material for making a triboelectric nanogenerator since it has a strong triboelectric behaviour and is just slightly less easy to lose electrons than wool in the triboelectric series [46]. Figure 2 depicts the schematic layout and operation of a typical silk fibroin-based triboelectric generator.

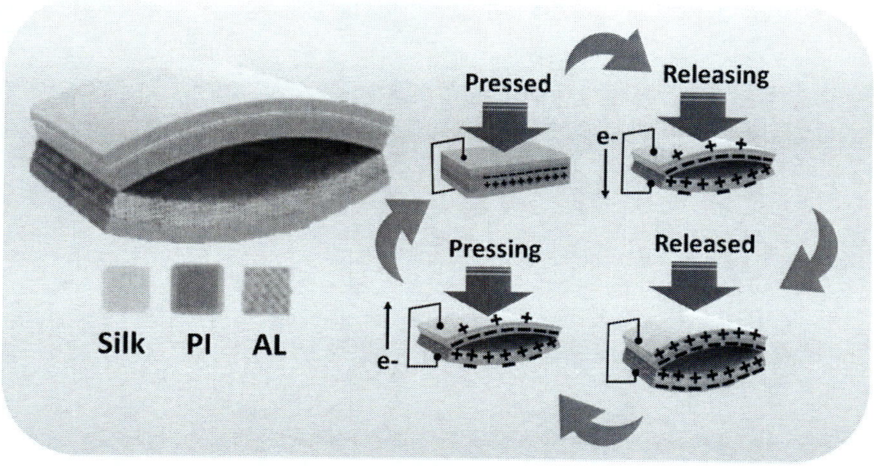

Figure 2. A schematic structure of silk-based triboelectric generator and its working principle.

The working principle can be seen in the figure as consisting of four steps:

1. By pressing the various electro affinity layers together, electrons can be moved to the side with greater electron affinity.
2. When the pressure is released, electrons move to the negatively charged side of the circuit through the circuit.
3. Upon complete release, an electric field is created between the two sides due to the opposite charge differential.
4. Reapplying pressure leads the electrons to flow in the other direction [47].

[1]

Once the two sides are joined, the process can be repeated, and with enough iterations, electricity can be produced.

Using silk's inherent triboelectric property, Kim et al. studied the impact of surface area shape on the amount of voltage generated between the two contacting surfaces [48].

The information about the triboelectric generator built using Silk fibroin is as follows and can be shown in Figure 2. One-half of the triboelectric nanogenerator was constructed by joining an electrospun nonwoven mesh or a cast film of B. mori silk to an aluminum foil. A polyimide film was sandwiched between two sheets of aluminium foil to create the other half. The electrospun nonwoven mat nanogenerator had a peak voltage of about 16 V, which was roughly 1.5 times higher than the peak voltage of the cast silk film nanogenerator. Like Fan et al. technology, the silk-based nanogenerator was capable of lighting LEDs, and when loaded with a resistance of 5 M, a peak power density of 4.3 mWm^2 was observed. The nanogenerator was effective and long-lasting, maintaining constant voltage after 25 000 cycles of deformation while charging a 10 F capacitor to 2 V in roughly 5 minutes. The potential to incorporate silk into a triboelectric nanogenerator and obtain effective performance by lighting LEDs is being demonstrated here for the first time.

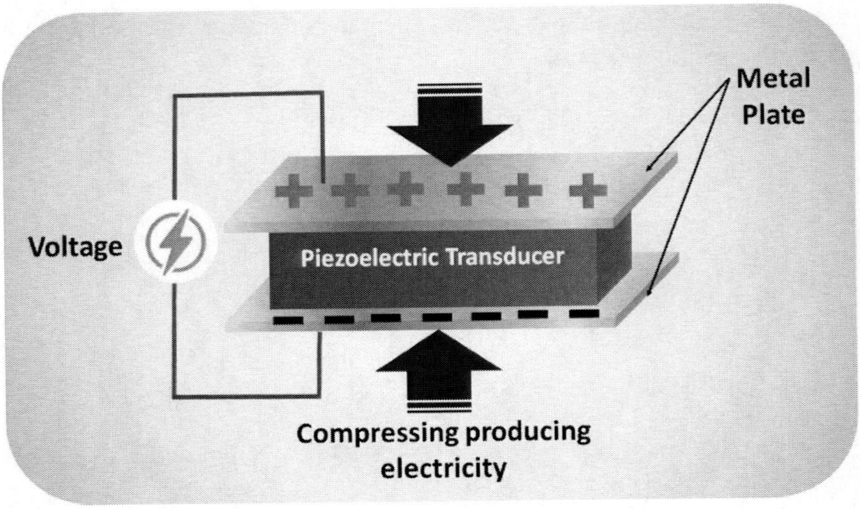

Figure 3. Schematic structure and principle of Piezoelectric generator.

Similarly, Liu et al. [49] looked at the viability of replacing Kim etalless.'s scalable electrospining with spray coating for the mass production of silk film-based triboelectric generators. In the experiment, 30 cm 36 cm PET-ITO films were spray coated with 15 weight per cent aqueous silk solutions for 5 to 15 seconds. At a load resistance of 55 M, the nanogenerators showed V up to 218 V and a peak power density of 68 Wm^{-2}. The triboelectric generators had a very long lifespan; after seven days, they were still produced 300 cycles per minute of continuous output voltage. These investigations have shown that the triboelectric property of silk can be exploited to produce sufficient electrical energy to power existing electrical equipment.

3.1.2. Piezoelectric Generators with SF

A piezoelectric generator is a power generation device that uses the principle of piezoelectricity to transform mechanical energy into electricity. The schematic structure and principle of piezoelectricity generation are shown in Figure 3. It consists of a piezoelectric material sandwitched between two metal plates.

The fundamental idea behind piezoelectricity is the production of an electrical potential difference due to the displacement of atoms in crystals with asymmetries in their lattice structure, on an atomic scale [50]. Silk is a semicrystalline material which is made up of sheets, and it can be distorted by external pressure, it may also be piezoelectric. Piezoelectric materials vibrate, which leads to polarization and depolarization actions that result in the conversion of mechanical energy into electrical energy. When a metal ball struck silk in a glass tube filled with neon gas in 1939, Harvey [51] showed how the piezoelectric effect in silk fibroin works by making it light. But it wasn't until the 2010s that significant research on silk-based piezoelectric materials started.

To power the implantable devices, Kim et al. [52] demonstrated Silk fibroin-based biodegradable composite-type nanogenerators with a controlled lifetime. The well-dispersed lead-free ferroelectric ($BaTiO_3$, $ZnSnO_3$, $Bi_{0.5}(Na_{0.82}K_{0.18})_{0.5}TiO_3$ and $K_{0.5}Na_{0.5}Nb_{0.995}Mn_{0.005}O_3$) nanoparticles are what make up the 2D thin film and 1D wire-type composites. Ag nanowires are employed to improve the nanoparticles' dispersion, while polyvinylpyrrolidone stops the Ag nanowires from tying together. Due to the most excellent piezoelectric coupling figure of merit, maximum output voltages and current densities of 2.2 V and 0.12 mA/cm^2 in the thin film and 1.8 V and 0.1 mA/cm^2 in the wire are obtained when 30 wt per cent KNN:Mn nanoparticles are well-dispersed in the solution. The glycerol content can also be used to alter the characteristics of water-soluble composite films for up to two days. Figure 4 shows schematic diagram of the fabrication process for the silk fibroin-based composite type nanogenerators with ferroelectric nanoparticles.

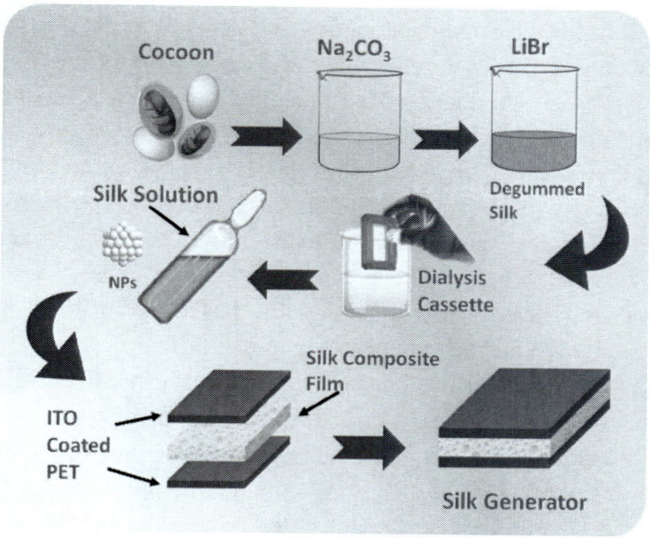

Figure 4. Schematic diagram of the fabrication process for the silk fibroin-based composite type nanogenerators with ferroelectric nanoparticles [52].

One such instance is a study from 2018 by Karan et al. [53]. They assessed the structure-dependent piezoelectric response of silk from the spider species Nephilapilipes and measured out-of-plane piezoelectric coefficients up to 0.36 pmV^{-1}. Then, by incorporating silk fibres into a PDMS film that was oriented in a single direction and sandwiching the composite film between two electrodes that could be connected to electronic equipment, they created a piezoelectric nanogenerator. The nanogenerator was then placed through various stresses and deformations, such as tapping with a finger, colliding with a falling metal block, stretching, and twisting. The maximum output voltage, current, and power density were 21.3 V, 0.68 A, and 3.23 Wcm^{-2}, respectively. The voltage was higher than that of other piezoelectric nano-generators made from viral protein, fish bladder, and fish scales [54-56]

Utilizing a composite made of silk and zinc oxide (ZnO), Lisa Sarkar et al. [57] produced a flexible nanogenerator (NG) by taking advantage of the materials' natural piezoelectric capabilities. Zinc oxide nanoparticles were created using a direct precipitation approach (NPs). The homogeneous silk-ZnO solution was then spin-coated to make the silk-ZnO composite film. Several standard characterization techniques were used to examine the composition and morphology of the silk-ZnO composite. Cell viability testing was also done as part of the composite film's biocompatibility investigation. The creation of a piezoelectric nanogenerator served to illustrate the usefulness of composites as prepared materials. For a particular ZnO concentration, this hybrid nanogenerator was able to produce a maximum open circuit voltage of 25 V (peak to peak value) in the bending condition. The bending angle of the nanogenerator's output response showed a strong association with it. With the nanogenerator, a peak output power density of 6.67 mW cm^3 was attained. The constructed prototype works well to generate energy from human movement and illuminate commercial red LEDs. Additionally, an experimental determination of the piezoelectric coefficient (d33) of the silk-ZnO composite film was made.

3.1.3. Hybrid Piezoelectric-Enhanced Triboelectric Nanogenerator with SF

Triboelectric and piezoelectric nanogenerators (TENG and PENG) allow for the effective conversion of mechanical energy to electricity and have sparked intense study over the last decade [58, 59]. TENGs use friction to convert mechanical energy to electricity via a combined effect of triboelectrification and electrostatic induction. It is typically made up of a single triboelectric pair with corresponding electrodes [60, 61]. PENGs convert mechanical energy to electricity by deforming a piezoelectric material sandwiched between two electrodes, causing a piezoelectric potential to be generated [62]. In reality, with appropriate structure designs, when an external force is applied, both contact friction and material deformation can occur concurrently. Thus, as with typical nanogenerators based exclusively on triboelectricity or piezoelectricity, the applied mechanical energy is under utilized.

An all-fiber hybrid piezoelectric-enhanced triboelectric nanogenerator is reported here, which was created by electrospinning silk fibroin and poly (vinylidene fluoride) (PVDF) nanofibers on conductive textiles. The hybrid nanogenerator has remarkable electrical performance, with a power density of 310 W/cm^2 due to the huge specific surface area of nanofibers and the extraordinary capacity of silk fibroin to donate electrons in triboelectrification. This unique all-fiber arrangement developed by Yinben Guo et al. [63] it has all the advantages of typical fabrics, including a flexible appearance and good air permeability, it can also be custom embedded in clothes in any desired size and shape. Furthermore, the device enables the identification of various types of bodily movements through a link between gesture and related electrical output. As a potential application sector, we envision an autonomous wearable fall alert microsystem for the elderly and others working in high-risk environments.

Compared to other textile-based nanogenerators, the output performance of produced TPNGs exhibits high power levels, with voltage, short-circuit current, and power density of 500 V, 12 A, and 0.31 mW/cm^2, respectively.

3.2. Thermoelectric Energy Conversion and Refrigerator

Thermoelectricity is the direct conversion of heat into electric energy or the opposite. The Seebeck effect ("power production mode"), produces electricity from a temperature gradient, and the Peltier effect ("refrigeration mode"), which produces a temperature gradient when electrical current is applied, are the two fundamental principles of thermoelectricity. To maximize the amount of power produced or the amount of heat absorbed, many of these couples are connected electrically in series and thermally in parallel to form a thermoelectric (TE) module or device, as shown in Figure 5. An essential thermoelectric couple comprises one leg made of n-type material and one leg made of p-type material.

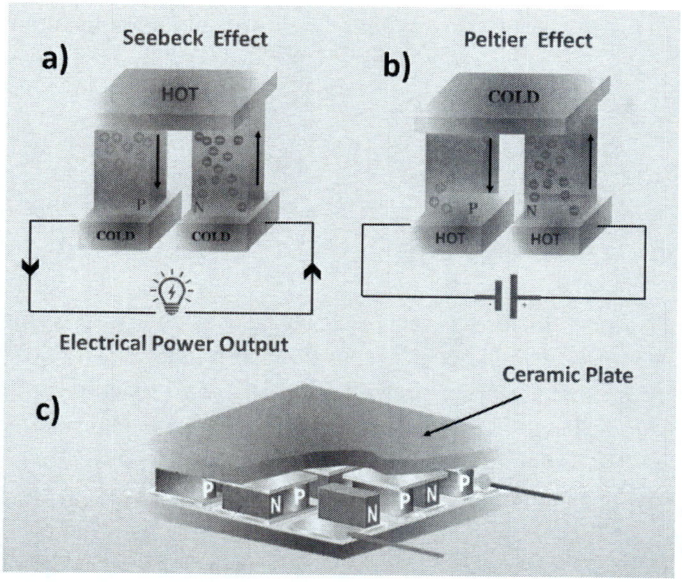

Figure 5. The fundamental thermoelectric module, which is composed of n-type and p-type materials, can function as (a) a power generator, (b) a refrigerator, or (c) a thermoelectric device.

Thermoelectricity, a fundamental phenomenon first observed in the nineteenth century, has been extensively employed in thermocouples to monitor temperature, still in the twentieth century it only found a few specialized uses in power production and refrigeration. The existing state-of-the-art bulk TE materials, such as Bi_2Te_3, PbTe, $CoSb_3$, and SiGe, have been in use for 20 to 30 years; nevertheless, the low ZT of these materials restricts the spread of thermoelectricity to more diverse applications. Due to advancements made in the burgeoning field of nanoscale materials (NTMs) over the previous ten years, this may soon alter. NTMs provide a way to fine-tune thermoelectric conversion efficiency beyond bulk materials thanks to quantum and classical scale effects. It is conceivable to expect that recent developments in NTMs, particularly innovative preparation techniques, novel materials, innovative concepts, and improved understanding, will result in eco-friendly, cost-effective thermoelectricity with higher performance and wider use of TE power generation and refrigeration[64, 65].

Dan-Liang Wen et al. [66] developed a wearable double-chain thermoelectric generator (DC-ThEG) to integrate thermal energy harvesting and capacitance-based multi-functional sensing. The suggested DC-ThEG was created by screen-printing double-chain thermocouples made of $Bi_2Te_{2.7}Se_{0.3}$ (n-type) and Sb_2Te_3 (p-type)-based thermoelectric inks on top of a PI substrate. Silk fibroin was then used to fill the space between the two chains to create a functional sensing layer. This allowed the DC-ThEG to demonstrate an output capability of 151 mV (at T = 50°C), and experimental confirmation of both the DC-outstanding ThEG's electrical output stability and mechanical stability. Additionally, it was successfully confirmed that the device can detect the presence of liquid-state water in the air and the temperature. Experimental evidence suggests that the aforementioned functions coexist while being worn. A 3.3 V output was obtained by collecting thermal energy as an appealing application example, which satisfies the energy supply needs of most wearable devices. Additionally, a commercial calculator was successfully driven by harvesting body heat with DC-ThEGs and storing the generated electricity in capacitors. In conclusion, this work's integration of energy harvesting and multiple-functional sensing opens up a wide range of potential applications for the suggested DC-ThEG.

4. Storage of Energy

4.1. Introduction to the Storage of Energy

The demand for electric vehicles has increased intensely. National governments are likewise moving toward cleaner, renewable energy sources than traditional ones based on fossil-fuels. Currently, batteries are the primary source of energy storage. specific applications necessitate the use of energy storage having a rapid charge and discharge mechanisms. The electrochemical supercapacitors can fill the gap in this case. They feature a high-power density along with high charge-discharge cycle endurance. Only the low energy density issue is a hindrance, which is also being improved due to research aimed at developing new nanostructured materials. Previously, carbon-based capacitors had an energy density of 5-10 Wh kg^{-1}, but this has now increased upto 40-50 Wh kg^{-1} [67] with the addition of new processes tunning materials and their properties. Soon batteries will be substituted with supercapacitors. In the Ragone plot Figure 6 depicts the recent location of energy and power density in energy storage devices.

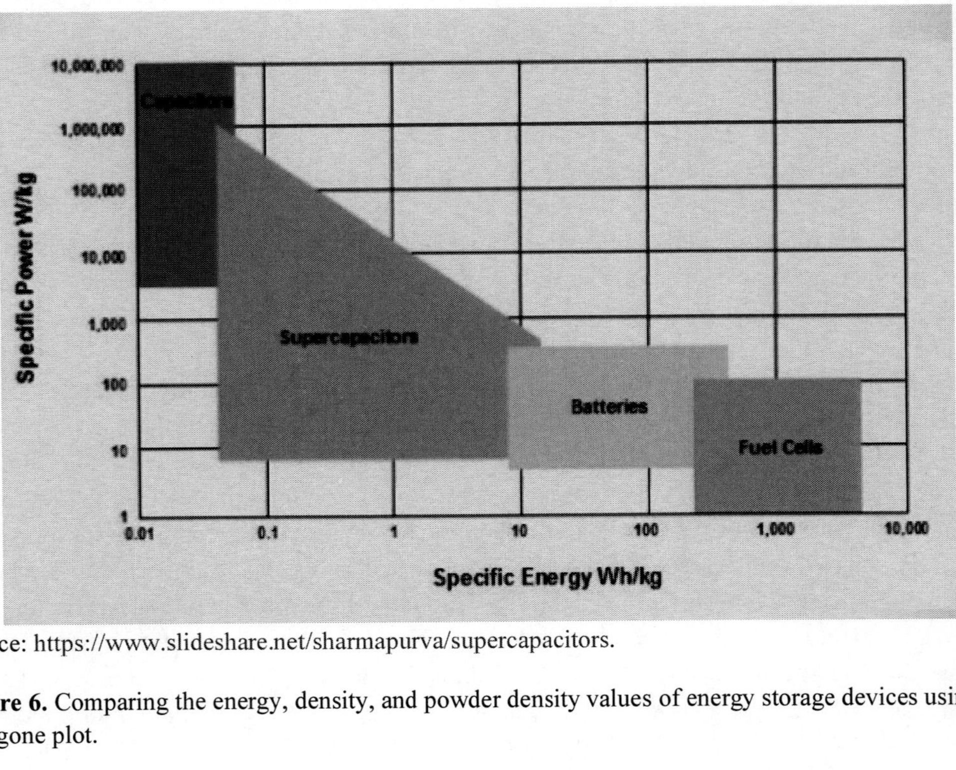

Source: https://www.slideshare.net/sharmapurva/supercapacitors.

Figure 6. Comparing the energy, density, and powder density values of energy storage devices using a Ragone plot.

A Ragone plot compares the efficiency of several energy storage systems. This graph depicts the specific energy (Wh/kg) vs the specific power (W/kg). The horizontal and vertical axes are often on a logarithmic scale, allowing for simple comparison of the performance of various energy storage devices. This was the first time a graph of this type was used to compare battery performance. It can, however, be used to compare various energy storage solutions.

4.2. The Basics of How a Supercapacitor Stores Energy

Capacitors are passive electrical components with two electrodes separated by a dielectric. The charge builds up on the electrodes' surfaces and is separated by the dielectric. The capacitance of a capacitor determines how much electrical energy it can store. Farads are used to measure capacitance (F). The ratio of charge Q to the voltage V delivered to the electrodes is known as capacitance C.

$$C=Q/V \tag{1}$$

C is directly proportional to the electrode surface area A and inversely proportional to the distance d between the electrodes in a typical capacitor:

$$C=\varepsilon_0\varepsilon_r A/d \tag{2}$$

where ε0 is the permittivity of empty space and r is the dielectric constant of the material between the electrodes. A few hundredths of a farad is the capacitance of conventional capacitors. This demonstrates how low the energy density is. The capacitor, however, has a very high-power density. They can expend all of their available energy in a relatively short amount of time without becoming hurt.

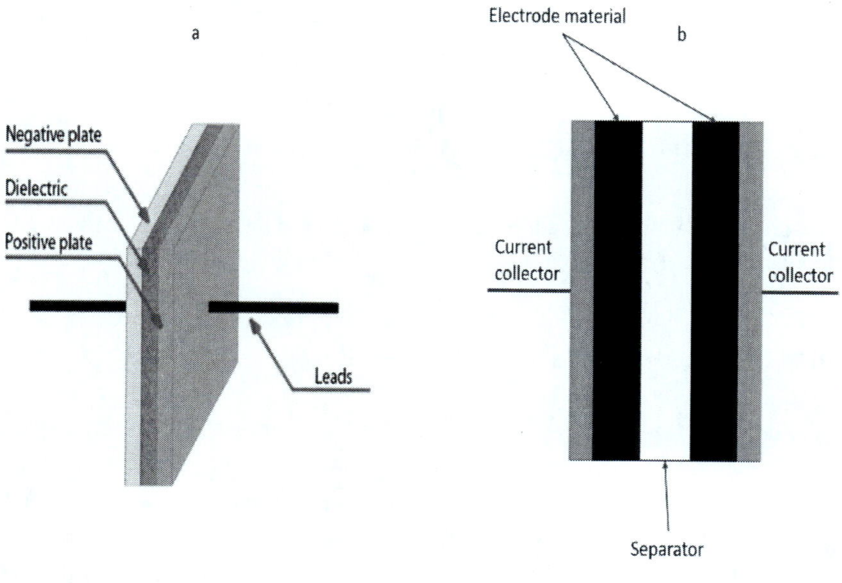

Figure 7. Construction of (a) a conventional capacitor and (b) a supercapacitor.

Batteries are the standard form of energy delivery and storage. Batteries can store more energy with this method, but at the expense of higher power usage. Redox mechanisms, which store charge, are not quick enough to match a capacitor's power density. Due to modern growth and societal energy demands, there is a greater need for devices with high energy and power densities. Electrochemical supercapacitors are a new technology that can close this gap. An electrochemical device known as a supercapacitor can store more charge than regular capacitors and deliver it faster than a battery. A supercapacitor is built differently from a standard capacitor, as shown in Figure 7. In a supercapacitor, an electrolyte-soaked separator separates two electrodes.

Although supercapacitors are built differently, their working principle is very similar to that of capacitors. Porous electrodes increase active surface area, while a thin electrolyte-electrode double layer, also known as the Helmholtz layer, drastically reduces distance. Using equation (2) to analyze these variables shows the possibility of a huge capacitance value.

Electrochemical double layer capacitors (EDLC), pseudocapacitors, and hybrid supercapacitors are the three types of supercapacitors. Figure 8 depicts the classification system. The actual and original supercapacitor is the EDLC supercapacitor. Pseudocapacitors are false supercapacitors with surface redox reaction that minimizes the performance characteristics of a supercapacitor, as the name implies. The first two classes combine to form the hybrid supercapacitor.

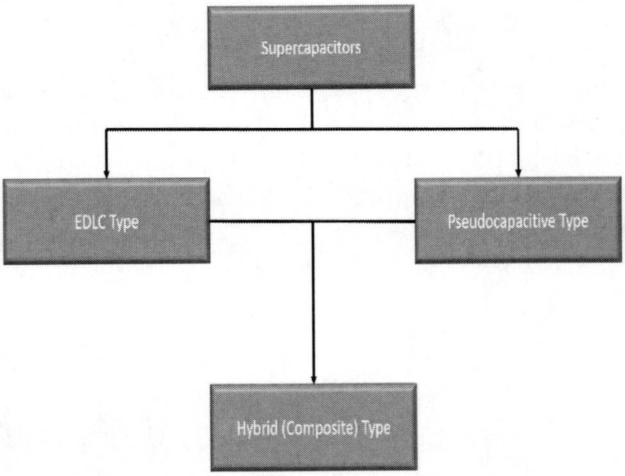

Figure 8. Classification of supercapacitors (EDLC-carbon-based high power density, Pseudocapacitive Metal oxide/hydroxide, polymer based- high energy density, hybrid (EDLC+ Pseudo).

4.3. Utilizing Silk-Based Products for Energy Storage

In addition to creating novel renewable materials for solar and alternative energy devices, there is an increasing demand for energy storage devices with high energy density, long life spans, and affordable material costs. [68] Technology for batteries and capacitors is currently dominated by alkali metals, particularly lithium. The demand for lithium, which is mainly used in lithium-ion batteries, more than quadrupled between 2015 and 2020, and it is anticipated to climb by more than 500% by 2030, reaching 2.2 million tonnes [69, 70]. This rise will only make the rising mineral problem worse when the world makes the switch to more environmentally friendly technologies.

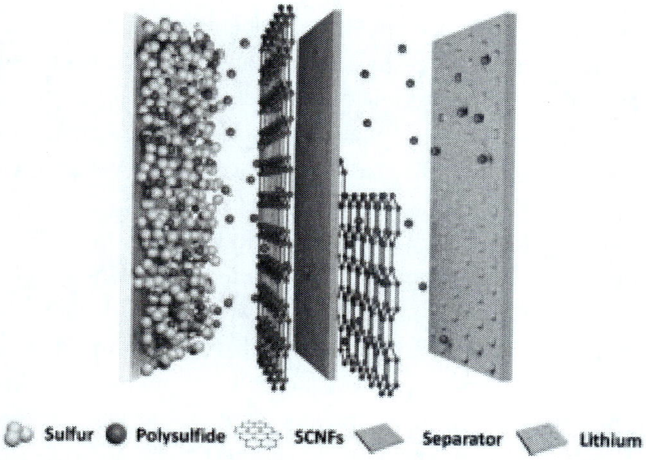

Figure 9. Schematic of a Li-S cell in assembly, with a sulphur cathode made of carbonized silk fibres a separator, an interlayer for the anode, and lithium metal. For the cathode interlayer (left half), a lower impedance was attained, but at the expense of the Li-metal being harmed by polysulfides after numerous cycles. In addition to providing fast electron transport, the addition of the carbonized silk layers between the cathode (on the left) and anode (on the right) also caught polysulfides during charging, preventing corrosion of the lithium interlayer [72].

4.3.1. Carbonized Silk is Used in Batteries and Supercapacitors

In a recent work by Wu et al. [69], regenerated silk nanofiber film was used as an interlayer between Li-S batteries. Figure 9 depicts a double layer of carbonized electrospun silk fibroins sandwiched between the S-cathode and Li-anode. The carbonized silk fibres (SCNFs) were woven into a continuous conductive network that functioned as a cathode and anode interlayer, provided quick electron pathways, and was stable for 200 cycles [69]. This network also offered a high discharge capacity. The battery's columbic efficiency and charge retention life was significantly increased, which suggests that Li-S battery stability may have been much enhanced.

Similarly, Houet al. demonstrated in an earlier study that porous nitrogen-doped carbon nanosheets made from B. mori silk may be utilized to create high-capacity battery anodes and supercapacitors. $FeCl_3$ and $ZnCl_2$ metal salts were used to treat the silk, and it was discovered that they were effective at both dissolving and regenerating natural silk as well as serving as carbonization agent activation sites [71]. The resulting nanosheets developed lamellar like layered structures, as would be expected from silk materials. This system's lamellar-layered structures had both micro and mesoporous, which gave them an extraordinarily high surface area. With a high storage capacity of 1865 mA h g^1 and only 9% stability loss after 10,000 cycles (conventional rechargeable batteries show total stability loss at about 10,000 cycles), this lamellar-layered structure paved the way for the use of organic silk-based batteries in commercial applications [71].

There are other options besides natural silk carbonization for anode/cathode interlayer. The integration of flexibility, biocompatibility, and degradability for energy storage/delivery systems for epidermal and implantable bioelectronics was recently studied by Pal et al.

According to the researchers, B. mori silk might be used in conjunction with conducting polymers like poly (3, 4-ethylenedioxythiophene), polystyrene sulfonate, or graphene biocomposite inks with photo patterning capabilities. These inks might be made into active electrodes and printed on flexible protein sheets in being circumstances with only water as a solvent, eliminating the necessity for many harsh metal ion solvents utilized in previous studies [73]. These patterned thin films displayed good capacitance, power density, and cycling stability over 500 cycles. The thin films were also discovered cytocompatible and completely disintegrated within a month, implying that they may be employed as biodegradable batteries [73].

Supercapacitors, as opposed to conventional lithium ion batteries, have far faster charge rates, can store much more energy, and have a much longer lifespan (100,000 to 1,000,000 million cycles versus 500 to 10,000 for lithium-ion batteries) [74]. These benefits are offset by the supercapacitors' slow self-discharge rate, which renders them unsuitable for long-term energy storage. Furthermore, supercapacitors suffer from gradual voltage loss, which causes the voltage output of capacitors to drop linearly with stored charge. Because of their wide operating temperature range, supercapacitors are still utilized in hybrid vehicles, in conjunction with batteries for hybrid or quick charging energy storage devices, as backup systems, and as power buffers [74, 75]. As a result, it's no surprise that supercapacitors have emerged as a viable study field for silk-based power sources and energy storage devices. One early work [76] focused on synthesising nitrogen-enriched carbon compounds from B. mori silk strands. The authors used heat to convert silk fibroin to nitrogen-containing activated carbon, which they then successfully used in electric double layers for supercapacitors [76]. In a subsequent investigation [77], porous nitrogen-doped carbon-derived films were produced from B. mori silk for use in potassium-ion hybrid capacitors. The resulting nitrogen-rich, porous carbonized silk anode exhibited a power density of 112.6 W kg^1 and could be fully charged in 7 minutes with a discharge period of 2.5 hours.

4.3.2. Self-Assembly of Silk-Graphene for Energy Storage

The main goal of earlier energy storage research was to transform a silk into carbonized silk for use as anodes in batteries or super capacitors. This highlights the adaptability and usefulness of silk as a material, but it also possesses a variety of other characteristics that make it excellent for energy extraction. One such example is the natural ability of silk proteins to self-assemble. According to Wang et al. research graphene and B. mori silk fibroin may be dissolved in KOH to make a solution that produces self-assembling structures that can be freeze-dried to create aerogels and carbonized to create graphene-silk composite materials [78]. The harvested graphene-silk nanofibril composite material demonstrated outstanding electrochemical stability with a capacitance retention ratio of 96% [78]. The composite material also obtained a high energy density, opening the door for its application in the low-cost fabrication of high-performance supercapacitors. Building on earlier work, Yun et al. produced carbon aerogels using silk proteins that had undergone regeneration and graphene oxide. The aerogels were created using a lyophilization/flash freezing process [79]. Silk-graphene composite nanoplates assembled into 3D cryogels due to strong bonds being established between the hydrophilic building blocks of silk proteins and the oxygen functional groups of graphene oxide during the lyophilization process. The materials were stable after 5000 cycles and exhibited a high capacitance of 298 Fg^1. Table 1 compares the energy and power densities and specific capacitance of various silk fibre-based supercapacitors.

Table 1. Performance of several silk fibre -based supercapacitors in terms of energy and power density, as well as specific capacitance

Sr. No	Supercapacitor Devices	Synthesis Method of Electrodes	Substrate	Morphology	Electrolyte	Capacitance		Energy Density (Wh Kg^{-1})	Power Density (W Kg^{-1})	Stability	Ref.
						Individual Electrode	Device				
1.	SC-DC//SC-DC silkworm cocoon carbons (SCs) degumming, (D) activated by $CaCl_2$.	Pretreatment and green activator	Nickel foam	Nanosheets	1 M Na_2SO_4		247 F gl at0.5 A gl	21.06 Wh kgl	225 W kgl	97% after 20 000 cycles at 20 A gl	[80]
2.	MnO_2-Coated carbon fabrics from silk (CS-MnO_2) CS-MnO_2 // (CS-MnO_2)	Hydrothermal method	Silk fabrics	Aligned/twisted	6 M KOH		105.00 F gl at 0.25 A gl	14.58 W h kgl	0.25 kW kgl	101.54% after 10 000 cycles	[81]
3.	Carbonized silk from HPMC-MoO_2 NP hydroxypropyl methyl cellulose	Chemical method	Nickel foam	Fibrous		456 F/g at 0.2 A/g	-	-	-	98.31% after 10 000 cycles	[82]
4.	Co-doped reduced graphene oxide (Co-rGO) Co-rGO porous silk fibroin composites film	Green pathways (microbial strain)	Si Substrate	Porous	EMIm+ BF4 - electrolyte		104 F gl at 0.5 A gl	28.31 Wh kgl	78.24 kW kgl	89% after 10,000 cycles,	[83]
5.	EGaIn/SF ink EGaIn/Silk fibroin (SF) ink Micro-supercapacitors (MSCs)	Chemical method	Polyester mesh screen	Hydrophobicity and rough surface	CMC/Na2SO4 gel electrolyte		35.72 mF cm− 2 at 25µA cm− 2	3.17 µWh cm^{-2}	0.053 mW cm^{-2}	99.2% after 3000 cycles	[84]
6.	(CSF/Ppy) carbonized silk fabric/Polypyrrole	Electro-depositing	Silk fabric	Interconnected fiber	1M Na_2SO_4	4040 $mFcm^{-2}$ at 2 $mAcm^{-2}$		6.88 $mWhcm^{-3}$ to 1.02 $mWhcm^{-3}$	0.04 Wcm^{-3} to 0.21 Wcm^{-3}	88.6% after 1500 Cycles	[85]

Table 1. (Continued)

Sr. No	Supercapacitor Devices	Synthesis Method of Electrodes	Substrate	Morphology	Electrolyte	Capacitance		Energy Density (Wh Kg^{-1})	Power Density (W Kg^{-1})	Stability	Ref.
						Individual Electrode	Device				
7.	PANI @ Ternary solvent (TS) silk fabric PANI@Without TS fabric	Chemical method	Silk fabric	Fibrous	1M H_2SO_4	4091.43 mF/cm ^{2} 1811.43 mF/cm ^{2}	-	-	-	99.54% after 3000 Cycles	[86]
8.	Carbonized silk cloth (SCS) CSC@Ti $_3$ C $_2$ T x core/shell electrodes	Chemical method	Carbon cloth	Nanoflakes	PVA-H $_2$ SO $_4$ hydrogel	362 mF/cm ^{2}	-	13 µWh/cm ^{2}	181 µW/cm ^{2} .	90.0% after 1000 Cycles	[87]
9.	Parallel-like fiber-shaped supercapacitor (PFSS) carboxyl-functionalized multi-walled carbon nanotube (F-MWCNT) PANI@F-MWCNT@silk	Chemical method	Silk yarn	Fibrous	1M H_2SO_4	95.5 mF cm−2 at 0.1 mA	-	1.22 mWh m^{-2}	127 mW m^{-2}	87% after 5000 cycles	[88]
10.	Asymmetrical Fiber-shaped supercapacitors (AFSSCs) AFSSCs-ICS PPy-MWCNTs-silk // graphene tube (HGT)	Chemical method	Silk	Sword-scabbard-like	H_2SO_4/PVA gel electrolyte		2.3 mF cm^{-1} at 0.5 mA	0.8 µWh cm^{-1}	33 mW cm^{-1},	81% after 3000 Cycles	[89]

Conclusion

Green electronics is a relatively new idea gathering a lot of interest in the scientific and industrial communities with the ultimate goal of using organic, biodegradable and self-sustainable technology to replace the standard inorganic battery-powered electronics. Due to its adaptable biodegradability and flexibility, silk is a green material that has been thoroughly investigated for usage in functional electronics. Due to its outstanding biocompatibility and controllable biodegradability, silk—a "green material" made from the plentiful Bombyxmori silkworm—has recently emerged as a leading contender in the quest for green bioelectronics

The flexibility of silk fibroin ranges from 43 kPa to 3.9 GPa and can be converted into various materials, including films, fibres, hydrogels, sponges, and capsules. In addition, the development of silk fibroin and its derivatives can help support large-scale production of green electronics that use silk by being simple to manufacture and using few harmful chemicals. As a result, silk has recently attracted much interest from the scientific community and it is being used to create functional bioelectronics like transient devices drug delivery systems, wearable epidermal sensors, smart textiles, bio-memristors, bio-photonic devices, thermoelectric sensors, and energy storage/harvesting devices. To harvest these forms of energy, several wearable device designs and working mechanisms have been developed recently. These include triboelectric nanogenerators (TENGs), piezoelectric nanogenerators (PENGs), thermoelectric generators (TEGs), biofuels cells (BFCs), solar cells (SCs), and hybrid generators (HGs). Silk-based energy storage devices like supercapacitors have been attempted for continuous nonstop operation of biological devices in health care are of prime importance. The elaborative presentation and in-depth discussion on silk fibroin-based energy conversion and storage of energy made in this chapter will be helpful to researchers in multi faculty interdisciplinary areas.

References

[1] Ashton, K. (2009). That 'Internet of Things' *Thing. RFID J.*, 22, 97–114.

[2] Gubbi, J., Buyya, R., Marusic, S., and Palaniswami, M. (2013). Internet of Things (Iot): A Vision, Architectural Elements, and Future Directions. *Future Gener. Comp. Sy*, 29, 1645–1660.

[3] Lee, I., and Lee, K. (2015). The Internet of Things (IoT): Applications, Investments, and Challenges for Enterprises. *Bus. Horiz.*, 58, 431–440.

[4] Atzori, L., Iera, A., and Morabito, G. (2010). The Internet of Things: A Survey. *Comput. Netw*, 54, 2787–2805.

[5] Wang, Z. L. (2019). Entropy Theory of Distributed Energy for Internet of Things. *Nano Energy*, 58, 669–672.

[6] Grey, C. P, and Tarascon, J. M. (2017). Sustainability and *in Situ* Monitoring in Battery Development. *Nat. Mater.*, 16, 45–56.

[7] Gu, P., Zheng, M. B., Zhao, Q. X., Xiao, X., Xue, H. G. and Pang, H. Rechargeable Zinc-Air Batteries: A Promising Way to Green Energy. *J. Mater. Chem. A*, 2017, 5, 7651–7666.

[8] He, Y. H., Matthews, B., Wang, J. Y., Song, L., Wang, X. X., and Wu, G. (2018). Innovation and Challenges in Materials Design for Flexible Rechargeable Batteries: From 1D to 3D. *J. Mater. Chem. A*, 6, 735–753.

[9] Kim, J., Campbell, A. S., de Avila, B. E., and Wang, J. (2019). Wearable Biosensors for Healthcare Monitoring. *Nat. Biotechnol.*, 37, 389–406.

[10] Liu, Y., Pharr, M., and Salvatore, G. A. (2017). Lab-on-Skin: A Review of Flexible and Stretchable Electronics for Wearable Health Monitoring. *ACS Nano*, 11, 9614−9635.

[11] Trung, T. Q, and Lee, N. E. (2016). Flexible and Stretchable Physical Sensor Integrated Platforms for Wearable Human-Activity Monitoringand Personal Healthcare, *Adv. Mater*., 28, 4338−4372.

[12] Zang, Y. P., Zhang, F. J., Di, C. A., and Zhu, D. B. (2015). Advances of Flexible Pressure Sensors toward Artificial Intelligence and Health Care Applications. *Mater. Horiz*., 2, 140−156.

[13] Hussain, A. M., Ghaffar, F. A., Park, S. I., Rogers, J. A., Shamim, A., and Hussain, M. M. (2015). Metal/Polymer Based Stretchable Antenna for Constant Frequency far-Field Communication in Wearable Electronics. *Adv. Funct. Mater*., 25, 6565−6575.

[14] Kim, J., Banks, A., Cheng, H., Xie, Z., Xu, S., Jang, K. I., Lee, J. W., Liu, Z., Gutruf, P., Huang, X., (2015). Epidermal Electronics with Advanced Capabilities in near-Field Communication. *Small*, 11, 906−912.

[15] Kim, J., Banks, A., Xie, Z., Heo, S. Y., Gutruf, P., Lee, J. W., Xu, S., Jang, K. I., Liu, F., Brown, G., et al. (2015). Miniaturized Flexible Electronic Systems with Wireless Power and near-Field Communication Capabilities, *Adv. Funct. Mater*., 25, 4761−4767.

[16] Son, D., Kang, J., Vardoulis, O., Kim, Y., Matsuhisa, N., Oh, J. Y., To, J. W., Mun, J., Katsumata, T., Liu, Y, (2018). An Integrated SelfHealable Electronic Skin System Fabricated via Dynamic Reconstruction of a Nanostructured Conducting Network. *Nat. Nanotechnol*., 13, 1057−1065.

[17] Liu, T., Ren, Y. Y., Shen, Y., Zhao, S. X., Lin, Y. H., and Nan, C. W. (2016). Achieving High Capacity in Bulk-Type Solid-State Lithium Ion Battery Based on $Li_{6.75}La_3Zr_{1.75}Ta_{0.25}O_{12}$ Electrolyte: Interfacial Resistance. *J. Power Sources*, 324, 349−357.

[18] Fu, J., Cano, Z. P., Park, M. G., Yu, A., Fowler, M., and Chen, Z. (2017). Electrically Rechargeable Zinc-Air Batteries: Progress, Challenges and Perspectives. *Adv. Mater*., 29, 1604685.

[19] Kang, D. H., Chen, M., andOgunseitan, O. A. (2013). Potential Environmental and Human Health Impacts of Rechargeable Lithium Batteries in Electronic Waste. *Environ. Sci. Technol*., 47, 5495− 5503.

[20] Li, Y., and Dai, H. (2014). Recent Advances in Zinc-Air Batteries. *Chem. Soc. Rev*., 43, 5257−5275.

[21] J. B., and Park, K. S. (2013). The Li-Ion Rechargeable Battery: A Perspective. *J. Am. Chem. Soc*., 135, 1167−1176.

[22] Lin, M. C., Gong, M., Lu, B., Wu, Y., Wang, D. Y., Guan, M., Angell, M., Chen, C., Yang, J., Hwang, B. J., and Dai, H. (2015). An Ultrafast Rechargeable Aluminium-Ion Battery. *Nature*, 520, 324−328.

[23] Placke, T., Kloepsch, R., Duhnen, S., and Winter, M. (2017). Lithium Ion, Lithium Metal, and Alternative Rechargeable Battery Technologies: The Odyssey for High Energy Density. *J. Solid State Electrochem*., 21, 1939−1964.

[24] Ponrouch, A., Frontera, C., Barde, F., and Palacin, M. R. (2016). Towards a Calcium-Based Rechargeable Battery. *Nat. Mater*., 15, 169−172.

[25] Starner, T. (1996). Human-Powered Wearable Computing. *IBM Syst. J*., 35, 618−629.

[26] Huang, L. B., Bai, G. X., Wong, M. C., Yang, Z. B., Xu, W., andHao, J. H. (2016). Magnetic-Assisted Noncontact Triboelectric Nanogenerator Converting Mechanical Energy into Electricity and Light Emissions. *Adv. Mater*., 28, 2744−2751.

[27] Szczupak, A., Halamek, J., Halamkova, L., Bocharova, V., Alfonta, L., and Katz, E. (2012). Living Battery - Biofuel Cells Operating *in Vivo* in Clams. *Energy Environ. Sci*., 5, 8891−8895.

[28] Kim, S., Vyas, R., Bito, J., Niotaki, K., Collado, A., Georgiadis, A., and Tentzeris, M. M. (2014). Ambient RF Energy-Harvesting Technologies for Self-Sustainable Standalone Wireless Sensor Platforms. *Proc. IEEE*. 1649−1666. (101).

[29] Li, Y., Grabham, N. J., Beeby, S. P., and Tudor, M. J. (2015). The Effect of the Type of Illumination on the Energy Harvesting Performance of Solar Cells. *Sol. Energy*, 111, 21−29.

[30] Tao, X. (2019). Study of Fiber-Based Wearable Energy Systems. *Acc. Chem. Res*., 52, 307−315.

[31] Zhang, H. L., Yang, Y., Su, Y. J., Chen, J., Hu, C. G., Wu, Z. K., Liu, Y., Wong, C. P., Bando, Y., and Wang, Z. L. (2013). Triboelectric Nanogenerator as Self-Powered Active Sensors for Detecting Liquid/Gaseous Water/Ethanol. *Nano Energy*, 2, 693−701.

[32] Zhang, K., Wang, X., Yang, Y., and Wang, Z. L. (2015). Hybridized Electromagnetic-Triboelectric Nanogenerator for Scavenging Biomechanical Energy for Sustainably Powering Wearable Electronics. *ACS Nano*, 9, 3521−3529.

[33] Wu, H., Huang, Y., Xu, F., Duan, Y., and Yin, Z. (2016). Energy Harvesters for Wearable and Stretchable Electronics: From Flexibility to Stretchability. *Adv. Mater.*, 28, 9881−9919.

[34] Yang, B., and Yun, K. S. (2012). Piezoelectric Shell Structures as Wearable Energy Harvesters for Effective Power Generation at Low-Frequency Movement. Sens. *Actuators A: Phys.*, 188, 427−433

[35] Tian, R., Wan, C., Wang, Y., Wei, Q., Ishida, T., Yamamoto, A., Tsuruta, A., Shin, W., Li, S., and Koumoto, K. (2017). A Solution-Processed TiS2/Organic Hybrid Superlattice Film Towards Flexible Thermoelectric Devices. *J. Mater. Chem. A*, 5, 564−570.

[36] Wan, C. L., Tian, R. M., Azizi, A. B., Huang, Y. J., Wei, Q. S., Sasai, R., Wasusate, S., Ishida, T., and Koumoto, K. (2016). Flexible Thermoelectric Foil for Wearable Energy Harvesting. *Nano Energy*, 30, 840−845.

[37] Chen, X. H., Yin, L., Lv, J., Gross, A. J., Le, M., Gutierrez, N. G., Li, Y., Jeerapan, I., Giroud, F., andBerezovska, A., et al. (2019). Stretchable and Flexible Buckypaper-Based Lactate Biofuel Cell for Wearable Electronics. *Adv. Funct. Mater.*, 29, 1905785.

[38] Ogawa, Y., Kato, K., Miyake, T., Nagamine, K., Ofuji, T., Yoshino, S., and Nishizawa, M. (2015). Organic Transdermal Iontophoresis Patch with Built-in Biofuel Cell. *Adv. Healthcare Mater.*, 4, 506−510.

[39] Bernechea, M., Miller, N. C., Xercavins, G., So, D., Stavrinadis, A., and Konstantatos, G. (2016). Solution-Processed Solar Cells Based on Environmentally Friendly $AgBiS_2$ Nanocrystals. *Nat. Photonics*, 10, 521−525.

[40] Duan, C., Zhang, K., Zhong, C., Huang, F., and Cao, Y. (2013). Recent Advances in Water/Alcohol-Soluble Pi-Conjugated Materials: New Materials and Growing Applications in Solar Cells. *Chem. Soc. Rev.*, 42, 9071−9104.

[41] Deng, F., Yue, X., Fan, X., Guan, S., Xu, Y., and Chen, J. (2019). Multisource Energy Harvesting System for a Wireless Sensor Network Node in the Field Environment. *IEEE Internet Things J.*, 6, 918−927.

[42] Habibzadeh, M., Hassanalieragh, M., Ishikawa, A., Soyata, T., and Sharma, G. (2017). Hybrid Solar-Wind Energy Harvesting for Embedded Applications: Supercapacitor-Based System Architectures and Design Tradeoffs. *IEEE Circuits Syst. Mag.*, 17, 29−63.

[43] Statistics, I. (2020). Internation Energy Agency, Key World Energy Statistics, Paris 2020.

[44] Shaw, P. E. (1917). Experiments on tribo-electricity. I.—Thetribo-electric series, *Proc. R. Soc. London, Ser. A*, 94, 16.

[45] Fan, F.-R., Tian, Z.-Q., and Wang, Z. L. (2012). Flexible triboelectric generator, *Nano Energy*, 1, 328.

[46] Wang, Z. L. (2013). Triboelectric Nanogenerators as New Energy Technology for Self-Powered Systems and as Active Mechanical and Chemical Sensors, *ACS Nano*, 7, 9533.

[47] Kim, H.J., Kim, J.H., Jun, K.W., Kim, J.H., Seung, W.C., Kwon, O. H., Park, J.Y., Kim, S.W., and Oh, I.K. (2016). Silk Nanofiber-Networked Bio-Triboelectric Generator: Silk Bio-TEG, *Adv. Energy Mater.* 6, 1502329.

[48] Ju Hyun Lee, Insang Yu, Seung Hyun, JinKon Kim, andUnyongJeong. (2017). Remarkable Increase in Triboelectrification by Enhancing the Conformable Contact and Adhesion Energy with a Film-Covered Pillar Structure, *Nano Energy*, 34.

[49] Liu, C., Li, J., Che, L., Chen, S., Wang, Z., and Zhou, X. (2017). Toward large-scale fabrication of triboelectric nanogenerator (TENG) with silk-fibroin patches film via spray-coating process, *Nano Energy*, 41, 359.

[50] Ikeda, T. (1996). *Fundamentals of Piezoelectricity*, Oxford University Press, Oxford, UK.

[51] Harvey, E. N. (1939). The Luminescence of Adhesive Tape, *Science*, 89, 460.

[52] Kyeong Nam Kima, Jinsung Chuna, Song A. Chaeb, Chang Won Ahnb, Ill Won Kimb, Sang-Woo Kimc, Zhong Lin Wangd, Jeong Min Baik. (2015). Silk fibroin-based biodegradable piezoelectric composite nanogenerators using lead-free ferroelectric nanoparticles, *Nano Energy*, 14, 87-94.

[53] Karan, S. K., Maiti, S., Kwon, O., Paria, S., Maitra, A., Si, S. K., Kim, Y., Kim, J. K., and Khatua, B. B. (2018). Nature driven spider silk as high energy conversion efficient bio-piezoelectric nanogenerator, *Nano Energy*, 49, 655.

[54] Ghosh, S. K., and Mandal, D. (2016). High-performance bio-piezoelectric nanogenerator made with fish scale. *Appl. Phys. Lett.*, 109, 103701.

[55] Ghosh, S. K., and Mandal, D. (2016). Efficient natural piezoelectric nanogenerator: electricity generation from fish swim bladder, *Nano Energy*, 28, 356.

[56] Lee, B. Y., Zhang, J., Zueger, C., Chung, W.J., Yoo, S. Y., Wang, E., Meyer, J., Ramesh, R., and Lee, S.W. (2012). Virus-based piezoelectric energy generation. *Nat. Nanotechnol.*, 7, 351.

[57] Lisa Sarkar, Mudigunda V Sushma, Bhavani Prasad Yalagala, Aravind Kumar Rengan, Shiv Govind Singh, and Siva Rama Krishna Vanjari. (2020). ZnO nanoparticles embedded silk fibroin-a piezoelectric composite for nanogenerator applications, *Nanotechnology*, 8, 33 (26).

[58] Ding, R., Zhang, X., Chen, G., Wang, H., Kishor, R., Xiao, J., Gao, F., Zeng, K., Chen, X., Sun, X. W., and Zheng, Y. (2017). High-performance piezoelectric nanogenerators composed of formamidinium lead halide perovskite nanoparticles and poly(vinylidene fluoride), *Nano Energy*, 37, 126-135.

[59] Zhu, G., Lin, Z. H., Jing, Q., Bai, P., Pan, C., Yang, Y., Zhou, Y., and Wang, Z. L. (2013). Toward large-scale energy harvesting by a nanoparticle-enhanced triboelectric nanogenerator, *Nano Lett.*, 13, 847-853.

[60] He, X., Zi, Y., Yu, H., Zhang, S. L., Wang, J., Ding, W., Zou, H., Zhang, W., Lu, C., and Wang, Z. L. (2017). An ultrathin paper-based self-powered system for portable electronics and wireless human-machine interaction, *Nano Energy*, 39, 328-336.

[61] Chen, X., Song, Y., Su, Z., Chen, H., Cheng, X., Zhang, J., Han, M., and Zhang, H. (2017). Flexible fiber-based hybrid nanogenerator for biomechanical energy harvesting and physiological monitoring, *Nano Energy* 38 43-50.

[62] Dagdeviren, C., Li, Z., and Wang, Z. L. (2017). Energy harvesting from the animal/human body for self-powered electronics, *Annu. Rev. Biomed. Eng.*, 19, 85-108.

[63] Yinben Guo, Xiao-Sheng Zhang, Ya Wang, Wei Gong, Qinghong Zhang, Hongzhi Wang, and Juergen Brugger (2018). All-Fiber Hybrid Piezoelectric-Enhanced Triboelectric Nanogenerator for Wearable Gesture Monitoring, *Nano Energy*, 48, 152-160.

[64] We, J. H., Kim, S. J., and Cho, B. J. (2014). Hybrid composite of screen-printed inorganic thermoelectric film and organic conducting polymer for flexible thermoelectric power generator. *Energy*, 73, 506–512.

[65] We, J. H., Kim, S. J., Kim, G. S., and Cho, B. J. (2013). Improvement of thermoelectric properties of screen-printed Bi_2Te_3 thick film by optimization of the annealing process. *J. Alloy. Compd.*, 552, 107–110.

[66] Dan-Liang Wen, Hai-Tao Deng, Xin Liu, Guo-Ke Li, Xin-Ran Zhang, and Xiao-Sheng Zhang, (2020). *Wearable multi-sensing double-chain thermoelectric generator, Microsystems & Nano-engineering*, 6, 68.

[67] Zheng, M., Xiao, X., Li, L., Gu, P., Dai, X., Tang, H., Hu, Q., Xue, H., and Pang, H. (2018). Hierarchically nanostructured transition metal oxides for supercapacitors, *Sci. China Mater.*

[68] Acar, C. (2018). A comprehensive evaluation of energy storage options for better sustainability, *Int. J. Energy Res.*, 42, 3732.

[69] Liu, W., Agusdinata, D. B., andMyint, S. W. (2019). Spatiotemporal patterns of lithium mining and environmental degradation in the Atacama Salt Flat, Chile Show affiliations, *Int. J. Appl. Earth Obs. Geoinf.*, 80, 145.

[70] Wu, K. S., Hu, Y., Cheng, Z. L., Pan, P., Jiang, L. Y., Mao, J. T., Ni, C. K., Gu, X. F., and Wang, Z. X. (2019). Hexagonal boron nitride monolayers on metal supports: Versatile templates for atoms, molecules and nanostructures, *J. Membr. Sci.*, 592, 8.

[71] Hou, J., Cao, C., Idrees, F., and Ma, X. (2015). Hierarchical Porous Nitrogen-Doped Carbon Nanosheets Derived from Silk for Ultrahigh-Capacity Battery Anodes and Supercapacitors, *ACS Nano*, 9, 2556.

[72] Wu, K. S., Hu, Y., Cheng, Z. L., Pan, P., Jiang, L. Y., Mao, J. T., Ni, C. K., Gu, X. F., and Wang, Z. X. (2019). Hexagonal boron nitride monolayers on metal supports: Versatile templates for atoms, molecules and nanostructures, *J. Membr. Sci.*, 592, 8.

[73] Pal, R. K., Kundu, S. C., and Yadavalli, V. K. (2018). Fabrication of Flexible, Fully Organic, Degradable Energy Storage Devices using Silk Proteins, *ACS Appl. Mater. Interfaces*, 10, 9620.

[74] Poonam, K. S., Arora, A., and Tripathi, S. K. (2019). Review of Supercapacitors: Materials and Devices. *J. Energy Storage*, 21, 801.

[75] Chotia, I., and Chowdhury, S. (2015) presented at 2015 IEEE Innovative Smart Grid Technol. – Asia (ISGT ASIA), Bangkok, Thailand November.

[76] Kim, Y. J., Abe, Y., Yanagiura, T., Park, K. C., Shimizu, M., Iwazaki, T., Nakagawa, S., Endo, M., and Dresselhaus, M. S. (2007). Easy preparation of nitrogen-enriched carbon materials from peptides of silk fibroins and their use to produce a high volumetric energy density in supercapacitors, *Carbon*, 45, 2116.

[77] Luo, H., Chen, M., Cao, J., Zhang, M., Tan, S., Wang, L., Zhong, J., Deng, H., Zhu, J., andLu, B. (2020). Cocoon Silk-Derived, Hierarchically Porous Carbon as Anode for Highly Robust Potassium-Ion Hybrid Capacitors, *Nano-Micro Lett*., 12, 113.

[78] Lihong Shi, Yang Huang, Lei Gao, andYuan Cheng. (2021). Application of Graphene in Coating Silk Fibril for Tunable Infrared Absorption, *Journal of Electronic Materials*, 50, 592–597.

[79] Yun, Y. S., Cho, S. Y., and Jin. H. J. (2014). Macromol.Carbon aerogels based on regenerated silk proteins and graphene oxide for supercapacitors, *Res*., 22, 509.

[80] Yang, J., Tan, Z., Chen, X., Liang, Y., Zheng, M., Hu, H., Xiao, Y. (2021). A mild method to prepare nitrogen-rich interlaced porous carbon nanosheets for high-performance supercapacitors. *Journal of Colloid and Interface Science*, 599, 381–389.

[81] Xia, T., Wang, Q., Wu, W., Ao, C., Zheng, Z., Lu, C., Zhang, W. (2021). Fabrication and characterization of MnO_2-Coated carbon fabrics from silk for shape-editable supercapacitors. *Journal of Alloys and Compounds*., 854, 157289.

[82] Liang, J., Zhang, X., Ji, Y., Chen, Z., Norton, M. L., Wang, Y., Cheng, G. (2021). Feeding silkworms with HPMC dispersed MoO_2 NPs: An efficient strategy to enhance the supercapacitance performance of carbonized silk. *Composites Science and Technology*, 215, 109025.

[83] Rath, T., Pramanik, N., and Kumar, S. (2017). High electrochemical performance flexible solid-state supercapacitor based on Co-doped reduced graphene oxide and silk fibroin composites. *Energy*, 141, 1982–1988.

[84] Zhang, Y., Cheng, Z., Ni, C., Wang, Z., Yu, Y., Zhai, X., and Hu, Y. (2022). Highly conductive EGaln/silk fibroin ink for graphene 3D array structure micro-supercapacitors. *Chemical Engineering Journal*, 428, 132084.

[85] Li, X., Sun, C., Cai, Z., and Ge, F. (2019). High-performance all-solid-state supercapacitor derived from PPy coated carbonized silk fabric. *Applied Surface Science*, 473, 967–975.

[86] Cai, H., Liu, Z., Xu, M., Chen, L., Chen, X., Cheng, L., and Dai, F. (2021). High performance flexible silk fabric electrodes with antibacterial, flame retardant and UV resistance for supercapacitors and sensors. *ElectrochimicaActa*., 390, 138895.

[87] Hu, M., Hu, T., Cheng, R., Yang, J., Cui, C., Zhang, C., and Wang, X. (2018). MXene-coated silk-derived carbon cloth toward flexible electrode for supercapacitor application. *Journal of Energy Chemistry*, 27(1), 161–166.

[88] Song, P., Tao, J., He, X., Sun, Y., Shen, X., Zhai, L., Li, B. (2020). Silk-inspired stretchable fiber-shaped supercapacitors with ultrahigh volumetric capacitance and energy density for wearable electronics. *Chemical Engineering Journal*, 386, 124024.

[89] Song, P., Xi, C., Premlatha, S., Shen, X., Ji, Z., Yan, Z., and Zhu, G. (2021). Sword/scabbard-shaped asymmetric all-solid-state supercapacitors based on PPy-MWCNTs-silk and hollow graphene tube for wearable applications. *Chemical Engineering Journal*, 411, 128522.

Chapter 7

Silk Fibroin as a Functional Biomaterial for Tissue Engineering

Vidya Murugesan[1], Shruthika Mundada[2] and Senthilkumar Rajagopal[2,*]

[1] Department of Chemistry and Biochemistry,
M. S. Ramaiah College of Arts, Science and Commerce, Bangalore, KA, India
[2] Department of Biotechnology,
School of Applied Sciences, REVA University, Bangalore, KA, India

Abstract

Silk is a biomaterial made up of functional proteins produced by insects such as flies, silkworms, scorpions, spiders, and mites. Silk produced by silkworms has been intensively researched for use in tissue engineering and wound healing. Silk is undeniably a natural biocompatible substance for humans, and it has long been used in medicinal procedures. The silk worm protein is made up of two proteins: fibroin and sericin. Silk fibroin accounts for over 70% of the weight of a cocoon and has a wide range of applications in textiles and biomedical applications due to its biocompatibility, non-toxicity, biodegradability, immunogenicity, and non-carcinogenicity. Silk fibroin has exceptional toughness and mechanical strength, while sericin has strong resistance to ultraviolet light and oxidation. Silk fibroin has been shown to promote wound healing by enhancing cell proliferation and growth, as well as migrating various types of cells involved in the wound healing process at various phases. Several silk variants have been studied for wound healing applications over the last several decades, including silk worm fibroin, silk sericin, recombinant silk materials, and native spider silk. In recent years, numerous technologies have been investigated and implemented to generate bio artificial skin grafts and bioactive wound dressings with the goal of leveraging silk's regenerative qualities.

In this chapter, we reveal the comprehensive insight of silk fibroin in various drug delivery systems, and also to present the current opportunities and requirements of silk fibroin, a functional biomaterial.

Keywords: bioartifical skin grafts, *Bombyx mori*, cocoons, immunogenicity, sericin, silk gum

* Corresponding Author's Email: senthilanal@yahoo.com.

In: Silk Fibroin
Editor: Shivaji H. Pawar
ISBN: 979-8-88697-402-7

1. Introduction

Silk bave, the silk fibre of the *Bombyx mori* silkworm, is single and continuous, and with spans of 700–1500 m reeled from its cocoon [1]. It's a type of natural biological composite material which is made up of two cores of fibroin filaments linked together by a sericin cementing layer [2], with each individual fibroin filament known as silk brin (sometimes called completely degummed silk) [3, 4]. The cross section of the silkworm silk bave and its brin is uneven, with feature sizes on the order of 20m and 10m [4], respectively. For centuries, Bombyx mori silkworms have been domestically bred and used in the textile industry [5]. Silk fibres are protein-based fibrous biopolymers made up of [Gly-Ser-Gly-Ala-Gly-Ala] xhexapeptides that form a stable anti-parallel beta-sheet structure with [Gly-Ala/Ser] dipeptides [6, 7]. The semi-crystalline structure of silk fibroin (SF) is due to strong hydrogen bonding between the beta-sheets, which make SF mechanically stronger to most bio-derived polymers [8, 9]. Because of its great biocompatibility, good oxygen and vapor permeability, and induction of minimal inflammatory reactions *in vivo*, SF has been proven to be a valuable material for the development of wound healing dressings [10].

The amino acid sequence Gly-Ser-Gly-Ala-Gly-Ala comprises the main molar mass structure of silk Fibroin, which crystallizes as the antiparallel β that improve the tensile strength and rigidity of silk fibroin. 70% of the β-sheet crystalline structure of SF, on the other hand, is made up of the protein structure surrounded by hydrophilic and non-repetitive amorphous areas [6, 11]. SF may be generated from a water-based solution at room temperature with a neutral pH, and it can be shaped into α-helix and random coils, although it has poor mechanical qualities [12]. Various treatment strategies have been evaluated in recent years to boost the strength of SF. Chemical, enzymatic, and physical cross-linking of SF hydrogels have specifically resulted in the production of β-sheet structure [13], because hydrogelation process result from transition of structure from random coil to β-sheet [14].

The crystallinity of fibroin can be changed to produce three unique morphologies: silk I, silk II, and silk III. Silk I, which is typically found in silkworm glands, possess the lowest crystallinity and the most random coils and α-helices [15]. These motifs produce a zigzag amorphous structure, which makes silk thermodynamically fragile and water-soluble [6,16]. Silk II, which is typically present in silk fibres, has the highest crystallinity and belongs to the monoclinic crystal system, with mostly antiparallel β-sheet components [15]. These β-sheet motifs are notable for having an asymmetrical structure, with Gly-hydrogen chains on one side and Ala-methyl chains on the other. Hydrophobic interactions between methyl groups and hydrogen atoms make silk II thermodynamically stable. Hence silk II has high mechanical and thermal resistances, stable and not soluble in water. Silk III is naturally found in films which are created from air-water interface of aqueous solution of fibroin, which is a crystalline fibroin polymorph [17].

Chemical and physical methods can interchange silk I and silk II, making it particularly adaptable in the generation of biomedical systems from fibroin. Water-soluble silk I is usually converted to insoluble silk II to produce useful platforms including hydrogels, scaffolds, micro- and nanoparticles and films. Three well-known methods utilized for this reason are temperature, organic solvents, and ionic solutions. Claudius Galenus of Pergamon treated gladiators with silk fibre as sutures in the first century, which was the earliest documented usage of silk in biomedical purposes [18]. It was, however, non-sterile silk. Joseph Lister initiated the first sterile silk suture into clinical practise in 1869, beginning a new era of silk-based medical supplies. The FDA recognized fibroin as a biomaterial in 1993 [19]. Fibroin is an excellent biomaterial for nanoparticle formulations and it was reported to possess excellent properties like biocompatibility, water solubility, structure modifiability and biodegradability.

SF fibres have shown to have excellent mechanical properties [20, 21]. Large break strain (4–26%), toughness (70–78 MJ m^{-3}) and ultimate strength (300–740 MPa) are among them [20]. Furthermore, the toughness of SF fibres has been reported to be higher than that of many synthetic fibres, including carbon fibre (25 MJ m^{-3}), Kevlar (50 MJ m^{-3}), and various collagens, such as tendon collagen (7.5 MJ m^{-3}) [22, 23]. Many studies have employed SF as a scaffold material for load-bearing Tissue engineering [TE] applications, especially in musculoskeletal TE, due to its robust mechanical characteristics [24]. With an objective of making use of the silk regenerative properties, plenty of strategies have been investigated and used to construct bioactive wound dressings and bio artificial skin grafts recently [25].

Moreover, using biocompatible and mechanically durable polymers with mild fabrication and processing conditions in such delivery systems is advantageous for preserving the bioactivity of the loaded APIs. As discussed earlier, SF meets all these requirements which make it a promising candidate for drug delivery [26, 27]. SF-based drug delivery systems can be fabricated by different methods, each resulting in a delivery system with unique properties such as modified release kinetics, stability, and other features which could be of benefit in various applications. Various types of SF-based drug delivery systems have been designed including nanofibers, hydrogels, films, lyophilized sponges, micro- and nanoparticles as well as SF-coated polymeric particles [28].

2. Biocompatible and Biodegradation

Fibroin is a highly biocompatible polymer for clinical usage since it is a protein with a basic chemical composition [19, 29]. Biocompatibility is a critical aspect in the effective use of scaffolds, since it allows cells to adhere and migrate to scaffold surfaces and migrate into the scaffold encountering differentiation and proliferation inside the scaffolds [30]. Hence, it's critical that biodegradation by-products are non-toxic and don't interfere with other tissue, organs, or activities when metabolized in the body. SF is an enzymatically biodegradable polymer that has not been found to elicit an immune response [29]. When enzymes are adsorbed onto the surface of the SF scaffold via surface-bonding domains, the degradation process begins. The enzymes next hydrolyze ester bonds to digest SF [29, 31, 32]. When compared to RSF silk scaffolds, native silk fibres degrade far more slowly. This is owing to the larger amount of β-sheet secondary structure in native silk fibres than in RSF structures [33]. As a result, the rate of SF deterioration is greatly reliant on the - β sheet secondary structures. Cytocompatibility studies proved that silk fibroin possess high compatibility with various cell lined like osteoblasts, endothelial cells, hepatocytes, mesenchymal stem cells and fibroblasts [34, 35].

3. Tissue Engineering

TE is an interdisciplinary field that combines science and technology to engineer tissues and organs *in vitro* using materials, cells and biochemical variables. Scaffold engineering has become a primary focus of research in the field of TE in order to mimic the extracellular matrix, which is one of the most challenging and difficult technological barriers. SF is regarded as a viable material for scaffold fabrication among the various materials examined. The ease of processing, excellent biocompatibility, tailorable degradability and remarkable mechanical properties silk fibroin have been investigated for fabrication of various articles such as porous matrices, nonwoven mats, films, hydrogels, for use in other tissue engineering applications. Figure 1 shows schematic of various human tissues engineering using silkworm-derived silk fibroin [36].

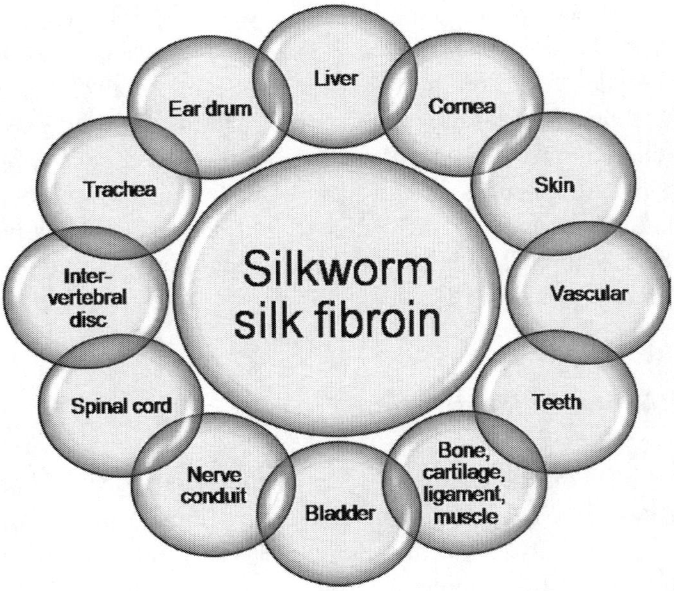

Figure 1. Schematic of various human tissues engineering using silkworm-derived silk fibroin. (Adopted from Kasoju and Bora, 2012) [36].

4. Regeneration of Cartilage Tissue

Cartilage is a non-innervated, avascular connective tissue that protects the subchondral bone from severe stresses in the joint and is essential for body movement. Unfortunately, the cartilage would degenerate over time due to its limited ability to self-repair after being damaged. Many surgical methods are currently being tried to restore cartilage, however, the results are not always satisfactory. As a result, TE offers a promising approach to cartilage healing [37]. Cai et al. developed a silk fibroin coating poly (D, L) lactic acid (PLA) film system and found that silk coating improved the PLA film's hydrophilicity, which strengthens the adhesion and proliferation of rat osteoblasts [38]. There have also been numerous other studies on the use of SF biomaterials to regenerate cartilage, with promising results [39, 40].

After three weeks of incubation the spatial cell arrangement and collagen type-II distribution in the artificial tissue construct mimic those of native articular cartilage described by Wang et al. after seeding hMSCs on 3D porous SF scaffolds generated. Also, they found that dexamethasone and TGFβ-3 have an important effect in enhancing chondrogenesis, proliferation, and survival [41]. The TGFβ-1@CS/BMP-2@SF hydrogel created in this study increased BMSC chondrogenic potential in vivo and in vitro by releasing TGF-β1 and BMP-2, presenting a unique therapeutic method for healing articular cartilage abnormalities in knee joints [42]. To test the feasibility of cartilage repair, Uebersax et al. insert IGF-I into SF scaffolds and find that the loaded group has a better capacity to stimulate hMSC chondrogenesis [43].

In vitro culture studies demonstrated that the genipincross-linked CS/SF sponges aided chondrocyte-like cells (ATDC5) in matrix synthesis, adhesion and proliferation [44]. Bhardwaj et al. discovered that SF/CS blended scaffolds aided in attachment, growth, and chondrogenic phenotype of bovine chondrocytes for 2 weeks *in vitro* culture, and biomechanical studies demonstrated that cell-seeded constructs had higher static and dynamic stiffness at high frequencies than non-cell-seeded controls [45]. Baek et al. and Jin et al. used microwave-induced argon plasma treatment to increase the cytocompatibility and hydrophilicity of nanofibrous SF mesh with human articular chondrocytes [46, 47]. Ghosh et al. established an SF-based *in vitro* model to show that hMSCs recognized small differences in morphology and stiffness of nanofibrous SF matrices, inducing migration and promoting aggregation, resembling chondrogenesis [48].

Morita et al. studied that how the frictional properties of cartilage regenerated from rabbit chondrocytes inoculated into SF sponges varied over time, and discovered that the friction coefficient of the regenerated cartilage decreased as the cultivation time increased, owing to the formation of a hydrophilic layer of ECM on the SF sponge surface [49]. Shangkai et al. investigated cartilage tissue engineering *in vitro* by growing chondrocytes on SF-sponge scaffolds in a bioreactor that allowed for mechanical stimulation, and then executed clinical feasibility investigations in rabbits [50]. The bioreactor settings were shown to favorably enable substantial increases in both DNA and GAG contents, as well as resulted in cartilage tissue maturation. The *in vitro* manufactured cartilage graft was subsequently utilized to successfully heal major damages in knee joints [50].

5. Engineering Ligament and Tendon Tissues

Ligament and tendon are made up of thick fibrous connective tissue which comprises mostly of fibrocytes and collagenous fibres. Ligaments are cord-like structures that support internal organs and connect bone to bone in joints to ensure appropriate articulation. Tendonsare connective tissue that connects muscles to other body components, most commonly bones, and transfer the mechanical power of muscle contraction to the bones [51, 52]. Because they lack a robust regeneration ability, they are frequently compromised during sports and other high-intensity activities, eventually resulting in severe joint dysfunction [53]. Tissue engineering has the potential to increase the quality of these tissues during the healing process, and it may be a more successful treatment than standard approaches. Cell-matrix composites with combination of biochemical cues and cells have been demonstrated to affect the tendon and ligament healing process.

While engineering anterior cruciate ligaments (ACL) in vitro, Altman et al. demonstrate that silk fibre matrices improve the adhesion, expansion, and differentiation of human progenitor bone marrow stem cells toward ligament-like tissue [54]. Chen et al. reported that coating silk fibres with Arg-Gly-Asp (RGD) significantly improved human bmSCs and ACL fibroblast adhesion, spreading, proliferation, and collagen matrix formation. As a result, it appears that modifying silk with RGD or collagen promotes structural and functional ligament repair [55]. Toh et al. develops a silk scaffold with knitted architecture for ligament tissue engineering, using an electro-spun nanofiber surface on knitted microfiber architecture, and shows that this method improves composite-material integrity in degradation resistance tests and promotes cell adhesion and proliferation *in vitro* [56]. The incorporation of silk fibres into the collagen matrix [57], and the loading of basic fibroblast growth factor (bFGF) and transforming growth factor (TGF-) in an optimal regimen within the scaffold [58], the coating of silk fibroin on electro-spun poly-lactic-co-glycolic acid (PLGA) nano-fibers [59], stimulates the mechanical and biochemical mechanisms for ligament tissue regeneration. A mature New Zealand white rabbit's Achilles tendon can also be repaired by A. pernyi [60]. The healing of the rabbit's Achilles tendon demonstrates the viability of employing non-mulberry silks in conjunction with mulberry silk as tendon scaffolds.

6. Bladder Tissue Engineering

The urinary bladder is a hollow muscular organ that can expand and contract (or stretchy) organ that rests on the pelvic floor and collects urine released by the kidneys before urination. A range of circumstances come across in urology lead to bladder dysfunction and they need reconstructive protocols. Transitional epithelial cells in the urinary bladders of New Zealand rabbits are well supported by silk fibroin films [61]. Short-length 1.5 cm defects are successfully repaired in rabbits using silk films [62]. Tian et al. used adipose-derived MSCs planted on a porous SF scaffold to heal urethral defects in New Zealand rabbits [63]. They discovered that rabbits implanted with MSC-seeded SF had a much-decreased incidence rate of post-operative urethral stricture and fistula. The manufacturing of SF-based woven patches adopting the gel spinning process was described, as well as their usage in bladder augmentation. Urinary incontinence due to stress is still a global issue that affects people of all ages. Zou et al. created a tissue engineered sling with a silk scaffold seeded with bmMSCs that was implanted into stress urine incontinence patients [64].

7. Bone Tissue Engineering

Bone is a specialized connective tissue made up primarily of calcified extracellular matrix including collagen type I and hydroxyapatite [65]. The bone tissue scaffolding material must assure matrix toughness and matrix deposition. Silk fibroin is a good choice in this case because of its great toughness and mechanical strength, as well as its biocompatibility [36]. One of the most well-studied tissue engineering technologies to date is silk fibroin-based bone tissue engineering. *In vitro* studies have revealed that RSF scaffolds stimulate osteogenic development of human mesenchymal stem cells (HMSC). In nude rat models, these constructs have been demonstrated to repair femoral lesions in vivo [35]. To improve osteogenic potential, RSF scaffolds are combined with additional biomaterials such as collagen or calcium phosphate-based inorganic components [66, 67]. Regenerated silk fibroin in combination with bone morphogenetic factors (BMP-7 and BMP 2 which are FDA approved growth factors together with HMSCs shown improved osteoblast, differentiation, adhesion, promoted bone formation and enhanced alkaline phosphatase activity in vivo [68, 69].

RSF was proposed by Ding as a carrier for loading DBM. This carrier may create stable porous structures and, when combined with bone marrow stem cells, has been proven to stimulate osteogenesis in mice (BMSC) [70]. RSF matrices that had been pre-incubated with osteoblasts in vitro and then implanted into mice demonstrated improved vascularization in vivo [70]. Furthermore, the vascular endothelial growth factor (VEGF) could produce neovascularization as well as enhance osteoblast differentiation [71, 72]. Farokhi used RSF–calcium phosphate–poly (lactic-co-glycolic acid) scaffolds to insert VEGF in this context. The scaffolds retained about 83 percent bioactivity following VEGF release for up to 28 days in vitro, according to the findings [73]. The development of neo-bone in rabbit defects following implantation for 10 weeks in an in vivo investigation [74]. According to Zhang, a sonicated silk hydrogel carrier containing VEGF and BMP-2 could enhance angiogenesis and osteogenesis after implantation for 12 weeks in maxillary sinus floor of rabbit [75].

8. Regeneration of Ear Drums

The tympanic membrane (TM) is a translucent structure between the outer and middle ear that receives sound waves while also protecting the middle ear. Middle ear infections or traumatic ruptures induced by mechanical stress and pressure explosions are the most common causes of TM perforations. If the rupture is not self-repaired within three months, it can develop into a chronic perforation, resulting in hearing loss and recurrent infections [76-78]. RSF has great qualities that make it a good material for tympanic membrane TE, allowing keratinocytes produced from human TM cells to develop and disseminate [79]. Human tympanic membrane keratinocytes adhered well to a silk fibroin membrane, and their growth kinetics were good [80]. RSF membranes, according to Allardyce, have excellent tensile strength, strong acoustic energy transmission capability which signifies that these membranes can be used to regenerate *in vivo* chronic TM perforations [81].

9. Regeneration of Skin and Wound Tissues

The skin is the human body's largest organ, and it serves as a barrier against infectious germs. It is only capable of limited self-healing. Skin integrity is lost when there is a lot of damage to the skin, which can lead to death [82, 83]. RSF biomaterials have been demonstrated to impact keratinocyte and fibroblast adhesion, and they are commonly used in TE skin regeneration [84]. In rabbit and porcine models, Zhang found that RSF films could be implanted into full-thickness skin defects, reducing healing time and improving skin regeneration when compared to conventional commercial wound dressings [9]. Staphylococcus aureus and Pseudomonas aeruginosa could be inhibited by RSF mats coated with antibacterial silver nanoparticles, which could be employed as antimicrobial wound dressings [85]. Cai used electrospinning to make RSF–chitosan scaffolds and observed that increasing RSF concentration increased mechanical strength and also antimicrobial activity along with enhancing chitosan concentration. Chitosan-RSF scaffolds were demonstrated to promote antimicrobial activity against Escherichia coli and stimulate cell proliferation [86].

Layering silk fibroin with collagen-I improves keratinocyte attachment and dispersion, whereas fibronectin coating promotes adhesion and dispersion of both keratinocytes and fibroblasts within the matrix [87]. Nano-fibrous silk-chitin [88], silk-collagen [89], and intermolecular cross-linked recombinant human-like collagen (RHLC) [90] with fibroin are other silk composites that have been effectively employed. Furthermore, in a full-thickness rat wound model, silk fibroin-alginate blended scaffolds sufficiently encourage re-epithelization [91]. Guang used a hydrogen-bonding approach to coat porous RSF scaffolds with chitosan to create 3D RSF–chitosan scaffolds, which he then implanted into a rat wound. The wound was totally fixed after 21 days with no teratogenic consequences or infections, according to their findings [92]. Non-mulberry RSF from Antheraea assama, on the other hand, has also been found to be a promising material for skin TE. This is due to the presence of the RGD peptide sequence, which facilitates cell adhesion [93]. These findings show that blended silk fibroin may have a better chance of regenerating skin tissue than pure silk fibroin.

10. Regeneration of Neural Tissues

The central nervous system (CNS) and the peripheral nervous system (PNS) are the two major divisions of the human nervous system (PNS). Minor injuries can be self-healed by the PNS, but major damage must be repaired surgically with nerve grafts obtained from elsewhere in the body. Silk fibroin promotes the survival of dorsal root ganglia and schwann cells while maintaining their normal phenotypic and functions [94]. Spider silk fibroin and B. mori silk fibroin is capable of bridging 13mm nerve gap within 12 weeks [95]. To improve the success of silk-based nerve grafts, researchers are currently investigating the direction of pores and the addition of neurotrophic agents for increased neuronal development.

11. Regeneration of Hepatic Tisssues

The liver is essential for glucose, protein, lipid, and vitamin metabolism. Hepatocytes are the main cellular component of hepatic tissue, and they are used in vitro to rebuild 3-D tissue or liver patches. Functionalized silk with lactose and cyanuric acid [96], silk fibroin blended films with collagen [97], recombinant human-like collagen [98], silk fibroin-collagen-heparin scaffolds [99], and silk micro-particle embedded PLA scaffolds [100] are among the silk composites for hepatic tissue engineering.

In rats, a mixture of silk fibroin and collagen causes minor inflammation in the omentum [101]. Furthermore, multi-scale scaffolds built on non-mulberry A. assama fibroin give enough mechanical support and aid cellular response [102].

12. Regeneration of Spinal Cord Tissue

One of the most popular treatments for spinal cord injuries is the transplantation of olfactory ensheathing cells (OECs). The development of silk-based spinal cord regeneration is still in its infancy. OEC cultivation on nano-fibrous silk fibroin indicates silk biomaterials' potential in this field [103]. The diameter of nanofibers has a regulatory effect on OEC proliferation; smaller diameters produce greater cellular responses than bigger diameters [104]. Modern tissue engineering's future lies in the regeneration and replacement of damaged tissue and organs.

13. Fluorescent Silk Fibroin

Because of its inherent fluorescence property obtained through the introduction of the fluorescence gene into silk cocoons, fluorescent SF has attracted a lot of attention for its potential use in biotechnological and medicinal applications. The incorporation of fluorescence properties into SF by gene insertion avoids the need to conjugate a fluorescently tagged protein like green fluorescent protein (GFP). As a result, the addition of fluorescent characteristics to SF may provide numerous advantages over fluorescently tagged proteins [105].

A method for producing functional silks that biologically introduce dyes into SF has been created by Tansil et al. 2011 [106]. Apart from that, the genetic engineering of silk proteins has garnered a lot of attention [107-110]. To develop green, fluorescent silkworm cocoons, a GFP is usually gene-targeted with the fibroin gene [110]. Transgenic silkworms have been used to mass-produce fluorescent silks in three colors (red, green, and orange) [111]. These fluorescent silks come in a variety of colors and could be used in medical applications or as a silk fibre for garments. Silkworms make yellow fluids and cocoons by attaching the yellow carotenoid pigment to the intracellular carotenoid-binding protein (among the pigments ingested into the silkworm body) (CBP). The Y (Yellow Blood) gene governs the uptake of carotenoids into the intestinal mucosa and the silk gland in the silkworm *Bombyx mori* and determines the color of the cocoon. Silk cocoons, in general, have a bright appearance which is because of the presence of flavonoids which include quercitin derivatives [112-114].

Tansil et al. were the first to demonstrate rhodamine B as a dye uptake into the SF. They administered silkworms a dye-containing diet and used enhanced characterization to corroborate their findings [106, 115] Silkworms were fed a blend of mulberry leaf powder and fluorescent compounds such rhodamine B, rhodamine 101, and rhodamine 110 to produce fluorescent silk cocoons and fibres. Silk genes have been used to create a variety of transformation vectors for the altering of silk [116]. Kim et al. generated a microcarrier by introducing jellyfish EGFP into exon 2 of the silkworm fibroin heavy chain gene. The green fluorescence is conserved, and the gene for green fluorescence is passed on to the next generation. They've also evolved transgenic silkworm which can express blue and red fluorescent protein [117].

14. Applications

To generate fluorescent SF materials for a variety of applications, SF has been combined with other materials or chemically/genetically changed. The use of fluorescent polymeric nanoparticles (PNPs) for bio imaging has received a lot of attention recently for non-invasive tracking [118]. Due to multiple substantial advantages over currently available detection methods, such as transmission electron microscopy (TEM) and scanning electron microscopy (SEM), the application of Two-Photon Fluorescence (TPF) microscopy for bio-imaging has sparked a lot of attention recently [119]. Silk films with TPF characteristics could be used as a film scaffold in biological imaging. In the future, functional fluorescent SF materials could be used in a variety of fields, including optoelectronics, biomedical science, and chemical science.

By loading spheres with the anthracycline DoX, Khalid et al. reported the drug release kinetics of nanodiamondSF-doxorubicins (NDSX) spheres. Degradable SF spheres precisely stabilize and discharge the drug, while ND provides a fluorescent imaging modality. The fluorescence of ND inside the spheres was used to track the degradability of SF and the subsequent release of DoX. These NDSX spheres could be used as dual-functional nanomaterials for drug delivery and subsequent imaging within biological structures, and they could be useful in targeted drug delivery [120].

15. Silk Fibroin-Based Flexible Electronics

Many outstanding properties of SF have been demonstrated, which include exceptional water solubility, biocompatibility, excellent optical transmittance, adjustable biodegradability, ease of processing, light weight and as well as good mechanical robustness, making it ideal for the construction of next-generation biocompatible flexible electronic devices [121-123]. 3D printing, Electro spinning, soft lithography, spin coating, freeze drying, ultrasonic induction and particulate leachingare some of the preparation technologies that have been used to process SF materials in various forms, such as silk sponges, silk films, silk fibres, and silk hydrogels, for various application areas. SF is valuable or even excellent as a substrate for wearable applications because of its superior mechanical qualities, optical properties (i.e., transparency), biological properties (i.e., biocompatibility), and variety of material morphologies (i.e., fibres, films, sponges, and hydrogels) [124].

Flexible electrodes, flexible sensors, flexible power supply devices, electronic skins, and other flexible and wearable devices based on various kinds of SF have been reported in the previous ten years. The development of wearable electronic devices, such as electronic skins, is now beset by numerous challenges, one of which is the device's strong conformal attachment to the biological surface [123].

Because it is a great precursor for chemical doping, SF has also been employed as a chemically dopable functional material to prepare flexible electrochemical sensors. Other flexible functional devices using SF as an active functional material, such as filters [125, 126], actuators [127], resistive switching memory devices [128, 129], OFETs [130] and OLEDs [131], have garnered considerable attention in addition to energy harvesters and sensors based on SF.

Due to a lack of pollutants in the preparation and operation of Nano network-based air filters created from electrospun biomaterials, they have been confirmed to be effective and are considered a green technology [132]. SF has gotten a lot of interest in the world of biological air filters since it not only has great biological qualities, but it also has a lot of functional groups, like hydroxyl groups, amide bonds and phenolic hydroxyl groups, that help catch various particles or chemicals [133]. Because of its high optical transmittance, SF has a lot of potential in the field of optical-electronic devices. Melikov presented a silk hydrogel (SH) lens for light-emitting diodes [134].

16. Fibroin Nanoparticles: A Promising Drug Delivery System

Fibroin is a major silk protein with good properties for drug delivery as a biomaterial. The development of fibroin nanoparticles (FNPs) for a wide range of biomedical applications has recently received a lot of attention [135]. These 1–1000-nm-diameter particles give several clinical benefits by encapsulating, entrapping, chemically binding the drug molecules or adsorbing. They have the capacity to (1) solubilize and/or improve their permeability, (2) preserve drugs from degradation, (3) target the disease site precisely (4) alter pharmacokinetics (5) modify drug release patterns, (6) selectively promote cellular absorption [15, 136, 137]. Sutures, coating devices, tissue regeneration, and drug delivery systems are just a few of the medical applications for fibroin, which is an FDA-approved polymer [138, 139].

Because of its suitability for nanoparticles (NP) formulation, fibroin has recently become more widely used in nanopharmaceuticals. FNPs, like pure fibroin, are extremely safe and biocompatible in a variety of applications, both *in vitro* and *in vivo* [140-142]. Zhang et al. 2007 reported that the first FNP formulation was developed by desolvation (nanoprecipitation, coacervation) [143]. Many approaches have been created since then, and they can be divided into two categories: top-down and bottom-up. Each approach has its own set of upsides and downsides, as well as variable sizes of FNPs; therefore, proper selection is critical in FNP formulation. In most top-down approaches, bulk materials (such as silk fibre or lyophilized regenerated fibroin) are size-reduced to FNPs via mechanical procedures like ball milling [144, 145]. Bottom-up approaches, on the other hand, entail the assembling of fibroin molecules into NPs [15].

FNPs have been shown to be exceedingly versatile in delivering a variety of therapeutic molecules, including proteins, enzymes, vaccines, and DNA, ranging in size from small to large molecules, hydrophobic to hydrophilic substances, and stable to labile pharmaceuticals. Furthermore, the molecular structure of fibroin is made up of a variety of amino acids with functional groups (amine, hydroxyl, carboxylic, thiol) that can be used to attach other biomolecules or antibodies as active targeting markers. To regulate the entrapped drug characteristics, these functional groups can be changed by coating with polymers [137]. FNPs demonstrate their efficacy as a drug delivery system by successfully incorporating a variety of natural chemicals, including quercitin [146], curcumin [147], triptolide [148], alpha mangostin [137], resveratrol [149] and celastrol [148]. Ding et al. loaded FNPs with triptolide and celastrol, two important drugs in traditional Chinese medicine [148].

Bovine serum albumin, platelet-derived growth factor (PDGF), Insulin [150, 151], vascular endothelial growth factor (VEGF) [152], asparaginase [153], bone morphogenetic protein 2

(BMP-2) [154], ß-glucosidase [155], naringinase [140], protease [156] and Pin1 isomerase [157] has been delivered via FNPs. Insulin may be conjugated covalently with FNPs at a significant recovery rate of 90–115 percent due to its capacity to be surface modified and crosslinked, boosting insulin in vitro half-life to more than 2.5 times [150]. Pin1 isomerase (a peptidyl prolyl cis–trans isomerase that can bind to the phosphoserine-proline or phosphothreonine-proline motifs of numerous proteins) was delivered directly into the cell cytoplasm by Kim et al. using FNPs encapsulated in cationic lipid [157]. These FNPs-lipid complexes successfully transported the enzyme with moderate cytotoxicity and high efficiency, resulting in an increase in Runx2 and Smad signalling, as well as the production of osteogenic marker genes and mineral deposition in Pin1-deficient cells. As a delivery mechanism, oligochitosan-fibroin NPs were developed as a delivery system for siRNA treatment in comparison with oligochotisan-siRNA polyplexus [158].

In an investigation, Mao et al. found that fibroin, in its hydrogel version; can improve the transdermal transport of curcumin-loaded polymeric NPs in a psoriatic mouse model [159]. Chomchalao et al. synthesized amphotericin B loaded FNPs in conjunction with the polymer PEG 400 as a ready-to-use eye drop for the treatment of fungal keratitis [160]. Besheli et al. successfully integrated the antibiotic vancomycin into FNPs for the treatment of local bone infections utilizing a desolvation process followed by complexation with fibroin scaffolds [161]. For the first time, Kim et al. used the spray-freeze-drying process to create cisplatin-loaded FNPs as a pulmonary drug delivery system for targeted lung cancer treatment [162].

Conclusion and Future Perspectives

Bombyx mori silkworm silk fibroin, in its native or regenerated forms, has proven to have remarkable properties suitable for sophisticated textile applications. The successful development of mechanically superior silk material is influenced not only by inherent molecular qualities, but also by in vivo and in vitro processing conditions, which have a significant impact on the structural organization and packing of the molecules. SF has been employed in the production of a number of drug delivery systems, including hydrogels, SF films, microparticles, and nanoparticles, employing a variety of fabrication processes. Each of these SF-based devices has demonstrated promise in a variety of biomedical applications. Experiments, simulations, and theoretical analysis have substantially improved our understanding of the mechanical properties of silk fibroin and facilitated its uses, providing a powerful driving force for the development of mechanically more robust silk. Mechanically improved silk is projected to have a slew of new functions that will make it a natural substitute for a variety of synthetic materials and open the door to a slew of new.

Despite its many advantages as a medication delivery system used in a variety of administration routes, fibroin still has some drawbacks that must be overcome. Because sericin may have immunogenic properties, the sericin removal process from silk fibres must be completed and adequate. The slow degradation of silk II crystalline antiparallel ß-sheet domains could be a disadvantage in some applications where the nanoparticulate carrier must be eliminated quickly and completely. Because fibroin is a protein, it may be proteolytically attacked by immune cells such as macrophages and giant cells, resulting in off-target drug release owing to granuloma development inside these cells [163]. The difficulties could be solved by coating or integrating fibroin or FNPs with polyethylene glycol or other hydrophilic polymers. Fibroin, particularly FNPs, has a high proclivity for serving as a delivery mechanism for a variety of therapeutic agents, including small molecule medicines, protein therapeutics, genes, and vaccines. Numerous FNP administration methods, including parenteral, oral, transdermal, ophthalmic, local bone implantation, and respiratory, have also been studied. Because of the beneficial features of FNP, further research should be focused on the less studied but potentially effective routes, such as ocular and respiratory. Last but not least, Finally, and perhaps most crucially, most FNP investigations are based on *in vitro* and *in vivo* experiments, implying that more clinical trials are needed to potentially bring FNPs to market.

References

[1] Zhao H-P, Shi H-J. Variability in mechanical properties of Bombyx mori silk. *Materials Science and Engineering: C*. 2007;27:675-83.

[2] Chen S, Liu M, Huang H, Cheng L, Zhao H-P. Mechanical properties of Bombyx mori silkworm silk fibre and its corresponding silk fibroin filament: A comparative study. *Materials & Design*. 2019;181:108077.

[3] Chen F, Porter D, Vollrath F. Silkworm cocoons inspire models for random fiber and particulate composites. *Physical Review E, Statistical, Nonlinear, and Soft Matter Physics*. 2010;82:041911.

[4] PÈrez-Rigueiro J, Viney C, Llorca J, Elices M. Silkworm silk as an engineering material. *Journal of Applied Polymer Science*. 1998;70:2439-47.

[5] Zhou CZ, Confalonieri F, Jacquet M, Perasso R, Li ZG, Janin J. Silk fibroin: structural implications of a remarkable amino acid sequence. *Proteins*. 2001;44 (2):119-22.

[6] Qi Y, Wang H, Wei K, Yang Y, Zheng RY, Kim IS, Zhang, KQ. A Review of Structure Construction of Silk Fibroin Biomaterials from Single Structures to Multi-Level Structures. *Int J Mol Sci*. 2017;18 (3).

[7] Liu B, Song Y-w, Jin L, Wang Z-j, Pu D-y, Lin S-q, Zhou, C, You, H, Ma, Y, Li, J, Yang, L, Sung, KLP, & Zhang, Y. Silk structure and degradation. *Colloids and Surfaces B: Biointerfaces*. 2015;131: 122-8.

[8] Tian D, Li T, Zhang R, Wu Q, Chen T, Sun P, Ramamoorthy, A. Conformations and Intermolecular Interactions in Cellulose/Silk Fibroin Blend Films: A Solid-State NMR Perspective. *J Phys Chem B*. 2017;121 (25):6108-16.

[9] Zhang W, Chen L, Chen J, Wang L, Gui X, Ran J, Xu, G, Zhao, H, Zeng, M, Ji, J, Qian, L, Zhou, J, Ouyang, H, & Zou, X. Silk Fibroin Biomaterial Shows Safe and Effective Wound Healing in Animal Models and a Randomized Controlled Clinical Trial. *Adv Healthc Mater*. 2017;6 (10).

[10] Ifkovits JL, Devlin JJ, Eng G, Martens TP, Vunjak-Novakovic G, Burdick JA. Biodegradable fibrous scaffolds with tunable properties formed from photo-cross-linkable poly (glycerol sebacate). *ACS Appl Mater Interfaces*. 2009;1 (9):1878-86.

[11] Zafar MS, Belton DJ, Hanby B, Kaplan DL, Perry CC. Functional material features of Bombyx mori silk light versus heavy chain proteins. *Biomacromolecules*. 2015;16 (2):606-14.

[12] Kundu B, Rajkhowa R, Kundu SC, Wang X. Silk fibroin biomaterials for tissue regenerations. *Adv Drug Deliv Rev*. 2013;65 (4):457-70.

[13] Ribeiro VP, Pina S, Costa JB, Cengiz IF, García-Fernández L, Fernández-Gutiérrez MDM, Paiva, OC., Oliveira, AL, San-Román, J, Oliveira, JM, & Reis, RL. Enzymatically Cross-Linked Silk Fibroin-Based Hierarchical Scaffolds for Osteochondral Regeneration. *ACS Appl Mater Interfaces*. 2019;11 (4):3781-99.

[14] Lin Y, Xia X, Shang K, Elia R, Huang W, Cebe P, Leisk, G, Omenetto, F, & Kaplan, DL. Tuning chemical and physical cross-links in silk electrogels for morphological analysis and mechanical reinforcement. *Biomacromolecules*. 2013;14 (8):2629-35.

[15] Pham DT, Saelim N, Tiyaboonchai W. Crosslinked fibroin nanoparticles using EDC or PEI for drug delivery: physicochemical properties, crystallinity and structure. *Journal of Materials Science*. 2018; 53:14087-103.

[16] Mottaghitalab F, Farokhi M, Shokrgozar MA, Atyabi F, Hosseinkhani H. Silk fibroin nanoparticle as a novel drug delivery system. *J Control Release*. 2015;206:161-76.

[17] Valluzzi R, Gido SP, Muller W, Kaplan DL. Orientation of silk III at the air-water interface. *Int J Biol Macromol*. 1999;24 (2-3):237-42.

[18] Holland C, Numata K, Rnjak-Kovacina J, Seib FP. The Biomedical Use of Silk: Past, Present, Future. *Advanced Healthcare Materials*. 2019;8:&NA;.

[19] Melke J, Midha S, Ghosh S, Ito K, Hofmann S. Silk fibroin as biomaterial for bone tissue engineering. *Acta Biomater*. 2016;31:1-16.

[20] Koh L-D, Cheng Y, Teng C-P, Khin Y-W, Loh X-J, Tee S-Y, Low, M, Ye, E, Yu, HD, Zhang, YW, & Han, MY. Structures, mechanical properties and applications of silk fibroin materials. *Progress in Polymer Science*. 2015;46:86-110.

[21] Sirichaisit J, Brookes VL, Young RJ, Vollrath F. Analyis of structure/property relationships in silkworm (Bombyx mori) and spider dragline (Nephila edulis) silks using Raman spectroscopy. *Biomacromolecules*. 2003;4 (2):387-94.

[22] Blamires SJ, Spicer PT, Flanagan PJ. Spider Silk Biomimetics Programs to Inform the Development of New Wearable Technologies. *Frontiers in Materials*. 2020;7.

[23] Gosline JM, Guerette PA, Ortlepp CS, Savage KN. The mechanical design of spider silks: from fibroin sequence to mechanical function. *J Exp Biol*. 1999;202 (Pt 23):3295-303.

[24] Ma D, Wang Y, Dai W. Silk fibroin-based biomaterials for musculoskeletal tissue engineering. *Materials Science and Engineering: C*. 2018;89:456-69.

[25] Vidya M, Rajagopal S. Silk Fibroin: A Promising Tool for Wound Healing and Skin Regeneration. *International Journal of Polymer Science*. 2021;2021:9069924.

[26] Wenk E, Merkle HP, Meinel L. Silk fibroin as a vehicle for drug delivery applications. *J Control Release*. 2011;150 (2):128-41.

[27] Winkler S, Wilson D, Kaplan DL. Controlling β-Sheet Assembly in Genetically Engineered Silk by Enzymatic Phosphorylation/Dephosphorylation. *Biochemistry*. 2000;39 (41):12739-46.

[28] Tomeh MA, Hadianamrei R, Zhao X. Silk Fibroin as a Functional Biomaterial for Drug and Gene Delivery. *Pharmaceutics*. 2019;11 (10):494.

[29] Cao Y, Wang B. Biodegradation of silk biomaterials. *Int J Mol Sci*. 2009;10 (4):1514-24.

[30] O'Brien FJ. Biomaterials & scaffolds for tissue engineering. *Materials Today*. 2011;14 (3):88-95.

[31] Nair LS, Laurencin CT. Biodegradable polymers as biomaterials. *Progress in Polymer Science*. 2007;32 (8):762-98.

[32] Arai T, Freddi G, Innocenti R, Tsukada M. Biodegradation of Bombyx mori silk fibroin fibers and films. *Journal of Applied Polymer Science*. 2004;91:2383-90.

[33] Hu Y, Zhang Q, You R, Wang L, Li M. The Relationship between Secondary Structure and Biodegradation Behavior of Silk Fibroin Scaffolds. *Advances in Materials Science and Engineering*. 2012;2012:185905.

[34] Jin H-J, Chen J, Karageorgiou V, Altman GH, Kaplan DL. Human bone marrow stromal cell responses on electrospun silk fibroin mats. *Biomaterials*. 2004;25 (6):1039-47.

[35] Meinel L, Hofmann S, Betz O, Fajardo R, Merkle HP, Langer R, Evans, CH, Vunjak-Novakovic, G, & Kaplan, DL. Osteogenesis by human mesenchymal stem cells cultured on silk biomaterials: comparison of adenovirus mediated gene transfer and protein delivery of BMP-2. *Biomaterials*. 2006;27 (28):4993-5002.

[36] Kasoju N, Bora U. Silk fibroin in tissue engineering. *Adv Healthc Mater*. 2012;1 (4):393-412.

[37] Chung C, Burdick JA. Engineering cartilage tissue. *Adv Drug Deliv Rev*. 2008;60 (2):243-62.

[38] Cai K, Yao K, Lin S, Yang Z, Li X, Xie H, Qing, T, & Gao, L. Poly (D,L-lactic acid) surfaces modified by silk fibroin: effects on the culture of osteoblast in vitro. *Biomaterials*. 2002;23 (4):1153-60.

[39] Chao P-HG, Yodmuang S, Wang X, Sun L, Kaplan DL, Vunjak-Novakovic G. Silk hydrogel for cartilage tissue engineering. *J Biomed Mater Res B Appl Biomater*. 2010;95 (1):84-90.

[40] Cheng G, Davoudi Z, Xing X, Yu X, Cheng X, Li Z, Deng, H, & Wang, Q. Advanced Silk Fibroin Biomaterials for Cartilage Regeneration. *ACS Biomaterials Science & Engineering*. 2018;4 (8):2704-15.

[41] Wang Y, Kim UJ, Blasioli DJ, Kim HJ, Kaplan DL. In vitro cartilage tissue engineering with 3D porous aqueous-derived silk scaffolds and mesenchymal stem cells. *Biomaterials*. 2005;26 (34):7082-94.

[42] Li Y, Liu Y, Guo Q. Silk fibroin hydrogel scaffolds incorporated with chitosan nanoparticles repair articular cartilage defects by regulating TGF-β1 and BMP-2. *Arthritis Research & Therapy*. 2021;23 (1):50.

[43] Uebersax L, Merkle HP, Meinel L. Insulin-like growth factor I releasing silk fibroin scaffolds induce chondrogenic differentiation of human mesenchymal stem cells. *J Control Release*. 2008;127 (1):12-21.

[44] Silva SS, Motta A, Rodrigues MT, Pinheiro AF, Gomes ME, Mano JF, Reis, RL, & Migliaresi, C. Novel genipin-cross-linked chitosan/silk fibroin sponges for cartilage engineering strategies. *Biomacromolecules*. 2008;9 (10):2764-74.

[45] Bhardwaj N, Nguyen QT, Chen AC, Kaplan DL, Sah RL, Kundu SC. Potential of 3-D tissue constructs engineered from bovine chondrocytes/silk fibroin-chitosan for in vitro cartilage tissue engineering. *Biomaterials*. 2011;32 (25):5773-81.

[46] Ki CS, Park YH, Jin H-J. Silk protein as a fascinating biomedical polymer: Structural fundamentals and applications. *Macromolecular Research*. 2009;17 (12):935-42.

[47] Jin S, Baek H, Woo Y, Lee M, Kim J-S, Park C, Park, YH, Rah, DK, Chung, KH, Lee, SJ, & Han, IH. Beneficial Effects of Microwave-Induced Argon Plasma Treatment on Cellular Behaviors of Articular Chondrocytes Onto Nanofibrous Silk Fibroin Mesh. *Macromolecular Research - MACROMOL RES*. 2009;17:703-8.

[48] Ghosh S, Laha M, Mondal S, Sengupta S, Kaplan DL. In vitro model of mesenchymal condensation during chondrogenic development. *Biomaterials*. 2009;30 (33):6530-40.

[49] Morita Y, Tomita N, Aoki H, Sonobe M, Wakitani S, Tamada Y, Suguro, T, & Ikeuchi, K. Frictional properties of regenerated cartilage in vitro. *J Biomech*. 2006;39 (1):103-9.

[50] Shangkai C, Naohide T, Koji Y, Yasuji H, Masaaki N, Tomohiro T, Yasushi, T. Transplantation of allogeneic chondrocytes cultured in fibroin sponge and stirring chamber to promote cartilage regeneration. *Tissue Eng*. 2007;13 (3):483-92.

[51] Tatara AM, Mikos AG. Tissue Engineering in Orthopaedics. *J Bone Joint Surg Am*. 2016;98 (13):1132-9.

[52] Kuo CK, Marturano JE, Tuan RS. Novel strategies in tendon and ligament tissue engineering: Advanced biomaterials and regeneration motifs. *Sports Med Arthrosc Rehabil Ther Technol*. 2010;2:20.

[53] Aslan H, Kimelman-Bleich N, Pelled G, Gazit D. Molecular targets for tendon neoformation. *J Clin Invest*. 2008;118 (2):439-44.

[54] Altman GH, Horan RL, Lu HH, Moreau J, Martin I, Richmond JC, Kaplan, DL. Silk matrix for tissue engineered anterior cruciate ligaments. *Biomaterials*. 2002;23 (20):4131-41.

[55] Chen J, Altman GH, Karageorgiou V, Horan R, Collette A, Volloch V, Colabro, T., & Kaplan, D. L. Human bone marrow stromal cell and ligament fibroblast responses on RGD-modified silk fibers. *J Biomed Mater Res A*. 2003;67 (2):559-70.

[56] Li M, Wu Z, Zhang C, Lu S, Yan H, Huang D, Ye, H. Study on porous silk fibroin materials. II. Preparation and characteristics of spongy silk fibroin materials. *Journal of Applied Polymer Science*. 2001;79:2192-9.

[57] Chen X, Qi YY, Wang LL, Yin Z, Yin GL, Zou XH, Ouyang, HW. Ligament regeneration using a knitted silk scaffold combined with collagen matrix. *Biomaterials*. 2008;29 (27):3683-92.

[58] Moreau JE, Chen J, Horan RL, Kaplan DL, Altman GH. Sequential growth factor application in bone marrow stromal cell ligament engineering. *Tissue Eng*. 2005;11 (11-12):1887-97.

[59] Sahoo S, Toh SL, Goh JC. PLGA nanofiber-coated silk microfibrous scaffold for connective tissue engineering. J *Biomed Mater Res B Appl Biomater*. 2010;95 (1):19-28.

[60] Fang Q, Chen D, Yang Z, Li M. In vitro and in vivo research on using Antheraea pernyi silk fibroin as tissue engineering tendon scaffolds. *Materials Science and Engineering: C*. 2009;29:1527-34.

[61] Liu CX, Liao YF, Li HL, Zheng SB. Cytocompatibility of silk fibroin film with rabbit urinary bladder transitional epithelial cells in vitro. *Journal of Southern Medical University*. 2008;28 (2):216-8.

[62] Liu C-x, Lin Y-y, Li H, Zheng S-b. Application of silk fibroin film for repairing rabbit urethral defect. *Journal of Southern Medical University*. 2007;27 2:184-7.

[63] Tian B, Song L, Liang T, Li Z, Ye X, Fu Q, Li, Y. Repair of urethral defects by an adipose mesenchymal stem cell-porous silk fibroin material. *Mol Med Rep*. 2018;18 (1):209-15.

[64] Zou XH, Zhi YL, Chen X, Jin HM, Wang LL, Jiang YZ, Yin, Z., & Ouyang, H. Mesenchymal stem cell seeded knitted silk sling for the treatment of stress urinary incontinence. *Biomaterials*. 2010;31 (18):4872-9.

[65] Griffith LG, Naughton G. Tissue engineering--current challenges and expanding opportunities. *Science*. 2002;295 (5557):1009-14.

[66] He P, Sahoo S, Ng KS, Chen K, Toh SL, Goh JC. Enhanced osteoinductivity and osteoconductivity through hydroxyapatite coating of silk-based tissue-engineered ligament scaffold. *J Biomed Mater Res A*. 2013;101 (2):555-66.

[67] Liu G, Zhang F, Zhang L, Wang Y, Fan L. Glass transition temperature of polystyrene determined by DSC revisited. *Acta Polymerica Sinica*. 2010;010:1065-9.

[68] Zhang Y, Fan W, Nothdurft L, Wu C, Zhou Y, Crawford R, Xiao, Y. In vitro and in vivo evaluation of adenovirus combined silk fibroin scaffolds for bone morphogenetic protein-7 gene delivery. *Tissue Eng Part C Methods*. 2011;17 (8):789-97.

[69] Kirker-Head C, Karageorgiou V, Hofmann S, Fajardo R, Betz O, Merkle HP, Hilbe, M, von Rechenberg, B, McCool, J, Abrahamsen, L, Nazarian, A, Cory, E, Curtis, M, Kaplan, D, & Meinel, L. BMP-silk composite matrices heal critically sized femoral defects. *Bone*. 2007;41 (2):247-55.

[70] Ghanaati S, Unger RE, Webber MJ, Barbeck M, Orth C, Kirkpatrick JA, Booms, P, Motta, A, Migliaresi, C, Sader, RA, & Kirkpatrick, CJ. Scaffold vascularization in vivo driven by primary human osteoblasts in concert with host inflammatory cells. *Biomaterials*. 2011;32 (32):8150-60.

[71] Unger RE, Sartoris A, Peters K, Motta A, Migliaresi C, Kunkel M, Bulnheim, U, Rychly, J, & James Kirkpatrick, C. Tissue-like self-assembly in cocultures of endothelial cells and osteoblasts and the formation of microcapillary-like structures on three-dimensional porous biomaterials. *Biomaterials*. 2007;28 (27):3965-76.

[72] Fuchs S, Jiang X, Schmidt H, Dohle E, Ghanaati S, Orth C, Hofmann, A, Motta, A, Migliaresi, C, & Kirkpatrick, CJ. Dynamic processes involved in the pre-vascularization of silk fibroin constructs for bone regeneration using outgrowth endothelial cells. *Biomaterials*. 2009;30 (7):1329-38.

[73] Farokhi M, Mottaghitalab F, Shokrgozar MA, Ai J, Hadjati J, Azami M. Bio-hybrid silk fibroin/calcium phosphate/PLGA nanocomposite scaffold to control the delivery of vascular endothelial growth factor. *Materials Science and Engineering: C*. 2014;35:401-10.

[74] Kanczler JM, Ginty PJ, White L, Clarke NMP, Howdle SM, Shakesheff KM, Oreffo, ROC. The effect of the delivery of vascular endothelial growth factor and bone morphogenic protein-2 to osteoprogenitor cell populations on bone formation. *Biomaterials*. 2010;31 (6):1242-50.

[75] Zhang W, Wang X, Wang S, Zhao J, Xu L, Zhu C, Zeng, D., Chen, J., Zhang, Z., Kaplan, D. L., & Jiang, X. The use of injectable sonication-induced silk hydrogel for VEGF165 and BMP-2 delivery for elevation of the maxillary sinus floor. *Biomaterials*. 2011;32 (35):9415-24.

[76] Villar-Fernandez MA, Lopez-Escamez JA. Outlook for Tissue Engineering of the Tympanic Membrane. *Audiology Research*. 2015;5 (1):117.

[77] Wang AY, Shen Y, Liew LJ, Wang JT, von Unge M, Atlas MD, Dilley, RJ. Rat model of chronic tympanic membrane perforation: Ventilation tube with mitomycin C and dexamethasone. *Int J Pediatr Otorhinolaryngol*. 2016;80:61-8.

[78] Teh BM, Marano RJ, Shen Y, Friedland PL, Dilley RJ, Atlas MD. Tissue engineering of the tympanic membrane. *Tissue Eng Part B Rev*. 2013;19 (2):116-32.

[79] Shen Y, Redmond SL, Teh BM, Yan S, Wang Y, Atlas MD, Dilley, RJ, Zheng, M., & Marano, RJ. Tympanic membrane repair using silk fibroin and acellular collagen scaffolds. *Laryngoscope*. 2013;123 (8):1976-82.

[80] Levin B, Redmond SL, Rajkhowa R, Eikelboom RH, Atlas MD, Marano RJ. Utilising silk fibroin membranes as scaffolds for the growth of tympanic membrane keratinocytes, and application to myringoplasty surgery. *J Laryngol Otol*. 2013;127 Suppl 1:S13-20.

[81] Allardyce BJ, Rajkhowa R, Dilley RJ, Xie Z, Campbell L, Keating A, Atlas, MD, von Unge, M, & Wang, X. Comparative acoustic performance and mechanical properties of silk membranes for the repair of chronic tympanic membrane perforations. *Journal of the Mechanical Behavior of Biomedical Materials*. 2016;64:65-74.

[82] Chua AW, Khoo YC, Tan BK, Tan KC, Foo CL, Chong SJ. Skin tissue engineering advances in severe burns: review and therapeutic applications. *Burns Trauma*. 2016;4:3.

[83] Farokhi M, Mottaghitalab F, Fatahi Y, Khademhosseini A, Kaplan DL. Overview of Silk Fibroin Use in Wound Dressings. *Trends Biotechnol*. 2018;36 (9):907-22.

[84] Zhang X, Reagan MR, Kaplan DL. Electrospun silk biomaterial scaffolds for regenerative medicine. *Advanced Drug Delivery Reviews*. 2009;61 (12):988-1006.

[85] Uttayarat P, Jetawattana S, Suwanmala P, Eamsiri J, Tangthong T, Pongpat S. Antimicrobial electrospun silk fibroin mats with silver nanoparticles for wound dressing application. *Fibers and Polymers*. 2012;13 (8):999-1006.

[86] Cai Z-x, Mo X-m, Zhang K-h, Fan L-p, Yin A-l, He C-l, Wang, H. Fabrication of Chitosan/Silk Fibroin Composite Nanofibers for Wound-dressing Applications. *International Journal of Molecular Sciences*. 2010;11 (9):3529-39.

[87] Min BM, Jeong L, Lee KY, Park WH. Regenerated silk fibroin nanofibers: water vapor-induced structural changes and their effects on the behavior of normal human cells. *Macromol Biosci*. 2006;6 (4):285-92.

[88] Yoo CR, Yeo IS, Park KE, Park JH, Lee SJ, Park WH, Min, BM. Effect of chitin/silk fibroin nanofibrous bicomponent structures on interaction with human epidermal keratinocytes. *Int J Biol Macromol*. 2008;42 (4):324-34.

[89] Yeo I-S, Oh J-E, Jeong L, Lee TS, Lee SJ, Park WH, Min, BM. Collagen-Based Biomimetic Nanofibrous Scaffolds: Preparation and Characterization of Collagen/Silk Fibroin Bicomponent Nanofibrous Structures. *Biomacromolecules*. 2008;9 (4):1106-16.

[90] Khor HL, Ng KW, Htay AS, Schantz JT, Teoh SH, Hutmacher DW. Preliminary study of a polycaprolactone membrane utilized as epidermal substrate. *J Mater Sci Mater Med*. 2003;14 (2):113-20.

[91] Roh DH, Kang SY, Kim JY, Kwon YB, Young Kweon H, Lee KG, Park, YH, Baek, RM, Heo, CY, Choe, J, & Lee, JH. Wound healing effect of silk fibroin/alginate-blended sponge in full thickness skin defect of rat. *J Mater Sci Mater Med*. 2006;17 (6):547-52.

[92] Guang S, An Y, Ke F, Zhao D, Shen Y, Xu H. Chitosan/silk fibroin composite scaffolds for wound dressing. *Journal of Applied Polymer Science*. 2015;132 (35).

[93] Gupta AK, Mita K, Arunkumar KP, Nagaraju J. Molecular architecture of silk fibroin of Indian golden silkmoth, Antheraea assama. *Sci Rep*. 2015;5:12706.

[94] Yang Y, Chen X, Ding F, Zhang P, Liu J, Gu X. Biocompatibility evaluation of silk fibroin with peripheral nerve tissues and cells in vitro. *Biomaterials*. 2007;28 (9):1643-52.

[95] Huang W, Begum R, Barber T, Ibba V, Tee NC, Hussain M, Arastoo, M, Yang, Q, Robson, LG, Lesage, S, Gheysens, T, Skaer, NJV, Knight, DP, & Priestley, JV. Regenerative potential of silk conduits in repair of peripheral nerve injury in adult rats. *Biomaterials*. 2012;33 (1):59-71.

[96] Gotoh Y, Niimi S, Hayakawa T, Miyashita T. Preparation of lactose-silk fibroin conjugates and their application as a scaffold for hepatocyte attachment. *Biomaterials*. 2004;25 (6):1131-40.

[97] Cirillo B, Morra M, Catapano G. Adhesion and function of rat liver cells adherent to silk fibroin/collagen blend films. *Int J Artif Organs*. 2004;27 (1):60-8.

[98] Hu K, Lv Q, Cui FZ, Feng QL, Kong XD, Wang HL, Huang, LY, & Li, T. Biocompatible Fibroin Blended Films with Recombinant Human-like Collagen for Hepatic Tissue Engineering. *Journal of Bioactive and Compatible Polymers*. 2006;21 (1):23-37.

[99] Lu Q, Zhang S, Hu K, Feng Q, Cao C, Cui F. Cytocompatibility and blood compatibility of multifunctional fibroin/collagen/heparin scaffolds. *Biomaterials*. 2007;28 (14):2306-13.

[100] Lv Q, Hu K, Feng Q, Cui F, Cao C. Preparation and characterization of PLA/fibroin composite and culture of HepG2 (human hepatocellular liver carcinoma cell line) cells. *Composites Science and Technology*. 2007;67:3023-30.

[101] She Z, Liu W, Feng Q. Self-assembly model, hepatocytes attachment and inflammatory response for silk fibroin/chitosan scaffolds. *Biomed Mater*. 2009;4 (4):045014.

[102] Kasoju N, Bhonde RR, Bora U. Preparation and characterization of Antheraea assama silk fibroin based novel non-woven scaffold for tissue engineering applications. *J Tissue Eng Regen Med*. 2009;3 (7):539-52.

[103] Qian Y, Shen Y, Lu Z, Fan Z, Liu T, Zhang J, Zhang, F. Biocompatibility of silk fibroin nanofibers scaffold with olfactory ensheathing cells. *Zhongguo Xiu Fu Chong Jian Wai Ke Za Zhi*. 2009;23 (11):1365-70.

[104] Shen Y, Qian Y, Zhang H, Zuo B, Lu Z, Fan Z, Zhang, P, Zhang, F & Zhou, C. Guidance of Olfactory Ensheathing Cell Growth and Migration on Electrospun Silk Fibroin Scaffolds. *Cell Transplantation*. 2010;19 (2):147-58.

[105] Lee OJ, Sultan MT, Hong H, Lee YJ, Lee JS, Lee H, Kim, SH, & Park, CH. Recent Advances in Fluorescent Silk Fibroin. *Frontiers in Materials*. 2020;7.

[106] Tansil NC, Li Y, Teng CP, Zhang S, Win KY, Chen X, Liu, XY, & Han, MY. Intrinsically colored and luminescent silk. *Adv Mater*. 2011;23 (12):1463-6.

[107] Kikuchi Y, Mori K, Suzuki S, Yamaguchi K, Mizuno S. Structure of the Bombyx mori fibroin light-chain-encoding gene: upstream sequence elements common to the light and heavy chain. *Gene*. 1992;110 (2):151-8.

[108] Yamao M, Katayama N, Nakazawa H, Yamakawa M, Hayashi Y, Hara S, Kamei, K, & Mori, H. Gene targeting in the silkworm by use of a baculovirus. *Genes Dev*. 1999;13 (5):511-6.

[109] Zhang F, Zhao Y, Chen X, Xu AY, Huang JT, Lu CD. Fluorescent Transgenic Silkworm. *Sheng Wu Hua Xue Yu Sheng Wu Wu Li Xue Bao* (Shanghai). 1999;31 (2):119-23.

[110] Zhao Y, Chen X, Peng WP, Dong L, Huang JT, Lu CD. Altering Fibroin Heavy Chain Gene of Silkworm Bombyx mori by Homologous Recombination. *Sheng Wu Hua Xue Yu Sheng Wu Wu Li Xue Bao* (Shanghai). 2001;33 (1):112-6.

[111] Iizuka T, Sezutsu H, Tatematsu K-I, Kobayashi I, Yonemura N, Uchino K, Nakajima, K, Kojima, K, Takabayashi, C, Machii, H, Yamada, K, Kurihara, H, Asakura, T, Nakazawa, Y, Miyawaki, A, Karasawa, S, Kobayashi, H, Yamaguchi, J, Kuwabara, N, … Tamura, T. Colored Fluorescent Silk Made by Transgenic Silkworms. *Advanced Functional Materials*. 2013.

[112] Tamura Y, Nakajima K, Nagayasu K, Takabayashi C. Flavonoid 5-glucosides from the cocoon shell of the silkworm, Bombyx mori. *Phytochemistry*. 2002;59 (3):275-8.

[113] Daimon T, Hirayama C, Kanai M, Ruike Y, Meng Y, Kosegawa E, Nakamura, M, Tsujimoto, G, Katsuma, S, & Shimada, T. The silkworm *Green b* locus encodes a quercetin 5-*O*-glucosyltransferase that produces green cocoons with UV-shielding properties. *Proceedings of the National Academy of Sciences*. 2010;107 (25):11471-6.

[114] Kusurkar T, Tandon I, Sethy N, Bhargava K, Sarkar S, Singh S, et al. Fluorescent silk cocoon creating fluorescent diatom using a "Water glass-fluorophore ferry." *Scientific Reports*. 2013;3:3290.

[115] Tansil NC, Li Y, Koh LD, Peng TC, Win KY, Liu XY, Han, MY. The use of molecular fluorescent markers to monitor absorption and distribution of xenobiotics in a silkworm model. *Biomaterials*. 2011;32 (36):9576-83.

[116] Tomita M. Transgenic silkworms that weave recombinant proteins into silk cocoons. *Biotechnol Lett*. 2011;33 (4):645-54.

[117] Kim SW, Yun E-Y, Choi K, Kim S-R, Park SW, Kang S-W, 강석우, & Goo, Tae-Won Expression of the blue fluorescent protein in fibroin H-chain of transgenic silkworm. *International Journal of Industrial Entomology*. 2014;52:25-32.

[118] Win KY, Teng CP, Ye E, Low M, Han MY. Evaluation of polymeric nanoparticle formulations by effective imaging and quantitation of cellular uptake for controlled delivery of doxorubicin. *Small*. 2015;11 (9-10):1197-204.

[119] He GS, Tan L-S, Zheng Q, Prasad PN. Multiphoton Absorbing Materials: Molecular Designs, Characterizations, and Applications. *Chemical Reviews*. 2008;108 (4):1245-330.

[120] Khalid A, Mitropoulos AN, Marelli B, Tomljenovic-Hanic S, Omenetto FG. Doxorubicin loaded nanodiamond-silk spheres for fluorescence tracking and controlled drug release. *Biomed Opt Express*. 2015;7 (1):132-47.
[121] Irimia Vladu M. @ Greeé Electronics: Biodegradable and Biocompatible Materials and Devices for Sustainable Future. *ChemInform*. 2014;45.
[122] Hakimi O, Knight D, Vollrath F, Vadgama P. Spider and mulberry silkworm silks as compatible biomaterials. *Composites Part B-Engineering*. 2007;38 (3):324-37.
[123] Sun Q, Qian B, Uto K, Chen J, Liu X, Minari T. Functional biomaterials towards flexible electronics and sensors. *Biosens Bioelectron*. 2018;119:237-51.
[124] Wen D-L, Sun D-H, Huang P, Huang W, Su M, Wang Y, Han, MD, Kim, B, Brugger, J, Zhang, HX, & Zhang, XS. Recent progress in silk fibroin-based flexible electronics. *Microsystems & Nanoengineering*. 2021;7 (1):35.
[125] Souzandeh H, Wang Y, Netravali AN, Zhong W-H. Towards Sustainable and Multifunctional Air-Filters: A Review on Biopolymer-Based Filtration Materials. *Polymer Reviews*. 2019;59 (4):651-86.
[126] Min K, Kim S, Kim S. Silk protein nanofibers for highly efficient, eco-friendly, optically translucent, and multifunctional air filters. *Scientific Reports*. 2018;8.
[127] Fengel CV, Bradshaw NP, Severt SY, Murphy AR, Leger JM. Biocompatible silk-conducting polymer composite trilayer actuators. *Smart Material Structures*. 2017;26:055004.
[128] Xing X, Chen M, Gong Y, Lv Z, Han ST, Zhou Y. Building memory devices from biocomposite electronic materials. *Sci Technol Adv Mater*. 2020;21 (1):100-21.
[129] Wang H, Zhu B, Wang H, Ma X, Hao Y, Chen X. Ultra-Lightweight Resistive Switching Memory Devices Based on Silk Fibroin. *Small*. 2016;12 (25):3360-5.
[130] Wang CH, Hsieh CY, Hwang JC. Flexible organic thin-film transistors with silk fibroin as the gate dielectric. *Adv Mater*. 2011;23 (14):1630-4.
[131] Fernández-Luna V, Fernández-Blázquez JP, Monclús MA, Rojo FJ, Daza R, Sanchez-deAlcazar D, Cortajarena, AL, & Costa, RD. Biogenic fluorescent protein–silk fibroin phosphors for high performing light-emitting diodes. *Materials Horizons*. 2020;7 (7):1790-800.
[132] Xu J, Liu C, Hsu PC, Liu K, Zhang R, Liu Y, Cui, Y. Roll-to-Roll Transfer of Electrospun Nanofiber Film for High-Efficiency Transparent Air Filter. *Nano Lett*. 2016;16 (2):1270-5.
[133] Gao X, Gou J, Zhang L, Duan S, Li C. A silk fibroin based green nano-filter for air filtration. *RSC Advances*. 2018;8 (15):8181-9.
[134] Melikov R, Press D, Kumar B, Dogru IB, Sadeghi S, Chirea M, Yılgör, İ, & Nizamoglu, S. Silk-hydrogel Lenses for Light-emitting Diodes. *Scientific Reports*. 2017;7:7258.
[135] Pham DT, Tiyaboonchai W. Fibroin nanoparticles: a promising drug delivery system. *Drug Deliv*. 2020;27 (1):431-48.
[136] Date AA, Hanes J, Ensign LM. Nanoparticles for oral delivery: Design, evaluation and state-of-the-art. *J Control Release*. 2016;240:504-26.
[137] Pham DT, Saelim N, Tiyaboonchai W. Alpha mangostin loaded crosslinked silk fibroin-based nanoparticles for cancer chemotherapy. *Colloids Surf B Biointerfaces*. 2019;181:705-13.
[138] Altman GH, Diaz F, Jakuba C, Calabro T, Horan RL, Chen J, Lu, H., Richmond, J., & Kaplan, D. L. Silk-based biomaterials. *Biomaterials*. 2003;24 (3):401-16.
[139] Numata K, Kaplan DL. Silk-based delivery systems of bioactive molecules. *Adv Drug Deliv Rev*. 2010;62 (15):1497-508.
[140] Wu J, Wang J, Zhang J, Zheng Z, Kaplan DL, Li G, Wang, X. Oral Delivery of Curcumin Using Silk Nano- and Microparticles. *ACS Biomaterials Science & Engineering*. 2018;4 (11):3885-94.
[141] Pham DT, Saelim N, Tiyaboonchai W. Paclitaxel loaded EDC-crosslinked fibroin nanoparticles: a potential approach for colon cancer treatment. *Drug Deliv Transl Res*. 2020;10 (2):413-24.
[142] Takeuchi I, Shimamura Y, Kakami Y, Kameda T, Hattori K, Miura S, Shirai, H., Okumura, M., Inagi, T., Terada, H., & Makino, K. Transdermal delivery of 40-nm silk fibroin nanoparticles. *Colloids Surf B Biointerfaces*. 2019;175:564-8.
[143] Zhang Y-Q, Shen W-D, Xiang R-L, Zhuge L-J, Gao W-J, Wang W-B. Formation of silk fibroin nanoparticles in water-miscible organic solvent and their characterization. *Journal of Nanoparticle Research*. 2007;9 (5):885-900.

[144] Rajkhowa R, Wang L, Kanwar J, Wang X. Fabrication of ultrafine powder from eri silk through attritor and jet milling. *Powder Technology*. 2009;191:155-63.

[145] Rajkhowa R, Wang L, Wang X. Ultrafine silk powder preparation through rotary and ball milling. *Powder Technology*. 2008;185:87-95.

[146] Lozano-Pérez AA, Rivero HC, Pérez Hernández MDC, Pagán A, Montalbán MG, Víllora G, Lozano-Pérez, AA, Rivero, HC, Pérez Hernández, M del C, Pagán, A, Montalbán, MG, Víllora, G, & Cénis, JL. Silk fibroin nanoparticles: Efficient vehicles for the natural antioxidant quercetin. *Int J Pharm*. 2017;518 (1-2):11-9.

[147] Gou S, Huang Y, Wan Y, Ma Y, Zhou X, Tong X, Huang, J, Kang, Y, Pan, G, Dai, F, & Xiao, B. Multi-bioresponsive silk fibroin-based nanoparticles with on-demand cytoplasmic drug release capacity for CD44-targeted alleviation of ulcerative colitis. *Biomaterials*. 2019;212:39-54.

[148] Ding B, Wahid MA, Wang Z, Xie C, Thakkar A, Prabhu S, Wang, J. Triptolide and celastrol loaded silk fibroin nanoparticles show synergistic effect against human pancreatic cancer cells. *Nanoscale*. 2017;9 (32):11739-53.

[149] Lozano-Pérez AA, Rodriguez-Nogales A, Ortiz-Cullera V, Algieri F, Garrido-Mesa J, Zorrilla P, Rodriguez-Cabezas, ME, Garrido-Mesa, N, Utrilla, MP, De Matteis, L, Martines de la Fuente, J, Cenis, JL, Galvez, J. Silk fibroin nanoparticles constitute a vector for controlled release of resveratrol in an experimental model of inflammatory bowel disease in rats. *Int J Nanomedicine*. 2014;9:4507-20.

[150] Yan H-B, Zhang Y-Q, Ma Y-L, Zhou L-X. Biosynthesis of insulin-silk fibroin nanoparticles conjugates and in vitro evaluation of a drug delivery system. *Journal of Nanoparticle Research*. 2009;11:1937-46.

[151] Xue X, Fu H, Lu S. Study on Antheraea perny Silk Fibroin Nanoparticles Carried Insulin. *Nano Research & Applications*. 2017;03.

[152] Kundu J, Chung YI, Kim YH, Tae G, Kundu SC. Silk fibroin nanoparticles for cellular uptake and control release. *Int J Pharm*. 2010;388 (1-2):242-50.

[153] Zhang Y-Q, Xiang RL, Yan HB, Chen XX. Preparation of silk fibroin nanoparticles and their application to immobilization of L-asparaginase. *Gaodeng Xuexiao Huaxue Xuebao/Chemical Journal of Chinese Universities*. 2008;29:628-33.

[154] Shi P, Abbah SA, Saran K, Zhang Y, Li J, Wong HK, Goh, JCH. Silk fibroin-based complex particles with bioactive encrustation for bone morphogenetic protein 2 delivery. *Biomacromolecules*. 2013;14 (12):4465-74.

[155] Cao TT, Zhou ZZ, Zhang YQ. Processing of β-glucosidase-silk fibroin nanoparticle bioconjugates and their characteristics. *Appl Biochem Biotechnol*. 2014;173 (2):544-51.

[156] Zhu L, Hu RP, Wang HY, Wang YJ, Zhang YQ. Bioconjugation of neutral protease on silk fibroin nanoparticles and application in the controllable hydrolysis of sericin. *J Agric Food Chem*. 2011;59 (18):10298-302.

[157] Kim WJ, Islam R, Kim BS, Cho YD, Yoon WJ, Baek JH, Woo, KM, & Ryoo, HM. Direct Delivery of Recombinant Pin1 Protein Rescued Osteoblast Differentiation of Pin1-Deficient Cells. *J Cell Physiol*. 2017;232 (10):2798-805.

[158] Shahbazi B, Taghipour M, Rahmani H, Sadrjavadi K, Fattahi A. Preparation and characterization of silk fibroin/oligochitosan nanoparticles for siRNA delivery. *Colloids Surf B Biointerfaces*. 2015; 136:867-77.

[159] Mao KL, Fan ZL, Yuan JD, Chen PP, Yang JJ, Xu J, ZhuGe, DL, Jin, BH, Zhu, QY, Shen, BX, Sohawon, Y, Zhao, YZ, & Xu, HL. Skin-penetrating polymeric nanoparticles incorporated in silk fibroin hydrogel for topical delivery of curcumin to improve its therapeutic effect on psoriasis mouse model. *Colloids Surf B Biointerfaces*. 2017;160:704-14.

[160] Pratthana C. Development of amphotericin B-loaded fibroin nanoparticles: a novel approach for topical ocular application. *Journal of Materials Science*. 2020;v. 55 (no. 12):pp. 5268-79-2020 v.55 no.12.

[161] Hassani Besheli N, Mottaghitalab F, Eslami M, Gholami M, Kundu SC, Kaplan DL, Farokhi, M. Sustainable Release of Vancomycin from Silk Fibroin Nanoparticles for Treating Severe Bone Infection in Rat Tibia Osteomyelitis Model. *ACS Applied Materials & Interfaces*. 2017;9 (6):5128-38.

[162] Kim SY, Naskar D, Kundu SC, Bishop DP, Doble PA, Boddy AV, Chan, HK, Wall, IB, & Chrzanowski, W. Formulation of Biologically-Inspired Silk-Based Drug Carriers for Pulmonary Delivery Targeted for Lung Cancer. *Sci Rep*. 2015;5:11878.

[163] Wang Y, Rudym DD, Walsh A, Abrahamsen L, Kim HJ, Kim HS, Kirker-Head, C, & Kaplan, DL. In vivo degradation of three-dimensional silk fibroin scaffolds. *Biomaterials*. 2008;29 (24-25):3415-28.

Chapter 8

Silk Fibroin - A Diagnostic Tool for Targeted Drug Delivery System

Priti R. Ghutepatil[1], Rujuta Barve Joshi[1] and Shivaji H. Pawar[2,3]

[1] Department of Applied Science,
Dr. D. Y. Patil Institute of Technology Pimpri, Pune, Maharashtra, India
[2] Centre for Innovative and Applied Research (CIAR),
T. C. College, Baramati, Maharashtra, India
[3] Center for Interdisciplinary Research,
D. Y. Patil University, Kolhapur, Maharashtra, India

Abstract

Drug delivery system has gained lot of attention because of its tremendous impact in healthcare industry. Amongst several other drug delivery methods, targeted drug delivery method is found to be dominant over other methods due to its potential to specifically target a drug to a region of the affected part. Nanomaterial based targeted drug delivery technique is a method of choice as it has a potential to overcome the limitations of other conventional drug delivery methods. Silk fibroin is one of the well-known materials for its use in targeted drug delivery system as it exhibits excellent properties like biocompatibility, biodegradability, optical transparency, fluorescence etc. The timely summary of silk-fibroin as a diagnostic tool for drug delivery system will provide a research perspective to promote the further improvement and development in the drug delivery system.

1. Introduction

Silk is a naturally derived polymer available in large quantities. It is produced by the process of sericulture or silk farming. In sericulture or silk farming process, the silkworms are cultivated and fed with mulberry leaves [1]. Silk fibres are collected from silkworm's cocoons. South Asia is one of the largest silk fabrics producers. The properties and structure of silk can be analysed using modern technologies and advanced tools in chemistry. Silk fibroin has been suggested as a platform for drug delivery in the form of films, hydrogels and porous 3D scaffolds. Due to hierarchical structure and versatility of silk fibroin, it has become immensely popular among

In: Silk Fibroin
Editor: Shivaji H. Pawar
ISBN: 979-8-88697-402-7

researchers as a promising biomaterial for pharmaceutical and biomedical applications [2, 3]. Many researchers have focused on silk fibroin as a biomaterial with customized properties and remarkable performance for various applications, such as water ultra-filtration systems, biosensors, wearable electronic devices, tissue-engineering and drug delivery systems [4-6].

Silk is secreted by spinning large number of arthropod lineages which is used to build nests, cocoons and webs [7, 8]. The domesticated silkworm, *Bombyxmori (B. mori)*, *Nephilaclavipes* and *Araneusdiadematus* are used to produce extensively characterized silk [9-13]. The cocoon of *B. mori* produces light chain and heavy chain fibroin proteins. These protein chains are coated with sericin proteins. Sericin proteins form the complex fibres of the cocoon case and hold the fibroin fibres together [14-16]. Spider silks are superior in mechanical strength than silkworm silk. *B. mori* silk is available in abundance as compared to spider silks and hence used in commercial industries. Before the hatching process, the eggs laid by silkworm are stored in an incubated and disinfected environment for ten days followed by nourishment of larvae with mulberry leaves for the period of 40-45 days. The larvae spin fibres to form cocoons and these cocoon fibres are untangled into commercial silk [17, 18]. Silk fibroin can be easily fabricated into different forms like nano-fibrous mats, sponges, micro-particles, nanoparticles and composite films for biomedical applications. The various methods are used to process silk fibroin solution and produce various types of delivery systems including hydrogels, scaffolds, films and nanoparticles [19, 20].

In the process of drug delivery, drug carrier plays a key role. The nanoparticles, microspheres, microcapsules, pills, emulsions etc can be used as drug carriers in different drug-controlled release systems. Amongst all, nanoparticles are being widely used as an effective carrier in drug delivery system. Around the 1970s, the nanoparticles were first used as drug carrier by Speiser and Birrenbach. Nanoparticles protect drugs from degradation, intracellular penetration and bioavailability, enhance biological stability and retention time, and reduce risks of toxicity [21-25].

1.1. Sources of Silk Fibroin

B. mori cocoons and *Nephilaclavipes* are commonly used sources to extract silk fibroin. *B. mori* (the domestic silk moth) is an insect from the moth family *Bombycidae*. The silkworm is the larva or caterpillar of a silk moth [26]. *Nephilaclavipes* is usually famed as the golden silk orb-weaver or banana spider. These spiders are excellent web builders, fabricating and using seven distinct types of silk [27].

1.2. Cocoon (Bombyx Mori) Silk

B. mori is a well-known producer of silk throughout the world [28]. Because of the well-equipped facilities, per annum around 400,000 tons of dry silkworm *B. mori* cocoons are available for the textile industry and hence for biomedical applications [29]. The raw silk fibre is prepared from 75 to 83% fibroin and 17 to 25% sericin. Fibroin is a core of two fibres with 10 to 25 μm in diameter and coated with sericin. This sericin acts as glue and holds fibroin fibres together. However, it may cause allergic and immunologic reactions in humans and hence has to be removed by undergoing degumming process [28, 31]. Silk fibroin fibres are made up of two proteins in a 1:1 ratio and joined by a single di-sulphide bond (light and heavy chain respectively). Silk fibroin is mostly made up of repetitive sequences of amino acids, such as serine (12%), glycine (43%) and alanine (30%) [32, 33]. It can set itself into two structural conformations like silk-I (α-helix) and silk-II (β-sheets) [32]. Silk-I can be dissolved in water, but silk-II has solubility in aqueous inorganic salt or concentrated acids because of their capability to break down hydrogen bonding for the stabilization of β-sheets silk-II [34-36].

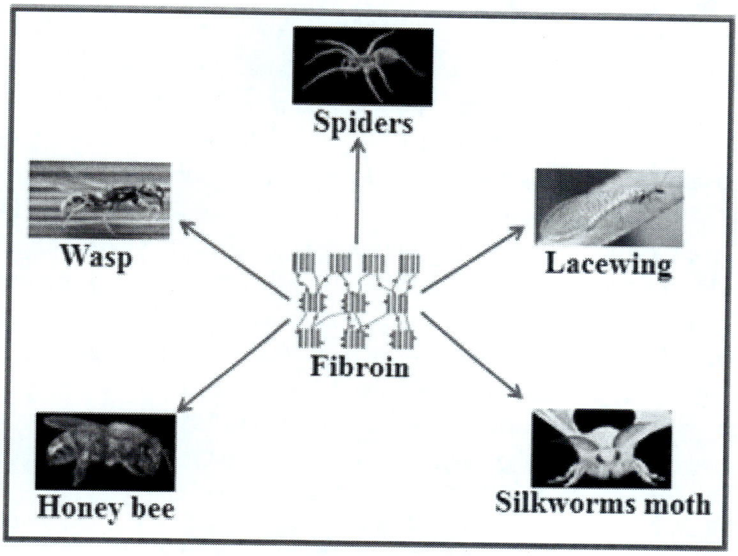

Figure 1. Sources of silk fibroin.

The preparation process of silk fibroin from *B. mori* cocoons is reported elsewhere. Firstly, the cocoons are boiled at 100°C in a Na_2CO_3 aqueous solution for 1 hour followed by boiling it in an alkaline soap solution at 100°C for 30 min. After washing them thoroughly with tap and distilled water, they are allowed to boil in distilled water at 100°C for 20 min. The complete process is repeated thrice. Eventually, the degummed silk is dried at room temperature and humidity for 48 h. Regenerated silk fibroin can be dissolved in an appropriate solvent system, such as 9.3M LiBr, CLiNS aqueous solution, Ajisawa's reagent ($CaCl_2$:H_2O:C_2H_5OH; 1:8:2.5 molar ratio), N-methyl morpholine-N oxide, calcium nitrate aqueous solution, hexafluoro-isopropanol (HFIP), and ionic liquids. Required time and temperature parameters can be adjusted depending up on the solvent and solubility power of the solvent. The aqueous solution of silk fibroin can then be obtained by dialyzing silk fibroin solution against pure, distilled water [37].

1.2.1. Nephila Clavipes

The process of obtaining spider silk from *Nephila Clavipes* is studied in detail. It is much safer than *B. mori* silk because it does not have a sericin coating and is obtained with almost no immunological response [38]. Spider silk is well-known for its remarkable properties like mechanical, thermal stability and high tensile strength [38, 39].

In *Nephila Clavipes* silk, two distinct proteins with similar sequences can be found namely; MaSp1 (major ampullatespidroin) which consists GGX motif (where X = Y, L, or Q), arranged in flexible 310-helical structures and MaSp2 (major ampullatespidroin) which has a GPGXX motif (where X = G, Q, L, or Y) in a repeated sequence and are arranged in elastic β-turn spiral or helical structures [40-42]. Glycine (G) (above 42%) and alanine (A) (above 25%) arethe main amino acids in the spidroin core. Additionally, other amino acid like Glutamine (Q), serine (S), leucine (L), valine (V) proline (PR), and tyrosine (Y) can be found in the core [40-42].

1.3. Structure and Properties of Silk Fibroin

In Silk fibroin, heavy-light complex is formed by forming a link between heavy and light chain via di-sulphide bonds [43-46]. There are two regions- crystalline region and amorphous region. The crystalline region consists of repetitive sequence of glycine-alanine interconnected with serine and tyrosine amino acids. However, amorphous region consists of amino acids like aspartic acid. P25 is a glycoprotein which includes Asn-linked oligosaccharide chains and is hydrophobically linked to heavy and light complex [43]. The heavy chain, light chain, and P25 are the three polypeptides which are found at molar ratio of 6:6:1 and forms cocoon of *B. mori* [43]. The amino acid sequence glycine, serine, alanine, valine, 15 other amino acid types with varying percentage forms the amino acid sequence. 75% of the heavy chain is occupied by the repeated units of Gly-X dipeptidemotif. Repeated sequence of hydrophobic residues of the dipeptide forms the stable antiparallel β-sheet crystallites. 70% of the Gly-X dipeptide motif region is occupied by the peptide sequenceGly-Ala-Gly-Ala-Gly-SerandGly-Ala-Gly-Ala-Gly-Tyr [47-53].

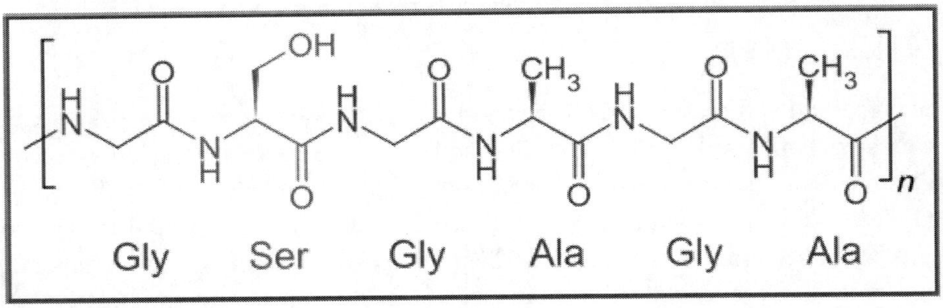

Figure 2. Chemical structure of silk fibroin.

1.4. Mechanical Properties

Significant properties like large break strain, ultimate strength and toughness make them a promising candidate for drug delivery system [54-56]. The toughness of SF fibres is found to be higher than other synthetic fibres like kevlar, carbon fibre, viscous rayon, synthetic rubber and natural materials like wool, cotton, resilin etc. [57,58]. Hence, many researchers are using SF as a scaffold material for load-bearing tissue engineering and biomedical applications [59]. These SF scaffolds are normally made from regenerated silk fibroin (RSF) solutions which are brittle and weak [60]. In order to ensure RSF has good mechanical properties, many different strategies have been implemented [61]. The mechanical properties of RSF produced silk scaffolds can be improved by using porogens [62] and 3D bioprinting [63] technologies. This allows careful handling during surgical procedures as and when required.

1.5. Biocompatibility

Biocompatibility plays a significant role in biomedical applications. The capability of biomaterials is to be in contact with a living system without showing any adverse effects. Hence, the monitoring of biocompatibility is one of the important tasks of the overall safety assessment of biomaterials as it is known to cause negligible immune reaction after implantation [64]. SF is known to be biologically inert and hence is a biocompatible natural polymer [65]. It is also known to have good blood compatibility making them suitable for *in-vivo* experiments and is also approved by FDA approved for its use as a suture material [66, 67]. In 1995, Minoura et al. [68] have successfully grown fibroblast cells on SF functionalized materials. The observed in-vivo inflammatory reaction in case of SF films is similar to that of collagen [69-72].

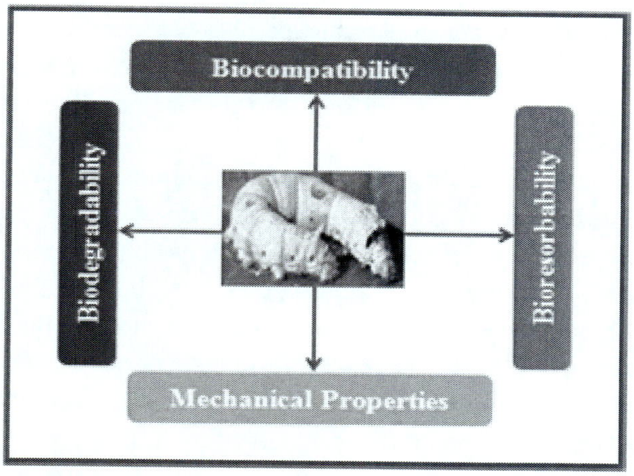

Figure 3. Properties of silk fibroin.

1.6. Biodegradability and Bioresorbability

Biodegradability and bioresorbability properties are the important essential properties for successful implementation of scaffold materials, as they should be replaced progressively with the patient's own cells [73]. Hence, it is required that by-products of biodegradation should be non-toxic and should not interfere with other tissue, organs, and functions during metabolization. SF does not cause an immunogenic response [74]. The degradation occurs after adsorption of enzymes onto the surface of the SF scaffold via surface-bonding domains. The enzymes then digest SF via hydrolysis of ester bonds [74-76]. Degradation of non-crystalline SF structures (hydrophilic blocks) in an enzyme solution results in hydrophobic crystal structures which are further dissolved in enzyme solutions [77]. SF can be proteolytic degraded through enzymes, such as α-chymotrypsin, protease-XIV and collagenase-IA [74-77]. As compared to α-chymotrypsin and collagenase-IA, Protease-XIV found to exhibit higher SF degradation activity. Hence, it is known to achieve the lowest average molecular weight of SF residues [77] and is used enzyme for silk degradation in abundance. Process of synthesis is found to affect the degradation process which results into different morphology of SF particles that dissolves in enzymes [78]. As the degradation products of SF are amino acids and peptides, they are easily absorbed in-vivo [77]. *In vivo* studies undertaken on SF porous scaffolds implanted in lewis rats showed that the scaffolds decomposed within 8 weeks. After 1 year, the implanted scaffolds were fully degraded, due to macrophage degradation [79]. This proves that SF scaffold is biodegradable as well as bioresorbable.

Since the RSF structures have a lower content of β-sheet secondary structure than native silk fibres, native silk fibres degrade at a much slower rate [80]. The rate of degradation of SF is thus dependent on the amount of β -sheet secondary structures present. [81-83].

2. Drug Delivery System

Drug delivery deals with the approaches, formulations, manufacturing techniques, storage systems, and technologies involved in transporting a pharmaceutical compound to its target site so that a desired therapeutic effect can be achieved [84]. It monitors the rate at which a drug is released and the location in the body where it is released. The line of treatment includes oral ingestion, topical, intramuscular, intravenous, rectal, inhalation routes etc. Drug targeting constitutes coordinated drug behaviour, targeting site and pharmaceutical carrier. The target could be a specific organ, a cell, or group of cells in chronic or acute condition demanding treatment with which the drug is going to interact. Drug carrier or a vehicle is a specially engineered substrate necessary for selective and effective transportation of the loaded drug toward diseased sites [85]. Ideally, drug carriers must be harmless, non-immunogenic; biochemically inert, biodegradable, biocompatible, and physico-chemically stable *in-vivo* and *in-vitro*, readily eliminated from the body, minimal drug leakage during transit and unrecognizable by the host's defence mechanisms. They should also have reasonably simple, reproductive and cost-effective preparation processes [86, 87]. Improving or designing several other techniques enhance the use of existing medications. There are five generations of drug delivery system starting from conventional drug delivery system (First generation) to the nanorobots, gene therapy based long term delivery system (Fifth generation) [88].

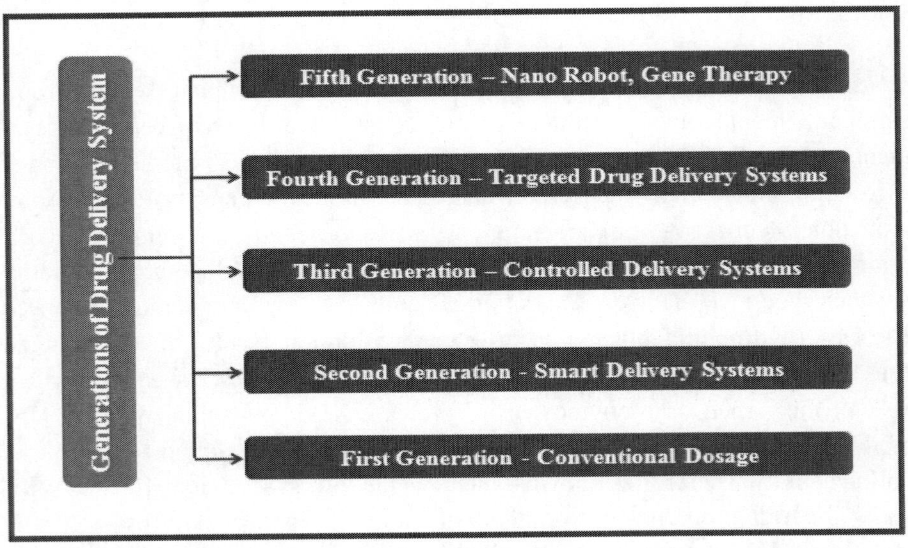

Figure 4. Generations of drug delivery system.

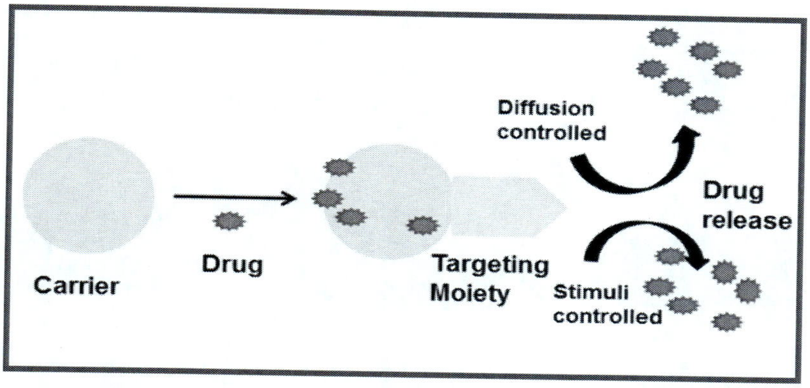

Figure 5. Drug delivery process.

By using First generation conventional drug delivery system for most therapeutic agents, only a small portion of the medication reaches the organ to be affected and major portion do not reach the affected part. This problem can be solved by using targeted drug delivery system, which is the fourth generation of drug delivery system. The concept of targeted drugs was first proposed in 1906 by scientist, Ehrlich. As a theoretical concept, it became most popular and found to be and strong alternative for effective and site-specific treatment. The basic principle lies in the fact that it is used to enhance the concentration of medication in the affected part by minimizing its effect in the unaffected part. It is based on a method that delivers a certain amount of a therapeutic agent for a prolonged period of time to a targeted diseased area within the body. This helps to maintain the required plasma and tissue drug levels in the body; thereby preventing any damage to the healthy tissue via the drug.

It provides drug potency, allows controlled drug release and ensures safety by reducing side-effects, maintaining sustained and consistent blood level, improving patient's compliance, enhancing bioavailability and decreasing dosing frequency. Targeted drug delivery can be used to treat many diseases, such as the cardiovascular diseases and diabetes. However, the most important application of targeted drug delivery is to treat cancerous tumours.

2.1. Properties of Ideal Targeted Drug Delivery [89]

1. It should be non-toxic, bio-degradable, bio-compatible and physico-chemical stable *in-vivo* and *in-vitro*.
2. It should be capable to deliver the drug to target cells or tissue or organ and should have uniform capillary distribution.
3. It should release the dug in a controlled and predictable manner for a suitable period of time.
4. It should efficiently maintain the drug concentration at the targeted site within the therapeutic window for prolong period of time.
5. Minimal drug losses due to leakage of the carrier system should be ensured.
6. Carrier used should be bio-degradable or and get readily eliminated from the body without showing any toxic interaction.
7. Its preparation should be easy or reasonably simple, reproductive and cost effective.

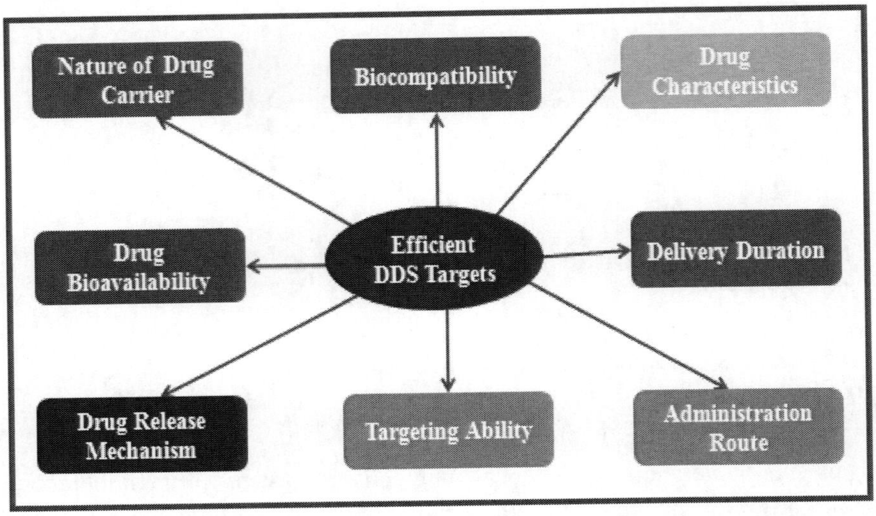

Figure 6. Efficient DDS targets.

2.2. Strategies of Drug Targeting

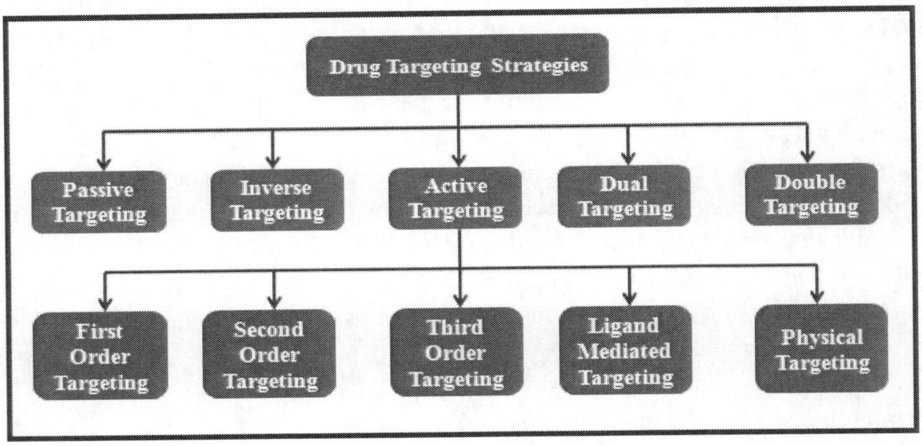

Figure 7. Drug targeting strategies.

2.2.1. Passive Targeting

Passive targeting targets systemic circulation. In this technique, drug targeting occurs because of the body's natural response to physico-chemical characteristics of the drug or drug-carrier system. It depends on the accumulation of drug(s) at areas targeting the site of interest. NPs are used as carriers in passive targeting and are directed to enter blood vessels present at the disease site, which provides the opportunity for significant drug accumulation at the target.

2.2.2. Inverse Targeting

Here, attempts are made to avoid passive uptake of colloidal carrier by RES and hence the process is referred to as inverse targeting. To achieve inverse targeting, RES normal function is suppressed by pre-injecting large amount of blank colloidal carriers or macromolecules like dextran sulphate. This approach leads to saturation of RES and suppression of defence mechanism.

2.2.3. Active Targeting

In this approach, carrier system bearing drug reaches to specific site on the basis of surface modification rather than natural uptake by RES. It mainly relies on the biological interface between target cells and the ligands attached to NPs. Various types of ligands such as, proteins, polysaccharides, nucleic acids, peptides, small molecules etc. have been employed for this purpose [90, 91]. It can be affected at different levels, 1. First order targeting - Restricted distribution of the drug carrier system to the capillary bed of a pre-determined target site, organ or tissue. 2. Second order targeting - The selective drug delivery to a specific cell type such as tumour cells (& not to the normal cells) 3. Third order targeting - Drug delivery specifically to the intracellular organelles of the target cells.

2.2.4. Ligand-Mediated Targeting

In his approach ligands are used as carrier surface group(s), which can selectively direct the carrier to the pre-specified site(s) housing the appropriate receptor units to serve as 'homing device' to the carrier/drug. Most of the carrier systems are colloidal in nature and can be specifically functionalized using various biologically relevant molecular ligands including anti-bodies, polypeptides, oligosaccharides, viral proteins and fusogenic residues. The ligands confer recognition and specificity upon drug carrier, endow them with an ability to approach the respective target selectivity and deliver the drug.

2.2.5. Dual Targeting

In this targeting approach carrier molecule itself have their own therapeutic activity and thus increase the therapeutic effect of drug.

2.2.6. Double Targeting

When temporal and spatial methodologies are combined to target a carrier system, then targeting may be called double targeting. Spatial placement relates to targeting drugs to specific organs tissues, cells or even subs cellular compartment, whereas temporal delivery refers to controlling the rate of drug delivery to target site.

2.2.7. Combination Targeting

Combination targeting systems are equipped with carriers, polymers and homing devices of molecular specificity that could provide a direct approach to target site.

2.3. Advantages and Disadvantages of Targeted Drug Delivery

2.3.1. Advantages

1. Simplified drug administration protocols.
2. Reduction of toxicity thereby reducing harmful systemic effects.
3. Proper administration of drug helps producing the desired effect.

4. Avoidance of hepatic first pass metabolism.
5. Enhances the absorption of target molecules such as peptides and particulates.
6. Dose is less compared to conventional drug delivery system.
7. No peak and valley plasma concentration.
8. Selective targeting to infections cells that compare to normal cells.

2.3.2. Disadvantages

1. Rapid clearance of targeted system.
2. Immune reactions against intravenous administered carrier system.
3. In-sufficient localization of targeted systems into tumour cells.
4. Diffusion and re-distribution of released drugs.
5. Requires highly sophisticated technology for the formulation.
6. Requires skill for manufacturing storage and administration.
7. Drug deposition at the target site may produce toxicity symptoms.
8. Difficult to maintain stability of dosage form.

2.4. Silk Fibroin-Based Targeted Drug Delivery

Silk-based delivery systems refer to the use of silk protein as a polymer for various drug delivery systems. Polymeric drug delivery systems are widely known to be a new efficient alternative to the conventional formulations because of their favourable, flexible characteristics and easy production at industrial scale [92]. Polymer therapeutics has the capability of working either as the bio-active molecule (polymer drug) or as the inert carrier to which a drug can be covalently linked. Some of these polymer composites include polymer-drug conjugates, polymer-antibody conjugates, polymer-DNA conjugate, dendrimers, polymeric micelles etc. [93]. Classification of polymers for drug delivery can be done on the basis of following points:

- Origin of the polymer which can: be synthetic, natural, or a combination of both;
- Chemical nature, biodegradability or non-biodegradability of polymer being used;
- Hydrophilic or hydrophobic nature of polymer [94,95].

However, each of these features has their own limitations, like (i) though they are biodegradable and available in abundance, reproduction and purification of natural polymers is difficult; (ii) High immunogenicity property of synthetic polymers prevents them from long-term usage; (iii) Non-biodegradable polymers should be removed by surgery once they release the drug at the targeted site.

Polymer acting as an inert carrier to which a drug can be conjugated is quite advantageous as it helps to improve the pharmaco-kinetic and pharmaco-dynamics properties of bio-pharmaceuticals through various ways such as, it increases the plasma half-life and decreases the immunogenicity, it boosts stability of biopharmaceuticals, improves solubility of low molecular weight drugs, and also has potential for targeted drug delivery system [96-98].

The nature of the drug, the target, and the disease state determine the selection of the carrier to be used. The details of some of these drug carriers are mentioned in Table 1.

Table 1. Advantages and disadvantages of various drug carriers

Drug carrier	Advantages	Disadvantages	Reference
Liposomes	Less toxic, small size, able to overcome bacterial resistance, bioavailability.	Unstable, once administered it cannot be removed, poor control over drug release, leakage of encapsulated drug during storage, low solubility.	[99]
Microspheres	Sustained control over drug release. Protection of drugs against environment, small and spherical size, enhance the biological half-life and also improve the bioavailability.	Controlled release rate of microspheres may vary due to intrinsic and extrinsic factors, toxicity, low drug loading is done in case of parenteral microspheres.	[100
Micelles	Small particle size, good thermos-dynamic solution stability, extended release of various drugs, ability to solubilize hydrophobic molecules.	Poor drug-loading efficiency, poor physical stability *in vivo*, and insufficient cellular interaction of neutral micelles with malignant cells.	[101]
Niosomes	Non-ionic, less toxic, osmotically active and stable, controlled drug release.	Poor skin permeability, breaking of vesicles, leakage of drug, aggregation, and fusion of vesicles.	[102]
Quantum dots	Possesses long-term, multiplexed, and quantitative imaging and detection, can be used as ultrasensitive and multicolour imaging of molecular targets.	Their delivery process across cells may be dangerous for the cell and may destroy it. In other cases, a QD may be toxic for cells.	[103]
Nanoparticles	Increases stability, delivers a higher concentration of pharmaceutical agent to a desired location.	Toxicity, low degradability, susceptible to oxidation, small size & large surface area can lead to particle aggregation, limited drug loading.	[104]
Dendrimers	High degree of loading of the active drug molecules, uniform size, water solubility, modifiable surface functionality and available internal cavities.	Low hydro solubility and high non-specific toxicity.	[105]
Silk Fibroin	High bio-compatibility, bio-degradability, bio-resorbility, high encapsulation capacity, low immunogenicity, controllable drug release kinetics and tuneable mechanical properties, non-toxic.		[106]

However, some of their drawbacks like long term toxicity to human and environmental biological system; non-biocompatibility limit their usage for the various biomedical applications and hence need to be overcome [107]. Replacing these materials by polymer based nanomaterials helps them to overcome such drawbacks. Utilizing polymer based nano-materials for the targeted drug delivery system results into sustained drug release, effective loading of drugs, precise targeting of the tumour site, enhanced biocompatibility, less toxicity, improved patient's compliance etc. Some of the well-known natural polymers such as gelatine, chitosan, alginates, albumin, starch; guar gum, fibroin etc. are the potential candidates exhibiting high bio-compatibility and bio-degradability [108]. One of these several well-known polymers, a unique natural protein viz; silk fibroin extracted from silkworm silk is known to be a potential biopolymer for various biomedical and pharmaceutical applications due to its excellent physiochemical properties which includes bio-compatibility, bio-degradability, bio-resorbability, low immunogenicity, high encapsulation capacity, controllable drug release kinetics and tuneable mechanical properties. It also has the potential to increase the stability of the loaded APIs [92, 109].

Silk protein has a unique combination of properties, which are rare among natural polymers. Presence of carboxyl and amino groups in SF allows for bio-functionalization with various bio-molecules or ligands which finds application in targeted drug delivery [110]. Many forms of SF including fibres, films, foams, tubes, gels and particles have been investigated for these purposes [111-114]. In order to enhance the physico-chemical, mechanical and bio-degradation properties of SF for specific applications, it has been blended with chitosan, cellulose, polyvinyl alcohol and polyethylene oxide [115-122].

Hydrogels can be synthesized by various techniques by using Hydrogels SF aqueous solution. The transition from solution to gel can be triggered by physio-chemical or chemical processes using natural polymers or synthetic reagents [123]. Tomeh et al. reported the use of curcumin-loaded gel scaffolds which not only improves process of protein adsorption and sustained release of curcumin, but also enhances bacterial growth inhibition by 6-fold against *S. aureus* [124].

Silk fibroin in the form of Films has also gained importance for its potential use as a bio-material in pharmaceutical formulations and tissue engineering applications [125]. Several techniques of preparing SF films are reported in literature [126-128]. Another study reports the structural changes with enhanced physical properties in SF when blended with other polymers such as sodium alginate (SA) before casting the film [129]. Hence, SF/SA blend can provide unique tuneable characteristics which can be beneficial in pharmaceutical applications.

Recently, nanoparticles have generated great excitement in the field of drug delivery because of their ability to provide new opportunities to overcome the limitations of conventional delivery methods with regards to the drugs. Nanoparticles are defined as a set of substances where at least one dimension is less than approximately 100 nm. They are of great interest because of its unique optical, magnetic, electrical and other properties emerging out at the nanoscale. Frequently used nanomaterials for drug delivery system includes liposomes, metal nanoparticles like gold and silver, magnetic nanoparticles, nanocapsules, micelles, nano-shell, Inorganic non-metallic nanoparticles like quantum dots, iron oxide, silicon, graphene etc. [130]. Seib et al. [131] designed lysosomotropic SF nanoparticles for pH-dependent release of the anticancer drug doxorubicin to overcome drug resistance. SF nanoparticles are largely employed for controlled release of the loaded drug at the site of action. Several methods of fabrication of SF nanoparticles are reported amongst which, the salting-out method is the most popular [132,133]. Addition of PVA results improve the morphological parameters of the SF-particles [134-136]. The properties of the SF particles can also be manipulated by blending with other polymers. Song et al. [137] have recently reported the manipulation of properties of SF particles by blending them with varying concentration of polyethyleneimine (PEI) which allows fine tuning of the drug delivery by controlling the size and zeta potential of the particles.

Along with outstanding environmental and thermal stabilities, SF also possesses excellent optical properties like refractive index of 1.546 and high light-guiding efficiency making them promising candidate in the field of green photonics [138-141]. In this context, the first SF-lasers were proposed by Omenetto and co-workers [142]. Besides several important findings in lasers and light guiding applications using various SF derivatives, limited research has been carried out on SF based Light Emitting Diodes (LED). In addition to it, SF is also found to exhibit excellent luminescent properties. Veronica et al. [143] proposed the fabrication of FP-SF based efficient down converting bio-hybrid light emitting diodes. Bio-hybrid LEDs are slowly replacing the rarer earth based inorganic phosphors due to their negative ecological impact, toxicity, visual and non-visual health effects under long exposure to blue light. Optical imaging is a powerful imaging method with the advantages of no radiation, no invasion, high resolution and good controllability, but its penetration is poor. Photostability is the basic property of the fluorescent SF for functional silk fibre essential for biomedical applications. Hence, if produced in different forms, they can be used in various biomedical fields. The multifunctional SF based materials with fluorescent can be used in *in-vitro* and *in-vivo* imaging purposes due to their excellent fluorescent properties and biocompatibility.

Fluorescence imaging is usually performed by using fluoresce in to generate fluorescence signals. Near-infrared fluorescence (NIRF) probes are widely used because of their strong penetrating power and safety. They have been used in small animal living imaging systems and clinical tumour transformation. At present, a large number of nano-drug carriers such as liposomes, metal, or non-metallic nanoparticles can enclose NIRF to achieve optical imaging of blood vessels [144-147]. Its application in cardio-vascular disease imaging has been paid more and more attention. McCarthy et al. [147] bound the group of near infrared light activated therapeutic (NILAT) with macrophage-targeted magnetic nanoparticles (MNPs) and prepared a kind of diagnostic and therapeutic nanoparticles.

Since past few decades, fluorescent SF is found to be a well-known bio-technological tool which helps in examining the function of a particular gene of interest by monitoring, visualizing, and measuring protein expression in living cells. Several methods have been reported by researchers for fabrication of fluorescent silk which possesses tremendous potential and offers biomedical applications, including drug delivery, bio-imaging, and sensing [148] due to its excellent fluorescence property achieved through the insertion of the fluorescence gene into silk cocoons. The fabrication of silk materials with microstructures, such as 3D PCs, micro lenses, and optical arrays, is a biocompatible method that provides unique opportunities for structurally coloured bio-degradable materials in microscale implantable bio-sensing and targeted therapeutics applications [149-151]. A set of bio-active diffractive optical elements micro fabricated using functionalized SF films are applied in sensing applications, including hydration sensing, biological concealment, therapeutic treatment, and in *in-vitro* and *in-vivo* drug-release monitoring upon degradation [152]. The addition of the fluorescence properties into SF via gene insertion escapes the conjugation step to insert a fluorescently tagged protein such as green fluorescent protein (GFP). GFP is gene-targeted with the fibroin gene to obtain green, fluorescent silkworm cocoons [153]. Thus, fluorescent SF possesses enormous potential for its use for biomedical applications, including drug delivery, bio-imaging, and sensing. Methodology for preparing a fluorescent SF solution and various forms like membrane, sponge, nanoparticle, film, hydrogel, nano-fibre, electrospun fibre etc. for biomedical applications was proposed for the first time by Kim et al. [154]. Mechanical properties of the fluorescent silk produced through feeding *B. mori* silkworms with carbon nano dots are found to be strengthened than that of the natural silk [155]. Incorporation of organic/inorganic molecules in the SF materials found to exhibit excellent optical properties making them promising candidate for different optical applications such as optical elements, nonlinear optics, light energy conversion, photochemical reactions, UV resistance, photo-catalysis, light-assisted synthesis, sensors, and bio-imaging [156].

Fluorescent dyes help in monitoring and visualizing the internalization of carriers or measuring the loading and release of drugs in controlled drug delivery systems [157, 158]. Khalid et al. [159] has demonstrated the drug release kinetics of the nano diamond-SF-doxorubicins (NDSX) spheres by loading spheres with anthracyclineDoX as it stabilizes and releases the drug in a precise manner and ND provides a fluorescent modality for imaging.

These NDSX spheres works as a dual-functional nanomaterials for drug delivery and succeeding imaging within biological structures and could be useful in targeted drug delivery [160].

Though extensive research is still being carried out for improvising the system of targeted drug delivery, above discussed development of polymer based targeted drug delivery system justifies the important significance of using silk fibroin as a potential tool for the targeted drug delivery system. During the process of targeted drug delivery utmost care should be taken that the process of releasing desired drug should be carried out in a controlled manner by controlling the rate of drug delivery and duration required for the process. For its implementation, third generation-controlled release drug delivery system can be used along with the targeted drug delivery system. The combination of the two systems will be a boon to the medical fraternity.

2.5. Controlled Release Drug Delivery System

Controlled release drug delivery systems refer to controlled release of the encapsulated API (Active Pharmaceutical Ingredient) in specific amounts over a certain period of time. It is necessary for maintaining the therapeutic concentrations of the drug in the blood or site of action for a longer period, to reduce the administration rate and the adverse drug reactions [161]. The widely used controlled release formulations are mainly composed of FDA approved synthetic polymers such as PEG and PLGA [162]. However, there exist certain restrictions on its utilization. Other natural polymers such as SF offer tuneable sustained release kinetics and stabilization of the loaded APIs because of which they have gained more attention for its use in controlled drug release systems.

2.5.1. Advantages and Disadvantages of Controlled or Sustained Release Drug Delivery [163]

2.5.1.1. Advantages

1. Decreased local and systemic side effects help in reducing gastro-intestinal irritation.
2. Better drug utilization helps in reducing the amount of drug used.
3. Helps in improving the efficiency in treatment, therapy optimization, uniformity in blood concentration, bio-availability of some drugs, patient compliance, less frequent dosing, reduced night-time dosing, reduced patient care time.
4. Reduction in fluctuation in drug level helps to have more uniform pharmacological response, lesser reduction in drug activity with chronic use.
5. Although the initial costing of treatment is greater than that of conventional dosage forms the average cost of treatment over an extended time period may be less.

2.5.1.2. Disadvantages

1. Due to incomplete release, increased first-pass metabolism and instability, insufficient residence time for complete release, site specific absorption, pH dependent stability,

etc systemic availability is found to decrease as compared to that of immediate release conventional dosage forms.

2. Poor *in-vitro* and *in-vivo* correlation.
3. Difficulty in drug retrieval is observed in case of toxicity, poisoning or hypersensitivity reactions.
4. Reduced potential for dose adjustment of drugs normally administered in varying strengths.

2.5.2. Mechanisms Involved for Controlling Drug Release

1. *Complexation*: It involves ion exchange resins or other adsorption agents.
2. *Matrix embedment*: It refers to designing matrix-based controlled release products with the help of long chain polymers or waxy lipids for embedding drugs in it which can be presented as part of the multi-layered tablets or tablet-in-tablet designs.
3. *Coated reservoir systems:* It is commonly used in multi-particulate dosage forms where diffusion controlling membranes are used to seal the drug reservoir, controlling the drug release rate etc.

2.5.3. Factors Responsible for Designing a Controlled Release Drug Delivery System

These parameters are broadly classified as formulation related and drug related.

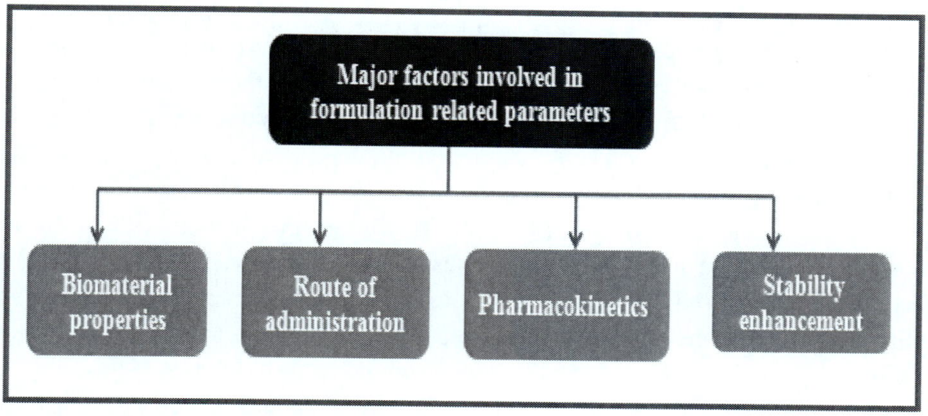

Figure 8. Major factors involved in formulation related parameters.

1. *Biomaterial properties*: Biocompatibility, surface chemistry, hydrophilicity, degradability, mechanical properties, etc.
2. *Route of administration*: Oral, renal, ocular, topical, parenteral, nasal, buccal, vaginal, rectal, etc.
3. *Pharmacokinetics*: Therapeutic index, therapeutic window, drug release kinetics, dosage forms, etc.
4. *Stability enhancement*: Small molecules, peptides, genes, growth factors, colloidal/non-colloidal particles, etc.

2.5.4. Major Factors Involved in Drug-Related Parameters Are [165]

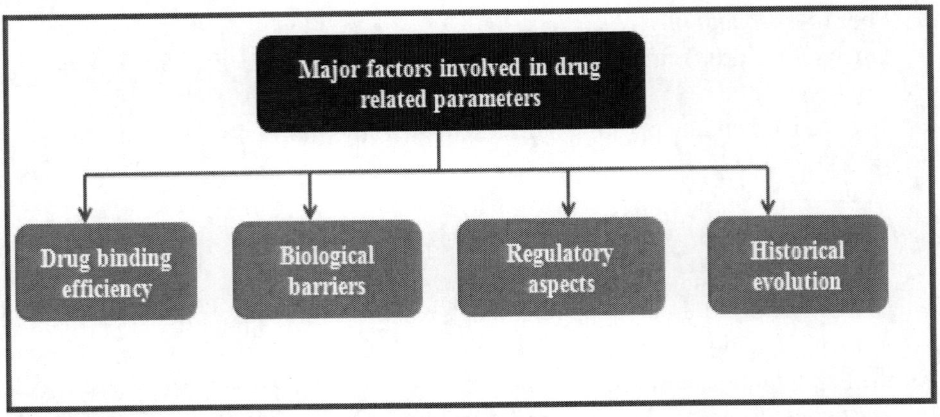

Figure 9. Major factors involved in drug related parameters.

1. Drug binding efficiency with plasma proteins.
2. Biological barriers– Mucosal, endothelial, epithelial.
3. Regulatory aspects – Animal model and clinical translation.
4. Historical evolution- Macro, micro and nano.

2.5.5. Classification of Controlled Release Drug Delivery Systems

Controlled release drug delivery systems are classified based on the mechanism of drug release from the dosage form into dissolution-controlled, diffusion-controlled, water penetration-controlled (osmotic pressure-controlled and swelling-controlled), chemically controlled and nanoparticle-based systems [166].

In dissolution-controlled release systems, drugs are either coated with or encapsulated within slowly dissolving polymeric membranes (reservoir systems) or matrices (monolithic systems). In reservoir systems, drugs are protected inside polymeric membranes with low solubility. Most of the conventional immediate-release tablets, pills and effervescent tablets are dissolution-controlled systems, where the rate-limiting step is dissolution [167, 168].

In diffusion-controlled release systems, drugs are trapped in and released via diffusion through inert water-insoluble polymeric membranes (reservoir systems) or polymeric matrices (monolithic systems) [166].

In monolithic or matrix-controlled delivery systems, the drug is either dissolved or dispersed homogenously throughout the polymer matrix. The drug release is through diffusion when the outside layer that is exposed to the solution gets dissolved first, allowing drugs to diffuse out of the matrix. In monolithic systems, where a drug is dissolved, drugs are loaded below the solubility limit. As the size of the matrix decreases, the drug released decreases. Here, the drug release is non-zero order. In monolithic systems where the drugs are dispersed in the polymer matrix, drugs are loaded above the solubility limit [166].

Water penetration controlled drug delivery systems are classified as osmotic pressure controlled drug delivery systems and swelling controlled drug delivery systems. The rate control is dependent on water penetration into the system. Osmotic drug delivery uses the osmotic pressure for controlled delivery of drugs by using osmogens. Osmosis refers to the process of movement of solvent from a lower concentration of solute towards a higher concentration of solute across the semi-permeable membrane [169].

In swelling controlled drug delivery systems, the drug is dispersed or dissolved in the hydrophilic polymer when in a glassy (hard and rigid) state [170].

Chemically controlled drug delivery systems change their chemical structure when exposed to the biological milieu. These are made of bio-degradable polymers which degrade in the body as a result of natural biological processes, eliminating the need to remove the delivery system after exhausting an active agent from the system [170]. These are classified into two types: polymer-drug dispersion system and polymer-drug conjugate systems.

In polymer drug dispersion systems, the drug is uniformly dispersed or dissolved in a bio-degradable polymer and released through degradation of polymers under physiological conditions. In polymer drug conjugate systems, the drug is chemically conjugated with the polymer either by covalent bonding or grafting to the backbone of the polymer. The drug releases through the cleavage of polymer-drug bonds under physiological conditions [171].

2.5.6. Evolution of the Controlled Release Dosage Forms

- *First-generation*: The first generation is of basics of controlled release dosage form drugs [1950–1980]. Mechanisms employed to accelerate the oral and transdermal formulation includes dissolution, osmosis, diffusion, and ion exchange. Amongst them, diffusion and dissolution-controlled systems are the most commonly used mechanisms of drug delivery. With these drugs, the correlation between *in-vitro* and *in-vivo* formulation was well understood and there were no biological barriers detected for this generation [172]. Also, good control of physico-chemical properties of delivery can be achieved.
- *Second-generation:* These are less successful. Unlike the first generation they have formulations for prolonged release using bio-degradable polymers for delivering proteins and peptides. During this period, pulmonary delivery systems were developed for delivering insulin. Due to its lesser bioavailability, it is delivered many times higher per dose than is required for the parenteral injection which results in adverse effects. In the last decade of the second generation, nanoparticles that target the gene and the tumour were studied [173].
- *Third generation*: The new technologies for drug delivery are delivery of poorly water soluble drugs, long term and non-invasive technology for delivering protein/nucleic acid/peptide, drug delivery to the targeted site using nanoparticles and the drug delivered through self-regulation [173].

Conclusion

Targeted drug delivery system is a world-wide accepted system amongst several other systems due to its effective and site-specific treatment. Naturally derived biopolymer; silk fibroin has been proven to be an ideal bio-material for targeted drug delivery system and an attractive natural polymer for drug delivery due to its excellent properties such as good bio-compatibility, modulated bio-degradability, large scale production, easy modification, and self-assembling property. Owing to the development of nanomaterial, SF based nanoparticles are also proven to be a promising carrier for delivery of various drugs, including small molecule drugs, protein drugs, and gene drugs, etc. Various forms of SF can be fabricated for the fruitful development of various types of drug delivery systems. Each of these SF-based systems exhibit promising features for different biomedical applications. It is observed that the fluorescent properties do not affect the biocompatibility of the fluorescent SF. Thus, fluorescent SF based nanomaterial has also gained ample attention for biomedical applications. Extensive research has been carried out on fluorescent SF for its applications in the biomedical and bio-technological fields. Controlled drug delivery systems are superior to the conventional method as they enhance bioavailability, maintain drug plasma level within the therapeutic window and expanse drug release. Controlled drug delivery improves stability, drug solubility and provides the specific delivery of drugs with an anticipated rate and mechanism to particular cells.

Hence, SF is found to be a suitable diagnostic tool for biomedical applications using targeted drug delivery system. There is a still huge scope for the development of SF based biomaterials by the fusion of various advanced technology to explore its applications in the biomedical field.

References

[1] Rockwood D. N., Preda R. C., Yücel T., Wang X, Lovett M. L., and Kaplan D. L. 2011. "Materials fabrication from Bombyxmori silk fibroin." *Nature Protocols* 6: 1612-1631.

[2] Altman, G. H., F. Diaz, C. Jakuba, T. Calabro, R. L. Ho- ran, J. S. Chen, H. Lu, J. Richmond and D. L. Kaplan. 2003. "Silk-Based Biomaterials," *Biomaterials,* 24(3): 401-416.

[3] Ciocci, M., Cacciotti, I., Seliktar, D., Melino, S. "Injectable silk fibroin hydrogels functionalized with microspheres as adult stem cells-carrier systems." *Int. J. Biol. Macromol.* 108: 960–971.

[4] Lammel, A. S., Hu, X., Park, S. H., Kaplan, D. L., Scheibel, T. R. 2010. "Controlling silk fibroin particle features for drug delivery. *Biomaterials*." 31: 4583–4591.

[5] Zhou, Z., Shi, Z., Cai, X., Zhang, S., Corder, S. G., Li, X., Zhang, Y., Zhang, G., Chen, L., Liu, M. 2017. "The Use of Functionalized Silk Fibroin Films as a Platform for Optical Diffraction-Based Sensing Applications." *Adv. Mater*. 29: 1–7.

[6] Wang, C., Li, X., Gao, E., Jian, M., Xia, K., Wang, Q., Xu, Z., Ren, T., Zhang, Y. 2016. "Carbonized Silk Fabric for Ultra stretchable, Highly Sensitive, and Wearable Strain Sensors." *Adv. Mater*. 28: 6640–6648.

[7] Porter, D., Vollrath, F. 2009. "Silk as a biomimetic ideal for structural polymers." *Adv. Mater*. 21: 487–492.

[8] Kaplan, D. L., Adams, W. W., Farmer, B., Viney, C. 1994. "Silk, Biology, Structure, Properties, and Genetics. In *Silk Polymers - Materials Science and Biotechnology*." Eds.; ACS Publications, Washington DC, USA, 544: 2–16.

[9] Pereira, R. F. P., Silva, M. M., de Zea Bermudez, V. 2015. "Bombyxmori Silk Fibres, An Outstanding Family of Materials." *Macromol. Mater. Eng*. 300: 1171–1198.

[10] Liu, Y. Q., Qin, L., Li, Y. P., Wang, H., Xia, R. X., Qi, Y. H., Li, X. S., Lu, C., Xiang, Z. H. 2010. "Comparative genetic diversity and genetic structure of three Chinese silkworm species Bombyxmori L. (Lepidoptera, Bombycidae), Antheraeapernyi Guérin-Meneville and Samiacynthiaricinidonovan (Lepidoptera, Saturniidae)." *Neotrop. Entomol*. 39: 967–976.

[11] MurugeshBabu, K. 2017. "Silk from silkworms and spiders as high-performance Fibres." In *Structure and Properties of High-Performance Fibres*. Elsevier Ltd., Amsterdam, The Netherlands, 327–366.

[12] Vepari C, Kaplan D. L. 2007. "Silk as a biomaterial." *Prog. Polym. Sci*. 32: 991–1007.

[13] Omenetto F. G., Kaplan D. L. 2010. "New opportunities for an ancient material." *Science* 329: 528–31.

[14] Kundu B, Rajkhowa R, Kundu S. C., Wang X. 2013. "Silk fibroin biomaterials for tissue regenerations." *Adv. Drug. Deliv. Rev.* 65: 457–70.

[15] Kazmierska K., Florczak A., Piekos K., Mackiewicz A., Dams-Kozlowska H. 2011. "Engineered spider silk, the intelligent biomaterial of the future, Part II." *Postepy. Hig. Med. Dosw.* 65: 389–96.

[16] Schacht K, Scheibel T. 2014. "Processing of recombinant spider silk proteins into tailor-made materials for biomaterial applications." *Curr. Opin. Biotechnol.* 29C :62–69.

[17] Heim M, Keerl D, Scheibel T. 2009. "Spider silk, from soluble protein to extraordinary Fibre." *Angew ChemInt Ed En.* 48: 3584–96.

[18] Wenk E., Merkle H. P., Meinel L. 2011. "Silk fibroin as a vehicle for drug delivery applications." *Journal of Controlled Release* 150, 128-141.

[19] Seib, F. P. 2018. "Reverse-engineered silk hydrogels for cell and drug delivery." *Ther. Deliv.* 9: 469–487.

[20] Ponrasu T., Ramachandran B., Kannan R., Muthuvijayan V. 2017. "Morin incorporated polysaccharide–protein (psyllium–keratin) hydrogel scaffolds accelerate diabetic wound healing in Wistarrats." *J. Biomed. Mater. Res. B. Appl. Biomater.* 105: 1401-1408.

[21] Kreuter, J. 2004. "Nanoparticles as drug delivery systems." *Encycl. Nanosci. Nanotechnol.* 7: 161–180.

[22] Birrenbach, G., Speiser, P. P. 1976. "Polymerized micelles and their use as adjuvants in immunology." *J. Pharm. Sci.* 65: 1763–1766.

[23] Alexis, F., Pridgen, E., Molnar, L. K., Farokhzad, O. C. 2008. "Factors affecting the clearance and bio-distribution of polymeric nanoparticles." *Mol. Pharm.* 5: 505–515.

[24] Kumari, A., Yadav, S. K., Yadav, S. C. 2010. "Biodegradable polymeric nanoparticles based drug delivery systems." *Colloids Surf.* B 75: 1–18.

[25] Elzoghby, A. O., Samy, W. M., Elgindy, N. A. 2012. "Albumin-based nanoparticles as potential controlled release drug delivery systems." *J. Control. Release* 157: 168–182.

[26] Arunkumar, K. P., Metta, M., Nagaraju, J. 2006. "Molecular Phylogeny of Silk moths Reveals the Origin of Domesticated Silk moth, Bombyx Mori from Chinese Bombyx Mandarina and Paternal Inheritance of Antheraea Proylei Mitochondrial DNA. *Mol. Phylogenet.* E 40: 419–427.

[27] Babb, P. L., Lahens, N. F., Correa-Garhwal, S. M., Nicholson, D. N., Kim, E. J., Hogenesch, J. B., Kuntner, M., Higgins, L., Hayashi, C. Y., Agnarsson, I. 2017. "The NephilaClavipes Genome Highlights the Diversity of Spider Silk Genes and Their Complex Expression. *Nat. Genet.* 49, 895–903.

[28] Sionkowska, A., Grabska, S., Lewandowska, K., Andrzejczyk, A. 2016. "Polymer Films Based on Silk Fibroin and Collagen—The Physico-Chemical Properties." *Mol. Cryst. Liq. Cryst.* 640, 13–20.

[29] Murugesh. B. K. 2017. "13- Silk from silkworms and spiders as high-performance Fibres," in *Structure and Properties of High-Performance Fibres*, G. Bhat, Ed., pp. 327–366, Wood head Publishing, Oxford.

[30] Puppi, D., Chiellini, F, Piras, A. M., Chiellini, E. 2010. "Polymeric Materials for Bone and Cartilage Repair." *Prog. Polym. Sci.* 35: 403–440.

[31] Melke, J., Midha, S., Ghosh, S., Ito, K., Hofmann, S. 2016. "Silk fibroin as biomaterial for bone tissue engineering." *Acta Biomater.* 31: 1–16.

[32] Sangkert, S., Meesane, J., Kamonmattayakul, S., Chai, W. L. 2016. "Modified Silk Fibroin Scaffolds with Collagen/Decellularized Pulp for Bone Tissue Engineering in Cleft Palate, Morphological Structures and Biofunctionalities." *Mater. Sci. Eng.* C 58: 1138– 1149.

[33] Hashimoto, T., Taniguchi, Y., Kameda, T., Tamada, Y., Kurosu, H. 2014. "Changes in the Properties and Protein Structure of Silk Fibroin Molecules in Autoclaved Fabrics." *Polym. Degrad. Stab.* 112: 20–26.

[34] Kundu, B., Rajkhowa, R., Kundu, S. C., Wang, X. 2013. "Silk Fibroin Biomaterials for Tissue Regenerations. *Adv. Drug Deliver. Rev.* 65: 457–470.

[35] Chen, J., Vongsanga, K., Wang, X., Byrne, N. 2014. "What Happens during Natural Protein Fibre Dissolution in Ionic Liquids. *Materials.* 7: 6158–6168.

[36] Kundu, B., Kurland, N. E., Yadavalli, V. K., Kundu, S. C. 2014. "Isolation and Processing of Silk Proteins for Biomedical Applications." *Int. J. Biol. Macromol* 70: 70–77.

[37] Qiang, L., Chuan-Bao, C., Ying, Z., He-Sun, Z. 2004. "Preparation and Properties of Insoluble Fibroin Films by Novel Method." *Chem. J. Chin. Univ*. 25: 1752–1755.

[38] Huemmerich, D., Helsen, C. W., Quedzuweit, S., Oschmann, J., Rudolph, R., Scheibel, T. 2004. "Primary Structure Elements of Spider Dragline Silks and Their Contribution to Protein Solubility." *Biochemistry* 43: 13604–13612.

[39] Zhao, Z., Li, Y., Xie, M. 2015. "B. Silk Fibroin Based Nanoparticles for Drug Delivery." *Ijmrs*16: 4880–4903.

[40] Hu, X., Vasanthavada, K., Kohler, K., McNary, S., Moore, A. M. F., Vierra, C. A. 2006. "Molecular Mechanisms of Spider Silk." *Cell. Mol. Life Sci*. 63: 1986–1999.

[41] Lewis, R. V. 2006. "Spider Silk, Ancient Ideas for New Biomaterials." *Chem. Rev*. 106: 3762–3774.

[42] Humenik, M., Smith, A. M., Scheibel, T. 2011. "Recombinant Spider Silks—Biopolymers with Potential for Future Applications. *Polymers."*3: 640.

[43] Inoue, S., Tanaka, K., Arisaka, F., Kimura, S.; Ohtomo, K.; Mizuno, S. 2000. "SF of B. mori is secreted, assembling a high molecular mass elementary unit consisting of H-chain, L-chain, and P25, with a 6,6,1 molar ratio."*J. Biol. Chem*. 275: 40517–40528.

[44] Tanaka, K., Inoue, S., Mizuno, S. 1999. "Hydrophobic interaction of P25, containing Asn-linked oligosaccharide chains, with the H-L complex of SF produced by B. mori." *Biochem. Mol. Biol*. 29: 269–276.

[45] Kundu, J., Chung, Y. I., Kim, Y. H., Tae, G., Kundu, S. C. 2010. "Silk fibroin nanoparticles for cellular uptake and control release." *Int. J. Pharm*. 388: 242–250.

[46] Mondal, M., Trivedy, K., Kumar, S. N. 2007. "The silk proteins, sericin and fibroin in silk worm, Bombyxmori Linn-A review." *Caspian J. Environ*. Sci. 5: 63–76.

[47] Zhou, C. Z., Confalonieri, F., Jacquet, M., Perasso, R., Li, Z. G., Janin, J. 2001. "Silk fibroin, Structural implications of a remarkable amino acid sequence." *Proteins Struct. Funct. Bioinform*. 44: 119–122.

[48] Asakura, T., Ito, T., Okudaira, M., Kameda, T. 1999. "Structure of Alanine and Glycine Residues of Samiacynthiaricini Silk Fibres Studied with Solid-State 15N and 13C NMR." *Macromolecules* 32: 4940–4946.

[49] Gage, L. P., Manning, R. F. 1980. "Internal structure of the silk fibroin gene of Bombyxmori-I The fibroin gene consists of a homogeneous alternating array of repetitious crystalline and amorphous coding sequences." *J. Biol. C*hem. 255: 9444–9450.

[50] Fedic, R., Zurovec, M., Sehnal, F. 2003. "Correlation between fibroin amino acid sequence and physical silk properties." *J. Biol. Chem*. 278: 35255–35264.

[51] Qi, Y., Wang, H., Wei, K., Yang, Y., Zheng, R. Y., Kim, I. S., Zhang, K. Q. 2017. "A Review of Structure Construction of Silk Fibroin Biomaterials from Single Structures to Multi-Level Structures." *Int. J. Mol. Sci*. 18: 237.

[52] Cho, S. Y. Yun, Y. S., Lee, S., Jang, D. Park, K. Y., Kim, J. K., Kim, B. H., Kang, K., Kaplan, D. L., Jin, H. J. 2015. "Carbonization of a stable β-sheet-rich silk protein into a pseudo graphitic pyroprotein." *Nat. Comm*un. 6: 7145.

[53] Cebe, P., Partlow, B. P., Kaplan, D. L., Wurm, A., Zhuravlev, E., Schick, C. 2017. "Silk I and Silk II studied by fast scanning calorimetry." *Acta Biomater*. 55: 323–332.

[54] Shao, Z., Vollrath, F. 2002. "Surprising strength of silkworm silk." *Nature* 418: 741.

[55] Sirichaisit, J., Brookes, V. L., Young, R. J., Vollrath, F. 2003. "Analyis of structure/property relationships in silkworm (Bombyxmori) andspider dragline (Nephilaedulis) silks using Raman spectroscopy." *Biomacromolecules*4: 387–394.

[56] Koh, L. D., Cheng, Y., Teng, C. P., Khin, Y. W., Loh, X. J., Tee, S. Y., Low, M., Ye, E., Yu, H. D., Zhang, Y. W., & Han, M. Y. 2015. "Structures, mechanical properties and applications of silk fibroin materials." *Prog. Polym. Sci*. 46: 86–110.

[57] Blamires, S. J., Spicer, P. T., Flanagan, P. J. 2020. "Spider Silk Biomimetics Programs to Inform the Development of New Wearable Technologies." *Front. Mater*. 7: 29.

[58] Gosline, J. M., Guerette, P. A., Ortlepp, C. S., Savage, K. N. 1999. "The mechanical design of spider silks, From fibroin sequence to mechanical function." *J. Exp. Biol*. 202: 3295–3303.

[59] Ma, D., Wang, Y., Dai, W. 2018. "Silk fibroin-based biomaterials for musculo skeletal tissue engineering." *Mater. Sci. Eng.* C 89: 456–469.

[60] Altman, G. H., Horan, R. L., Lu, H. H., Moreau, J, Martin, I., Richmond, J. C., Kaplan, D. L. 2002. "Silk matrix for tissue engineered anterior cruciate ligaments." *Biomaterials* 23: 4131–4141.

[61] Zhang, C., Zhang, Y., Shao, H., Hu, X. 2016. "Hybrid Silk Fibres Dry-Spun from Regenerated Silk Fibroin/Graphene Oxide Aqueous Solutions." AC*S Appl. Mater. Interfaces* 8: 3349–3358.

[62] Nazarov, R.; Jin, H. J.; Kaplan, D. L. 2004. "Porous 3-D Scaffolds from Regenerated Silk Fibroin." *Biomacromolecules*5: 718–726.

[63] Gregory, D. A., Zhang, Y., Smith, P. J., Zhao, X., Ebbens, S. J. 2016. "Reactive Inkjet Printing of Biocompatible Enzyme Powered SilkMicro-Rockets." *Small* 12: 4048–4055.

[64] O'Brien, F. J. 2011. "Biomaterials & scaffolds for tissue engineering." *Mater. To*day. 14: 88–95.

[65] Melke, J., Midha, S., Ghosh, S., Ito, K., Hofmann, S. 2016. "Silk fibroin as biomaterial for bone tissue engineering." *Acta Biomate*r. 31: 1–16.

[66] Sakabe, H., Ito, H., Miyamoto, T., Noishiki, Y., Ha, W. S. 1989. "*In vivo* blood compatibility of regenerated silk fibroin." *Sen'IGakkaishi* 45: 487–490.

[67] Valluzzi, R., Gido, S. P., Muller, W., Kaplan, D. L. 1999. "Orientation of silk III at the air-water interface." *Int. J. Biol. Macromol.* 24: 237–242.

[68] Minoura, N., Aiba, S. I., Higuchi, M., Gotoh, Y., Tsukada, M., Imai, Y. 1995. "Attachment and Growth of Fibroblast Cells on Silk Fibroin." *Biochem. Biophys. Res. Commun.* 208: 511–516.

[69] Kim, J. Y., Yang, B. E., Ahn, J. H., Park, S. O., Shim, H. W. 2014. "Comparable efficacy of silk fibroin with the collagen membranes for guided bone regeneration in rat calvarial defects." *J. Adv. Prosthodont.* 6: 539–546.

[70] Santin, M., Motta, A., Freddi, G., Cannas, M. 1999. "*In vitro* evaluation of the inflammatory potential of the silk fibroin." *J. Biomed. Mater. Res.* 46: 382–389.

[71] Panilaitis, B., Altman, G. H., Chen, J., Jin, H.-J., Karageorgiou, V., Kaplan, D. L. 2003. "Macrophage responses to silk." *Biomaterials* 24: 3079–3085.

[72] Meinel, L., Hofmann, S., Karageorgiou, V., Kirker-Head, C., McCool, J, Gronowicz, G. Zichner, L., Langer, R., Vunjak-Novakovic, G., Kaplan, D. L. 2005. "The inflammatory responses to silk films *in vitro* and *in vivo*."*Biomaterials* 26: 147–155.

[73] Babensee, J. E., Anderson, J. M., McIntire, L. V., Mikos, A. G. 1998. "Host response to tissue engineered devices." *Adv. Drug. Deliv. Rev.* 33: 111–139.

[74] Cao, Y., Wang, B. 2009. "Biodegradation of silk biomaterials." *Int. J. Mol. Sci.* 10, 1514–1524.

[75] Nair, L. S., Laurencin, C. T. 2007. "Biodegradable polymers as biomaterials." *Prog. Polym. Sci.* 32: 762–798.

[76] Arai, T., Freddi, G., Innocenti, R., Tsukada. 2004. "M. Biodegradation of Bombyxmori silk fibroin Fibres and films." *J. Appl. Polym. Sci.* 91: 2383–2390.

[77] Li, M.; Ogiso, M.; Minoura, N. Enzymatic degradation behavior of porous silk fibroin sheets. *Biomaterials* 2003, 24, 357–365.

[78] Horan, R. L., Antle, K., Collette, A. L., Wang, Y., Huang, J., Moreau, J. E., Volloch, V., Kaplan, D.L., Altman, G. H. 2005. "*In vitro* degradationof silk fibroin." Biomaterials 26: 3385–3393.

[79] Wang, Y., Rudym, D. D., Walsh, A., Abrahamsen, L., Kim, H. J., Kim, H. S., Kirker-Head, C., Kaplan, D. L. 2008. "*In vivo* degradation of three-dimensional silk fibroin scaffolds." Biomaterials 29: 3415–3428.

[80] Hu, Y., Zhang, Q., You, R., Wang, L., Li, M. 2012. "The Relationship between Secondary Structure and Biodegradation Behavior of SilkFibroin Scaffolds." Adv. Mater. Sci. Eng. 1–5.

[81] Jin, H. J., Park, J., Karageorgiou, V.; Kim, U. J., Valluzzi, R., Cebe, P., Kaplan, D. L. 2005. "Water-Stable Silk Films with Reduced β-Sheet." Content. Adv. Funct. Mater. 15: 1241–1247.

[82] Lu, Q., Hu, X., Wang, X. Kluge, J. A. Lu, S., Cebe, P. Kaplan, D. L. 2010. "Water-insoluble silk films with silk I structure." ActaBiomater. 6: 1380–1387.

[83] Wang, J. N., Liu, Z. W., Yang, Y. X., Huang, H. Y. 2012. "Enzymatic degradation behaviour of silk fibroin Fibre treated by ray irradiation." Text. Res. J. 82: 1799–1805.

[84] Mohammad Z., Zeeshan A., Faisal S., Md Wasim H., Suhail A., Sahar I., Mohd S., Nazma K. 2017. "Vesicular drug delivery system used for liver diseases." *World J Pharm Sci.* 5(4): 28–35.

[85] Manish G., Vimukta S. 2011. "Targeted drug delivery system, a review." *Res J Chem Sci.* 1(2): 135–138.

[86] Gujral S., Khatri S. 2013. "A review on basic concept of drug targeting and drug carrier system." *IJAPBC.* 2(1).

[87] Bhargav E, Madhuri N, Ramesh K, Manne A, Ravi V. 2013. "Targeted drug delivery - a review." *World J Pharm Pharm Sci.* 3(1),150–169.

[88] Kumar A., Nautiyal U., Kaur C., Goel V., Piarchand N. 2017. "Targeted drug delivery system, current and novel approach." *Int J Pharm Med Res.* 5448–454.

[89] Mahajan H. S. Sanjay B. Patil, Surendra Gattani, Bhanudas Kuchekar. 2007. "Targeted Drug delivery system." *Pharma Times* 39 (2).

[90] Yoo J, Park C, Yi G, Lee D, Koo H. 2019. "Active targeting strategies using biological ligands for nanoparticle drug delivery systems." *Cancers.* 11: 640.

[91] Mishra N, Pant P, Porwal A, Jaiswal J, Samad MA, Tiwari S. 2016. "Targeted drug delivery, a review." *Am. J. Pharm. Tech. Res.* 6(1).

[92] Luo, Z., Li, J., Qu, J., Sheng, W., Yang, J., Li, M. 2019. "Cationized Bombyxmori silk fibroin as a delivery carrier ofthe VEGF165–Ang-1 coexpression plasmid for dermal tissue regeneration." *J. Mater. Chem. B* 7: 80–94.

[93] Duncan R. 2003. "The dawning era of polymer therapeutics." *Nat Rev Drug Discov.* 2: 347-360.

[94] Angelova N and Hunkeler D. 1999. Rationalizing the design of polymeric biomaterials" *Trends Biotechnol.* 17: 409-421.

[95] Pillai O, Panchagnula R. 2001. "Polymers in drug delivery." *Curr. Opin. Chem. Biol.* 5.447.

[96] Lockman PR, Koziara JM, MumperRJ, Allen DD. 2004."Nanoparticlesurface charges alter blood-brain barrier integrity and permeability." *J. Drug. Target.* 12: 635.

[97] Hobel S, Aigner A. 2010. "Polyethylenimine (PEI)/siRNA-Mediated Gene Knockdown *in Vitro* and *in Vivo*." *Methods. Mol. Biol.* 623: 283-297.

[98] Kodaira H, Tsutsumi Y, Yoshioka Y, Kamada H, Kaneda Y, et al. 2004. *Biomaterials* 25, 4309-4315.

[99] Jain, A., and Jain, S. K. 2018. "Stimuli-responsive smart liposomes in cancer targeting." *Curr. Drug Targets* 19: 259–270.

[100] Lammel A, Schwab M, Slotta U, Winter G, Scheibel T. 2008. "Processing conditions for the formation of spider silk microspheres." *Chem. Sus. Chem.* 1:413–416.

[101] Cagel, M., Tesan, F. C., Bernabeu, E., Salgueiro, M. J., Zubillaga, M. B., Moretton, M. A., & Chiappetta, D. A. 2017. "Polymeric mixed micelles as nanomedicines: achievements and perspectives." *Eur. J. Pharm. Biopharm.* 113: 211–228.

[102] Qumbar, M., Ameeduzzafar., Imam, S. S., Ali, J., Ahmad, J., and Ali, A. 2017. "Formulation and optimization of lacidipine loaded niosomal gel for transdermal delivery: in-vitro characterization and in-vivo activity." *Biomed. Pharmaco. Ther.* 93: 255–266.

[103] Liu, P., Wei, Z., Cheng, K., Liao, X., Qin, X., Feng, Y., Rao, J., Chen, J., Wang, T., Liu, L., Huang, J., & Liu, M. 2020. "Self-healable flexible luminescent films basedon silk fibroin and graphene quantum dots." *OSA Continuum* 3: 1021-1030.

[104] Vimbela, G. V., Ngo, S. M., Fraze, C., Yang, L., and Stout, D. A. 2017. "Antibacterial properties and toxicity from metallic nanomaterials. *Int. J. Nanomedi*cine 12: 3941–3965.

[105] He, H., Yuan, Q., Bie, J., Wallace, R. L., Yannie, P. J., Wang, J., Lancina, M. G., Zolotarskaya, O. Y., Korzun, W., Yang, H., & Ghosh, S. 2018. Development of mannose functionalized dendrimeric nanoparticles for targeted delivery to macrophages: use of this platform to modulate atherosclerosis." *Transl. R*es. 193: 13–30.

[106] Altman, G. H., Diaz, F., Jakuba, C., Calabro, T., Horan, R. L., Chen, J., Lu, H., Richmond, J., & Kaplan, D. L. 2003. "Silk-based biomaterials." *Biomaterials* 24:401–16.

[107] Matteis V. D. 2017. "Exposure to inorganic nanoparticles: routes of entry, immune response, biodistribution and *in vitro/in vivo* toxicity evaluation." *Toxics* 5:29.

[108] Patra, J. K., Das, G., Fraceto, L. F., Campos, E. V. R., Rodriguez-Torres, M. del P., Acosta-Torres, L. S., Diaz-Torres, L. A., Grillo, R., Swamy, M. K., Sharma, S., Habtemariam, S., & Shin, H. S. 2018. "Nano based drug delivery systems: recent developments and future prospects." *J. Nanobiotechnol* (2018) 16:1-33.

[109] Song, W., Gregory, D. A., Al-janabi, H., Muthana, M., Cai, Z., Zhao, X. 2019. "Magnetic-silk/polyethyleneimine core-shell nanoparticles for targeted gene delivery into human breast cancer cells." *Int. J. Pharm.* 555:322–336.

[110] Vepari, C., Kaplan, D. L. 2007. "Silk as a Biomaterial." *Prog. Polym. Sci.* 32: 991–1007.

[111] Lu, Q., X. Hu, X. Q. Wang, J. A. Kluge, S. Z. Lu, P. Cebe and D. L. Kaplan, 2010. "Water-Insoluble Silk Films with Silk I Structure," *Acta Biomaterialia*, 6 (4): 1380-1387.

[112] Sultan, M. T., Lee, O. J., Kim, S. H., Ju, H. W., and Park, C. H. 2018. "Silk fibroin in wound healing process." *Adv. Exp. Med. Biol.* 1077: 115–126.

[113] Elzoghby, A. O., W. M. Samy, N. A. Elgindy. 2012."Protein-based nanocarriers as promising drug and gene delivery systems." *Journal of Controlled Release* 161: 38-49.

[114] Malafaya, P. B., G. A. Silva, R. L. Reis. 2007. "Natural–origin polymers as carriers and scaffolds for biomolecules and cell delivery in tissue engineering applications." *Advanced Drug Delivery Reviews* 59: 207-233.

[115] Mano, J., G. Silva, H. S. Azevedo, P. Malafaya, R. Sousa, S. Silva, L. Boesel, J. M. Oliveira, T. Santos, A. Marques. 2007. "Natural origin biodegradable systems in tissue engineering and regenerative medicine, present status and some moving trends." *Journal of the Royal Society Interface* 4: 999-1030.

[116] Hosseinkhani, H., M. Hosseinkhani, H. Kobayashi. 2006. "Design of tissue-engineered nano scaffold through self-assembly of peptide amphiphile." *Journal of Bioactive and Compatible Polymers* 21: 277-296.

[117] Hosseinkhani, H., T. Kushibiki, K. Matsumoto, T. Nakamura, Y. Tabata. 2006. "Enhanced suppression of tumor growth using a combination of NK4 plasmid DNA-PEG engrafted cationized dextran complex and ultrasound irradiation." *Cancer Gene Therapy* 13: 479-489.

[118] Hosseinkhani, H., T. Aoyama, O. Ogawa, Y. Tabata. 2003. "Ultrasound enhances the transfection of plasmid DNA by non-viral vectors." *Current Pharmaceutical Biotechnology* 4: 109-122.

[119] M. Konishi, Y. Tabata, M. Kariya, H. Hosseinkhani, A. Suzuki, K. Fukuhara, M. Mandai, K. Takakura, S. Fujii. 2005. "*In vivo* anti-tumor effect of dual release of cisplatin and adriamycin from biodegradable gelatin hydrogel." *Journal of Controlled Release*, 103: 7-19.

[120] Tabata Y. 2003. "Tissue regeneration based on growth factor release." *Tissue Engineering* 9: 5-15.

[121] Mottaghitalab, F., M. Farokhi, V. Mottaghitalab, M. Ziabari, A. Divsalar, M. A. Shokrgozar. 2011. "Enhancement of neural cell lines proliferation using nano-structured chitosan/poly (vinylalcohol) scaffolds conjugated with nerve growth factor, Carbohydrate." *Polymers* 86: 526-535.

[122] Shokrgozar, M. A., F. Mottaghitalab, V. Mottaghitalab, M. Farokhi. 2011. "Fabrication of porous chitosan/poly (vinyl alcohol) reinforced single-walled carbon nanotube nanocomposites for neural tissue engineering", *Journal of Biomedical Nanotechnology*, 7: 276-284.

[123] Ahmed, E. M. 2015. "Hydrogel: Preparation, characterization, and applications: A review." *J. Adv. Res.* 6: 105–121.

[124] Tomeh, M. A., Hadianamrei, R.; Zhao, X. 2019. "A review of curcumin and its derivatives as anticancer agents." *Int. J. Mol. Sci.* 20: 1033.

[125] Lee, M. C., Kim, D. K., Lee, O. J., Kim, J.-H., Ju, H. W., Lee, J. M., Moon, B. M., Park, H. J., Kim, D. W., Kim, S. H., & Park, C. H. 2016. "Fabrication of silk fibroin film using centrifugal casting technique for corneal tissue engineering." *J. Biomed. Mater. Res. Part B* 104: 508–514.

[126] Sagnella, A., Pistone, A., Bonetti, S., Donnadio, A., Saracino, E., Nocchetti, M., Dionigi, C., Ruani, G., Muccini, M., Posati, T., Benfenati, V., & Zamboni, R. Effect of different fabrication methods on the chemo-physical properties of silk fibroin films and on their interaction with neural cells. *RSC Adv.* 2016, 6, 9304–9314.

[127] Terada, D., Yokoyama, Y., Hattori, S., Kobayashi, H., Tamada, Y. The outermost surface properties of silk fibroin films reflect ethanol-treatment conditions used in biomaterial preparation. *Mater. Sci. Eng. C* 2016, 58, 119–126.

[128] Jiang, C., Wang, X., Gunawidjaja, R., Lin, Y. H., Gupta, M. K., Kaplan, D. L., Naik, R. R., Tsukruk, V. V. 2007. "Mechanical Properties of Robust Ultrathin Silk Fibroin Films." *Adv. Funct. Mater.* 17: 2229–2237.

[129] De Moraes, M. A., Silva, M. F., Weska, R. F., Beppu, M. M. 2014. "Silk fibroin and sodium alginate blend: Miscibility and physical characteristics." *Mater. Sci. Eng.* C 40: 85–91.

[130] Pablo Vega-Vásquez 1, Nathan S. Mosier 1 and Joseph Irudayaraj. 2020. "Nanoscale Drug Delivery Systems: From Medicine to Agriculture." *Frontiers in Bioengineering and Biotech*nology 8 (79): 1-16.

[131] Philipp Seib, F. *AIMS Bioengineering*, Volume 4, Issue 2, 239-258.

[132] Rachasit Jeencham, Manote S., Waree T. 2019. "Preparation and characterization of chitosan/regenerated silk fibroin (cs/rsf) films as a biomaterial for contact lenses-based ophthalmic drug delivery system." *International Journal of Applied Pharmaceutics* 11: 1-10.

[133] Lammel, A. S., Hu, X., Park, S. H., Kaplan, D. L., Scheibel, T. R. 2010. "Controlling silk fibroin particle features for drug delivery." *Biomaterials* 31: 4583–4591.

[134] Shi, P., Goh, J. C. H. 2012. "Self-assembled silk fibroin particles: Tunable size and appearance." *Powder Technol.* 215–216: 85–90.

[135] Wang, X., Yucel, T., Lu, Q., Hu, X., Kaplan, D. L. 2010. "Silk nanospheres and microspheres from silk/pva blend films for drug delivery." *Biomaterials* 31: 1025–1035.

[136] Li, A. B., Kluge, J. A., Guziewicz, N. A., Omenetto, F. G., and Kaplan, D. L. 2015. "Silk-based stabilization of bio-macromolecules." *J. Control. Release* 219, 416–430.

[137] Song, W., Gregory, D. A., Al-janabi, H., Muthana, M., Cai, Z., & Zhao, X. 2019. "Magnetic-silk/polyethyleneimine core-shell nanoparticles for targeted gene delivery into human breast cancer cells." *Int J Pharm* 555:322–36.

[138] Perry, H., A. Gopinath, D. L. Kaplan, L. D. Negro and G. Omenetto. 2008. "Nano- and Micro patterning of Optically Transparent, Mechanically Robust, Biocompatible Silk Fibroin Films." *Adv. Mater.* 20: 3070–3072.

[139] Bucciarelli, A., V. Mulloni, D. Maniglio, R. K. Pal, V. K. Yadavalli, A. Motta and A. Quaranta. 2018. "A comparative study of the refractive index of silk protein thin films towards biomaterial based optical devices." *Opt. Mater.,* 78: 407–414.

[140] Tao, H., D. L. Kaplan and F. G. Omenetto. 2012. "Silk materials- a road to sustainable high technology." *Adv. Mater.*, 24: 2824–2837.

[141] Parker, B. S. T., P. Domachuk, J. Amsden, J. Bressner, A. Lewis, D. L. Kaplan and F. G. Omenetto. 2009. "Biocompatible Silk Printed Optical Waveguides" *Adv. Mater.*, 21: 2411–2415.

[142] Toffanin, S., S. Kim, S. Cavallini, M. Natali, V. Benfenati, S. Toffanin, S. Kim, S. Cavallini, M. Natali, M. Mucciniand F. G. Omenetto2012. *Appl. Phys. Lett.* 101: 091110154.

[143] Fernández-Luna, V., Fernández-Blázquez, J. P., Monclús, M. A., Rojo, F. J., Daza, R., Sanchez-deAlcazar, D., Cortajarena, A. L., & Costa, R. D. 2020. "Biogenic fluorescent protein–silk fibroin phosphors for high performing light-emitting diodes." *Mater. Horiz.*, 7: 1790 -1800.

[144] Weissleder, R., and Ntziachristos, V. 2003. "Shedding light onto live molecular targets." *Nat. Med.* 9, 123–128.

[145] Setua, S., Menon, D., Asok, A., Nair, S., and Koyakutty, M. 2010. "Folate receptor targeted, rare-earth oxide nanocrystals for bi-modal fluorescence and magnetic imaging of cancer cells." *Biomaterials* 31: 714–729.

[146] Sevick-Muraca, M. E. 2012. "Translation of near-infrared fluorescence imaging technologies, emerging clinical applications." An*nu. Rev. Med.* 63: 217–231.

[147] McCarthy, J. R., Jaffer, F. A., and Weissleder, R. 2006. "A macrophage-targeted theranostic nanoparticle for biomedical applications." *Small* 2: 983–987.

[148] Zhao, Y., Chen, X., Peng, W. P., Dong, L., Huang, J. T., and Lu, C. D. 2001. "Altering fibroin heavy chain gene of silkworm *B. mori* by homologous recombination." *Sheng Wu Hua Xue Yu Sheng Wu Wu Li Xue Bao* (Shanghai) 33: 112–116.

[149] Kim, D. W., Lee, O. J., Kim, S. W., Ki, C. S., Chao, J. R., Yoo, H., Yoon, S., Lee, J. E., Park, Y. R., Kweon, H., Lee, K. G., Kaplan, D. L., & Park, C. H.2015. "Novel fabrication of fluorescent silk utilized in biotechnological and medical applications." *Biomaterials* 70: 48–56.

[150] MacLeod, J., and Rosei, F. 2013. "Sustainable sensors from silk." *Nat. Mater.* 12: 98–100.

[151] Palermo, G., Barberi, L., Perotto, G., Caputo, R., DeSio, L., Umeton, C., and Omenetto, F. G. 2017. "Conformal silk-azobenzene composite foroptically switchable diffractive structures." *ACS Appl. Mater. Interfaces* 9: 30951–30957.

[152] Zhou, Z., Shi, Z., Cai, X., Zhang, S., Corder, S. G., Li, X., Zhang, Y., Zhang, G., Chen, L., Liu, M., Kaplan, D. L., Omenetto, F. G., Mao, Y., Tao, Z., & Tao, T. H. 2017. The use of functionalized silk fibroin films as a platform for optical diffraction-based sensing applications. *Adv. Mater.* 29: 1605471.

[153] Zhao, Y., Chen, X., Peng, W. P., Dong, L., Huang, J. T., and Lu, C. D. 2001. "Altering fibroin heavy chain gene of silkworm *Bombyxmori* by homologous recombination." *Sheng Wu HuaXue Yu Sheng Wu Wu Li XueBao* (Shanghai) 33: 112–116.

[154] Kim, D. W., Lee, O. J., Kim, S. W., Ki, C. S., Chao, J. R., Yoo, H., Yoon, S., Lee, J. E., Park, Y. R., Kweon, H., Lee, K. G., Kaplan, D. L., & Park, C. H 2015. "Novel fabrication of fluorescent silk utilized in biotechnological and medical applications." *Biomaterial*s 70: 48–56.

[155] Chu, M., and Liu, G. 2008. "Fluorescent silkworm silk prepared via incorporation of green, yellow, red, and near-infrared fluorescent quantum dots." *IEEE Trans. Nanotechnol.* 7: 308–315.

[156] Fernández-Luna, V., Fernández-Blázquez, J. P., Monclús, M. A., Rojo, F. J., Daza, R., Sanchez-deAlcazar, D., Cortajarena, A. L., & Costa, R. D. 2020. "Biogenic fluorescent protein–silk fibroin phosphors for high performing light-emitting diodes." *Mater. Horiz.* 7: 1790—1800.

[157] Wang, X., Hu, X., Daley, A., Rabotyagova, O., Cebe, P., and Kaplan, D. L. 2007. "Nanolayer biomaterial coatings of silk fibroin for controlled release." *J. Control Release* 121: 190–199.

[158] Ren, D., Kratz, F., and Wang, S. W. 2011. "Protein nanocapsules containing doxorubicin as a pH-responsive delivery system." *Small* 7: 1051–1060.

[159] Khalid, A., Lodin, R., Domachuk, P., Tao, H., Moreau, J. E., Kaplan, D. L., Omenetto, F. G., Gibson, B. C., & Tomljenovic-Hanic, S 2014. "Synthesis and characterization of biocompatible nanodiamond-silk hybrid material." *Biomed. Opt. Express* 5: 596–608.

[160] Khalid, A., Mitropoulos, A. N., Marelli, B., Tomljenovic-Hanic, S., and Omenetto, F. G. 2016. "Doxorubicin loaded nano-diamond silk spheres for fluorescence tracking and controlled drug release." *Biomed.* Opt. Express 7: 132–147.

[161] Gad, S. C. 2008. "Semisolid dosages: Ointments, creams and gels." In *Pharmaceutical Manufact*uring *Handbook: Production and Processes*; John Wiley & Sons: Hoboken, NJ, USA, Volume 5, pp. 267–312.

[162] Adepu, S., Gaydhane, M. K., Kakunuri, M., Sharma, C. S., Khandelwal, M.; Eichhorn, S. J. 2017. "Effect of micropatterning induced surface hydrophobicity on drug release from electrospun cellulose acetate nanofibers." *Appl. Surf. Sci.* 426, 755–762.

[163] Navin Dixit, Sheo Dutt Maurya and Bhanu P. S. Sagar. 2013. "Navin Dixit, Sheo Dutt Maurya, Bhanu P. S. Sagar." *Indian Journal of Research in Pharmacy and Biotech*nology. Vol. 1, 305-310.

[164] Fenton, O. S., Olafson, K. N., Pillai, P. S., Mitchell, M. J., Langer, R. 2018. "Advances in biomaterials for drug delivery." *Adv. Mater.* 30,1705328 pp. 1-29.

[165] Hardenia, A., Maheshwari, N., Hardenia, S. S., Dwivedi, S. K., Maheshwari, R., Tekade, R. K. 2019. "Scientific rationale for designing controlled drug delivery systems." In *Basic Fundamentals of Drug Delivery*; Elsevier: Amsterdam, The Netherlands. pp. 1–28.

[166] Srikonda, S., Kotamraj, P., Barclay, B. 2006. "Osmotic controlled drug delivery systems." *Des. Control. Release Drug Deliv. Syst.* 1, 203.

[167] Wang, Z., Shmeis, R. A. 2006. "Dissolution controlled drug delivery systems." *Des. Control. Release Drug Deliv. Systems*. 139–172.

[168] Siepmann, J., Siegel, R. A., Siepmann, F. 2012. "Diffusion controlled drug delivery systems." In *Fundamentals and Applications of Controlled Release Drug Delivery*; Springer: Berlin/Heidelberg, Germany, pp. 127–152.

[169] Patil, P. B., Uphade, K. B., Saudagar, R. B. 2018. "A review: Osmotic drug delivery system." *Pharma Sci. Monit.* 9, 2.

[170] Siepmann, J., Siepmann, F. 2012. "Swelling controlled drug delivery systems." In *Fundamentals and Applications of Controlled Release Drug Delivery;* Springer: Berlin/Heidelberg, Germany, pp. 153–170.

[171] Shivakalyani Adepu and Seeram Ramakrishna. 2021. "Controlled Drug Delivery Systems: Current Status and Future Directions." *Molecules* 26, 5905 pp. 1-45.

[172] Yun, Y. H., Lee, B. K., Park, K. 2015. "Controlled drug delivery: Historical perspective for the next generation." *J. Control. Release* 219,2–7.

[173] Park, K. 2014. "Controlled drug delivery systems: Past forward and future back." *J. Control. Release* 190, 3–8.

Chapter 9

Silk Fibroin Nanoparticles as a Promising Tool for Controlled Drug Delivery

S. J. Rajoba[1], L. D. Jadhav[2], S. H. Pawar[3,4] and R. D. Kale[1,*]

[1] Department of Physics,
Tuljaram Chaturchand College, Baramati, India
and Jaysingpur College, Jaysingpur Dist Kolhapur, India
[2] Department of Physics, Rajaram College, Kolhapur, India
[3] Centre for Innovative and Applied Research (CIAR),
T. C. College, Baramati, Maharashtra, India
[4] Center for Interdisciplinary Research,
D. Y. Patil University, Kolhapur, Maharashtra, India

Abstract

Silk fibroins are spun by several species of Lepidoptera and Arthropoda to build structures external to their bodies such as cocoons and webs. It is one of the strongest natural fibers, and this strength can be attributed to the chemical structure of the protein itself. Fibroin is a silk protein that possesses ideal properties such as mechanical strength, slow degradation profile, and aqueous processability, a biomaterial for various biomedical applications like sutures, coatings for cell culture, matrices, and 3D scaffolds for ligament, bone, cartilage, fat, vasculature engineering, etc. Silk fibroin (SF) differs widely in composition, structure, and properties depending on the specific source. Recently, the development of fibroin nanoparticles (FNPs) for various biomedical applications has been extensively studied. FNP has gained attention in the field of drug delivery due to its excellent biocompatibility, unique morphology, and structure. FNPs can encapsulate different types of therapeutic compounds, including small and big molecules, proteins, enzymes, vaccines, and genetic materials because of their versatility and chemical modifiability. FNPs can deliver different drugs targeted for various parts of the body for a particular period. Several administrative routes of parenteral, oral, transdermal, ocular, orthopaedic, and respiratory can be adapted for controlled drug delivery.

In the present chapter, we summarize basic information on the silk fibroin origin and characteristics, followed by a discussion on the FNPs preparation and characterization methods. Medical applications as a drug delivery system of FNPs are in-depth explored.

[*] Corresponding Author's Email: rajendra_kale@yahoo.com.

In: Silk Fibroin
Editor: Shivaji H. Pawar
ISBN: 979-8-88697-402-7

Finally, the challenges and suggested solutions to overcome challenges, as well as the future perspectives of these systems are discussed.

Keywords: silk fibroin, nanoparticles, drug delivery

1. Introduction

Drug delivery aims to deliver appropriate amounts of therapeutic agents to the diseased sites with the right concentration to improve the therapeutic effect of drugs and reduce their adverse effects. Furthermore, the purposes of drug delivery systems are the improvement in cellular uptake, the control of drug release, the enhancement of drug bioavailability, and the reduction of the drug degradation rate. Recently, to achieve these goals various drug delivery systems such as mesoporous silica nanoparticles (NPs), liposomes, and polymeric NPs have been designed for multiple drugs and released in a controlled manner [1, 2]. Among them recently nanoparticles have attracted a myriad of attention because of their various features like modifiable platforms, tunable size, ability to act as a modifiable platform, and high surface-to-volume ratio. Moreover, they can deliver hydrophilic and hydrophobic drug molecules [3]. Therefore, principles of nanotechnology are applied to design drug delivery systems, which improves therapeutic efficiency and preserves the properties of bioactive molecules [4]. Among the various nanoparticles, polymeric NPs have gained more attention due to their biocompatibility, desirable biodegradability, and ease of functionalization [5].

Several polymeric nanoparticles including synthetic (polyesters, polyorthoesters, polyanhydrides, polyphosphazenes, and polyphosphoesters, etc.) and natural polymer (alginates, chitosan, collagen, dextran, pullulan and gelatin) are used for drug delivery system [6]. Synthetic polymers have limitations in application due to some of their intrinsic properties and processing requirements [7]. In addition, the nature of material plays a significant role in several parameters such as loading amount, kinetic of release and destiny of drug in the body [8]. Therefore natural polymers are attractive alternatives due to their properties such as higher biocompatibility, biodegradability, valuable encapsulation capacity, and controlled release property [9]. Because of these properties, biodegradable polymeric nanoparticles are widely used in drug delivery systems.

Silk Fibroin (SF) is a natural polymer produced by several insects and spiders, therefore it has a wide range in sequence, morphology and properties. SF extracted from the cocoons of the silkworms (Bombyx mori) is one of the attractive natural polymers and it is approved by US FDA. Recently compared with other silk, silkworms silk is widely used for drug delivery applications because it shows unique self-assembling ability, intrinsic anti-inflammatory activity, wound-healing capacity, high drug encapsulation efficiency, and the ability to undergo lysosomal environment-responsive drug release [9, 10]. In addition, silk fibroin nanoparticles (SFNPs) can efficiently load small-molecule drugs, proteins, and nucleic acids through surface adsorption, physical encapsulation, and chemical coupling. Therefore, SFNPs increase the efficiency of drug uptake capacity and prevent drug degradation.

1.1. Conventional Carrier vs. Nanocarrier

Conventional drug delivery systems are based on orally administered or injectable drugs. The downside of a conventional drug delivery system is low bioavailability, rapid clearance, and high toxicity. To reduce its side effect, optimize its efficiency and enhance patient conformity, there is a need to develop alternative delivery methods and carriers. Nanoparticles have a large surface area which allows making the surface modification. In addition, properties of nanoparticles like particle size, surface morphology, etc. are modified according to the application. Therefore, recently nano technology-based drug delivery system is increasingly considered to be a future drug delivery system because it resolves challenges like bioavailability due to less solubility, poor intestinal absorption mechanism by degradation, deficient delivery mechanism at the target site, therapeutic potency, side effects and plasma inconsistency of drugs. However, nanoparticles have advantages like prolonged drug half-life and improved solubility of the hydrophobic drug. To release the drug at the right concentration at the targeted site nanoparticles are modified or functionalized. When nanoparticles are used to deliver drugs, they are known as nanocarriers. Depending on the application nanoparticles are transformed into functionalized nanoparticles.

1.2. Parameter to Nanoparticle Drug Delivery System

Drug delivery systems depend on the interaction between nanomaterials and cells. The size (nano or micro) and shape (spherical or rod) of nanoparticles govern the interaction between them. Based on its application or uptake mechanism particle size is controlled through the experimental parameters of the synthesis technique. To attempt this issue, J. A. Champion et al. [2] reported that the transport of particles in the body is affected by particle parameters like shape, diameter, particle velocity, diffusion and adhesion to walls in blood vessels, airways and intestine. In addition, the shape of particles affects these properties but in more complex ways. Spherical particles move easier due to their inherent symmetry, in contrast, non-spherical particles may align or tumble in the presence of flow. Alignment or tumbling issues will be increased when particles flow through filtering organs, such as the liver or spleen. This reveals that shape and size are important factors for an effective drug delivery system.

1.3. pH Sensitive System

pH sensitive drug delivery systems drugs are delivered in response to pH in the body. It is used to deliver drugs at predetermined rates for predefined periods, which are used to overcome the drawbacks of conventional drug delivery systems. This system is more attractive because it delivers the drug at a specific time as per the pathophysiological need of the disease; it results in improved patient therapeutic efficiency and compliance. Therefore, pH sensitive release systems should be promising in the future.

2. Polymeric Nanosystems for Drug Delivery

A polymeric drug delivery system has defined as a formulation or a device that enables the introduction of a therapeutic substance into the body. It improves its safety and efficacy by controlling the rate, time, and place of release of drugs in the body [11]. Recently polymer science with nanotechnology has introduced a new dimension in the field of drug delivery, food preservation, packaging, tissue engineering, nano-implants and biomedical engineering, etc. For drug delivery application various polymers such as synthetic and natural have been used for constructing biodegradable polymeric nano and micro particles. In the drug delivery system natural biodegradable polymers like protein based (albumin, gelatin, soy, collagen) and polysaccharides (chitosan, agarose, dextran, hyaluronic acid, alginate, carrageenan, cyclodextrin) are widely used due to its attractive properties such as its abundance in nature, biodegradability, biocompatibility and lower toxicity. However, it is challenging to use it as a drug carrier because of its broad molecular weight distributions and batch-to-batch variability [12].

2.1. Silk-Based Materials

The term 'silk' refers to a wide range of fiber spun by several species of Lepidoptera larvae (silkworms, spiders, scorpions, mites and flies) and Arthropoda to build an external structure to the body like cocoons and webs. A property like composition, structure, etc. depends on a specific source. Bombyx mori silkworms produce cocoons made of a fascinating protein-based material called silk fibroin (SF). SF is one of the strongest natural fibers, and this strength can be attributed to the chemical structure of the protein itself. Recently SF has been widely tested for drug delivery because of its excellent properties such as biocompatibility, biodegradability, low immunogenicity excellent mechanical properties, slow degradation profile, and aqueous processability [13]. Various physiochemical properties of SF are briefly discussed in the following section.

2.2. Structure and Physicochemical Properties of SF

2.2.1. Structure

Normally, native silk fiber consists of two types of self-assembled proteins: fibroin and sericin. These two proteins contain the same 18 amino acids such as glycine, alanine and serine in different amounts. The core fibroins are encased in a coat of sericin.

The amino acid sequence of SF contains repetitive glycine-alanine-glycine-alanine-glycine-serine (GAGAGS) repeats which self-assemble into an anti-parallel b-sheet structure (Figure 1). These b-sheets are highly crystalline and essentially crosslink the protein through strong intra- and inter-molecular hydrogen bonds, as well as strong Van der Waals interactions between stacked b-sheets, giving the material robust mechanical properties. The primary structure of SF is shown in Figure 1.

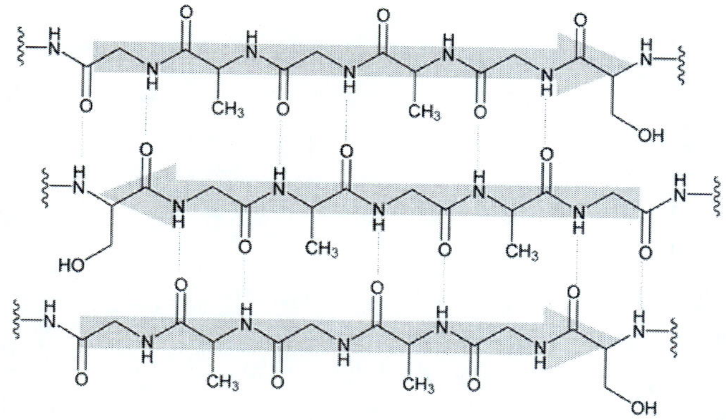

Figure 1. Primary structure of SF.

In an aqueous solution of silk fibroin, molecules lose amorphous structures due to their intrinsic electrostatic repulsion. Meanwhile, water tends to couple with these silk fibroin molecules and forma layer of hydration film. Upon the external treatment, β-sheets structures are formed, resulting in the self-assembly of molecular chains and the formation of SFNPs. Generally, SFNPs are produced by the self-assembly method. Self-assembly is a thermodynamic process and is determined by molecular aggregation, which can be modulated by external environmental factors. Under certain external parameters like metal ion, low temperature, organic solvent, and ultrasound, the soluble, and irregular Silk can be transformed into non-soluble Silk. On the other hand, under certain conditions (e.g., acid, ROS, enzyme, and hyperthermia), the β-sheet structures of Silk undergo a conformational reversion to amorphous structures of Silk. Therefore, the transformation between the crystal structures of silk fibroin is a complex and multi-factorial regulated process, which is fundamental to the multi-responsive property of SFNPs [14].

2.3. Properties of Silk Fibroin

2.3.1. Mechanical Properties

Silk as a fiber possesses certain mechanical qualities in terms of a balanced modulus as well as elongation ability contributing to the level of toughness. Compared with other natural biopolymers silk fibroin shows excellent mechanical properties such as modulus, ductility, breaking strength, and elongation. Silk also possesses a higher strength-to-density ratio up to ten times higher than that of steel. Table 1 shows that comparison of mechanical properties which provide a remarkable combination of strength and toughness [15-17].

Table 1. Tensile properties of silk polymeric fibers

S. N.	Source organisms	Tensile strength	Tensile modulus (g/den)	Breaking strain (%)
1.	Bombyx mori	4.3-5.2	84-121	11-20
2.	Antheraea mylitta	2.5-4.5	66-70	20-25
3.	Coscinocerahercules	5	85	15
4.	Antheraea oculea	3	55	14
5.	Hyalophoragloveri	2.5	45	20

It is also important that the excellent mechanical properties of the natural silk fibers are not yet reproduced by regenerated silk fibroin materials due to the lack of organized hierarchical secondary structure. There are several reports which show that there are scopes to significantly improve the strength of regenerated silk products. Such investigations are still at the conceptual stage.

2.3.2. Biocompatibility

Silk sutures over a period have had a long successful history therefore silk is known as biocompatible material. Recorded incidents of deferred hypersensitivity of silk sutures suggest the presence of a silk-gum like protein called the sericin. Another investigation was conducted stating that the sericin based materials have shown no established proof to indicate sericin as a source of adverse effects [18, 19]. However, these established reports provided a broad acceptance that adequately degummed sterile silk fibroin products with excellent biocompatibility capacity. For general surgical applications benefits and concerns with the use of silk for biomedical applications are given as follows [20].

Benefits:

1. The novel mechanical properties of some silks are superior to any other natural fiber and rival many high performance fibers.
2. Natural fiber with a long standing history of use in clinical applications.
3. The ability to process silks in aqueous solutions for subsequent formation of films and other material formats, with relatively simple insolubilization via exposure to alcohols and other environmental factors.
4. Easily chemically modified with surface decorations, such as adhesion sites or cytokines, due to the availability of amine and acid side chains on some of the amino acids.
5. Genetically tailorable composition and sequence to moderate specific features, such as molecular weight, crystallinity, solubility.
6. Slow rates of degradation *in vitro* and *in vivo*, this is particularly useful in biodegradable scaffolds in which slow tissue growth is desirable.

Concerns:

1. Adequate removal of contaminating sericin from silkworm silk to avoid biocompatibility problems.
2. Slow degradation of crystalline (b-sheet) regions.
3. Aborted proteolytic attack by macrophages and giant cells leading to encapsulation and the formation of a granuloma.
4. Potential sensitization to silk fibroin results in an allergic response upon exposure to the biomaterial.

2.3.3. Biodegradation

The term biodegradability is often used to discuss the disintegration of silk materials. According to the definition, biodegradability is the degradability of an implantable polymer by biological elements giving fragments, which can move away from the site through fluid transfer but not necessarily from the body. On the other hand, biosorption is the total elimination of the initial foreign material either through filtration or metabolization of the degraded bio-products [21]. Generally, the biodegradation behavior has two steps, as explained above. At first, silk biomaterials are adsorbed by different enzymes, which demands that the enzymes must find binding domains on the materials' surface. After that, silk biomaterials are digested by enzymes. The final wastes of silk fibroins are the corresponding amino acids, which are easily absorbed *in vivo*. That is one of the advantages of silk biomaterials used in the biomedical field [22].

To develop a successful formulation of nanoparticle-based drug carriers, it is necessary to consider their biodegradation rate. Many factors affect the release rate of nanoparticles, including desorption of surface-bound/adsorbed drug molecules, diffusion through the nanoparticle matrix, diffusion through the polymer wall (in the case of nano capsules), erosion of nanoparticle matrix, and a combined erosion/diffusion process. Generally, synthetic polymers such as PLA, PGA, and PLGA produce acidic compounds as by-products of hydrolytic degradation, and this is potentially undesirable at the targeted sites. However, SF usually undergoes proteolytic degradation resulting in non-toxic byproducts [23]. In addition, the characteristics of silk biodegradation behaviors vary with different enzymes. The degradation of the silk-based system is said to be initiated by cells *in-vitro*, these *in-vitro* studies have shown a various comparison of biodegradability rate of silk-based materials and structures. Synthetic biomaterials over time tend to lose certain mechanical properties during degradation. While silk materials can retain their strength over an extended period that is an advantage, especially in tissue engineering requiring a slow rate of degradability and load bearing properties [24].

The hydrophobic domain plays an important role in the molecular assembly of the proteins in the silk fibers [25]. These domains are responsible for insolubility, the high strength and the thermal stability of the silk fibers. It leads to the formation of β-sheet secondary structure. This outcome reveals that the biodegradability and biocompatibility of silk fibroin are governed by its.

2.3.4. Solubility Properties of Silk Fibroin

The silk fibroin is water soluble in its α-helical and random coil forms. Solubility can be maintained over days and even weeks depending on the storage temperature, pH and concentration of silk solution. Hence, silk based systems can be prepared using water based solutions under mild manufacturing conditions such as room temperature, neutral pH and without the application of high shear force. Such conditions are favorably exploited for loading sensitive drugs into silk implants [26, 27].

Fibroin dissolves majorly in a concentrated aqueous solution of acids such as a formic, sulphuric and concentrated aqueous, biological and aqueous organic solution of salts. Fibroin has specific dissolution features in various systems associated with its molecular structure. The dissolution of fibroin in salt solutions is caused due to the interaction of the solvent ions altogether with the functional group of the fibroin macromolecules. The intramolecular and intermolecular hydrogen bonds present in the fibroin structure are reported disrupted as a result of the nucleophilic savage by the anion. The interaction of the solvent ions with the polar and charged groups of pendant chains of fibroin causes the breakage of the hydrogen bonds between the macromolecules [28].

3. Synthesis Methods of Silk Fibroin Nanoparticles

SFNPs have been extensively studied as particulate carriers in the pharmaceutical and medical fields because they show promise as drug delivery systems due to their controlled and sustained release properties and biocompatibility with tissue and cells.

The synthesis of SFNPs is based on balancing the attractive and repulsive forces in the protein. During synthesis, nanoparticles undergo conformational changes depending on their composition, concentration, cross linking, and preparation conditions such as pH, ionic strength, and type of solvent. Therefore, SFNP is prepared by several methods, these methods are classified into four main groups viz. direct mixing, *in-situ* synthesis and silkworm feeding method. In the first method, NPs are added to SF solution and the resultant mixture can be used to form film, nanofiber, hydrogel, etc. In the second method, SF materials are put into a precursor solution of NPs. In the third type, silk solution is electro spun and in fourth type of method silkworms are fed with diets containing NPs and the SF-NPs composite fibers will be naturally spun.

Each type of method mentioned above has its advantages and disadvantages and are used for different applications. These four types of methods are discussed in the following sections.

3.1. Desolvation Method

The desolvation process is a widely used technique to prepare protein-based nanoparticles because of its comparatively mild conditions. In this technique, the protein is dissolved in a solvent and then gradually extracted into a non-solvent phase. Several parameters of SFNPs such as the size of nanoparticles, morphology, etc. are governed by the solubility of proteins in solvents, solvent polarity, pH, ionic strength, and presence of electrolytes [29]. Here, organic solvents such as acetone and ethanol have been used as anti-solvents for the preparation of protein nanoparticles. It is observed that particle size and diffusion of the solvent into the antisolvent phase decrease with an increase in anti-solvent, which limits the growth of nanoparticles [30, 31]. pH of the resultant solution is also one of the important parameters because it is reported that higher pH values produce smaller nanoparticles [32]. In short, the addition of desolvating agent leads to conformation changes in protein structure resulting in precipitation of the protein. A schematic representation of the desolvation method for preparing SFNPs is shown in Figure 2.

Figure 2. Schematic of the desolvation method for preparing silk fibroin.

3.2. Salting Out Method

The salting out method provides better crystalline, phase pure material, chemically uniform and more homogeneous powder size distribution due to homogeneous mixing of starting material at a molecular level. In this method, proteins have hydrophilic and hydrophobic parts. Hydrophobic parts can interact with the water molecules and allow proteins to form hydrogen bonds with the surrounding water molecules. With the increase of the salt concentration, the salt ions attract some of the water molecules, resulting in the removal of the water barrier between protein molecules and the increase of the protein-protein interactions. Therefore, the protein molecules aggregate together by forming hydrophobic interactions with each other and precipitating from the solution [30]. Schematic representation of the salting out method for synthesis of Silk fibroin nano particle is shown in Figure 3.

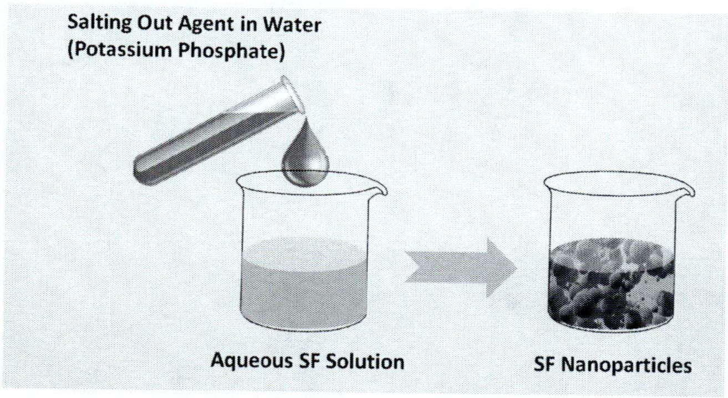

Figure 3. Schematic representation of the salting out method for synthesis of Silk fibroin nano particle.

3.3. Electrospraying Method

The basic principle involved in electro spraying is the liquid flowing out of a capillary nozzle, which is maintained at high electric potential, is forced by the electric field to be dispersed into small droplets [33, 34]. The schematic diagram of the electro spraying method for preparing SF nanoparticles is shown in Figure 4.

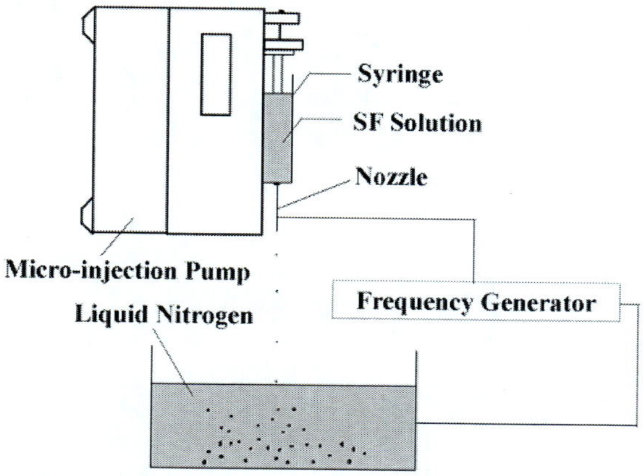

Figure 4. Schematic representation of preparatory steps of silk fibroin nanopartilces by Electrospraying method [34].

3.4. Silkworm Feeding Method

Recently, silk fiber, metal or metal oxide composites are successfully obtained by directly feeding silkworms with diets containing nano materials, such as metal NPs (copper, iron, silver, etc.), metal-oxide NPs (titanium dioxide, iron oxide, etc.) and carbon-based NPs (carbon nano tubes, graphene) [30]. Biological modifications of the silk fibroin (silk) material have broad applications, in textiles, biomedical materials and other industrial materials because it significantly improved tensile strength and strain at break. In addition, it is economical to incorporate nanoparticles to the biosynthesis of silk fibroin by adding them to silkworm larval diets [35]. Schematic representation of silkworm feeding method is shown in Figure 5.

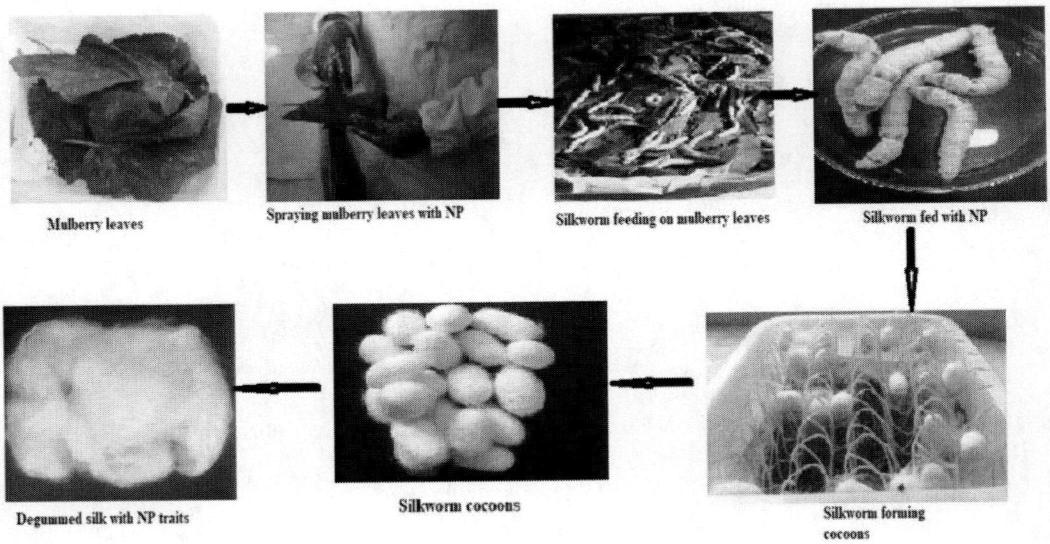

Figure 5. Schematic representation of silkworm feeding method [36].

Above mentioned types have its advantages because of their properties and applications. Composites prepared from directly mixing perfectly integrate SF and NPs, so it is easier to perform their functions as a whole, especially for bulk materials such as scaffold and hydrogel in tissue engineering. *In-situ* synthesis demonstrates the role of SF as a template to mediate the formation of NPs, which can promote the specific applications of composites related to NPs. The silkworm feeding method provides a natural way to yield SF fibers doped with NPs, but in most cases, their applications are limited by the fiber form. Therefore, researchers can choose different methods according to their needs.

4. Recent Advances in Drug Delivery System

Therapeutic agents such as proteins/peptides, nucleic acids, anticarcinogens, and other drugs have the drawbacks of low bioavailability, rapid clearance, and high toxicity. Drug delivery systems (DDS) are capable of designing to increase the bioavailability of drugs, control drug delivery and maintain the drug intact transport to the site of action while avoiding the non-diseased host tissues. In the early 1970s, Controlled drug delivery systems (DDS) aimed to deliver drugs at predetermined rates and predefined periods. In recent two decades naturally occurring biodegradable polymers are widely used in drug delivery due to their attractive properties such as abundance in nature, biodegradability, biocompatibility, lower toxicity, etc. In addition, polymeric drug delivery systems enable the introduction of therapeutic substances into the body. It improves its safety and efficacy by controlling the rate, time, and place of release of drugs in the body. This polymer is classified into two types one is protein based polymer (albumin, gelatin, soy, collagen) and another is polysaccharides (chitosan, agarose, dextran, hyaluronic acid, alginate, carrageenan and cyclodextrin). Though, its commercialization is still challenging due to its broad molecular weight distributions and batch-to-batch inconsistency. Therefore, it is important to explore new candidates from this family of polymers for specific targets and sustained drug release. The common polymers used in drug delivery system are briefly discussed in the following section.

5. Polymers in Drug Delivery System

For drug delivery applications polymers must be biocompatible, biocompatibility means the ability of a material to act with an appropriate host response in a specific application. In addition, they have properties like biodegradable with the formation of non-harmful byproducts, such as non-toxic alcohols, acids and other easily eliminated low molecular weight products. They can indeed contribute to the drug release as a result of their erosion/degradation, in addition to drug diffusion through the polymeric material. The biodegradable polymer must meet very specific requirements such as [37, 38]:

1. Biocompatibility backbone of the polymer and its degradation products.
2. Mechanical strength sufficient to meet the needs of specific applications.
3. Degradability with degradation kinetics matching a biological process such as wound healing.
4. Processibility using available equipment.
5. Solubility in various solvents.
6. Chemical, structural and application versatility.
7. Economically acceptable shelf life.

Recently Polymers become a fundamental part of drug delivery systems due to their improved pharmacokinetic properties. They have better circulation time than conventional small drug molecules thus targeting tissue more specifically. Because of the aforementioned benefits polymers plays a major role in the drug delivery system, the role of various polymers is summarized in the following section.

5.1. Polysaccharides

Polysaccharides are high molecular weights derivatives of monosaccharides with repeating units. Natural polysaccharides have been investigated for application in the drug delivery industry as well as in biomedical fields. The modified polymer has found its application as a support material for gene delivery, cell culture, and tissue engineering. Nowadays, natural polymers have been modified to obtain novel biomaterials for controlled drug delivery applications. After surface modification, they are used as drug delivery systems surface modification of their structure. Chitosan, alginate, cellulose and starch are generally used polysaccharides in the drug delivery system.

5.2. Chitosan

Chitosan is one of the cationic polysaccharides derived from natural chitin. Chitosan is the basic part of marine crustacean shells such as crabs, shrimps, lobsters and prawns in the form of chitin. Chitosan is a linear copolymer composed of glucosamine and N-acetyl glucosamine units, via β-(1, 4) linkages, namely 2-amino-2-deoxy-β-d-glucan. The chemical structure of chitosan is shown in Figure 6 [11].

Chitosan is one of the attractive materials for pharmaceutical applications because it is highly biocompatible and biodegradable with great potentials such as non-toxicity, high drug loading ability and high charge density [39].

One major advantage of chitosan NPs is their quick uptake by cells due to their surface amine groups. Protonation of these amine groups on chitosan glucosamine monomers is facilitated at pH below6.5, which confirms the cationic properties of chitosan [40]. This enables it to interact with anionic components such as cell surface macromolecules and nucleic acids. The high positive charge on chitosan gives it mucoadhesive properties making it an ideal candidate for the delivery of drugs to mucosal tissues [41].

5.3. Sodium Alginate

Alginate is a naturally occurring anionic polysaccharide and it can be used for many biomedical applications due to its mild gelation by the addition of divalent cations, low toxicity, biocompatibility and relatively low cost [42]. Alginates show different structural and chemical properties depending on the seasonal and growth conditions of the algae. The drug delivery from alginate hydrogels is a pH-dependent phenomenon leading to targeted delivery. Alginate shrinks in a low pH environment (gastric environment) and the encapsulated drugs are not released in the stomach. On reaching the intestinal tract the pH increases and so does the degree of swelling of the polymer which facilitates its rapid disintegration and the subsequent release of the drug [43]. The intrinsic properties of the calcium-alginate gel beads make them one of the most widely used carriers for controlled release.

5.4. Starch

Starch has a broad range of applications in pharmaceutical and biomedical research due to its unique physicochemical and functional characteristics and it can be isolated from natural resources such as rice, wheat, corn and potato. Starches are generally used in controlled drug release as a carrier and characteristic applications such as bone repair and replacement. The solubility, enzymatic digestibility, rheological and morphological characteristics, swelling and mechanical properties are the key factors to choose starches for these applications. Chemical modifications are an important tool to increase the reactive sites to interact with the drug molecules and consequently modified starches can easily be metabolized in the human body and are biocompatible [44].

5.5. Proteins

Proteins are high molecular weight compounds and are commonly used to make drug delivery systems. Animal originated proteins such as silk fibroin; collagen, gelatin and albumin are widely used for drug delivery systems.

5.6. Albumin

Human blood plasma contains albumin as the amplest protein (35e50 g/L human serum) and has a molar mass of 66.5 kDa. Albumin based NPs have received significant attention due to their biological origin, biodegradability, lack of toxicity, easy accessibility, water solubility, non-immunogenicity and most importantly their ability to accumulate in the tumor sites [44]. Albumins are of different types based on their sources. The one obtained from egg white is ovalbumin, from bovine serum is bovine serum albumin (BSA) and from human serum is human serum albumin (HSA). It is also available in soybeans, milk and grains. BSA and HSA have been widely used as carriers for nano drug formulations.

6. Silk-Based Drug Delivery System

Silk fibroin (SF) proteins are the promising natural fibers that help to address the slow and sustained drug delivery, due to their mechanical hardness, processing malleability, self-assembly, biocompatibility and ecological therein show considerable benefit for several therapeutic intrusions. This biomaterial can be used in the form of hydrogels, electro spun fibers, films and three dimensional (3-D) scaffolds. Mainly two species, namely silkworm (Bombyx mori) silk and spider (Nephila clavipes) dragline silk are used for making nanofibers by various methods.

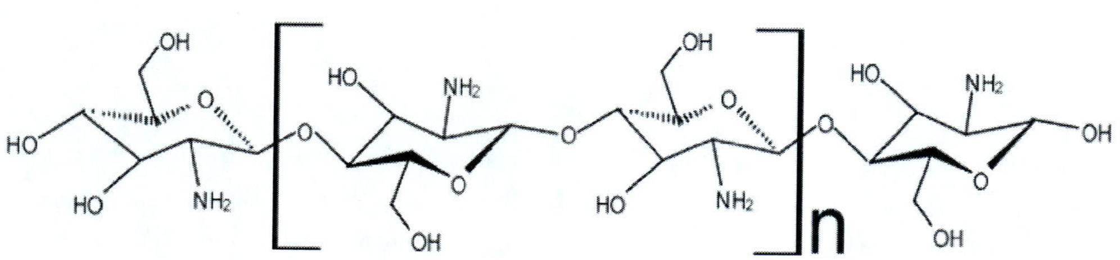

Figure 6. The chemical structure of chitosan [11].

6.1. Drug Incorporation

There are two ways to incorporate drugs into the SF drug delivery system one is during SF processing and another after SF processing. Generally, drugs are dissolved or mixed into the SF solution. Otherwise, the drug may be included post-fabrication, such as by adsorption or covalent coupling of the drug to the pre-fabricated SF system [45, 46].

During these processes, the challenging part is to make sure that there is no negative impact of the fabrication process on the integrity and biological potency of the drug.

Here, the drug would not be exposed to harsh fabrication conditions because the capacity of a drug to bind with an SF matrix depends on its physicochemical properties. It reveals that SFNP's allow to bound drugs under proper conditions. It allows the SF matrix to be prepared in advance and stored stably under suitable conditions for a long time.

The covalent coupling of drugs to the SF matrix can sustain its release compared to simple adsorption. However, harsh coupling conditions may result in a loss of activity and drug. Moreover, additional washing steps have to be included in the process to remove side products and additives.

6.2. Drug Release

Generally, the drug release rate is governed by the nature of the incorporated drug and polymer matrix. It is observed that increasing the drug molecular weight of drug it reduces the drug release rate [47]. In addition, the drug release rate also depends on the interaction of drugs with SFNPs. Recently, SFs are coated with microparticles and liposome's to enhance cellular recognition and prolong release. It is observed that coated SF slowed down the decomposition of the alginate microparticles as compared to the corresponding uncoated microparticles [48].

6.3. Application of SF as a Drug Delivery System

Recently, SF drug delivery systems focused on tissue regeneration applications because of the attractive properties of SF such as superior mechanical strength and its long-term biodegradability. Mostly, SF drug delivery devices have been mainly used for *in vitro* and *Vivo* bone and cartilage tissue engineering, vascular regeneration *in vivo* using a porcine model, and *in vitro* nerve regeneration [49].

Conclusion

Recently nanoscience and technology is widely used in the medical field for targeted drug delivery applications, which results to maximize therapeutic benefit (efficacy) and minimizing toxicity (safety). Among the various materials, polymeric nanoparticle-based drug delivery systems have gained more attention and demonstrated several advantages over conventional macro drug delivery systems because it has promising, viable, and safe option for the controlled release of drugs. In addition, it reduces the dosing frequency due to its amazing physicochemical properties such as the ability to target specific organs and tissues, increase bioavailability, solubility and permeability of potent drugs, etc. in the current decade Silk fibroin has become an attractive natural polymer for drug delivery because of its flexible qualities such as good biocompatibility, modulated biodegradability, large scale production, easy modification, and self-assembling property. Therefore, SF has been successfully used to control the release rate of biomolecules in a sustained manner with high stability. On the other hand, there are some challenges like the properties of silk may vary between both species and individuals of the same species, inconsistencies in the degumming process may render the quality control of SF delivery systems, etc.

References

[1] H. Rahmani, A. Fattahi, K. Sadrjavadi, S. Khaledian, Y. Shokoohinia, "Preparation and Characterization of Silk Fibroin Nanoparticles as a Potential Drug Delivery System for 5-Fluorouracil," *Adv. Pharm. Bull.* 9(4) (2019) 601-608.

[2] Julie A. Champion, Yogesh K. Katare, Samir Mitragotri, "Particle shape: A new design parameter for micro- and nanoscale drug delivery carriers," *Journal of Controlled Release* 121 (2007) 3-9.

[3] S. Wang, T. Xu, Y. Yang, Z. Shao, "Colloidal stability of silk fibroin nanoparticles coated with cationic polymer for effective drug delivery," *ACS Appl. Mater. Interfaces*, https://doi.org/10.1021/acsami.5b05335.

[4] Y. Ma, B. S. B. Canup, X. Tong, F. Dai, B. Xiao, "Multi-Responsive Silk Fibroin-Based Nanoparticles for Drug Delivery," *Front. Chem.* 8:585077. https://doi.org/10.3389/fchem.2020.585077.

[5] F. Mottaghitalab, M. Farokhi, M. A. Shokrgozar, F. Atyabi, H. Hosseinkhani, "Silk fibroin nanoparticle as a novel drug delivery system," *Journal of Controlled Release*, https://doi.org/10.1016/j.jconrel.2015.03.020.

[6] Z. Zhao, A. Chen, Y. Li, J. Hu, X. Liu, J. Li, Y. Zhang, G. Li, Z. Zheng, "Fabrication of silk fibroin nanoparticles for controlled drug delivery," *J. Nanopart. Res.* 14 (2012) 736, https://doi.org/10.1007/s11051-012-0736-5.

[7] O. Gianak, E. Pavlidou, C. Sarafidis, V. Karageorgiou, E. Deliyanni, "Silk Fibroin Nanoparticles for Drug Delivery: Effect of Bovine Serum Albumin and Magnetic Nanoparticles Addition on Drug Encapsulation and Release," *Separations* 5 (2018) 25, https://doi.org/10.3390/separations 5020025.

[8] J. Jacob, o. T. Haponiuk, S. Thomas, S. Gopi, "Biopolymer based nanomaterials in drug delivery systems: A review," *Materials Today Chemistry* 9 (2018) 43-55.

[9] R. Murphy, D. L. Kaplan, "Biomedical applications of chemically-modified silk fibroin," *J. Mater. Chem.*, 19 (2009) 6443-6450.

[10] H. Mori, M. Tsukada, "New silk protein: modification of silk protein by gene engineering for production of biomaterials," *Reviews in Molecular Biotechnology* 74 (2000) 95-103.

[11] Y. K. Sung, S. W. Kim, "Recent advances in polymeric drug delivery systems," *Biomaterials Research*, (2020) 24;12, https://doi.org/10.1186/s40824-020-00190-7.

[12] George, P. A. Shah, P. S. Shrivastav, "Natural biodegradable polymers based nano-formulations for drug delivery: A review," *International Journal of Pharmaceutics* (2019).

[13] E. Wenk, H. P. Merkle, L. Meinel, "Silkfibroin as a vehicle for drug delivery applications," *Journal of Controlled Release* 150 (2011) 128-141.

[14] Ya Ma, B. S. B. Canup, X. Tong, F. Dai, B. Xiao, "Multi-Responsive Silk Fibroin-Based Nanoparticles for Drug Delivery," https://doi.org/10.3389/fchem.2020.585077.

[15] Banani Kundu, Rangam Rajkhowa, Subhas C. Kundu, Xungai Wang "Silk fibroin biomaterials for tissue regenerations," *Advanced Drug Delivery Reviews* 65 (2013) 457-470.

[16] R. Rajkhowa, V. B. Gupta, V. K. Kothari, "Tensile stress-strain and recovery behavior of Indian silk fibers and their structural dependence," *J. Appl. Polym. Sci.* 77 (2000) 2418-2429.

[17] N. Reddy, Y. Yang, "Morphology and tensile properties of silkfibers produced by un-common saturniidae," *Int. J. Biol. Macromol.* 46 (2010) 419-424.

[18] T. Giese, M. Arslan, N. M. Pugno, M. J. Buehler "Nano confinement of spider silk spider fibrils begets superior strength, extensibility, and toughness," *Nano Letter* (2011) 11, 5038-5046.

[19] Y. Zhang, H. Wang, H. Shao, X. Hu "Antheraea pernyi silk fiber: a potential resource for artificially spinning spider dragline silk," *Journal of Biomedicine and Biotechnology* 90 (2010) 1123-1130.

[20] G. H. Altman, F. Diaz, C. Jakuba, T. Calabro, R. L. Horan, J. Chen, H. Lu, J. Richmond, D. L. Kaplan, "Silk-based biomaterials," *Biomaterials* 24 (2003) 401-416.

[21] B. Kundu, R. Rajkhowa, S. C. Kundu, X. Wang "Silkfibroin biomaterials for tissue regenerations," *Advanced Drug Delivery Reviews* 65 (2013) 457-470.

[22] Yang Cao, Bochu Wang, "Biodegradation of Silk Biomaterials," *Int. J. Mol. Sci.*, 10 (2009) 1514-1524, doi:10.3390/ijms10041514.

[23] Fatemeh Mottaghitalab, Mehdi Farokhi, Mohammad Ali Shokrgozar,Fatemeh Atyabi, Hossein Hosseinkhani, "Silk fibroin nanoparticle as a novel drug delivery system," 206 (2015) 161-176. https://doi.org/10.1016/j.jconrel.2015.03.020.

[24] Y. Wang, D. D. Rudym, A. Walsh, L. Abrahamsen, H. J. Kim, C. Kirker-Head., D. L. Kaplan, "In vivo degradation of three-dimensional silk fibroin scaffolds," *Biomaterials*, 29, 3415-3428.

[25] J. Huang, C. W. P. Foo, D. L. Kaplan, "Biosynthesis and applications of silk-like and collagen-like proteins," *Polym. Rev.*, 47 (2007), 29-62.

[26] V. Karageorgiou, L. Meinel, S. Hofmann, A. Malhotra, V. Volloch, D. Kaplan, "Bone morphogenetic protein-2 decorated silk fibroin films induce osteogenic differentiation of human bone marrow stromal cells," *J. Biomed. Mater. Res. Part A*, 71A (2004) 528-537.

[27] V. Karageorgiou, M. Tomkins, R. Fajardo, L. Meinel, B. Snyder, K. Wade, J. Chen, G. Vunjak-Novakovic, D. L. Kaplan "Porous silk fibroin 3-D scaffolds for delivery of bone morphogenetic protein-2 *in vitro* and *in vivo*," *J. Biomed. Mater. Res. Part A* 78A (2006) 324-334.

[28] T. R. Dawsey, C. L. McCormick, "Lithium chloride/dimethylacetemide solvent for Cellulose," *Journal of Macromolecular Science* 30 (1990) 405-440.

[29] Warangkana L ohcharoenkal, Liying Wang, Yi Charlie Chen, Yon Rojanasakul, "Protein Nanoparticles as Drug Delivery Carriers for Cancer Therapy," *Bio Med Research International* (2014) http://dx.doi.org/10.1155/2014/180549.

[30] Zheng Zhao, Yi Li, Mao-Bin Xie, "Silk Fibroin-Based Nanoparticles for Drug Delivery," *International Journal of Molecular Sciences* 16 (2015) 4880-4903 https://doi.org/10.3390/ijms16034880.

[31] Lohcharoenkal, W., Wang L., Chen Y. C., Rojanasakul, Y. Protein nanoparticles as drug delivery carriers for cancer therapy. *Biomed. Res. Int.* 2014, 2014, https://doi.org/10.1155/2014/180549.

[32] Pinto Reis, C., Neufeld, R. J., Ribeiro, A. J., Veiga, F. Nanoencapsulation I. Methods for preparation of drug-loaded polymeric nanoparticles. *Nanomedicine* 2006, 2, 8-21.

[33] Lan Cheng, Huiming Huang, Shaoyong Chen, Wenlu Wang, Fangyin Dai, Hongping Zhao, "Characterization of silkworm larvae growth and properties of silk fibres after direct feeding of copper or silver nanoparticles," *Materials & Design* (2017) https://doi.org/10.1016/j.matdes.2017.04.096.

[34] Jing Qu, Yu Liu, Yanni Yu, Jing Li, Jingwan Luo, Mingzhong Li, "Silk fibroin nanoparticles prepared by electrospray as controlled release carriers of cisplatin," *Materials Science & Engineering C* (2014) https://doi.org/10.1016/j.msec.2014.08.034.

[35] Yun-Hu Zhang, Mei-Juan Shi, Kai-Le Li, Rui Xing, Zhuo-Hua Chen, Xue-Dong Chen, Yong-Feng Wang, Xiao-Fei Liu, Xin-Yin Liang, Yang-Hu Sima, Shi-Qing Xu, "Impact of Adding Glucose-Coated Water-Soluble Silver Nanoparticles to the Silkworm Larval Diet on Silk Protein Synthesis and Related Properties," *Journal of Biomaterials Science* https://doi.org/10.1080/09205063.2019.1692642.

[36] Sandra Senyo Fometu, Guohua Wu, Lin Ma, Joan Shine Davids, "A review on the biological effects of nanomaterials on silkworm (Bombyx mori)," *Beilstein J. Nanotechnol.* 12 (2021) 190-202. https://doi.org/10.3762/bjnano.12.15.

[37] J. Safari, Z. Zarnegar, "Advanced drug delivery systems: Nanotechnology of health design A review," *Journal of Saudi Chemical Society* (2013) http://dx.doi.org/10.1016/j.jscs.2012.12.009.

[38] George, P. A. Shah, P. S. Shrivastav, "Natural biodegradable polymers based nano-formulations for drugdelivery: A review," https://doi.org/10.1016/j.ijpharm.2019.03.011.

[39] J. Jacob, O. T. Haponiuk, S. Thomas, S. Gopi, "Biopolymer based nanomaterials in drug delivery systems: A review," *Materials Today Chemistry* 9 (2018) 43-55.

[40] M. Jean, M. Alameh, D. De Jesus, M. Thibault, M. Lavertu, V. Darras, M. Nelea, M. D. Buschmann, A. Merzouki, "Chitosan-based therapeutic nanoparticles for combination gene therapy and gene silencing of *in vitro* cell lines relevant to type 2 diabetes," *Eur. J. Pharm. Sci.* 45 (2012) 138-149.

[41] E. Nagarajan, P. Shanmugasundaram, V. Ravichandiran, A. Vijayalakshmi, B. Senthilnathan, K. Masilamani, "Development and evaluation of chitosan based polymeric nanoparticles of an antiulcer drug lansoprazole," *J. Appl. Pharm. Sci.* 5 (2015) 20-25.

[42] T. Lin, P. Zhao, Y. Jiang, Y. Tang, H. Jin, Z. Pan, H. He, V. C. Yang, Y. Huang, "Blood brain-barrier-penetrating albumin nanoparticles for biomimetic drug delivery via albumin-binding protein pathways for antiglioma therapy," *ACS Nano* 10 (2016) 9999-10012

[43] F. F. Azhar, A. Olad, "A study on sustained release formulations for oral delivery of 5-fluorouracil based on alginate–chitosan/montmorillonite nano composite systems," *Appl. Clay Sci.* 101 (2014) 288-296.

[44] L. Kaur, J. Singh, Q. Liu "Starchea potential biomaterial for biomedical applications, in: M.R. Mozafari (Ed.), Nanomaterials and Nanosystems for Biomedical Applications," Springer, Dordrecht, (2007) 83-98.

[45] C. Kirker-Head, V. Karageorgiou, S. Hofmann, R. Fajardo, O. Betz, H. P. Merkle, M. Hilbe, B. von Rechenberg, J. McCool, L. Abrahamsen, A. Nazarian, E. Cory, M. Curtis, D. Kaplan, L. Meinel, BMP-silk composite matrices heal critically sized femoral defects, *Bone* 41 (2) (2007) 247-255.

[46] V. Karageorgiou, L. Meinel, S. Hofmann, A. Malhotra, V. Volloch, D. Kaplan, Bone morphogenetic protein-2 decorated silkfibroin films induce osteogenic differentiation of human bone marrow stromal cells, *J. Biomed. Mater. Res. A* 71 (3) (2004) 528-537.

[47] S. Hofmann, C. T. Foo, F. Rossetti, M. Textor, G. Vunjak-Novakovic, D. L. Kaplan, H. P. Merkle, L. Meinel, Silk fibroin as an organic polymer for controlled drug delivery, *J. Control. Release* 111 (1-2) (2006) 219-227.

[48] X. Wang, E. Wenk, X. Hu, G. R. Castro, L. Meinel, X. Wang, C. Li, H. Merkle, D. L. Kaplan, Silk coatings on PLGA and alginate microspheres for protein delivery, *Biomaterials* 28 (28) (2007) 4161-4169.

[49] X. Wang, X. Zhang, J. Castellot, I. Herman, M. Iafrati, D. L. Kaplan, Controlled release from multilayer silk biomaterial coatings to modulate vascular cell responses, *Biomaterials* 29 (7) (2008) 894-903.

Chapter 10

Silk Fibroin: Advancement in Wound Healing Technologies

S. S. Rohiwal[1,*], Z. Ellederova[1] and A. P. Tiwari[2]

[1] Laboratory of Cell Regeneration and Plasticity,
Institute of Animal Physiology and Genetics,
The Czech Academy of Sciences, Libechov, Czech Republic
[2] Department of Stem Cell and Regenerative Medicine,
Center for Interdisciplinary Research,
D. Y. Patil University, Kolhapur, India

Abstract

Silk is produced by numerous organisms such as insects like silkworms, spiders, flies, scorpions, and mites. Silk is extensively synthesized by silkworms and studied for its applications in tissue engineering and wound healing. Silk fibroin (SF) is widely used in bio-medical fields, including wound healing, due to its robust biocompatibility, biodegradability, less immunogenic, non-carcinogenic, and non-toxic properties. SF promote wound healing by enhancing the cell growth, proliferation, and migration of different cells involved in the other phase of the wound healing process. Several morphologic forms of SF are tailored in various forms, such as solutions, hydrogels, sponges and films that have been successfully used for therapeutic purposes, such as wound dressings to induce healing. In this book chapter, we examine several inherent properties of silk that aid in the acute events of the healing process.

A detailed insight into the progress of SF to be used as an extensive biomaterial in different forms for wound healing potential.

Keywords: silk fibroin, wound healing, biomaterial, scaffold

1. Introduction

Silk is a polymeric biomaterial produced from numerous insects like silkworms, spiders, mites, flies, scorpions and bees, used extensively for biomedical applications [1]. Silk has been

* Corresponding Author's Email: rohiwalsonali@gmail.com.

In: Silk Fibroin
Editor: Shivaji H. Pawar
ISBN: 979-8-88697-402-7

utilized as a potential biomaterial for engineering and regenerative medicine due to its biocompatibility, biodegradability, mechanical properties, and wound healing for skin regeneration [2]. The structural properties and sequence of silk fibres vary depending upon the source. The silk is mainly produced by two silkworms: no mulberry (wild variety) and domesticated mulberry. Silkworms develop silk cocoons by spinning silk fibres around them, which accommodate them during their period of metamorphism. This silk is engineered into various forms like films, fibres, sponges, hydrogels, and tubes to be utilized further upon application [3-5]. Silk fibres differ from species to species due to substantial variations in their amino acid sequence and their biological and physico-chemical properties [6]. Silk is a versatile biopolymer widely explored in many applications such as bone engineering, dental care, skin regeneration and regenerative medicine [7]. The fibre portion of silk fibroin (SF) is extracted from the cocoons of silkworms and isolated from the domesticated silkworm, Bombyx mori, i.e., most widely used and studied in terms of physical properties, chemical composition, and biological properties [8]. Silk-based material allows tissue ingrowth, does not evoke a long-term inflammatory response, and allows [9]. After *in vivo* transplantation of the material, it induces only a mild initial inflammatory response which subsides with time, thus indicating immune-compatible properties [10]. SF-based have unique structural and incredible mechanical properties compared to other biopolymer-based biomaterials. The degradation rate can be tailored depending on scaffold form, cross-linking agent types of blending etc., from weeks to years after *in vivo* implantation. Therefore, silk fibroin can be used in several biomedical applications and play a key role in tissue engineering, regenerative medicine, drug delivery, bone regeneration, skin regeneration, corneal, dental and cardiovascular implantation [11].

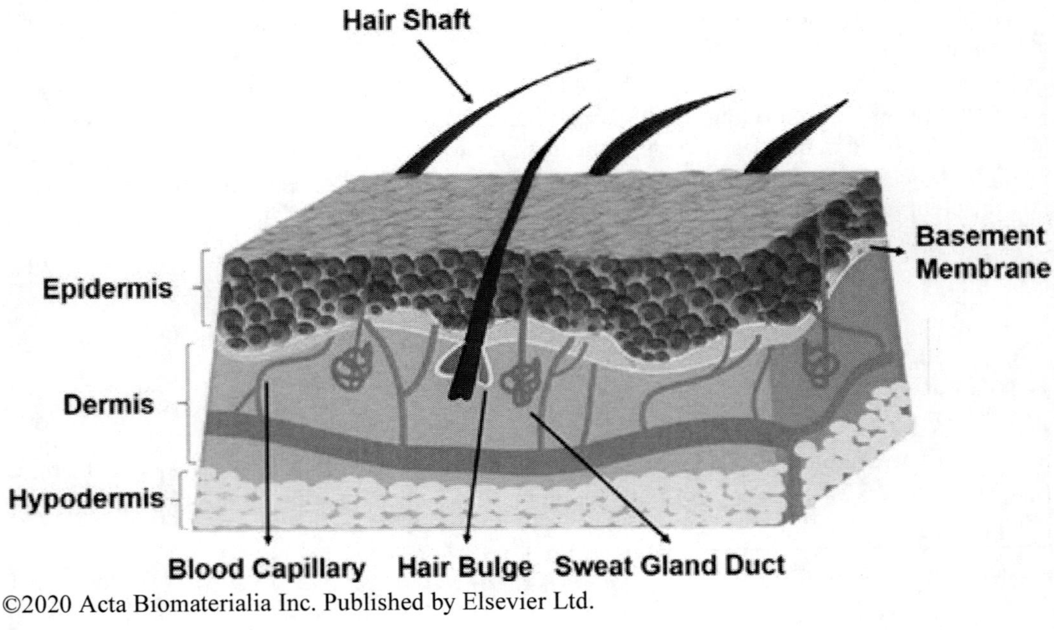

Figure 1. The three-dimensional (3D) structure of the skin explores all three layers (epidermis, dermis and hypodermis), sweat glands, blood capillaries and hair shafts protecting the body from the external environment and infectious pathogens [8].

Silk fibroin has also been widely explored for skin repair due to its unique properties like enabling hemostatic, excellent oxygen permeability and low inflammatory potential [12]. The characteristics property of silk enhances cell migration and proliferation, ultimately leading to accelerated wound healing [13]. It also promotes the adhesion of human keratinocytes and fibroblasts and enhances the deposition of type I collagen *in vitro* [14]. Several studies have been conducted by researchers that proved the efficacy of silk fibroin to be effectively used as wound dressings and bioartificial skin grafts [15, 16]. Furthermore, silk protein was safe under acute dermal irritation, acute dermal toxicity, and skin sensitization [17]. Skin is the largest organ of our body. It performs a vital role in protecting the human body from the environment. Thus, preventing the internal organs by providing barrier properties towards the external environment and harmful pathogens, dehydration, and infectious agents (Figure 1) [18].

Phases of Wound Healing
i) Homeostasis or coagulation beginning immediately after injury
ii) Inflammation which begins shortly after that
iii) Proliferation which starts within days of the injury and encompasses the major healing processes
iv) Remodeling in which scar tissue formation takes place and which may last up to a year or more

Figure 2. The phases of the wound healing mechanism.

2. Wound Healing Mechanism with Silk Fibroin As a Suitable Biomaterial

Silk fibroin has been a potential wound dressing material for treatment for more than one hundred years. Wound healing is a biological process in all body cells, tissues and organs. It is a dynamic and complex process that involves several types of cells, extracellular matrix (ECM) components, growth factors and enzymes like proteinases. Although the healing process is continuous, it is randomly divided into several phases to help understand the physiological courses in the wound and adjacent tissue. Wound healing has commonly been divided into three or four sequential and overlapping events that begin with haemostasis to stop bleeding and prevent excessive blood loss. Followed by inflammation and proliferation, which encompasses the recruitment of various types of cells like macrophages, neutrophils, fibroblasts, endothelial cells, and keratinocytes [19]. Healing is a complex process involving coordinated interactions and a sequence of carefully and specifically regulated steps and events associated with the appearance of various cell types in the wound during distinct phases of the wound healing process [20]. These numerous cells play a vital role in secreting multiple growth factors cytokines, forming blood vessels and developing tissue granulation and re-epithelialization, leading to a closer wound cavity. The wound healing mechanism is shown in Figure 3.

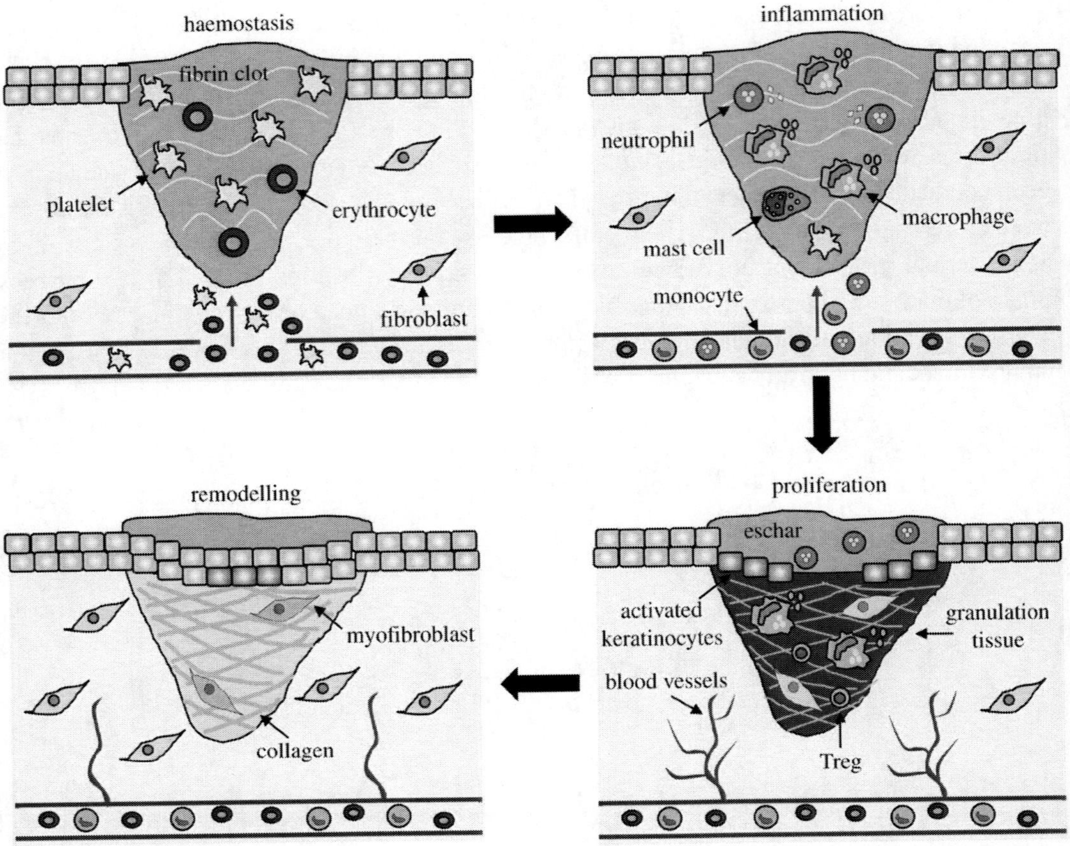

Figure 3. The diagrammatic representation of the wound healing mechanism [38].

Healing is a complex process involving coordinated interactions between discrete biological and immunological systems. It consists of a cascade of precisely regulated events that correlate with the presence of various cell types in the wound during different phases of the healing process. Although the process of healing is perpetual, it is divided into different phases to aid understanding of the physiological processes in the wound and encompasswound and encompassing tissue. Although the process of healing is perpetual, it is divided into different phases to aid understanding of the physiological processes that are happening in the wound and encompassing tissue. The final stage of healing starts when the matrix is remodelled into a new matrix, gradually leading to completely regenerated skin or semi/non-functional scar tissue [19].

2.1. Homeostasis

Immediately after injury, coagulation and homeostasis take place in the wound. The principal aim of these mechanisms is to prevent exsanguinations. It is a way to protect the vascular system, keeping it intact and providing a matrix for invading cells that are needed in the later phases of healing [21]. The silk fibres were also found to have some surface properties for haemostatic activity, the ability of silk threads to bind with proteins, and helping in the clotting cascade mechanism [22]. Some researchers also reported the thrombogenic effect of black braided silk sutures that aid in binding with blood platelets and fibrin during implantation.

Moreover, the thrombotic response was also reduced when the wax-coated silk threads were used. To make the silk an anticoagulant material, different surface functionalization/ modifications were used, like sulfation and heparinization, because the non-treated silk was reported to clot the blood [23]. The collagen exposed during the wound formation activates both the cascade pathways, intrinsic and extrinsic, and initiates the inflammatory phase. Once the injury occurs, the neuronal reflex mechanism and injured vessels compress immediately due to the compression of vascular smooth muscle cells in the circular muscle layer. This constriction is strong enough to prevent blood loss from arteriole with a diameter of 0.5 cm. As blood spills into the injury site, the blood components and platelets come in contact with exposed collagen and other extracellular matrix components. This contact triggers the cell membranes to release potent components such as vasoconstrictors (thromboxane A2 and prostaglandins 2-α) of clotting factors from the platelets and the formation of a blood clot of fibronectin, fibrin, vitronectin and thrombospondin. The fibrin clot is scaffolding for arriving cells, such as neutrophils, monocytes, fibroblasts, and endothelial cells. It also serves to concentrate the cytokines and growth factors. Some researchers reported using thin films made of regenerated silk solution that aids in binding with fibrinogen, confirming the haemostatic property of silk firoin even in the regenerated form [24]. These interactions of silk and clotting factors also help as a carrier that releases inflammatory factors to lead the healing process toward its next phase. Based on this natural interaction between silk and fibrinogen, researchers also developed porous sponges supplemented with coagulants, namely, fibrinogen and thrombin were mixed with silk solution [25]. The blood clot and platelets trapped within it are essential for haemostasis and provide a provisional matrix for cell migration in the subsequent haemostatic and inflammatory phases. These molecules act as promoters in the wound healing cascade by activating and attracting neutrophils, macrophages, endothelial cells, and fibroblasts. Platelets also contain vasoactive amines, such as serotonin, that are stored in dense bodies and cause vasodilation and increased vascular permeability, leading to fluid extravasation in the tissue that results in oedema, which, in turn, potentiates itself during the following inflammatory phase [26].

2.2. Inflammatory Phase

Our body regulates the prevention of foreign substance invasion by triggering the inbuilt immune mechanism. These foreign substances are not only harmful pathogens and microorganisms but could also be biomaterials that are implanted in case of organ transplant. The inflammatory response is induced once the graft is implanted into the host system. The humoral and cellular inflammatory phase is divided into two phases early inflammatory response and late inflammatory phase [28]. The immune reaction either subsides in a later phase or becomes chronic due to long-term immunogenicity. Several researchers studied silk as an organ transplantation material and suggested that it does not provoke long-term inflammation; it elicits an initial immune response that subsides in the later phases, showing signs of graft acceptance [27, 28].

2.2.1. Early Inflammatory Response

This phase starts during the coagulation and, shortly after that, the early inflammatory functions. So, in this phase of wound healing, silk-based dressings and grafts can act as a source to accelerate the wound healing process because of overlapping inflammatory responses [29]. Some mild inflammatory responses may arise after implantation of silk-based matrices at initial points or implantation help in the activation of macrophages, thereby stimulating them to secrete chemokines and growth factors [27,29]. This activation includes the formation of multinuclear giant cells and the recruitment of immune cells and fibroblast [29]. This phase of wound healing activates the complement cascade and initiates molecular events, leading to infiltration of the wound site by neutrophils, whose primary function is to prevent infection [30]. The neutrophils begin to be attracted to the wound site within 24-36 h of injury by chemo-attractive agents, including TGF-β, complement components such as C3a and C5a, and formyl methionyl peptides produced by bacteria and platelet products [31]. The neutrophils destroy and remove bacteria, foreign particles and damaged tissues through phagocytosis with the help of oxygen-free radicals and proteolytic enzymes. Due to alterations in the regulation of surface adhesion molecules, neutrophils become sticky and, through a process of margination, begin to adhere to the endothelial cells in the post-capillary venules surrounding the wound [28]. Then the neutrophils roll and move along the surface of the endothelium, being pushed forward by blood flow; selectins mediate this pathway

Similarly, chemokines which are secreted by endothelial cells, help in the activation of the adhesion system mediated by integrin [32]. Neutrophil activity gradually changes and is eliminated from the wounded site before the progression of the following healing step. Redundant cells are destroyed by apoptosis with the help of macrophages [30, 33].

2.2.2. Late Inflammatory Phase

This phase starts after 48-72 h after injury; macrophages impart a significant role [28, 34]. These cells are attracted to the wounded site by a myriad of chemo attractive agents, including clotting factors, complement components, cytokines such as PDGF, TGF-β, leukotriene B4 and platelet factor IV, as well as elastin and collagen breakdown products, macrophages have a longer lifespan than neutrophils and continue to work at a lower pH. TGF-β and other mediators such as TGF-α, heparin-binding epidermal growth factor, fibroblast growth factor [FGF], collagenase, activating keratinocytes, fibroblasts and endothelial cells play a fundamental role and act as regulatory cells [35]. Lastly, lymphocytes attract to the wounded site for almost 72 h by the action of interleukin-1 (IL-1), complement components and immunoglobulin G (IgG) breakdown products. For the production of extracellular matrix and collagenase, regulation of IL-1 plays an important role [36], which afterwards is needed for collagen remodelling. The detailed diagrammatic mechanism of wound healing is shown in Figure 3. It was observed that after long-term implantation of silk sponges, silk material does not induce foreign body multinucleated giant cells for a prolonged duration; thus, it is safe to use for implantation. These porous silk sponges have a less population of immune cells post 4-weeks implantation compared to 2 weeks [37].

2.3. Proliferative Phase

The proliferative phase starts on the third day after wounding and lasts about two weeks. It is characterized by the migration of fibroblast cells and deposition of the newly synthesized extracellular matrix, which acts as a provisional network composed of fibrin and fibronectin. This phase of wound healing can be seen as an abundant granulation tissue formation. In the proliferative phase, epithelialization, angiogenesis, granulation tissue formation, and collagen deposition are the principal steps in this building portion of wound healing. This process occurs early in wound repair. For wound healing, the basement membrane should remain intact and helps epithelial cells migrate in an upward pattern. The epithelial progenitor cells remain intact, and within 2-3 days, normal layers of the epidermis arise. TNF- α stimulates the process of angiogenesis marked by endothelial cell migration and capillary formation. The migration of capillaries into the wound bed is critical for proper wound healing. For proper wound healing, proper nutrients supplied by capillary are required, especially for the granulation phase. This failure to epithelial cells on the skin edge begins proliferating and sending out projections to renew a defensive barrier against fluid losses bacterial invasion. The activated platelets and the macrophages produce epidermal growth factor (EGF) and TGF-α, which stimulate epithelial proliferation and chemotaxis [39]. Epitheliazation begins shortly after wounding and is first stimulated by inflammatory cytokines.

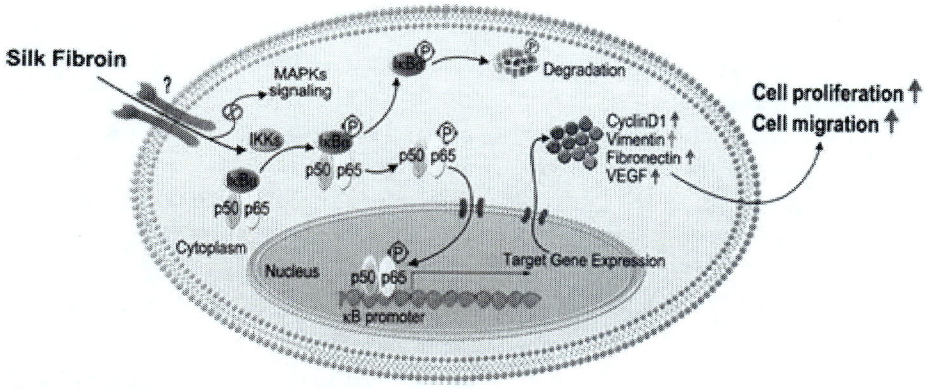

Figure 4. The role of SF in enhancing cell migration and cell proliferation in the wound by the NF-κB signalling pathway [13].

IL-1 and TNF- α up-regulate keratinocyte growth factor (KGF) gene expression in fibroblasts. In turn, fibroblasts synthesize and secrete KGF-1, KGF-2, and IL-6, which simulate neighbouring keratinocytes to immigrate to the wound site, proliferate, and differentiate in the epidermis [39, 40]. The main part of the proliferative phase is granulation tissue formation; fibroblasts migrate to the wound site from the surrounding tissue and become activated, synthesizing collagen and proliferating. Platelet-derived growth factors (PDGF) and EGF signal these fibroblast cells, which are derived from platelets and macrophages. Fibroblasts amplify the autocrine and paracrine signalling by PDGF expression. The wound fibroblasts will begin and start synthesizing collagen and transform into myofibroblasts for wound contraction. In response to PDGF, fibroblasts begin synthesizing a provisional matrix composed of collagen type III, glycosaminoglycans, and fibronectin [41]. A study reveals that SF protein stimulates cell migration by activating MEK, JNK, and PI3K signalling pathways when a starch assay has been performed on keratinocytes [42]. The wound healing property was proved in another, showing that the BmSF is attributed to the NF-κB signalling via microarray analysis and scratch assay. Park et al. proved not only the influence of silk on cell migration but also showed that it promotes healing by regulating the expression of well-known markers of cell proliferation, i.e., vimentin, cyclin D1, VEGF, and fibronectin, as shown in Figure 4 [13].

It was observed that by significantly tailoring the physical properties of silk, the development of pre-vascularized skin grafts could be obtained using silk for wound healing applications.

2.4. Maturation and Remodeling

This phase starts after the 8th day after injury and may last through 1 year. This phase of wound healing is most important, which features collagen deposition in an organized and well-mannered network. The collagen synthesis will continue 4 to 5 weeks after the wounding. The collagen is oriented parallel to the uninjured skin and is thinner. Over time, the initial collagen threads are reabsorbed, deposited thicker, and organized along the stress lines. The collagen found in granulation tissue is biochemically different from collagen from uninjured skin. The hydroxylation and glycosylation of lysine residues are higher in granulation tissue, and this increase in glycosylation correlates with the thinner fibre size. There will always be a difference between the organization of collagen of injured and uninjured skin, and the strength of the wound will never return the same. According to the literature, at 1 week, the wound has only 3% of its final strength; at three weeks, 30%; and at 3 months (and beyond), approximately 80 percent [43]. Silk has proven the migratory cell effects as previously described via NF-κB signalling pathways, which are attributed to enhanced re-epithelialization. Silk nanofibers stimulate the formation of an epithelial layer over an *in vitro* wound model formed of collagen gel [44].

Additionally, in *in vivo* wound healing model, expression of cytokeratin 10 (CK10) and cytokeratin 14 (CK14) markers was observed in the regenerated epithelial layer [45]. This expression of markers was observed in the newly formed epidermis in the wounds treated with silk matrices in the differentiation of keratinocytes. In conclusion, it can be said that silk provides an effective platform and artificial cell conducive microenvironment that aids in the migration of keratinocytes and accelerates the early-stage re-epithelialization of wounds.

3. Different Forms of Silk Fibroin for Wound Healing

SF solutions have been used to fabricate various forms such as solution, fibres, powder, sponges, hydrogels, and films via different treatments.

3.1. Silk Fibroin Solution

SF solution can be formulated for different types of creams, ointments and other products for various wound healing purposes. By degumming the silk cocoons, SF solution can be obtained in an aqueous solution of 0.02 M Na_2CO_3 by boiling for 40 min at 95°C, followed by washing it with distilled water to eliminate the glue-like sericin proteins yields silk fibroin solution [46] It was then solubilized with CaCl2, ethanol, and H2O (at a molar ratio of 1:2:8) for 50 min at

98°C followed by filtration to purify and subsequent dialysis for three days over the membrane to obtain silk fibroin solutions [13]. To test the wound healing ability, NIH3T3 cells were used to perform *in vitro* scratch wound assay and partial excision skin wound model with rats for *in vivo* evaluation of silk solution. SF in solution state significantly induced wound healing in a scratch wound model *in vitro* and *in vivo* [13]. and *in vivo* [13].

3.2. Hydrogels

Hydrogels are water-swollen 3D polymer networks with a high swelling ratio in an aqueous solution. The hydrogel can be cross-linked through physical or chemical methods. These are excellent for implementing cell seeding and encapsulation in developing tissue engineering applications. Hydrogels have been extensively applied in wound treatment because of their exclusive properties of induction of cell attachment, growth, proliferation, migration, and extracellular matrix (ECM) production, providing mechanical strength [47]. Silk can be blended with other polymers such as collagen for a wound healing treatment, possesses a suitable interfacial adhesion, and displays enhanced mechanical properties [48]. Moreover, hydrogel biocompatibility also allows them to conserve cell viability.

3.3. Films

Wound healing can be enhanced with silk fibroin blends in the form of well-designed and straightforward films. Silk fibroin films offer tremendous advantages of being easily obtainable, sterilizable, transparent, and permitting tissue regeneration during the wound healing process; thus, they could be clinically suitable for wound treatment. There are two methods to fabricate SF films spin coating and vertical deposition. In the case of spin coating, SF solution and ethanol are alternately coated onto the substrates. After being employed in the wounded area, these films showed faster healing with a lower inflammatory response than conventional porcine-based wound dressings. Similar results have been obtained by researchers stating the qualities of SF films as a novel wound healing material are very efficient with epidermal growth factors for the acute wound [49].

3.4. Sponges

Sponges are porous and are significant in tissue engineering materials and regeneration. SF solutions have also been utilized to fabricate porous sponges that provide a network of linked pores with a large surface area. It provides a predetermined three-dimensional volume that allows better surface area for tissue ingrowth and cell adhesion.

These interconnected areas of the porous structure of sponges closely mimic physiological environments *in vivo* [50]. SF porous sponges can be obtained using porogens, gas forming, electrospun fibre, freeze-drying, and freeze-drying/foaming. For wound healing applications, SF sponge and SF as a hybrid sponge with other polymers have been successfully applied due to the fragile nature of sponges [50].

Conclusion and Future Perspectives

Silk fibroin, due to its incredible properties like biocompatibility, biodegradability, user-friendly nature, easy availability, and low immune response to host tissue, has attracted intense interest in to be used in wound healing applications. It can be tailored in various forms such as solutions, films, hydrogels, and sponges. It is efficiently used in therapeutic practices to enhance skin and tissue regeneration and stimulate wound healing via cell proliferation, migration and differentiation. The cell-matrix interaction is one of the most important parameters that regulate the wound healing process. In connection, interactions between silk fibroin and cells play an important role in healing cutaneous wounds. Many challenges are yet to be explored besides having numerous literature focused on this biomaterial. Thus, additional studies are needed to progress the potential effect of SF for different wound healing applications, especially on the mechanistic basis of SF for wound healing applications.

Acknowledgements

The authors would like to acknowledge the Czech Academy of Sciences - Strategy AV21 (VP29) Towards Precision Medicine and Gene Therapy.

References

[1] Altman, G. H., Diaz, F., Jakuba, C., Calabro, T., Horan, R. L., Chen, J., Lu, H., Richmond, J., Kaplan, D. L. Silk-Based Biomaterials. *Biomaterials* 2003, 24, 401-416, https://doi.org/10.1016/S0142-9612(02)00353-8.

[2] Gil, E. S., Panilaitis, B., Bellas, E., Kaplan, D. L. Functionalized Silk Biomaterials for Wound Healing. *Advanced Healthcare Materials* 2013, 2, 206-217, https://doi.org/10.1002/adhm.201200192.

[3] Wittmer, C. R., Claudepierre, T., Reber, M., Wiedemann, P., Garlick, J. A., Kaplan, D., Egles, C. Biomedical Applications: Multifunctionalized Electrospun Silk Fibers Promote Axon Regeneration in the Central Nervous System (Adv. Funct. Mater. 22/2011). *Advanced Functional Materials* 2011, 21, 4202-4202, https://doi.org/10.1002/adfm.201190103.

[4] Wharram, S. E., Zhang, X., Kaplan, D. L., McCarthy, S. P. Electrospun Silk Material Systems for Wound Healing. *Macromolecular Bioscience* 2010, 10, 246-257, https://doi.org/10.1002/mabi.200900274.

[5] Lawrence, B. D., Marchant, J. K., Pindrus, M. A., Omenetto, F. G., Kaplan, D. L. Silk Film Biomaterials for Cornea Tissue Engineering. *Biomaterials* 2009, 30, 1299-1308, https://doi.org/10.1016/j.biomaterials.2008.11.018.

[6] Kundu, S. C., Kundu, B., Talukdar, S., Bano, S., Nayak, S., Kundu, J., Mandal, B. B., Bhardwaj, N., Botlagunta, M., Dash, B. C., Chitrangada Acharya, Ananta K. Ghosh. Nonmulberry Silk Biopolymers. *Biopolymers* 2012, 97, 455-467, https://doi.org/10.1002/bip.22024.

[7] Holland, C., Numata, K., Rnjak-Kovacina, J., Seib, F. P. The Biomedical Use of Silk: Past, Present, Future. *Advanced Healthcare Materials* 2019, 8, 1800465, https://doi.org/10.1002/adhm.201800465.

[8] Chouhan, D., Mandal, B. B. Silk Biomaterials in Wound Healing and Skin Regeneration Therapeutics: From Bench to Bedside. *Acta Biomaterialia* 2020, 103, 24-51, https://doi.org/10.1016/j.actbio.2019.11.050.

[9] Zhang, W., Chen, L., Chen, J., Wang, L., Gui, X., Ran, J., Xu, G., Zhao, H., Zeng, M., Ji, J., Li Qian, Jianda Zhou, Hongwei Ouyang, Xiaohui Zou. Silk Fibroin Biomaterial Shows Safe and Effective

Wound Healing in Animal Models and a Randomized Controlled Clinical Trial. *Advanced Healthcare Materials* 2017, 6, 1700121, https://doi.org/10.1002/adhm.201700121.
[10] Thurber, A. E., Omenetto, F. G., Kaplan, D. L. *In Vivo* Bioresponses to Silk Proteins. *Biomaterials* 2015, 71, 145-157, https://doi.org/10.1016/j.biomaterials.2015.08.039.
[11] Gholipourmalekabadi, M., Sapru, S., Samadikuchaksaraei, A., Reis, R. L., Kaplan, D. L., Kundu, S. C. Silk Fibroin for Skin Injury Repair: Where Do Things Stand? *Advanced Drug Delivery Reviews* 2020, 153, 28-53, https://doi.org/10.1016/j.addr.2019.09.003.
[12] Gil, E. S., Panilaitis, B., Bellas, E., Kaplan, D. L. Functionalized Silk Biomaterials for Wound Healing. *Advanced Healthcare Materials* 2013, 2, 206-217, https://doi.org/10.1002/adhm.201200192.
[13] Park, Y. R., Sultan, Md. T., Park, H. J., Lee, J. M., Ju, H. W., Lee, O. J., Lee, D. J., Kaplan, D. L., Park, C. H. NF-KB Signaling Is Key in the Wound Healing Processes of Silk Fibroin. *Acta Biomaterialia* 2018, 67, 183-195, https://doi.org/10.1016/j.actbio.2017.12.006.
[14] Min, B. M., Lee, G., Kim, S. H., Nam, Y. S., Lee, T. S., Park, W. H. Electrospinning of Silk Fibroin Nanofibers and Its Effect on the Adhesion and Spreading of Normal Human Keratinocytes and Fibroblasts *in Vitro*. *Biomaterials* 2004, 25, 1289-1297, https://doi.org/10.1016/j.biomaterials.2003.08.045.
[15] Janani, G., Kumar, M., Chouhan, D., Moses, J. C., Gangrade, A., Bhattacharjee, S., Mandal, B. B. Insight into Silk-Based Biomaterials: From Physicochemical Attributes to Recent Biomedical Applications. *ACS Applied Bio Materials* 2019, 2, 5460-5491, https://doi.org/10.1021/acsabm.9b00576.
[16] Vidal Yucha, S. E., Tamamoto, K. A., Nguyen, H., Cairns, D. M., Kaplan, D. L. Human Skin Equivalents Demonstrate Need for Neuro-Immuno-Cutaneous System. *Advanced Biosystems* 2019, 3, 1800283, https://doi.org/10.1002/adbi.201800283.
[17] Padol, A., Jayakumar, K., Shridhar, N., Swamy, Narayana H. D., Swamy, N., Mohan, K. Safety Evaluation of Silk Protein Film (A Novel Wound Healing Agent) in Terms of Acute Dermal Toxicity, Acute Dermal Irritation and Skin Sensitization. *Toxicology International* 2011, 18, 17, https://doi.org/10.4103/0971-6580.75847.
[18] Clark, R. A. F., Ghosh, K., Tonnesen, M. G. Tissue Engineering for Cutaneous Wounds. *Journal of Investigative Dermatology* 2007, 127, 1018-1029, https://doi.org/10.1038/sj.jid.5700715.
[19] Martin, P. Wound Healing - Aiming for Perfect Skin Regeneration. *Science* (1979) 1997, 276, 75-81, https://doi.org/10.1126/science.276.5309.75.
[20] Attinger, C. E., Janis, J. E., Steinberg, J., Schwartz, J., Al-Attar, A., Couch, K. Clinical Approach to Wounds: D??Bridement and Wound Bed Preparation Including the Use of Dressings and Wound-Healing Adjuvants. *Plastic and Reconstructive Surgery* 2006, 117, 72S-109S, https://doi.org/10.1097/01.prs.0000225470.42514.8f.
[21] Robson, M. C., Steed, D. L., Franz, M. G. Wound Healing: Biologic Features and Approaches to Maximize Healing Trajectories. *Current Problems in Surgery* 2001, 38, A1-140, https://doi.org/10.1067/msg.2001.111167.
[22] Dahlke, H., Dociu, N., Thurau, K. Thrombogenicity of Different Suture Materials as Revealed by Scanning Electron Microscopy. *Journal of Biomedical Materials Research* 1980, 14, 251-268, https://doi.org/10.1002/jbm.820140307.
[23] Zamani, M., Khafaji, M., Naji, M., Vossoughi, M., Alemzadeh, I., Haghighipour, N. A Biomimetic Heparinized Composite Silk-Based Vascular Scaffold with Sustained Antithrombogenicity. *Scientific Reports* 2017, 7, 4455, https://doi.org/10.1038/s41598-017-04510-1.
[24] Santin, M., Motta, A., Freddi, G., Cannas, M. *In Vitro* Evaluation of the Inflammatory Potential of the Silk Fibroin. *Journal of Biomedical Materials Research* 1999, 46, 382-389, https://doi.org/10.1002/(SICI)1097-4636(19990905)46:3<382::AID-JBM11>3.0.CO;2-R.
[25] Teuschl, A. H., Zipperle, J., Huber-Gries, C., Kaplan, D. L. Silk Fibroin Based Carrier System for Delivery of Fibrinogen and Thrombin as Coagulant Supplements. *Journal of Biomedical Materials Research Part A* 2017, 105, 687-696, https://doi.org/10.1002/jbm.a.35940.
[26] Li, J., Chen, J., Kirsner, R. Pathophysiology of Acute Wound Healing. *Clinics in Dermatology* 2007, *25*, 9-18, https://doi.org/10.1016/j.clindermatol.2006.09.007.

[27] Janani, G., Kumar, M., Chouhan, D., Moses, J. C., Gangrade, A., Bhattacharjee, S., Mandal, B. B. Insight into Silk-Based Biomaterials: From Physicochemical Attributes to Recent Biomedical Applications. *ACS Applied Bio Materials* 2019, 2, 5460-5491, https://doi.org/10.1021/acsabm.9b00576.

[28] Hart, J. Inflammation 1: Its Role in the Healing of Acute Wounds. *Journal of Wound Care* 2002, 11, 205-209, https://doi.org/10.12968/jowc.2002.11.6.26411.

[29] Thurber, A. E., Omenetto, F. G., Kaplan, D. L. *In Vivo* Bioresponses to Silk Proteins. *Biomaterials* 2015, 71, 145-157, https://doi.org/10.1016/j.biomaterials.2015.08.039.

[30] Broughton, G., Janis, J. E., Attinger, C. E. Wound Healing: An Overview. *Plastic and Reconstructive Surgery* 2006, 117, 1e-S-32e-S, https://doi.org/10.1097/01.prs.0000222562.60260.f9.

[31] Robson, M. C., Steed, D. L., Franz, M. G. Wound Healing: Biologic Features and Approaches to Maximize Healing Trajectories. *Current Problems in Surgery* 2001, 38, A1-140, https://doi.org/10.1067/msg.2001.111167.

[32] Singh, S., Young, A., McNaught, C. E. The Physiology of Wound Healing. *Surgery (Oxford)* 2017, 35, 473-477, https://doi.org/10.1016/j.mpsur.2017.06.004.

[33] Diegelmann, R. F. Wound Healing: An Overview of Acute, Fibrotic and Delayed Healing. *Frontiers in Bioscience* 2004, 9, 283, https://doi.org/10.2741/1184.

[34] Ennis, W. J., Meneses, P. Wound Healing at the Local Level: The Stunned Wound. *Ostomy Wound Manage* 2000, 46, 39S-48S; quiz 49S-50S.

[35] Ramasastry, S. S. Acute Wounds. *Clinics in Plastic Surgery* 2005, 32, 195-208, https://doi.org/10.1016/j.cps.2004.12.001.

[36] Singh, S., Young, A., McNaught, C. E. The Physiology of Wound Healing. *Surgery (Oxford)* 2017, 35, 473-477, https://doi.org/10.1016/j.mpsur.2017.06.004.

[37] Rnjak-Kovacina, J., Wray, L. S., Golinski, J. M., Kaplan, D. L. Arrayed Hollow Channels in Silk-Based Scaffolds Provide Functional Outcomes for Engineering Critically Sized Tissue Constructs. *Advanced Functional Materials* 2014, 24, 2188-2196, https://doi.org/10.1002/adfm.201302901.

[38] Wilkinson, H. and Hardman M. Wound healing: cellular mechanisms and pathological outcomes Article: Nanotechnology-Driven Therapeutic Interventions in Wound Healing: Potential Uses and Applications 2017, 3,3, 163-175, https://doi.org/10.1098/rsob.200223.

[39] Xia, Y. P., Zhao, Y., Marcus, J., Jimenez, P. A., Ruben, S. M., Moore, P. A., Khan, F., Mustoe, T. A. Effects of Keratinocyte Growth Factor-2 (KGF-2) on Wound Healing in an Ischaemia-Impaired Rabbit Ear Model and on Scar Formation. *The Journal of Pathology* 1999, 188, 431-438, https://doi.org/10.1002/(SICI)1096-9896(199908)188:4<431::AID-PATH362>3.0.CO;2-B.

[40] Pierce, G. F., Mustoe, T. A., Altrock, B. W., Deuel, T. F., Thomason, A. Role of Platelet-Derived Growth Factor in Wound Healing. *Journal of Cellular Biochemistry* 1991, 45, 319-326, https://doi.org/10.1002/jcb.240450403.

[41] Martínez-Mora, C., Mrowiec, A., García-Vizcaíno, E. M., Alcaraz, A., Cenis, J. L., Nicolás, F. J. Fibroin and Sericin from Bombyx Mori Silk Stimulate Cell Migration through Upregulation and Phosphorylation of C-Jun. *PLoS ONE* 2012, 7, e42271, https://doi.org/10.1371/journal.pone.0042271.

[42] Forrest, L. Current Concepts in Soft Connective Tissue Wound Healing. *British Journal of Surgery* 2005, 70, 133-140, https://doi.org/10.1002/bjs.1800700302.

[43] Schneider, A., Wang, X. Y., Kaplan, D. L., Garlick, J. A., Egles, C. Biofunctionalized Electrospun Silk Mats as a Topical Bioactive Dressing for Accelerated Wound Healing. *Acta Biomaterialia* 2009, 5, 2570-2578, https://doi.org/10.1016/j.actbio.2008.12.013.

[44] Gil, E. S., Panilaitis, B., Bellas, E., Kaplan, D. L. Functionalized Silk Biomaterials for Wound Healing. *Advanced Healthcare Materials* 2013, 2, 206-217, https://doi.org/10.1002/adhm.201200192.

[45] Park, Y. R., Ju, H. W., Lee, J. M., Kim, D. K., Lee, O. J., Moon, B. M., Park, H. J., Jeong, J. Y., Yeon, Y. K., Park, C. H. Three-Dimensional Electrospun Silk-Fibroin Nanofiber for Skin Tissue Engineering. *International Journal of Biological Macromolecules* 2016, 93, 1567-1574, https://doi.org/10.1016/j.ijbiomac.2016.07.047.

[46] Kamoun, E. A., Kenawy, E. R. S., Chen, X. A Review on Polymeric Hydrogel Membranes for Wound Dressing Applications: PVA-Based Hydrogel Dressings. *Journal of Advanced Research* 2017, 8, 217-233, https://doi.org/10.1016/j.jare.2017.01.005.

[47] de Moraes, M. A., Paternotte, E., Mantovani, D., Beppu, M. M. Mechanical and Biological Performances of New Scaffolds Made of Collagen Hydrogels and Fibroin Microfibers for Vascular Tissue Engineering. *Macromolecular Bioscience* 2012, 12, 1253-1264, https://doi.org/10.1002/mabi.201200060.

[48] Padol, A., Jayakumar, K., Shridhar, N., Swamy, Narayana H. D., Swamy, N., Mohan, K. Safety Evaluation of Silk Protein Film (A Novel Wound Healing Agent) in Terms of Acute Dermal Toxicity, Acute Dermal Irritation and Skin Sensitization. *Toxicology International* 2011, 18, 17, https://doi.org/10.4103/0971-6580.75847.

[49] Ma, D., Wang, Y., Dai, W. Silk Fibroin-Based Biomaterials for Musculoskeletal Tissue Engineering. *Materials Science and Engineering: C* 2018, 89, 456-469, https://doi.org/10.1016/j.msec.2018.04.062.

[50] Sultan, Md. T., Jeong, J. Y., Seo, Y. B., Lee, O. J., Ju, H. W., Park, H. J., Lee, Y. J., Lee, J. S., Kim, S. H., Park, C. H. Fabrication and Characterization of the Porous Duck's Feet Collagen Sponge for Wound Healing Applications. *Journal of Biomaterials Science, Polymer Edition* 2018, 29, 960-971, https://doi.org/10.1080/09205063.2017.1367636.

Chapter 11

Regulation of Cytokines in Wound Healing by Silk Fibroin

Deepinder Sharda[1], Komal Attri[1,2] and Diptiman Choudhury[1,2,*]

[1] School of Chemistry and Biochemistry,
Thapar Institute of Engineering and Technology,
Patiala, Punjab, India

[2] Centre of Excellence in Emerging Materials (CEEMS),
Thapar Institute of Engineering and Technology and Virginia Tech,
Patiala, Punjab, India

Abstract

Wound healing is a complex process involving a cascade of inflammatory and anti-inflammatory physiological events. Some key regulatory factors involved in this process include NF-kβ signaling, Wnt and Notch signaling and JAK/STAT signaling. In diabetes, a disease that has affected five hundred thirty-seven million people globally (2019 IDF report), these key factors get involved and cause problems, including retinopathy, neuropathy, damaged blood vessels, and impaired wound healing. Around 15-25% of diabetic patients develop foot ulcers, of which 6% require hospitalization, and of the hospitalized ones, 15-20% suffer from lower limb amputation. The normal healing process gets delayed in diabetic patients due to the accumulation of pro-inflammatory cytokines such as TNF-α, IFN-ϒ, IL-1β, and IL-6, resulting in the free radical generation and prolonged inflammatory phase. Various factors, like proteins, polyphenols, etc., regulate cytokine accumulation and shift from pro-inflammatory to anti-inflammatory signaling. One such protein is silk fibroin, a natural fibrous biomaterial obtained from silkworms, spiders, and bees and exhibits wound healing ability in normal and diabetic conditions. It influences NF-kβ signaling by inducingprotein expression levels ofIKKα, IKKβ, p65, cyclin D1, fibronectin, vimentin, VEGF, and the degradation ofIκBα. Silk fibroin also influences the anti-apoptotic and MAPK signaling pathways and drives healing.

Further, it is highly biocompatible, non-toxic, non-carcinogenic, sustained biodegradability, low antigenicity, oxygen permeability, and excellent mechanical properties such as high tensile strength and elasticity, which provides skeleton support to the wound. It is used in various morphological forms, including solutions, scaffolds,

* Corresponding Author's Email: diptiman@thapar.edu.

In: Silk Fibroin
Editor: Shivaji H. Pawar
ISBN: 979-8-88697-402-7

hydrogels, sponges, blend films, hydrocolloid dressings, and as the carrier for the delivery of drugs, growth factors, and other bioactive agents to the wound site. The chapter describes various morphological forms and mechanisms of cytokine regulation influencing the wound healing mechanism by silk fibroin.

Keywords: diabetes, wound healing, silk fibroin, wound dressings, cytokines, scaffolding, TNF-α, IFN-ϒ, IL-1β, IL-6, IKKα, IKKβ, p65, Cyclin D1, fibronectin, vimentin, VEGF, NF-kβ signaling, MAPK pathway

1. Introduction

Despite extensive development and research in this field, chronic wounds have remained a significant challenge for scientists and clinicians worldwide. Skin damage is the loss of tissue continuity due to any means, including chemical, mechanical, electrical, or radiations [1, 2]. It can be acute or chronic and internal or external [3, 4]. Globally, around 40 million people are affected by chronic wounds, which may last for three months, while acute injuries can last only up to fourteen days. Chronic wounds pose a severe economic burden of 19.8 billion till 2021 globally. They are estimated to increase up to 24.8 billion in 2024, thus becoming a major challenging issuewhichbecomes more harmful when combined with diabetes and other ailments especially in old age [5, 6, 7, 8].

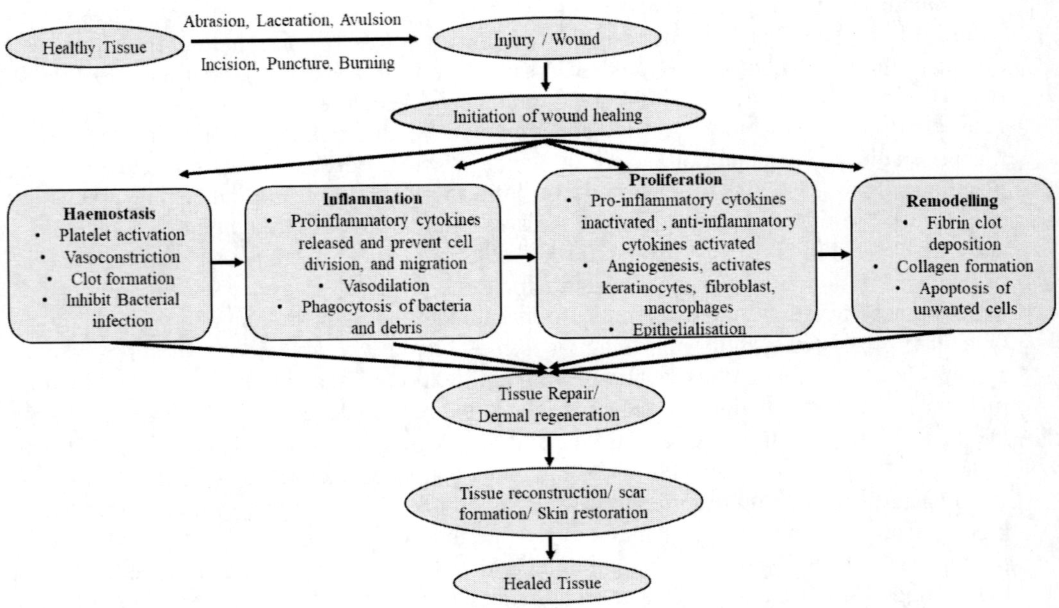

Figure 1. This flow chart shows the wound healing process, involving different causes of the wound, initiation of skin repair and regeneration followed by the phases involved in the healing process with the detailed study of each stage with its characteristic features related to wound care.

The skin constitutes the largest organ of the human body, which prevents the entry of harmful organisms into the body by acting as a barrier and have a self-healing ability to a certain extent. Loss of skin integrity occurs in case of extensive skin damage, which may eventually lead to death. The epidermis and dermis are the two main components of skin having hairs and hormonal glands in the upper epidermal layer with their roots deep inside the dermis, making the skin a complex organ and its healing even more challenging [9]. Wound healing involves a cascade of steps divided into three different phases: hemostasis/inflammation, proliferation, and remodeling. Different cell types, including leukocytes, fibroblasts, and keratinocytes, regulate cytokines, chemokines, and growth factors that interact and promote wound healing at each stage [10, 11], as shown in Figure 1.

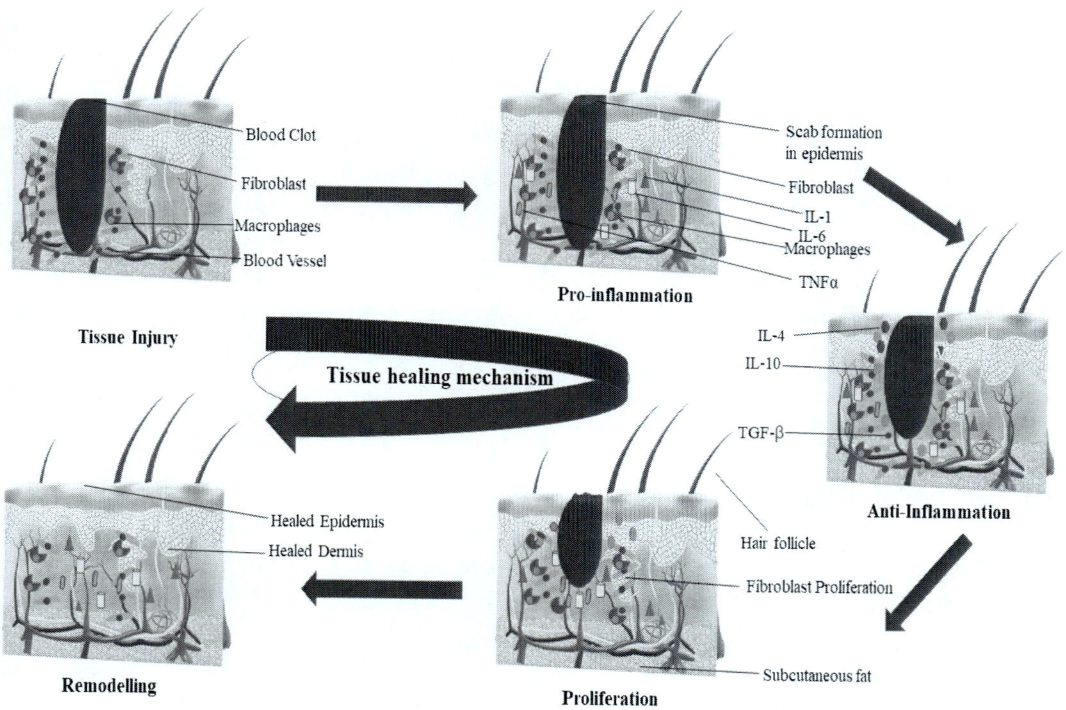

Figure 2. The diagrammatic presentation of wound healing phases: inflammatory, proliferative, and remodeling after the deep tissue injury. Cytokines, macrophages, and growth factors are shown in scab formation at the injury site after blood clotting, followed by fibroblast proliferation and the healed wound.

Briefly, the healing process involves clot formation during the initial hemostasis phase by activating platelets that release growth factors and chemokines. After that, the platelets help in barrier formation against micro-organisms, thus preventing them from affecting the wound and promoting the organization of the matrix responsible for the migration of cells for wound contraction [12]. In the proliferation phase, there is an accumulation of cells, connective tissues, and several other factors, such as cytokines, growth factors, and angiogenesis factors, across the wound. In the remodeling phase, there is resynthesis of the extracellular matrix to maintain the balance between the death of existing cells and the formation of new cells [13]. In the case of chronic or diabetic wounds, there is a persistent expression of cytokines responsible for inflammation, also called pro-inflammatory cytokines, which cause an interruption in normal wound healing and are accountable for impaired or non-healing of wounds[14, 15] and the entire mechanism of healing is shown in Figure 2. The healing process is delayed due to multiple factors directly or indirectly associated with the wound surface. The local factors to the wound area that hinder the healing include desiccation, necrosis, maceration, infection or abnormal presence of bacteria around the wound, pressure, trauma, and edema. On the other side, the systemic factors that are in little or no direct relationship with the affected surface but still interfere with healing are the patient's age, nutritional intake, body type, previous chronic ailment record, insufficient vascular system, immunosuppression, and radiation therapies, etc. [16]. Diabetes further adds several complications, including gangrene, amputations, ulceration, infection, and dehiscence of the wound after operation [17].

To overcome all these healing-related issues, many natural and artificial materials are utilized including, polymers, proteins, polyphenols, etc. Numerous nano-formulations based on such materials are synthesized for better wound healing. One such natural protein is silk fibroin obtained from various insects and is widely in practice for its tissue regeneration and wound healing applications [18]. Silk fibroin is a constituent of silk and is coated with silk sericin on the outside and is widely known to promote healing by enhancing cell growth and proliferation and regulating different kinds of cells involved in various phases of wound healing. We will discuss the role of cytokines, silk fibroin effect on cytokines that are crucial for wound healing, the signaling mechanism followed for healing using silk fibroin, and the different silk fibroin-based nano-formulations essential for the healing process in the wound area.

2. Role of Cytokines in Wound Healing

2.1. Pro-Inflammatory Cytokines

One of the first factors produced in response to damaged skin is pro-inflammatory cytokines which regulate the immune cell functioning for skin regeneration and affect fibroblasts and keratinocytes [19]. Low production of these factors causes wound infection, thus, altering their routine healing [20]. The major pro-inflammatory cytokines involved in skin regeneration are tumor necrosis factor (TNF), interleukin-1(IL-1), and interleukin-6 (IL-6) [21]. When there is an increase in the expression of pro-inflammatory cytokines, there is a delay in wound healing [22]. On the other hand, less production of such cytokines promotes skin wound infection and, thus, prevents normal wound healing [23]. The important cytokines are briefly described below.

2.1.1. Interleukin-6 (IL-6)

It was discovered in 1986 and is a pleiotropic cytokine that is a member of the IL-6 family of cytokines along with IL-11, IL-27, IL-35, and IL-39 [24]. The core protein size is nearly 20 kDa and consists of 212 amino acids, including 28 amino acid signal peptides [25]. It is a multifunctional cytokine that acts as a mediator of inflammation and lymphocyte differentiation and initiates its functioning after binding with its receptor, which has two subunits, gp130 and IL-6R, of which gp130 is expressed in many cells. At the same time, IL-6R is restricted to hepatocytes and immune system cells [26]. The receptor gets homodimerized when IL-6 binds to it and activates Janus-family kinases that comprise JAK1, JAK2, and TYK2, which further causes dimerization and translocation of STAT3 into the nucleus inducing transcription of IL-6 dependent genes after its phosphorylation [27]. The activation of STAT3 prevents apoptosis, causes differentiation and proliferation of cells, and further induces inflammation of wounds [28]. IL-6 also plays a role in regulating leukocyte infiltration, angiogenesis, and collagen accumulation [29]. The IL-6-STAT-3 pathway is involved in epithelial-mesenchymal transition, tumor migration, cancer stemness, and induction of vascular endothelial growth factor and hypoxia-inducible factor-1α (HIF-1α), which are the primary key factors for angiogenesis [30, 31, 32, 33]. Further, IL-6-STAT-3 targets the oncogenic miR-21 and miR-181b, thus, activating the NF-kβ pathway [34]. The IL-6 effect, in the case of wound healing, is significantly linked with the activation of fibroblasts, which further promotes the migration of cells in the wound and promotes the pro-regenerative factors expression in the injury [35]. Thus, IL-6 (Figure 3a) is functional in cutaneous wound healing, while other family members play a compensatory role in the regulation of wound contraction [36].

2.1.2. Interleukin-1 (IL-1)

It is one of the most ancient cytokines with 31-33 kDa size, whose signaling pathway is associated with activating a cascade involving Toll-like receptors [37]. In case of skin damage, it is intensely expressed along with other pro-inflammatory cytokines. It is present in two distinct forms, IL-1α and IL-1β, and sends signals via IL-R1, and IL-1RAcP receptor complex, which contains a TIR domain that activates the classicalNF-kβ cascade and MAP kinases [38]. Apart from this, two more regulatory pathways negatively affect IL-1 mediated signal transduction, of which IL-1Ra competes with IL-1 for binding to IL-1R1 and has no triggering effect on the signaling cascade after binding to the receptor. In contrast, the IL-1R2 receptor lacks the TIR domain and fails to activate the signaling pathway after binding with the ligand [39]. The absence of IL-1Ra decelerates skin regeneration, and its recombinant analog increases corneal wound healing in diabetic patients [40]. IL-1 negatively affects hair follicle growth in humans and mice and is essential for wound infection prevention, particularly in oral wounds [41]. The recombinant IL-1β is efficient in treating septic wounds in human beings. Further, IL-1 helps the skin stem cells to adapt to the inflammatory response, thus enhancing wound healing. Also, IL-1 (Figure 3b) mediated signal transduction is induced by the local inflammatory response to injury or TLR ligands, thus causing epigenetic rearrangement in skin stem cells, facilitating faster wound healing [42].

2.1.3. TumorNecrosis Factor (TNF)

It is a multifunctional cytokine of 17.3 kDa size. It can induce inflammation, protect tissues from infection, help to maintain the structure of secondary lymphoid organs, sustains immune regulation, and helps in the pathogenesis of numerous autoinflammatory and autoimmune disorders, including rheumatoid arthritis, multiple sclerosis, etc. due to being the critical inductor of inflammatory processes [43]. TNF is expressed in various cells such as stromal cells, endothelial cells, adipocytes, neurons, andimmune system cells and starts functioning from the postnatal period after childbirth due to the invasion of skin with microbiota [44]. It exists as a homotrimer on the cell surface in the transmembrane form, which converts into soluble form after proteolysis by TNF-α converting enzyme. In target cells, the intracellular signal transmission is induced by TNF after binding to its receptors, namely, TNFR1 and TNFR2. The first one is present in almost all cells, and the second is present in neurons, oligodendrocytes, and hematopoietic cells. These receptors further activate NFKβ and AP1 transcription factors that help in cell proliferation and survival, immune regulation, and expression of pro-inflammatory cytokines, including TNF itself [45, 46]. The induction of systemic inflammation is mediated by the TNF-TNFR1 signaling pathway, which plays a role in defense against pathogens present within the cells [47, 48]. There is an increase in the rate of wound healing when recombinant TNF administers the wound, but the rate is decreased if TNF is inhibited. Similarly, TNF (Figure 3c) derived from macrophages enhances skin appendage restoration after damage and activates hair follicle growth [49, 50].

2.1.4. Interferons (IFNs)

Interferon family consists of three distinct types: Type 1, which consists of IFNα, IFNβ, and others (Figure 3d). Type 2 consists of IFNγ, and Type 3 consists of IFNλ [51]. Plasmacytoid dendritic cells mainly produce Type 1 IFNs that act as antiviral cytokines, while Type 3 IFNs play a significant role in activating pro-inflammatory macrophages [52, 53]. Type 1 and Type 2 IFNs trigger the immune response at damaged sites, and their activity plays a role in treating wounds, just as in IL-1. They were also involved in TNF and NO production in cutaneous wounds by macrophages and triggered the immune response. Further, they induce higher production of TGF-β1 and VEGF, which are crucial for myofibroblast differentiation and collagen production along with blood vessel growth, respectively [54, 55]. IFNγ also decreases immune cell infiltration at the wound site and initiates more intense angiogenesis [56].

2.2. Anti-Inflammatory Cytokines

These are the series of immunoregulatory molecules that control pro-inflammatory cytokines' responses. These act in relation to specific inhibitors of cytokines and soluble cytokine receptors for regulating the human immune response.

2.2.1. Interleukin-10 (IL-10)

It is a homodimeric cytokine with 30 kDa size and 130 amino acids produced by different human cells, including T Cells, monocytes, macrophages, and keratinocytes, after an injury in human skin [57]. The tetramer receptor complex consists of two molecules, each of IL-10R1 and IL-10R2, and causes phosphorylation and dimerization of STAT3 after binding with dimerized IL-10. It enhances the functioning ability of macrophages and monocytes and decreases the production of pro-inflammatory cytokines [58, 59]. It plays an essential role in fetal development as a potent anti-inflammatory cytokine, thus preventing the maternal immune system from responding to the fetus and its antigens, which act as foreign particles to the mother [60]. IL-10 (Figure 3e) also prevents the excessive deposition of collagen linked with scar formation after being stimulated by pro-fibrotic agents [61]. It initiates scarless wound healing in postnatal tissue like it does in the fetus. Further, it re-establishes integrity of tissues in skin, tendon, and myocardium after injury; regulates extracellular matrix, modulates functioning of fibroblast and stem cells, and acts as an anti-scarring agent in case of excess fibrosis [62].

2.2.2. Interleukin-4 (IL-4)

It is a 20 kDa cytokine with numerous functions and is secreted by mast cells, T lymphocytes, and basophilic polymorphonuclear cells. It shows pleiotropic biological effects in hematopoietic and non-hematopoietic cells, mainly fibroblasts, by stimulating the extracellular matrix synthesis. It increases collagen production and the production of extracellular matrix macromolecules by human fibroblasts from different tissues. IL-4 (Figure 3f) initiates differentiation of myofibroblast, chemotaxis of fibroblast, and decreases biosynthesis of matrix metalloproteinase, which collectively indicates its involvement in connective tissue formation during both normal and pathologic wound healing. Its expression is highly enhanced in normal wound healing and decreases after wound contraction [63, 64].

2.2.3. Transforming Growth Factor (TGF-β)

It is a superfamily and involves the signaling pathway consisting of TGF-β (Figure 3g) type 1, 2, and 3, Activin A, B, and C, which are bone morphogenic genes, growth differential factors, and anti-mullerian hormones [65]. TGF-β1 is released at an early stage of healing and begins the recruitment of inflammatory cells at the wound site. When granulation tissue is formed, TGF-β1 enhances the expression of proteins, including fibronectin, collagen type I and III, and VEGF [65]. Further, it enhances the fibroblast's contraction to stimulate wound closure, enhances keratinocyte migration, and increases blood supply to the wound via improving the angiogenic properties of endothelial cells [66, 67]. Similarly, TGF-β2 is involved in recruiting fibroblasts and immune cells from wound edges to the wound site, which causes the formation of granulation tissue, angiogenesis, and collagen production [65]. It is also helpful in expressing and organizing collagen and other key ECM components during the healing process [68]. The TGF-β3 reduces the deposition of collagen type I by restricting the differentiation of myofibroblast and promoting collagen degradation by MMP – 9. All this collectively reduces scar formation [69].

All these cytokines come together to heal the wound, decrease the persistent pain and prevent any microbiota infection around the injured organ of the body. The mechanism of cytokine interaction for healing is discussed in the next part.

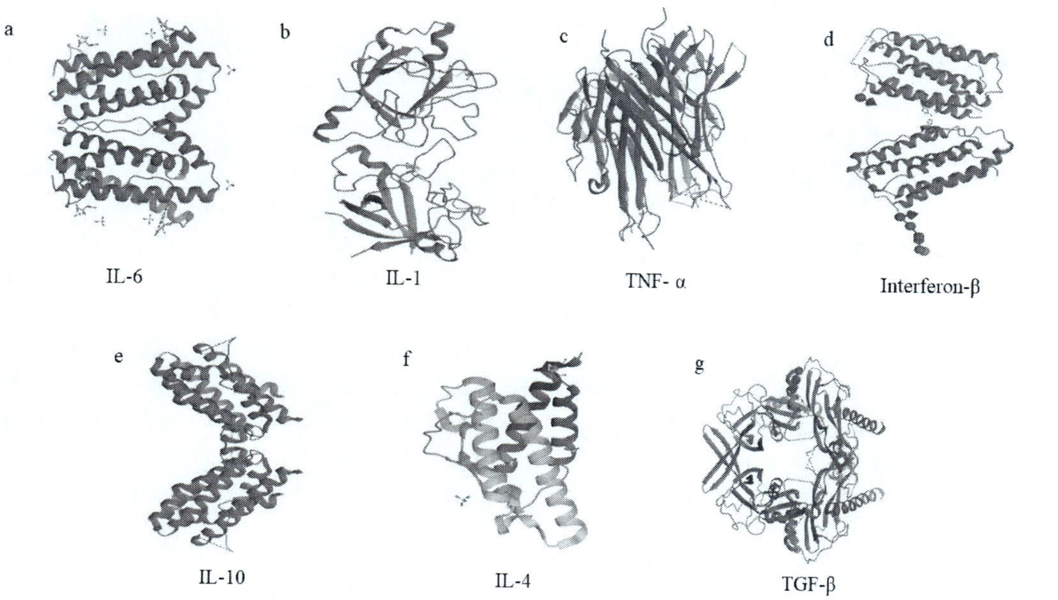

Figure 3. Different pro-inflammatory and anti-inflammatory cytokines which plays a crucial role in wound healing. The proinflammatory cytokines involved are a) IL-6 (PDB ID- 1ALU) b) IL-1 (PDB ID-1ILR) c) TNF- α (PDB ID- 2TNF) d) Interferon-β (PDB ID- 1AU1) and the anti-inflammatory cytokines involved are e) IL-10 (PDB ID- 2H24) f) IL-4 (PDB ID- 2B8U) g) TGF-β (PDB ID- 1TGK).

3. The Cascade of StepsFollowed by Pro-Inflammatory and Anti-Inflammatory Cytokines for Wound Healing

Both pro-inflammatory cytokines and anti-inflammatory cytokines play a crucial regulatory role in wound healing. There is the release of pro-inflammatory cytokines, including interleukin-6 (IL-6), IL-1, IL-12, IL-18, tumor necrosis factor (TNF), and interferon-gamma (IFN-γ), which causes tissue inflammation after the wound [42]. The neurons get signals for releasing IL-6 and IL-1βpro-inflammatory cytokines from the macrophages and monocytes [70]. The apoptosis and pyroptosis occur after activating oxidative stress andinnate immunity by IFN-γ, IL-1β, and TNF-α [71]. Different cytokines play different roles. IL-12 is responsible for stimulating the production of TNF-α and IFN-γ. It reduces the production of the anti-inflammatory cytokine IL-4, which is responsible for inhibiting IFN-γ by activating STAT-3 signaling. IFN-γ acts as an activator that activates the defense mechanism against the micro-organisms or pathogens in the infected area after activating the macrophages by stimulatingSTAT1 [72, 73].

Further, to defend against pathogens, IL-18 activates T cells and natural killer cells. In the case of prolonged expression of pro-inflammatory cytokines, there is a delay in wound healing due to damage caused to tissues [74]. TNF is activated by IL-1, which causes damage to cells. Also, overproduction of IL-1β causes inflammation to neuronal tissue, damages neuro-muscular junction, and delays healing [75]. IL-6 is secreted by macrophages in the presence of pathogenic organisms, which mediates the Toll-Like Receptor-9 responsible for killing foreign particles. In prokaryotic organisms, unmethylated DNA activates the TLR-9 pathway, killing pathogens at the wound site and causing mitochondrial DNA spillage. IL-12 causes inhibition of angiogenesis due to the overexpression of IFN-γ mediated interferon-gamma-induced protein 10 (CXCL-10 or IP-10) [76, 77].

The most crucial step for the initiation of wound healing is the beginning of inflammation. In the case of the non-diabetic wound, there is the release of anti-inflammatory cytokines, including IL-4, IL-10, IL-13, IL-11, transforming growth factor-beta (TGF-β), and interferon-alpha (IFN-α) [78]. Activation of the MAPK pathway is prolonged in non-diabetic wounds. It leads to activation of MAPK phosphate enzyme activation, which acts as a negative regulator of MAPK/p38 and JNK pathways causing further negative regulation of TNF-a production. There is an increase in anti-inflammatory cytokine IL-10 due to the dephosphorylation of the MAPK/p38 pathway. It is a homodimeric cytokine released by macrophages and monocytes, which inducesTGF- β - signaling responsible for cell division [79]. IL-4, IL-13, and IL-10 further stimulate the synthesis of fibrinogen, collagen, and extracellular matrix, which is crucial for healing [80]. IL-4, when getting secreted by mast cells, macrophages, and inflamed T-Cells, causes activation of wound repairing pathway Janus kinase/signal transducer and transcription-6 (Jak/STAT6).

Further, immune cells, adipocyte cells, and epithelial cells secrete L-1RA cytokine responsible for inhibiting pro-inflammatory IL-1β cytokine after binding with the interleukin-1 receptor. There is a delay in wound healing due to the prolonged inflammatory phase because of the deregulation of TNF-α and IL-1β [81]. The cells of bone marrow secrete IL-11 Cytokine, which has an anti-inflammatory effect and inhibits the secretion ofIL-1 and TNF-α by causing inhibition of NF-kβ P50/P65 and enhancing the expression of inhibitory NF-kβ P50/P50 synthesis in monocytes/macrophage cells [82]. There is an impairment of the transition of cytokines in diabetic wounds, but the transition is balanced in the case of non-diabetic injuries [77], as is shown in Figure 4.

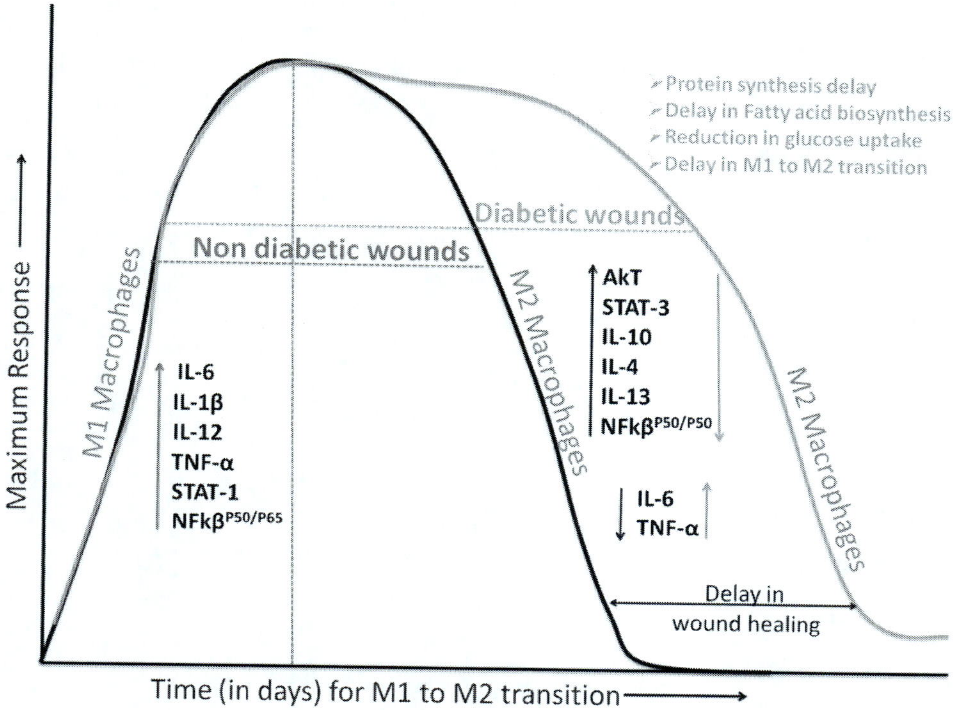

Figure 4. The bell-shaped graph shows the role of M1 and M2 macrophages in the wound healing process. The M1 macrophages lead to wound inflammation, whereas M2 macrophages heal the wound by decreasing the effect of inflammation. In diabetic patients, there is a prolonged expression of M1 macrophages compared to non-diabetic patients, thus delaying the transition from M1 to M2 macrophages, thus delaying the entire healing process.

Different molecules have been present in nature since time immemorial that affect the release and expression of different cytokines in the human body organs during injury and thus prevent blood loss and skin infection by enhancing the wound closure rate by distinct signaling mechanisms. One such compound is silk fibroin, which significantly impacts human beings due to its utilisation in various applications ranging from textile to tissue engineering, particularly wound healing, by controlling different signaling mechanisms involved in the wound healing process. We will study it in the next part.

4. Silk Fibroin: A Natural Polymer for Wound Healing

Silk fibroin is a naturally obtained, advanced, protein-based biomacromolecule derived from multiple sources such as non-mulberry and mulberry silkworm *Bombyxmori*, spiders, and insects [83]. Silk fibroin consists of 90% of amino acids. It has 5507 amino acid repeats formed by six repetitive units, including glycine, serine, and alanine, in different percentages and arranged as (Gly-Ala-Gly-Ala-Gly-Ser)$_n$(Figure 5a), which gets attached in antiparallel order to form β- sheets (Figure 5b) [84, 85]. It is a major structural protein of silk and comprises the core, thus, providing mechanical strength, and is surrounded by a glue-like coating of sericin protein. Each silk fiber has fibroin and sericin filaments in 2:1 proportion [86]. Each fibroin consists of nanofibrils with a 3.5nm diameter, which intertwines, leading to bigger fibril units having much stronger interactions and diameters of 20-200 nm, and are thus called microfibrils. There were hydrophobic repetitive units and hydrophilic non-repetitive units, and the heavy chain has distinct N and C terminals. It is attached to the light chain through a disulfide bond via a C terminal [87], as shown in Figure 6.

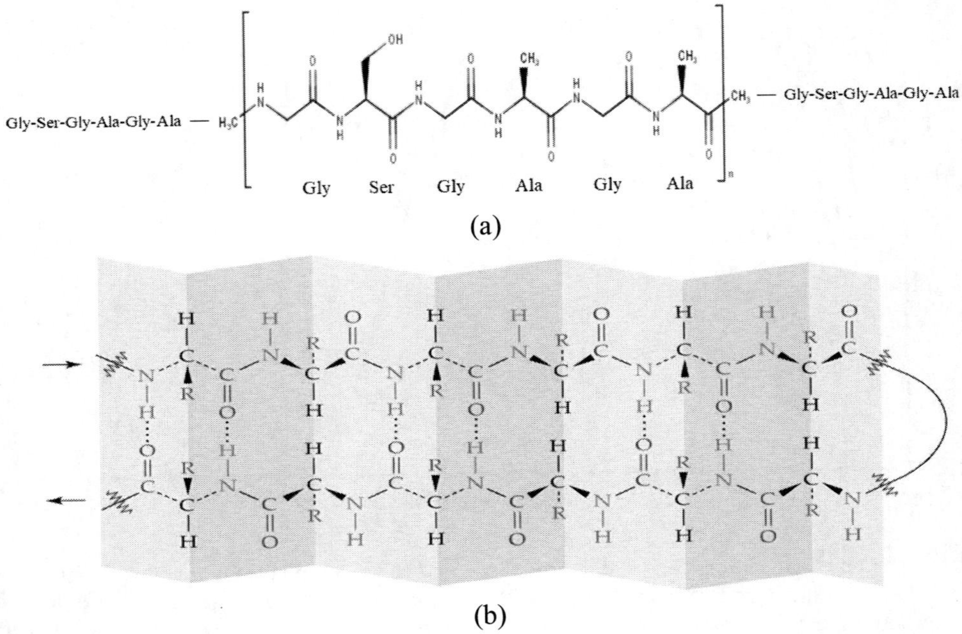

Figure 5. (a) The chain structure indicating the repetitive amino acid sequence in silk fibroin protein; (b) The beta-sheet structure of silk fibroin in which the amino acid chains are attached in an antiparallel manner via hydrogen bonds to form a βsheet-like structure.

Further, it has both amorphous and crystalline domains. The amorphous domains consist mainly of aspartic acid, a bulky side chain amino acid that provides elasticity and flexibility to the fiber, whereas the crystalline domain has glycine and alanine units distributed via motifs having tyrosine and serine, which provides strength and toughness to the fibroin [18].

There are mainly two types of silks – mulberry and non-mulberry silk. Mulberry silk has polypeptide chains having a molecular weight in the range of 200-350 kDa with repetitive hydrophobic parts interrupted by hydrophilic groups, and the primary structure has an H-L complex made up of heavy (-H) and light (-L) chain polypeptides linked together by a disulfide linkage at C terminus of the heavy chain. The other factor contributing to its overall integrity is the P25 glycoprotein, which is non-covalently connected to H-L chains. The H-chain comprises mainly glycine, serine, and alanine and is responsible for the fibrous properties of the protein. The hydrophobic domain of the H-chain forms antiparallel β- sheets. The L-chain consists of isoleucine, leucine, valine, and other acidic amino acids that are relatively hydrophilic and provide a nonfibrous effect to the fibroin [88, 89]. H-fibroin, L-fibroin, and P25 are present in the ratio of 6:6:1 in mulberry silk [90, 91].

On the other hand, in non-mulberry silk, P25 and L-chain are absent. Instead, they only have H-chain homodimers with a molecular weight of 330 kDa formed by individual proteins of 160 kDa. Further, they exhibit a higher ratio of alanine to glycine and poly-alanine blocks, which assists in β-sheets formation. Also, they show higher basic/acidic, polar/non-polar, hydrophilic/hydrophobic, and bulky/non-bulky ratios of different amino acids in their structure [92, 93, 94].

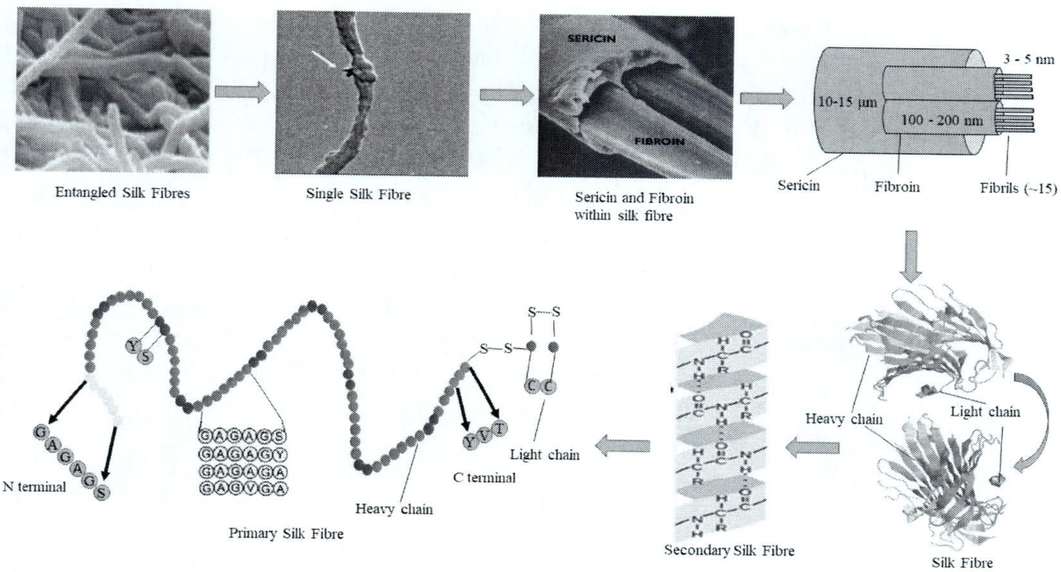

Figure 6. The general structure of silk protein in which silk fibroin is coated with silk sericin from the outer side and contains microfibrils. After degumming the silk protein, silk fibroin can be obtained, consisting of repetitive units of six amino acids that constitute hydrophobic domains separated from each other by hydrophilic domains. The disulfide linkage connects heavy and light chains through the C-terminal of the heavy chain.

It possesses outstanding mechanical properties, little immunogenicity, high adaptability, high biocompatibility, gradual disintegration, and the ability to modify itself with respect to surroundings which add to better integration [95, 96, 97]. All these structural features make silk fibroins widely popular in many applications in the textile industry and biomedical applications. It can be further utilized to make gels, solutions, sponges, films, and scaffolds for skin healing. It can adhere to cells and cell lines and let them proliferate and differentiate on its surface [18, 98]. Further, silk fibroin has been explored for skin healing because of its permeability to water and oxygen, low potential for inflammations, and hemostatic properties [99]. It is also widely used for bone regeneration, drug delivery, cartilage regeneration, cornea replacement, etc. It enhances wound healing and tissue regeneration by increasing cell proliferation and differentiation [100]. The regulation of pro and anti-inflammatory cytokines by silk fibroin and its formulations is explained below.

5. Regulation of Pro and Anti-Inflammatory Cytokines by Silk Fibroin

Silk fibroin effectively increases the proliferation of human fibroblasts and keratinocytes, increases collagen deposition and production just like in normal skin, and enhances the formation of a new epithelial layer over the wound site, promoting the wound healing process [101, 102]. Silk fibroin matrix is found to initially decrease the expression level of pro-inflammatory cytokines IL-6 and IL-1α till day seven and rapidly increase till day 21. These cytokines bind to their receptors, IL-6R and IL-1R1, respectively, which further enhance the signaling of NF-kβ, JAK, and MAPK pathways, leading to enhanced proliferation and inflammation and, thus, better wound healing. Similarly, the expression of TGF-β1, which helps in hypertrophic scar formation, is rapidly enhanced and peaked on day 21 of the injury. Thus, it accelerates scarless wound healing by regulatingTGF-β1 expression in the case of burn wounds [102]. Also, silk fibroin-based nano-formulations are found to decrease the expression of pro-inflammatory cytokines, including IL-1β, IL-6, and TNFα, which can prevent the synthesis of growth hormones that influence protein synthesis crucial for healing wounds [103, 104, 105, 106]. IL-6 modulates the expression of TGF-β and TGF-βR2 receptors in the skin, which further triggers the migration of keratinocytes to the wound region, thus enhancing the provisional matrix deposition and promoting healing [107]. Therefore, silk fibroin suppresses the release of pro-inflammatory cytokines during the initial stage of the inflammatory phase, thus resulting in a protective effect on cells and tissues during wound healing [108]. Further, the Silk fibroin matrix enhances the IL-10 cytokine level in the wound area from 7 days of injury to 21 days which is an anti-inflammatory cytokine and acts on the AMPK and Akt pathway and enhances cell growth and angiogenesis [102], as explained in Figure 7.

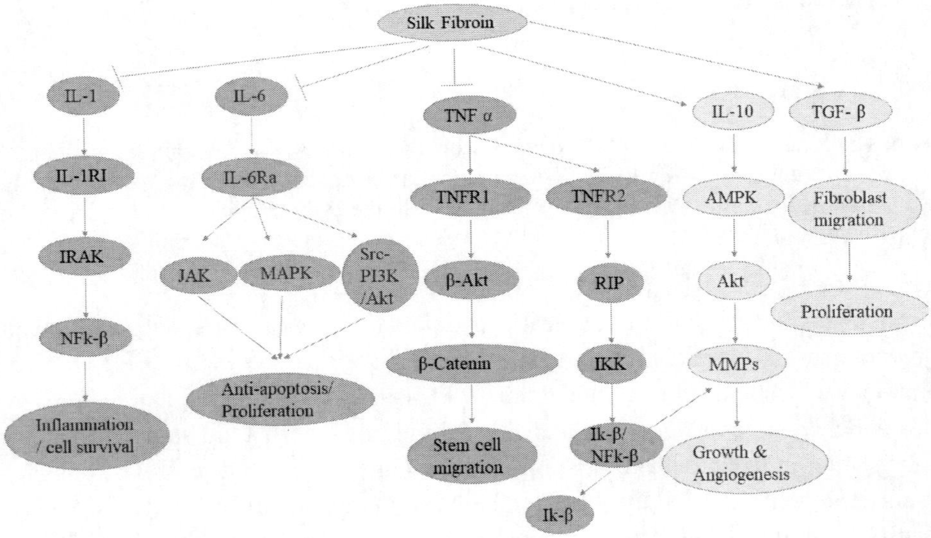

Figure 7. The role of silk fibroin in regulating different pro-inflammatory and anti-inflammatory cytokines for enhancing the wound healing process.

6. Regulation of Wound Healing by Silk Fibroin-Mediated NF-kβ Signaling

Silk fibroin is very effective in wound healing both *in vivo* and *in vitro*, but the actual mechanism behind the whole process depends on NF-kβ signaling. As we have discussed, silk fibroin enhances cell proliferation and cell growth. It has been tested *in vitro* on NIH3T3 cells in which fibroblast cell adhesion and proliferation was enhanced by activating a particular pathway. The migration of primary HDF cells is also increased [109, 110]. To study the molecular mechanism, the microarray gene profiling analysis was done after treating cells with silk fibroin solution, and it was found that there is an upregulation of NF-kβ genes in silk fibroin treated cells compared with control. Further, western blotting confirmed the involvement of NF-kβ signaling in wound healing [100].

The expression of genes involved in cellular activities such as cell adhesion, differentiation, migration, and survival is influenced by NF-kβ, which is a transcription factor. Two different pathways can be followed to activate this transcription factor, including the canonical pathway called the classical pathway or the non-canonical pathway, also called the alternate pathway [111, 112]. The classical pathway mainly participates in wound healing by enhancing fibroblast migration and activating different receptors, including toll-like receptors, interleukin-1 receptors, tumor necrosis factor receptors, and antigen receptors for the functioning of NF-kβ signaling [113, 114]. Silk fibroin mainly acts on the canonical wound healing pathway by activating various proteins and enhancing the expression of different subunits of NF-kβ signaling that include p65, IKK α, and IKKβ. Further, it decreases the expression of IKβα but upregulated the toll-like receptors and tumor necrosis factor receptors [100]. Also, the expression of the large number of proteins involved in different wound healing stages is controlled by NF-kβ signaling. For instance, there are growth factors and pro-inflammatory cytokines in the inflammatory phase, such as Epidermal Growth Factor (EGF), Transforming Growth Factor (TGF), IL-1β, and IL-10, while for the proliferative phase, Cyclin D1, fibronectin, vimentin, and VEGF are essential. Fibronectin also acts in the remodeling phase. Silk fibroin increases the protein expression level of factors involved in proliferation, such as Cyclin D1, fibronectin, vimentin, and VEGF. Still, there is not much effect of silk fibroin on EGF and TGF [115, 100]. Thus, it was concluded that NF-kβ signaling plays a crucial role in wound healing *in vivo* and *in vitro* after getting activated by silk fibroin and works mainly in the proliferative and regeneration phase of wound recovery, as shown in Figure 8.

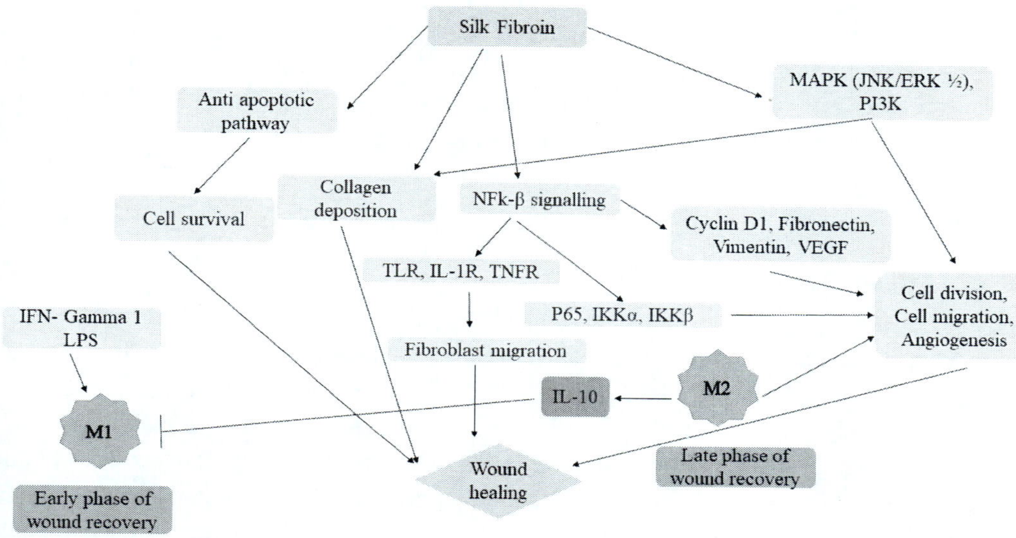

Figure 8. The interlinked signaling pathways followed by silk fibroin for treating wounds. The pathways involved include NF-kβ signaling, anti-apoptotic pathway, and MAPK pathway, which regulate the healing process by enhancing the transition from M1 to M2 macrophages through different mechanisms.

7. Other Pathways Involved in Wound Healing by Silk Fibroin

AKT/mTOR signaling and MAPK pathway significantly affect wound healing [116, 117]. Silk fibroin stimulates the migration of cells by activating extracellular signal-related kinases ½ (ERK 1/2) and c-Jun N-terminal kinases ½ (JNK1/2) [118]. The MEK, JNK, and PI3K inhibitors prevent the migration of cells and c-Jun expression. It also induces phosphorylation of c-Jun and JNK ½, which further leads to the activation of various downstream target genes, including PAI-1, a type-1 plasminogen activator inhibitor whose elevated levels protect proteolytic degradation of ECM proteins which accelerate wound healing. It further increases cell migration, wound re-epithelialization, and successful cellular attachment [119, 120]. JNK pathway also promotes fibroblast migration in the wound areas [121]. LY294002 inhibits silk fibroin-induced migration and c-Jun protein expression at the leading edge of cell migration. It prevents the phosphorylation of the c-Jun protein, thus indicating the role of PI3K in fibroin-induced c-Jun expression during cellular migration in the healing process [122]. Also, silk fibroin enhances collagen deposition in a much more ordered and regular manner in the wound area, which leads to healing without scarring, as wound closure depends on the quantity, organization, manner, and timing of collagen deposition [123].

8. Silk Fibroin and Its Different Formulations Used for Wound Healing

Wound healing involves the migration and proliferation of cells, with fibroblasts being the key components working from the initial phase until complete epithelialization. Initial studies have shown that silk fibroin plays a crucial role in cell migration and growth [54] and supports the secretion of fibroblasts and keratinocytes [9]. It increases the proliferation of skin fibroblasts, growth and proliferation of NIH3T3 cells, and cell migration of primary HDF cells [124, 125]. Collagen-1 coating of silk fibroin improves the attachment and dispersion of keratinocytes while layering silk fibroin with fibronectin enhances the attachment and dispersion of fibroblasts and keratinocytes within the matrix [126]. Various formulations were made using silk fibroin alone and in combination with different molecules, and it was found that blended ones may have more benefits than pure silk fibroin [9]. Some of the major forms and their application are discussed here briefly.

8.1. Silk Fibroin Hydrogels

Hydrogels have been extensively used for various biomedical applications, including wound healing [127], and are mainly obtained from natural polymers such as chitosan, collagen, and alginate, hyaluronic acid, and silk fibroin [18]. They are three-dimensional networks of polymers with a high swelling ratio and enhance the mechanical strength of hydrogels obtained from different polymers [128]. Fibroin fibers have more excellent toughness and extensibility than polymers such as elastin [129]. The presence of β-sheet crystal structure is responsible for the stiffness and strength of silk fibroin, making it even more potent than steel. In contrast, the inter-sheet Van der Waals interactions and hydrogen bonding provide enormous stability to the fibroin [130]. A pure silk fibroin gel was developed with high strength and biocompatibility, which allows them to sustain the viability of cells [131]. Another hydrogel containing silk and L-proline was tested *in vitro* and found to be effective in wound healing [132].

8.2. Silk Fibroin Nanoscaffolds

The electrospinning technique makes two-dimensional and pseudo-three-dimensional scaffolds that further provide a surface for cell adhesion, growth, skin regeneration, and wound closure [133, 134]. The nanofibrous scaffolds enhance the differentiation and proliferation of human mesenchymal cells *in vitro* [135]. The concentration of silk fibroin in spinning fluid directly affects the biological and mechanical properties of electrospun silk fibroin mats. The size also affects wound healing efficiency. The smaller the size, the greater the effect on extracellular matrix formation and skin proliferation [136]. The scaffolds having a two-dimensional structure make cells adapt to artificial morphology, which affects the genes involved in adhesion and cell migration and prevents scarring or contracture during cutaneous wound healing [137, 138].

8.3. Silk Fibroin Sponges/Blended Films

Sponges and films were made using silk fibroin for tissue engineering as they provide a large surface area inside the pores for cell adhesion and growth [139, 140]. Silk fibroin-based sponge was made using a freeze dryer to control the release of neurotensin-loaded gelatin microspheres for diabetic foot ulcers, demonstrating a sound delivery system by enhancing fibroblast accumulation, tissue granulation, and wound closure in the diabetic foot [141]. Compared to typical porcine-based wound dressings, silk films were made to treat full-thickness skin wounds, which show faster healing and less inflammatory response [142].

8.4. Silk Fibroin Solution

The silk fibroin solution was tested for both *in vivo* and *in vitro* models for wound healing activity. The solution was used as an eye drop for healing the corneal epithelium [143, 100].

8.5. Silk Fibroin Nanoparticles

Silk solutions are utilized in making microparticles and nanoparticles by following different techniques, including freeze-drying, spray drying, jet breaking, freeze-thawing, and self-assembly. These particles are of great interest as they can play a dual role by acting as growth factor carriers and increasing scaffolds' mechanical properties for treating wounds [94].

Further, silk fibroin-based biomaterials are employed for different tissue regenerations in the human body, which include vascular tissue, neural tissue, skin tissue, bone tissue, cartilage tissue, ligament and tendon tissue, hepatic tissue, cardiac tissue, spinal cord tissue, ocular tissue, tracheal tissue, bladder tissue, inter vertebrae tissue and eardrum tissue [94] (Figure 9).

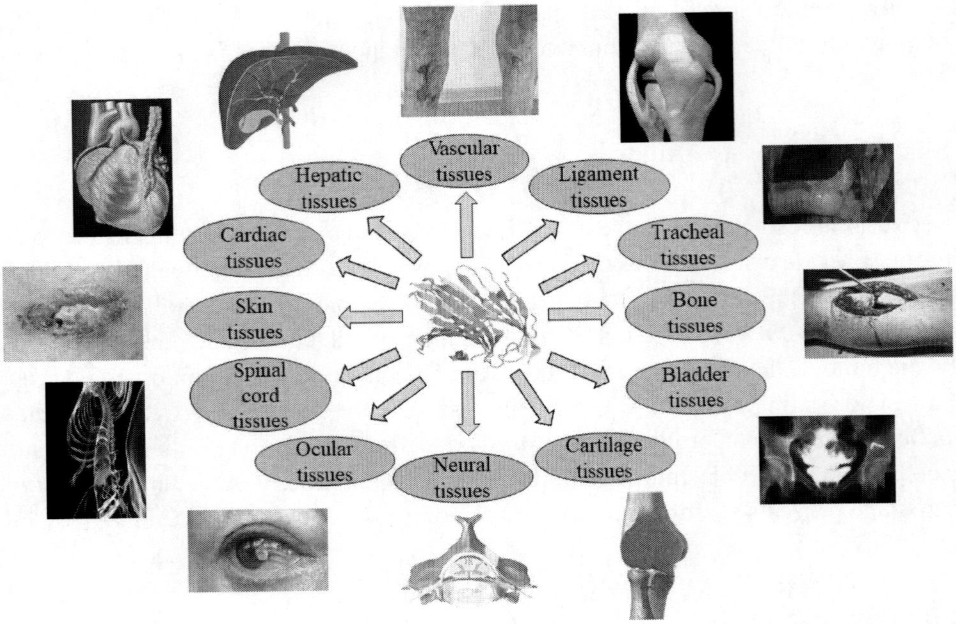

Figure 9. The application of silk fibroin in the regeneration of different body tissues.

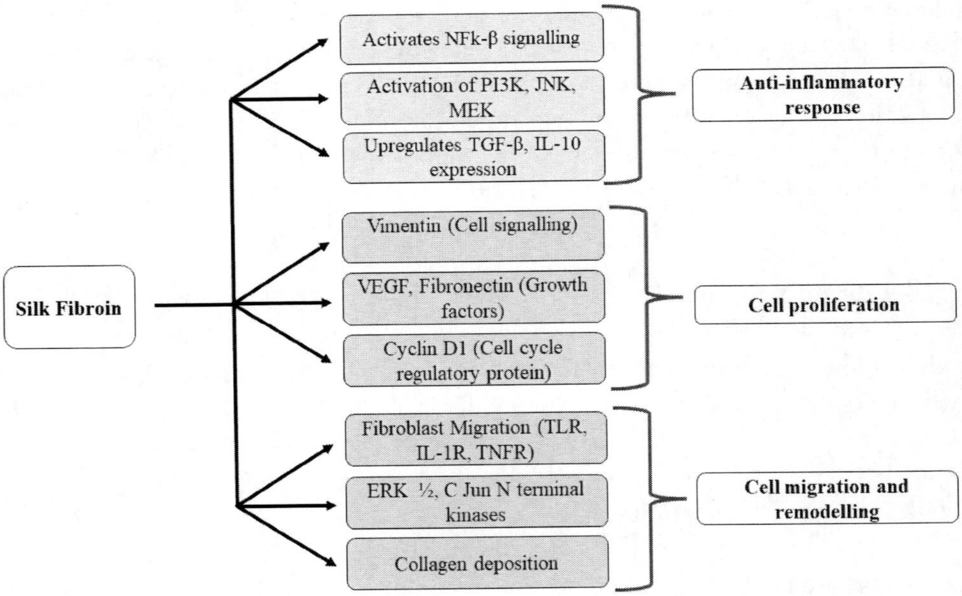

Figure 10. The role of silk fibroin in different wound healing phases, including inflammatory, proliferative, and remodeling phases, by regulation of different cytokines, growth factors, cell cycle regulatory proteins, and signaling pathways involved.

Conclusion

Wound healing is a complex process involving different stages. Several natural compounds are available worldwide, having wound healing applications that are yet to be explored in detail.

Silk fibroin is one of them and is among the most ancient polymers obtained naturally. It has attracted scientists worldwide for its role in numerous biomedical applications, including wound healing and tissue regeneration. Its wide range of applications in healing is due to its biocompatible nature, slow biodegradation, easy and cheap availability, and less inflammatory response to host tissue. Moreover, silk fibroin is found to be a promising biomaterial available in various forms, including scaffolds, sponges, blended films, nanoparticles, microparticles, and solutions, as a result of which silk fibroin can be employed alone or in the form of blends to treat both *in vitro* and *in vivo* wounds. Silk fibroin-activated NF-kβ signaling effectively treats injuries in proliferative and remodeling phases by enhancing cell proliferation, differentiation, adhesion, migration, and anti-inflammatory response by regulating different cytokines, growth factors, and cell signaling through distinct mechanisms, as shown in Figure 10.

Further research is needed to find out the elaborated mechanism of wound treatment using silk fibroin to fabricate a new array of silk fibroin-based biomaterials for treating different kinds of wounds. Also, silk fibroin is very brittle, easily fragmented, and thus, challenging to make sheets of uniform thickness for making scaffolds etc. Therefore, more detailed studies are required to enhance the therapeutic potential of silk fibroin and silk fibroin derivatives for wound care. Silk fibroins loaded with or without antimicrobial agents need to be explored for future clinical trials. Furthermore, more emphasis must be placed on studying the interaction of organic and inorganic compounds with silk fibroin as different functional groups in silk fibroin can make covalent bonds with inorganic phases to make scaffolds of interest with specific designs. Therefore, intense research is required to explore the interaction between different groups and the mechanism of action of silk fibroin on wound healing. Here in this chapter, the authors have discussed the role of cytokines and signaling pathways in wound recovery by taking the instance of silk fibroin as a model natural polymer. The silk fibroin effect on pro-inflammatory and anti-inflammatory cytokines, which can promote wound recovery by activating the transition from M1 to M2 macrophages, is discussed. This indicates that silk fibroin can be of great virtue for designing advanced wound healing agents both for diabetic and non-diabetic conditions.

References

[1] Diegelmann, R.F., M.C. Evans, Wound healing: an overview of acute, fibrotic and delayed healing, *Front. Biosci.* 9 (2004) 283–289. https://doi.org/10.2741/1184.

[2] Friedstat, J., D.A. Brown, B. Levi, Chemical, Electrical, and Radiation Injuries, *Clin. Plast. Surg.* 44 (2017) 657–669. https://doi.org/10.1016/J.CPS.2017.02.021.

[3] Frykberg, R.G., J. Banks, Challenges in the Treatment of Chronic Wounds, *Adv. Wound Care.* 4 (2015) 560. https://doi.org/10.1089/WOUND.2015.0635.

[4] Cardona, A.F., S.E. Wilson, Skin and Soft-Tissue Infections: A Critical Review and the Role of Telavancin in Their Treatment, *Clin. Infect. Dis.* 61 (2015) S69–S78. https://doi.org/10.1093/CID/CIV528.

[5] Guimarães, I., S. Baptista-Silva, M. Pintado, A.L. Oliveira, Polyphenols: A Promising Avenue in Therapeutic Solutions for Wound Care, *Appl. Sci* .11 (2021) 1230. https://doi.org/10.3390/APP11031230.

[6] Kant, V., B.L. Jangir, V. Kumar, A. Nigam, V. Sharma, Quercetin accelerated cutaneous wound healing in rats by modulation of different cytokines and growth factors, *Growth factors,* 38 (2020) 105-119. https://doi.org/10.1080/08977194.2020.1822830. (2020).

[7] Armstrong, D.G., A.J.M. Boulton, S.A. Bus, Diabetic Foot Ulcers and Their Recurrence, *N. Engl. J. Med.* 376 (2017) 2367–2375. https://doi.org/10.1056/NEJMRA1615439.

[8] Kalan, L.R., M.B. Brennan, The role of the microbiome in nonhealing diabetic wounds, *Ann. N. Y. Acad. Sci.* 1435 (2019) 79–92. https://doi.org/10.1111/NYAS.13926.

[9] Zhang,X., M.R. Reagan, D.L. Kaplan, Electrospun silk biomaterial scaffolds for regenerative medicine, *Adv. Drug Deliv. Rev.* 61 (2009) 988–1006. https://doi.org/10.1016/J.ADDR.2009.07.005.

[10] Schreml, S., R.M. Szeimies, L. Prantl, M. Landthaler, P. Babilas, Wound healing in the 21st century, *J. Am. Acad. Dermatol.* 63 (2010) 866–881. https://doi.org/10.1016/J.JAAD.2009.10.048.

[11] Martin, P., R. Nunan, Cellular and molecular mechanisms of repair in acute and chronic wound healing, *Br. J. Dermatol.* 173 (2015) 370–378. https://doi.org/10.1111/BJD.13954.

[12] Jeffrey, J., H. Bridget, Wound healing: part I. *Basic science,* 133 (2014) 199e-207e. https://doi.org/10.1097/01.prs.0000437224.02985.f9.

[13] Wang, P.H., B.S. Huang, H.C. Horng, C.C. Yeh, Y.J. Chen, Wound healing, *J. Chinese Med. Assoc.* 81 (2018) 94–101. https://doi.org/10.1016/J.JCMA.2017.11.002.

[14] Dumville, J.C., B.A. Lipsky, C. Hoey, M. Cruciani, M. Fiscon, J. Xia, Topical antimicrobial agents for treating foot ulcers in people with diabetes, *Cochrane Database Syst. Rev.* 6 (2017). https://doi.org/10.1002/14651858.CD011038.PUB2.

[15] Uccioli, L., V. Izzo, M. Meloni, V. Ruotolo, L. Giurato, Nonhealing foot ulcers in diabetic patients: general and local interfering conditions and management options with advanced wound dressings, *J. Wound Care.* 24 (2015) 35–42. https://doi.org/10.12968/JOWC.2015.24.SUP4B.35.

[16] Thomas Hess, C., Checklist for factors affecting wound healing., *Adv. Skin Wound Care.* 24 (2011) 192. https://doi.org/10.1097/01.ASW.0000396300.04173.EC.

[17] Greenhalgh, D.G., Wound healing and diabetes mellitus, *Clin. Plast. Surg.* 30 (2003) 37–45. https://doi.org/10.1016/S0094-1298(02)00066-4.

[18] Vidya, M., S. Rajagopal, Silk Fibroin: A Promising Tool for Wound Healing and Skin Regeneration, *Int. J. Polym. Sci.* 2021 (2021). https://doi.org/10.1155/2021/9069924.

[19] Larouche, J., S. Sheoran, K. Maruyama, M.M. Martino, Immune regulation of skin wound healing: mechanisms and novel therapeutic targets, *Adv. Wound Care.* 7 (2018) 209–231. https://doi.org/10.1089/wound.2017.0761.

[20] Wu, G., B. Zhu, X. Hong, P. Luo, Z.Xia, Role of cytokines in host defense against Staphylococcus aureus skin infection, *Histol. Histopathol.* 32 (2017) 761-766. https://doi.org/10.14670/HH-11-867.

[21] Turabelidze, A., L.A. Dipietro, Inflammation and wound healing, *Endod. Top.* 24 (2011) 26–38. https://doi.org/10.1111/ETP.12012.

[22] Zhao, R., H. Liang, E. Clarke, C. Jackson, M. Xue, Inflammation in chronic wounds, *Mdpi.Com.* 17 (2016) 2085. https://doi.org/10.3390/ijms17122085.

[23] Wu, G., B. Zhu, X. Hong, P. Luo, Z.X.-H. Histopath, Role of cytokines in host defense against Staphylococcus aureus skin infection, *Histol. Histopath.* 32 (2017) 761-766. https://doi.org/10.14670/HH-11-867.

[24] Hirano, T., IL-6 in inflammation, autoimmunity and cancer, *Int. Immunol.* 33 (2021) 127–148. https://doi.org/10.1093/INTIMM/DXAA078.

[25] Tanaka, T., M. Narazaki, T. Kishimoto, IL-6 in Inflammation, Immunity, and Disease.*Cold Spring Harb. Perspect. Biol.* 6 (2014) 16295–16296. https://doi.org/10.1101/CSHPERSPECT.A016295.

[26] Taga, T., M. Hibi, Y. Hirata, K. Yamasaki, K.Yasukawa, T. Matsuda, T. Hirano, T. Kishimoto, Interleukin-6 triggers the association of its receptor with a possible signal transducer, gp130, *Cell* 58 (1989) 573-581. https://doi.org/10.1016/0092-8674(89)90438-8.

[27] Wegenka, U.M., J. Buschmann, C. Lütticken, P.C. Heinrich, F. Horn, Acute-phase response factor, a nuclear factor binding to acute-phase response elements, is rapidly activated by interleukin-6 at the posttranslational level, *Mol. Cell. Biol.* 13 (1993) 276–288. https://doi.org/10.1128/MCB.13.1.276-288.1993.

[28] Rose-John, S., IL-6 trans-signaling via the soluble IL-6 receptor: importance for the pro-inflammatory activities of IL-6, *Int. J. Biol. Sci.*8 (2012)1237-1347. https://doi.org/10.7150/ijbs.4989.

[29] Lin, Z.-Q., T. Kondo, Y. Ishida, T. Takayasu, N. Mukaida, Essential involvement of IL-6 in the skin wound-healing process as evidenced by delayed wound healing in IL-6-deficient mice, *J. Leukoc. Biol.* 73 (2003) 713–721. https://doi.org/10.1189/JLB.0802397.

[30] *Wound Healing Center at Inova Loudoun treats complex wound and ostomy cases* | Inova Newsroom, (n.d.). https://www.inovanewsroom.org/ilh/2017/05/wound-healing-center-at-inova-loudoun-treats-complex-wound-and-ostomy-cases/ (accessed June 25, 2021).

[31] Barbieri, I., S. Pensa, T. Pannellini, E.Quaglino, D. Maritano, M. Demaria, A.Voster, J. Turkson, F. Cavallo, C.J. Watson, P. Provero, P. Musiani, V. Poli, Constitutively active Stat3 enhances neu-mediated migration and metastasis in mammary tumors via upregulation of Cten, *Cancer Res.* 70 (2010) 2558-2567. https://doi.org/10.1158/0008-5472.CAN-09-2840.

[32] Islam, M., S. Sharma, T.N. Teknos, RhoC regulates cancer stem cells in head and neck squamous cell carcinoma by overexpressing IL-6 and phosphorylation of STAT3, *PLoS One.* 9 (2014).e88527. https://doi.org/10.1371/JOURNAL.PONE.0088527.

[33] Yeh, H., W. Lai, H. Chen, H. Liu, W.Su, Autocrine IL-6-induced Stat3 activation contributes to the pathogenesis of lung adenocarcinoma and malignant pleural effusion, *Oncogene.* 25 (2006) 4300-4309. https://doi.org/10.1038/sj.onc.1209464.

[34] Löffler, D., K. Brocke-Heidrich, G.Pfeifer, C. Stocsits, J. Hackermuller, A.K. Kretzschamr, R. Burger, M. Gramatzki, C. Blumert, K. Bauer, H. Cvijic, A.K. Ullmann, P.F. Stadler, F. Horn, Interleukin-6–dependent survival of multiple myeloma cells involves the Stat3-mediated induction of microRNA-21 through a highly conserved enhancer, *Blood,* 110 (2007) 1330-1333. https://doi.org/10.1182/blood-2007-03-081133.

[35] Luckett-Chastain, L. R., R. M. Gllucci, Interleukin (IL)-6 modulates transforming growth factor-β expression in skin and dermal fibroblasts from IL-6-deficient mice, *Br. J. Dermatol.* 161 (2009) 237–248. https://doi.org/10.1111/j.1365-2133.2009.09215.x.

[36] Nosenko, M.A., S.G. Ambaryan, M.S. Drutskaya, Proinflammatory Cytokines and Skin Wound Healing in Mice, *Mol. Biol.* 53 (2019) 653–664. https://doi.org/10.1134/S0026893319050121/TABLES/4.

[37] Dinarello, C. A., Overview of the IL-1 family in innate inflammation and acquired immunity.*Immunol Rev.* 281 (2017) 8–27. https://doi.org/10.1111/imr.12621.

[38] Palomo, J., D. Dietrich, P. Martin, G. Palmer, C.Gabay, The interleukin (IL)-1 cytokine family–Balance between agonists and antagonists in inflammatory diseases, *Cytokine,* 76 (2015) 25-37. https://doi.org/10.1016/j.cyto.2015.06.017.

[39] Dripps, D. J., B. J. Brandhuber, R.C. Thompson, S.P. Eisenberg, Interleukin-1 (IL-1) receptor antagonist binds to the 80-kDa IL-1 receptor but does not initiate IL-1 signal transduction, *J Biol. Chem.* 266 (1991) 10331-10336. https://doi.org/10.1016/S0021-9258(18)99230-6.

[40] Ishida, Y., T. Kondo, A. Kimura, K. Matsushima, N. Mukaida, Absence of IL-1 receptor antagonist impaired wound healing along with aberrant NF-κB activation and a reciprocal suppression of TGF-β signal pathway, *Am Assoc Immnol.* 176 (2006) 5598-5606. https://doi.org/10.4049/jimmunol.176.9.5598.

[41] Yan, C., N. Gao, H. Sun, J. Yin, P. Lee, L. Zhou, X. Fan, F, Yu, Targeting imbalance between IL-1β and IL-1 receptor antagonist ameliorates delayed epithelium wound healing in diabetic mouse corneas, *Am. J. Pathol.* 186 (2016) 1466-1480. https://doi.org/10.4049/jimmunol.176.9.5598.

[42] Lin, E., F.B. Vincent, J. Sahhar, G.S. Ngian, R. Kandane-Rathnayake, R. Mende, E.F. Morand, T. Lang, J. Harris, Analysis of serum interleukin(IL)-1α, IL-1β and IL-18 in patients with systemic sclerosis, *Clin. Transl. Immunol.* 8 (2019) e1045. https://doi.org/10.1002/CTI2.1045.

[43] Feldmann, M., Development of anti-TNF therapy for rheumatoid arthritis, *Nat. Rev. Immunol.* 2 (2002) 364-371. https://doi.org/10.1038/nri802.

[44] Yuzhakova, D., M. Shirmanova, A. A. Bocharov, I. V. Astrakhantseva, E. A.Vasilenko, E. N. Gorshkova, M. S. Drutskaya, E. V. Zagaynova, S. A. Nedospasov, A. A. Kruglov, Microbiota induces expression of tumor necrosis factor in postnatal mouse skin, *Biochemistry (Mosc)* 81 (2016) 1303–1308. https://doi.org/10.1134/S0006297916110080.

[45] Kankaanranta, H., P. Ilmarinen, X. Zhang, I.M. Adcock, A. Lahti, P.J. Barnes, M.A. Giembycz, M.A. Lindsay, E. Moilanen, Tumour necrosis factor-α regulates human eosinophil apoptosis via ligation of

TNF-receptor 1 and balance between NF-κB and AP-1, *PLoS One*. 9 (2014). https://doi.org/10.1371/JOURNAL.PONE.0090298.

[46] Doberstein, S., F. Erard, B. Ryffel, D.E. Eivazi, J. Kung, D.-H.T. Nguyen, S.K.L. Janot, P.M. Steed, C. O'brien, J. Zalevsky, T. Secher, S.A. Ezhevsky, Dominant-negative inhibitors of soluble TNF attenuate experimental arthritis without suppressing innate immunity to infection, *Am Assoc Immnol.* 179 (2007) 1872–1883. https://doi.org/10.4049/jimmunol.179.3.1872.

[47] Kalliolias, G., L. B. Ivashkiv, TNF biology, pathogenic mechanisms and emerging therapeutic strategies, *Nat. Rev. Rheumatol.* 12 (2016) 49-62. https://doi.org/10.1038/nrrheum.2015.169.

[48] Allie, N., S. Grivennikov, R. Keeton, N.Hsu, M. Baurigolt, N. Court, C. Fremond, V. Yeremeev, Y. Shebzukhov, B. Ryfell, S. A. Nedospasov, V. F. J. Quesniaux, M. Jacobs, Prominent role for T cell-derived tumour necrosis factor for sustained control of Mycobacterium tuberculosis infection, *Sci Rep.* 3 (2013) 1809. https://doi.org/10.1038/srep01809.

[49] Wang, X., H. Chen, R. Tian, Y.Zhang, M. S. Drutskaya, C. Wang, J. Ge, Z. Fan, D. Kong, X. Wang, T. Cai, Y. Zhou, J. Wang, J. Wang, S. Wang, Z. Qin, H. Jia, Y. Wu, J. Liu, Y. Jiang, Y. Wu, Macrophages induce AKT/β-catenin-dependent Lgr5+ stem cell activation and hair follicle regeneration through TNF, *Nat Commun,* 27 (2017) 14091. https://doi.org/10.1038/ncomms14091.

[50] Chen, C., L. Wang, M. V. Plikus, T. X. Jiang, P. J. Murray, R. Ramos, C. F. Guerrero-Juarez, M. W. Hughes, O. K. Lee, S. Shi, R. B. Widelitz, A. D. Lander, C. M. Chuong,Organ-level quorum sensing directs regeneration in hair stem cell populations, *Cell.* 161 (2015) 277-290. https://doi.org/10.1016/j.cell.201502.016.

[51] Negishi, H., T. Taniguchi, H.Yanai, The interferon (IFN) class of cytokines and the IFN regulatory factor (IRF) transcription factor family, *Cold Spring Harb. Perspect. Biol.*10 (2018). https://doi.org/10.1101/cshperspect.a028423.

[52] McNab, F., K. Mayer-Barber, A. Sher, A. Wack, A. O'Garra, Type I interferons in infectious disease, *Nat. Rev. Immunol.* 15 (2015) 87-103. https://doi.org/10.1038/nri3787.

[53] Schoenborn, J., C. B. Wilson, Regulation of interferon-γ during innate and adaptive immune responses, *Adv. Immunol.* 96 (2007) 41-101. https://doi.org/10.1016/S0065-2776-(07)96002-2.

[54] Gregorio, J., S. Meller, C. Conrad,A.D.Nardo, B. Homey, A. Lauerma, N. Arai, R.L. Gallo, J. Digiovanni, M. Gilliet, Plasmacytoid dendritic cells sense skin injury and promote wound healing through type I interferons, *J. Exp. Med,* 207 (2010) 2921-2930. https://doi.org/10.1084/jem.20101102.

[55] Schäffer, M., M. Bongartz, W. Hoffmann, R. Viebahn, Regulation of nitric oxide synthesis in wounds by IFN-γ depends on TNF-α, *J. Investig. Surg.* 19 (2006) 371–379. https://doi.org/10.1080/08941930600985710.

[56] Iwakura, Y., N. Mukaida, Y. Ishida, T. Kondo, T. Takayasu, The essential involvement of cross-talk between IFN-γ and TGF-β in the skin wound-healing process, *Am Assoc Immnol.* 172 (2004) 1848–1855. https://doi.org/10.4049/jimmunol.172.3.1848.

[57] An, L., G.Q. Dong, Q. Gao, Y. Zhang, L.W. Hu, J.H. Li, Y. Liu, Effects of UVA on TNF-α, IL-1β, and IL-10 expression levels in human keratinocytes and intervention studies with an antioxidant and a JNK inhibitor, *Photodermatol. Photoimmunol. Photomed.* 26 (2010) 28–35. https://doi.org/10.1111/J.1600-0781.2009.00481.X.

[58] Fortunato, S., R. Menon, S. J. Lombardi, The effect of transforming growth factor and interleukin-10 on interleukin-8 release by human amniochorion may regulate histologic chorioamnionitis, *Am. J. Obstet. Gynecol.* 179 (1998) 794-799.

[59] Moore, K.W., R. De Waal Malefyt, R.L. Coffman, A. O'Garra, Interleukin-10 and the interleukin-10 receptor, *Annu. Rev. Immunol.* 19 (2001) 683–765. https://doi.org/10.1146/ANNUREV.IMMUNOL.19.1.683.

[60] Tinsley, J.H., S. South, V.L. Chiasson, B.M. Mitchell, Interleukin-10 reduces inflammation, endothelial dysfunction, and blood pressure in hypertensive pregnant rats, *Am. J. Physiol. - Regul. Integr. Comp. Physiol.* 298 (2010). https://doi.org/10.1152/AJPREGU.00712.2009.

[61] Shi, J.H., H. Guan, S. Shi, W.X. Cai, X.Z. Bai, X.L. Hu, X. Bin Fang, J.Q. Liu, K. Tao, X.X. Zhu, C.W. Tang, D.H. Hu, Protection against TGF-β1-induced fibrosis effects of IL-10 on dermal fibroblasts and its potential therapeutics for the reduction of skin scarring, *Arch. Dermatol. Res.* 305 (2013) 341–352. https://doi.org/10.1007/S00403-013-1314-0.

[62] King, A., S. Balaji, L.D. Le, T.M. Crombleholme, S.G. Keswani, Regenerative Wound Healing: The Role of Interleukin-10, *Adv. Wound Care.* 3 (2014) 315. https://doi.org/10.1089/WOUND.2013.0461.

[63] Salmon-Ehr, V., L. Ramont, G. Godeau, P. Birembaut, M. Guenounou, P. Bernard, F.X. Maquart, Implication of Interleukin-4 in Wound Healing, *Lab. Investig.* 2000 808. 80 (2000) 1337–1343. https://doi.org/10.1038/labinvest.3780141.

[64] Serezani, A.P.M., G. Bozdogan, S. Sehra, D. Walsh, P. Krishnamurthy, E.A. Sierra Potchanant, G. Nalepa, S. Goenka, M.J. Turner, D.F. Spandau, M.H. Kaplan, IL-4 impairs wound healing potential in the skin by repressing fibronectin expression, *J. Allergy Clin. Immunol.* 139 (2017) 142. https://doi.org/10.1016/J.JACI.2016.07.012.

[65] Barrientos, S., O. Stojadinovic, M.S. Golinko, H. Brem, M. Tomic-Canic, Growth factors and cytokines in wound healing, *Wound Repair Regen.* 16 (2008) 585–601. https://doi.org/10.1111/J.1524-475X.2008.00410.X.

[66] Evrard, S.M., C. Daudigier, L. Mauge, D. Israël-Biet, C.L. Guerin, I. Bieche, J.C. Kovacic, A.M. Fischer, P. Gaussem, D.M. Smadja, The profibrotic cytokine transforming growth factor-β1 increases endothelial progenitor cell angiogenic properties, *J. Thromb. Haemost.* 10 (2012) 670–679. https://doi.org/10.1111/J.1538-7836.2012.04644.X.

[67] Pakyari, M., A. Farrokhi, M.K. Maharlooei, A. Ghahary, Critical Role of Transforming Growth Factor Beta in Different Phases of Wound Healing, *Adv. Wound Care.* 2 (2013) 215–224. https://doi.org/10.1089/WOUND.2012.0406.

[68] Thompson, H.G.R., J.D. Mih, T.B. Krasieva, B.J. Tromberg, S.C. George, Epithelial-derived TGF-β2 modulates basal and wound-healing subepithelial matrix homeostasis, *Am. J. Physiol. - Lung Cell. Mol. Physiol.* 291 (2006). https://doi.org/10.1152/AJPLUNG.00057.2006.

[69] Hosokawa, R., K. Nonaka, M. Morifuji, L. Shum, M. Ohishi, TGF-beta 3 decreases type I collagen and scarring after labioplasty, *J. Dent. Res.* 82 (2003) 558–564. https://doi.org/10.1177/154405910308200714.

[70] Laudisi, F., R. Spreafico, M. Evrard, T.R. Hughes, B. Mandriani, M. Kandasamy, B.P. Morgan, B. Sivasankar, A. Mortellaro, Cutting edge: the NLRP3 inflammasome links complement-mediated inflammation and IL-1β release, *J. Immunol.* 191 (2013) 1006–1010. https://doi.org/10.4049/JIMMUNOL. 1300489.

[71] Lamkanfi, M., V.M. Dixit, Mechanisms and Functions of Inflammasomes, *Cell.* 157 (2014) 1013–1022. https://doi.org/10.1016/J.CELL.2014.04.007.

[72] Rigante, D., The Broad-Ranging Panorama of Systemic Autoinflammatory Disorders with Specific Focus on Acute Painful Symptoms and Hematologic Manifestations in Children., *Mediterr. J. Hematol. Infect. Dis.* 10 (2018) e2018067–e2018067. https://doi.org/10.4084/MJHID.2018.067.

[73] Qiao, Y., E. Giannopoulou, C. Chan, S. Park, S.Gong, J. Chen, X. Hu, O. Elemento, L. B. Ivashkiv, Synergistic activation of inflammatory cytokine genes by interferon-γ-induced chromatin remodeling and toll-like receptor signaling, *Immunity,* 39 (2013) 454-469. https://doi.org/10.1016/j.immuni.2013.08.009.

[74] Bernardo, M.E., W.E. Fibbe, Mesenchymal stromal cells: sensors and switchers of inflammation, *Cell Stem Cell.* 13 (2013) 392–402. https://doi.org/10.1016/J.STEM.2013.09.006.

[75] Zhang, J.Z., Z. Liu, J. Liu, J.X. Ren, T.S. Sun, Mitochondrial DNA induces inflammation and increases TLR9/NF-κB expression in lung tissue, *Int. J. Mol. Med.* 33 (2014) 817–824. https://doi.org/10.3892/IJMM.2014.1650.

[76] Lee, E.Y., Z.H. Lee, Y.W. Song, CXCL10 and autoimmune diseases, *Autoimmun. Rev.* 8 (2009) 379–383. https://doi.org/10.1016/J.AUTREV.2008.12.002.

[77] Kaur, P., D. Choudhury, Modulation of Inflammatory Dynamics by Insulin to Promote Wound Recovery of Diabetic Ulcers, *Wound Heal.* (2020). https://doi.org/10.5772/INTECHOPEN.92096.

[78] Ma, T.Y., G.K. Iwamoto, N.T. Hoa, V. Akotia, A. Pedram, M.A. Boivin, H.M. Said, TNF-α-induced increase in intestinal epithelial tight junction permeability requires NF-κB activation, *Am. J. Physiol. - Gastrointest. Liver Physiol.* 286 (2004). https://doi.org/10.1152/AJPGI.00173.2003.

[79] Vieira, I.R., R.K. Fernandes, D.R. Rodrigues, C.M. Gorgulho, R. Kaneno, Â.M.V.C. Soares, TLR9 stimulation induces increase in fungicidal activity of human dendritic cells challenged with

Paracoccidioides brasiliensis, *Med. Mycol.* 56 (2018) 911–915. https://doi.org/10.1093/MMY/MYX126.

[80] Duan, X.Z., H. Zhuang, M. Wang, H.W. Li, J.C. Liu, F.S. Wang, Decreased numbers and impaired function of circulating dendritic cell subsets in patients with chronic hepatitis B infection (R2), *J. Gastroenterol. Hepatol.* 20 (2005) 234–242. https://doi.org/10.1111/J.1440-1746.2004.03529.X.

[81] Dinarello, C.A., Immunological and inflammatory functions of the interleukin-1 family, *Annu. Rev. Immunol.* 27 (2009) 519–550. https://doi.org/10.1146/ANNUREV.IMMUNOL.021908.132612.

[82] Sultani, M., A.M. Stringer, J.M. Bowen, R.J. Gibson, Anti-inflammatory cytokines: important immunoregulatory factors contributing to chemotherapy-induced gastrointestinal mucositis, *Chemother. Res. Pract.* 2012 (2012) 1–11. https://doi.org/10.1155/2012/490804.

[83] Silva, S.S., E.M. Fernandes, S. Pina, J. Silva-Correia, S. Vieira, J.M. Oliveira, R.L. Reis, 2.11 Polymers of Biological Origin, *Compr. Biomater. II.* (2017) 228–252. https://doi.org/10.1016/B978-0-12-803581-8.10134-1.

[84] Zhang, Y., A. Atala, Regenerative Medicine of the Bladder, *Princ. Regen. Med.* (2019) 1263–1279. https://doi.org/10.1016/B978-0-12-809880-6.00072-2.

[85] Elzoghby, A.O., M.M. Elgohary, N.M. Kamel, Implications of Protein- and Peptide-Based Nanoparticles as Potential Vehicles for Anticancer Drugs, *Adv. Protein Chem. Struct. Biol.* 98 (2015) 169–221. https://doi.org/10.1016/BS.APCSB.2014.12.002.

[86] Koh, L. D., Y. Cheng, C. Teng, Y. Khin, X. Loh, S. Tee, M. Low, E. Ye, H. Yu, Y. Zhang, M. Han, Structures, mechanical properties and applications of silk fibroin materials, *J. Prog. Polym. Sci.* 46 (2015) 86-110. https://doi.org/10.1016/j.progpolymsci.2015.02.001.

[87] Ling, S., W. Chen, Y. Fan, K. Zheng, K. Jin, H. Yu, M. J. Buehler, D. L. Kaplan, Biopolymer nanofibrils: Structure, modeling, preparation, and applications, *J. Prog. Polym. Sci.* 85 (2018) 1-56. https://doi.org/10.1016/j.progpolymsci.2018.06.004.

[88] Qi, Y., H. Wang, K. Wei, Y. Yang, R. Zheng, I. S. Kim, K. Zhang, A review of structure construction of silk fibroin biomaterials from single structures to multi-level structures, *Int. J. Mol. Sci.* 18 (2017) 237. https://doi.org/10.3390/ijms18030237.

[89] Shimura, K., Chemical composition and biosynthesis of silk proteins, *Experientia.* 39 (1983) 455–461. https://doi.org/10.1007/BF01965160.

[90] Inoue, S., K. Tanaka, F. Arisaka, S. Kimura, K. Ohtomo, S. Mizuno, Silk fibroin of Bombyx mori is secreted, assembling a high molecular mass elementary unit consisting of H-chain, L-chain, and P25, with a 6: 6: 1 molar ratio, *J. Biol. Chem.* 275 (51) 40517-40528. https://doi.org/10.1074/jbc.M006897200.

[91] Kundu, B., S. C. Kundu, Osteogenesis of human stem cells in silk biomaterial for regenerative therapy, *J.Prog. Polym. Sci.* 35 (2010) 1116-1127. https://doi.org/10.1016/j.progpolymsci.2010.04.004.

[92] Freddi, G., Y. Gotoh, T. Mori, I. Tsutsui, M. Tsukada, Chemical structure and physical properties of antheraea assama silk, *J. Appl. Polym. Sci.* 52 (1994) 775–781. https://doi.org/10.1002/APP.1994.070520608.

[93] Sen, K., M. Babu K, Studies on Indian Silk. I. Macrocharacterization and Analysis of Amino Acid Composition, *J. Appl. Polym. Sci.* 92 (2004) 1080–1097. https://doi.org/10.1002/APP.13609.

[94] Kundu, B., R. Rajkhowa, S.C. Kundu, X. Wang, Silk fibroin biomaterials for tissue regenerations, *Adv. Drug Deliv. Rev.* 65 (2013) 457–470. https://doi.org/10.1016/J.ADDR.2012.09.043.

[95] Kundu, B., R. Rajkhowa, S. C. Kundu, X.Wang, Silk fibroin biomaterials for tissue regenerations, *Adv. Drug. Deliv.Rev.* 65 (2013) 457-470. https://doi.org/10.1016/j.addr.2012.09.043.

[96] Rodriguez, M., J. Brown, J. Giordano, S. J. Lin, F. G. Omenetto, D. L. Kaplan, Silk based bioinks for soft tissue reconstruction using 3-dimensional (3D) printing with in vitro and in vivo assessments, *Biomaterials,* 117 (2017) 105-115. https://doi.org/10.1016/j.biomaterials.2016.11.046.

[97] Thurber, A. E., F. G. Omenetto, D. L. Kaplan, In vivo bioresponses to silk proteins, *Biomaterials.* 71 (2015) 145-157. https://doi.org/10.1016/j.biomaterials.2015.08.039.

[98] Luan, X., Y. Wang, X. Duan, Q. Duan, M. Li, S. Lu, H. Zhang, X. Zhang, Attachment and growth of human bone marrow derived mesenchymal stem cells on regenerated Antheraea pernyi silk fibroin films, *Biomed. Mater.* 1 (2006) 181-187. https://doi.org/10.1088/1748-6041/1/4/001.

[99] Zhang, W., L. Chen, J. Chen, L. Wang, X. Gui, J. Ran, G. Xu, H. Zhao, M. Zeng, J. Ji, L. Qian, J. Zhou, H. Ouyang, X. Zou, Silk Fibroin Biomaterial Shows Safe and Effective Wound Healing in Animal Models and a Randomized Controlled Clinical Trial, *Adv. Healthc. Mater.* 6 (2017). https://doi.org/10.1002/ADHM.201700121.

[100] Park, Y.R., M.T. Sultan, H.J. Park, J.M. Lee, H.W. Ju, O.J. Lee, D.J. Lee, D.L. Kaplan, C.H. Park, NF-κB signaling is key in the wound healing processes of silk fibroin, *Acta Biomater.* 67 (2018) 183–195. https://doi.org/10.1016/J.ACTBIO.2017.12.006.

[101] Min, B., G. Lee, S. Kim, Y. Nam, T. Lee, W. H. Park, Electrospinning of silk fibroin nanofibers and its effect on the adhesion and spreading of normal human keratinocytes and fibroblasts in vitro, *Biomaterials.* 25 (2004) 1289-1297. https://doi.org/10.1016/j.biomaterials.2003.08.045.

[102] Ju, H.W., O.J. Lee, J.M. Lee, B.M. Moon, H.J. Park, Y.R. Park, M.C. Lee, S.H. Kim, J.R. Chao, C.S. Ki, C.H. Park, Wound healing effect of electrospun silk fibroin nanomatrix in burn-model, *Int. J. Biol. Macromol.* 85 (2016) 29–39. https://doi.org/10.1016/J.IJBIOMAC.2015.12.055.

[103] Lan, Y., W. Li, Y. Jiao, R. Guo, Y. Zhang, W. Xue, Y. Zhang, Therapeutic efficacy of antibiotic-loaded gelatin microsphere/silk fibroin scaffolds in infected full-thickness burns, *Acta Biomater.* 10 (2014) 3167–3176. https://doi.org/10.1016/J.ACTBIO.2014.03.029.

[104] Farokhi, M., F. Mottaghitalab, Y. Fatahi, A. Khademhosseini, D.L. Kaplan, Overview of Silk Fibroin Use in Wound Dressings, *Trends Biotechnol.* 36 (2018) 907–922. https://doi.org/10.1016/J.TIBTECH.2018.04.004.

[105] Arkhipova, A.Y., M.A. Nosenko, N. V. Malyuchenko, R. V. Zvartsev, A.M. Moisenovich, A.S. Zhdanova, T. V. Vasil'eva, E.A. Gorshkova, I.I. Agapov, M.S. Drutskaya, S.A. Nedospasov, M.M. Moisenovich, Effects of fibroin microcarriers on inflammation and regeneration of deep skin wounds in mice, *Biochem.* 81 (2016) 1251–1260. https://doi.org/10.1134/S0006297916110031.

[106] Kim, D.W., W.S. Eum, S.H. Jang, J. Park, D.H. Heo, S.H. Sheen, H.R. Lee, H. Kweon, S.W. Kang, K.G. Lee, S.Y. Cho, H.J. Jin, Y.J. Cho, S.Y. Choi, A transparent artificial dura mater made of silk fibroin as an inhibitor of inflammation in craniotomized rats: Laboratory investigation, *J. Neurosurg.* 114 (2011) 485–490. https://doi.org/10.3171/2010.9.JNS091764.

[107] Naomi, R., J. Ratanavaraporn, M.B. Fauzi, Comprehensive review of hybrid collagen and silk fibroin for cutaneous wound healing, *Materials* (Basel). 13 (2020). https://doi.org/10.3390/MA13143097.

[108] Sultan, M.T., O.J. Lee, S.H. Kim, H.W. Ju, C.H. Park, Silk Fibroin in Wound Healing Process, *Adv. Exp. Med. Biol.* 1077 (2018) 115–126. https://doi.org/10.1007/978-981-13-0947-2_7.

[109] Yamada, H., Y. Igarashi, Y. Takasu, H. Saito, K. Tsubouchi, Identification of fibroin-derived peptides enhancing the proliferation of cultured human skin fibroblasts, *Biomaterials.* 25 (2004) 467–472. https://doi.org/10.1016/S0142-9612(03)00540-4.

[110] Sheikh, F.A., H.W. Ju, B.M. Moon, H.J. Park, J.H. Kim, S.H. Kim, O.J. Lee, C.H. Park, A comparative mechanical and biocompatibility study of poly(ε-caprolactone), hybrid poly(ε-caprolactone)–silk, and silk nanofibers by colloidal electrospinning technique for tissue engineering, *Journal of Bioactive and Compatible Polymers,*29 (2014) 500–514. https://doi.org/10.1177/0883911514549717.

[111] Pahl, H.L., Activators and target genes of Rel/NF-kappaB transcription factors, *Oncogene.* 18 (1999) 6853–6866. https://doi.org/10.1038/SJ.ONC.1203239.

[112] Song, Q., Q. Gou, Y. Xie, Z. Zhang, C. Fu, Periplaneta americana Extracts Promote Skin Wound Healing via Nuclear Factor Kappa B Canonical Pathway and Extracellular Signal-Regulated Kinase Signaling, Evidence-Based Complement. *Altern. Med.* 2017 (2017). https://doi.org/10.1155/2017/5821706.

[113] Hoesel, B., J.A. Schmid, The complexity of NF-κB signaling in inflammation and cancer, *Mol. Cancer.* 12 (2013). https://doi.org/10.1186/1476-4598-12-86.

[114] Schmid, J.A., A. Birbach, IkappaB kinase beta (IKKbeta/IKK2/IKBKB)--a key molecule in signaling to the transcription factor NF-kappaB, *Cytokine Growth Factor Rev.* 19 (2008) 157–165. https://doi.org/10.1016/J.CYTOGFR.2008.01.006.

[115] Dal Prà, I., P. Petrini, A. Charini, S. Bozzini, S. Farè, U. Armato, Silk Fibroin-Coated Three-Dimensional Polyurethane Scaffolds for Tissue Engineering: Interactions with Normal Human Fibroblasts, *Tissue Eng.* 9 (2003) 1113–1121. https://doi.org/10.1089/10763270360728026.

[116] Xing, W., W. Guo, C. Zou, T. Fu, X. Li, M. Zhu, J. Qi, J. Song, C. Dong, Z. Li, Y. Xiao, P. Yuan, H. Huang, X. Xu, Acemannan accelerates cell proliferation and skin wound healing through AKT/mTOR signaling pathway, *J. Dermatol. Sci.* 79 (2015) 101-109. https://doi.org/10.1016/j.jdermsci.2015.03.016.

[117] Thuraisingam, T., Y. Xu, K. Eadie, M. Heravi, M. Guiot, R. Greemberg, M. Gaestel, D. Radzioch, MAPKAPK-2 signaling is critical for cutaneous wound healing, *J. Invest. Dermatol.* 130 (2010) 278-286. https://doi.org/10.1038/jid.2009.209.

[118] Martínez-Mora, C., A. Mrowiec, E.M. García-Vizcaíno, A. Alcaraz, J.L. Cenis, F.J. Nicolás, Fibroin and sericin from Bombyx mori Silk stimulate cell migration through upregulation and phosphorylation of c-Jun, *PLoS One.* 7 (2012). https://doi.org/10.1371/JOURNAL.PONE.0042271.

[119] Czekay, R., C. Wilkins-Port, S. P. Higgins, J. Freytag, J. M. Overstreet, R. M . Klein, C. E. Higgins, R. Samarakoon, P. J. Higgins, PAI-1: an integrator of cell signaling and migration, *Int. J. Cell Biol.* 2011 (2011). https://doi.org/10.1155/2011/562481.

[120] Martínez-Mora, C., A. Mrowiec, E.M. García-Vizcaíno, A. Alcaraz, J.L. Cenis, F.J. Nicolás, Fibroin and Sericin from Bombyx mori Silk Stimulate Cell Migration through Upregulation and Phosphorylation of c-Jun, *PLoS One.* 7 (2012) e42271. https://doi.org/10.1371/JOURNAL.PONE.0042271.

[121] Javelaud, D., J. Laboureau, E. Gabison, F. Verrecchia, A. Mauviel. Disruption of basal JNK activity differentially affects key fibroblast functions important for wound healing, *J. Biol. Chem.* 278 (2003) 24624-24628. https://doi.org/10.1074/jbc.M301942200.

[122] Desai, L. P., S. R. White, C. M. Waters, Cyclic Mechanical Stretch Decreases Cell Migration by Inhibiting Phosphatidylinositol 3-Kinase-and Focal Adhesion Kinase-mediated JNK1 Activation, *J. Biol. Chem.* 285 (2010) 4511-4519. https://doi.org/10.1074/jbc.M084335.

[123] Eming, S.A., P. Martin, M. Tomic-Canic, Wound repair and regeneration: Mechanisms, signaling, and translation, *Sci. Transl. Med.* 6 (2014). https://doi.org/10.1126/SCITRANSLMED.3009337.

[124] Petrini, P., C. Parolari, M.C. Tanzi, Silk fibroin-polyurethane scaffolds for tissue engineering, *J. Mater. Sci. Mater. Med.* 12 (2001) 849–853. https://doi.org/10.1023/A:1012847301850.

[125] Chiarini, A., P. Petrini, S. Bozzini, I.D. Pra, U. Armato, Silk fibroin/poly(carbonate)-urethane as a substrate for cell growth: in vitro interactions with human cells, *Biomaterials.* 24 (2003) 789–799. https://doi.org/10.1016/S0142-9612(02)00417-9.

[126] Min, B.M., L. Jeong, K.Y. Lee, W.H. Park, Regenerated silk fibroin nanofibers: water vapor-induced structural changes and their effects on the behavior of normal human cells, *Macromol. Biosci.* 6 (2006) 285–292. https://doi.org/10.1002/MABI.200500246.

[127] Drury, J. L., D. J. Mooney, Hydrogels for tissue engineering: scaffold design variables and applications, *Biomaterials.* 24 (2003) 4337-4351. https://doi.org/10.1016/s0142-9612(03)00340-5.

[128] Kapoor, S., S. C. Kundu, Silk protein-based hydrogels: promising advanced materials for biomedical applications, *Acta. Biomater.* 31 (2016) 17-32. https://doi.org/10.1016/j.actabio.2015.11.034.

[129] Hakimi, O., D. Knight, F. Vollrath, P.Vadgama, Spider and mulberry silkworm silks as compatible biomaterials, *Compos. B. Eng.* 38 (2007) 324-337. https://doi.org/10.1016/j.compositesb.2006.06.012.

[130] Keten, S., Z. Xu, B. Ihle, M. J. Buehler. Nanoconfinement controls stiffness, strength and mechanical toughness of β-sheet crystals in silk, *Nature Materials,*9 (2010) 359-367. https://doi.org/10.1038/NMAT2704.

[131] Gu, X., X. Chen, X. Tang, Z. Zhou, T. Huang, Y. Yang, J. Ling, Pure-silk fibroin hydrogel with stable aligned micropattern toward peripheral nerve regeneration, *Nanotechnol. Rev.* 10 (2021) 10–19. https://doi.org/10.1515/NTREV-2021-0002/HTML.

[132] Thangavel, P., Ramachandran, B., Kannan, R., & Muthuvijayan, V. (2017). Biomimetic hydrogel loaded with silk andl-proline for tissue engineering and wound healing applications: SF and LP Loaded Hybrid Hydrogels for Wound Healing.*Journal of Biomedical Materials Research Part B: Applied Biomaterials,*105(6), 1401–1408. https://doi.org/10.1002/jbm.b.33675.

[133] Lee, O. J., H. W. Ju, J. H. Kim, J. M. Lee, C. S. Ki, J. Kim, B. M. Moon, H. J. Park, F. A. Sheikh, C. H. Park, Development of artificial dermis using 3D electrospun silk fibroin nanofiber matrix, *J. Biomed. Nanotechnol.* 10 (2014) 1294-1303. https://doi.org/10.1166/jbn.2014.1818.

[134] Ige, O. O., L. E. Umoru, S.Aribo, Natural products: a minefield of biomaterials, *International Scholarly Research Articles.* 2012 (2012) 20. https://doi.org/10.5402/2012/983062.

[135] Meka, S. R. K., V. Agarwal, K.Chatterjee, *In situ* preparation of multicomponent polymer composite nanofibrous scaffolds with enhanced osteogenic and angiogenic activities, *Mater Sci. Eng. C. Mater. Biol. Appl.* 94 (2019) 565-579. https://doi.org/10.1016/j.msec.201809.055.

[136] Hodgkinson, T., X.F. Yuan, A. Bayat, Electrospun silk fibroin fiber diameter influences in vitro dermal fibroblast behavior and promotes healing of ex vivo wound models, *J. Tissue Eng.* 5 (2014). https://doi.org/10.1177/2041731414551661.

[137] Suarez, E., F. Syed, T.A. Rasgado, A. Walmsley, P. Mandal, A. Bayat, Skin equivalent tensional force alters keloid fibroblast behavior and phenotype, *Wound Repair Regen.* 22 (2014) 557–568. https://doi.org/10.1111/wrr.12215.

[138] Suarez, E., F. Syed, T. Alonso-Rasgado, P. Mandal, A. Bayat, Up-regulation of tension-related proteins in keloids: knockdown of Hsp27, α2β1-integrin, and PAI-2 shows convincing reduction of extracellular matrix production, *Plant Reconstr. Surg.* 131 (2013) 158e-173e. https://doi.org/10.1097/PRS.0b013e3182789b2b.

[139] Liu, H., H. Fan, Y. Wang, S. L. Toh, J C. H. Goh., The interaction between a combined knitted silk scaffold and microporous silk sponge with human mesenchymal stem cells for ligament tissue engineering, *Biomaterials.* 29 (2008) 662-674. https://doi.org/10.1016/j.biomaterials.2007.10.035.

[140] Shen, W., X. Chen, J. Chen, Z. Yin, B. C.Heng, W. Chen, H. Ouyang. The effect of incorporation of exogenous stromal cell-derived factor-1 alpha within a knitted silk-collagen sponge scaffold on tendon regeneration, *Biomaterials.* 31 (2010) 7239-7249. https://doi.org/10.1016/j.biomaterials.2010.05.040.

[141] Liu, J., L. Yan, W. Yang, Y. Lan, Q. Zhu, H. Xu, C.Zheng, R. Guo, Controlled-release neurotensin-loaded silk fibroin dressings improve wound healing in diabetic rat model, *Bioact. Mater.* 4 (2019) 151-159. https://doi.org/10.1016/j.bioactmat.2019.03.001.

[142] Sugihara, A., K. Sugiura, H. Morita, T. Ninagawa, K. Tubouchi, R. Tobe, M. Izumiya, T. Horio, N. G. Abraham, S. Ikehara, Promotive effects of a silk film on epidermal recovery from full-thickness skin wounds, *Proc. Soc. Exp. Biol. Med.* 225 (2000) 58-64. https://doi.org/10.1046/j.1525-1373.2000.22507.x.

[143] Abdel-Naby, W., B. Cole, A. Liu, J. Liu, P. Wan, R. Schreiner, D.W. Infanger, N.B. Paulson, B.D. Lawrence, M.I. Rosenblatt, Treatment with solubilized Silk-Derived Protein (SDP) enhances rabbit corneal epithelial wound healing, *PLoS One.* 12 (2017). https://doi.org/10.1371/JOURNAL.PONE.0188154.

Chapter 12

Silk Fibroin/PVA/ZnO Nanocomposites Wound Dressings

Priyanka Patil[1,2] and Shivaji H. Pawar[2,3*]

[1] Sigma Institute of Science Bakrol, Vadodara, Gujrat, India
[2] Center for Interdisciplinary Research,
D. Y. Patil University, Kolhapur, Maharashtra, India
[3] Centre for Innovative and Applied Research (CIAR),
T. C. College, Baramati, Maharashtra, India

Abstract

Impaired wound recovery can fail to local hypoxia or tissue necrosis and ultimately result in abnormality or even death. Several factors can influence the wound healing environment, including bacterial or fungal infections, different disease states, desiccation, edema, and even systemic viral infections such as COVID-19. Silk fibroin, the fibrous structural-protein component in silk, has emerged as a promising treatment for these impaired processes by promoting functional tissue regeneration. Silk fibroin's dynamic properties allow for customizable nanoarchitectures, which can be tailored for effectively treating various wound healing impairments. Wound dressing materials designed from biocompatible and biodegradable polymersare widely used in wound healing.

The present investigation deals withthe preparation of a unique blend of Silk Fibroin (SF)-Polyvinyl Alcohol (PVA) and ZnO nanoparticles (NPs) composite film (SF-PVA-ZnO NPs). The comparison of blend of SF- PVA and SF-PVA-ZnO Ps composite films were studied. The physical-chemicalcharacterizations of synthesized ZnO NPs and prepared composite films (SF-PVA and SF-PVA-ZnO) were studied. The field emission scanning electron microscopy confirmed the ZnO NPs embedded into SF-PVA composite film. The SF-PVA-ZnO NPs composite film showed enhanced mechanical property due to ZnO NPs. Antibacterial activity of the prepared composite filmsreveals that embedded ZnO NPs shows excellent antibacterial activity against wound infection-causing microorganisms. SF-PVA-ZnO composite film showed enhanced swelling behavior and faster blood clotting ability compared to control SF-PVA composite film. *In-vitro* cytocompatibility study exhibits the nontoxic nature of the synthesized SF-PVA-ZnO composite film. These studies confirm the designed composite film holds a huge potential to be used in dressing material for infected wound.

---

* Corresponding Author's Email: shpawar1946@gmail.com.

In: Silk Fibroin
Editor: Shivaji H. Pawar
ISBN: 979-8-88697-402-7

Graphical Abstract

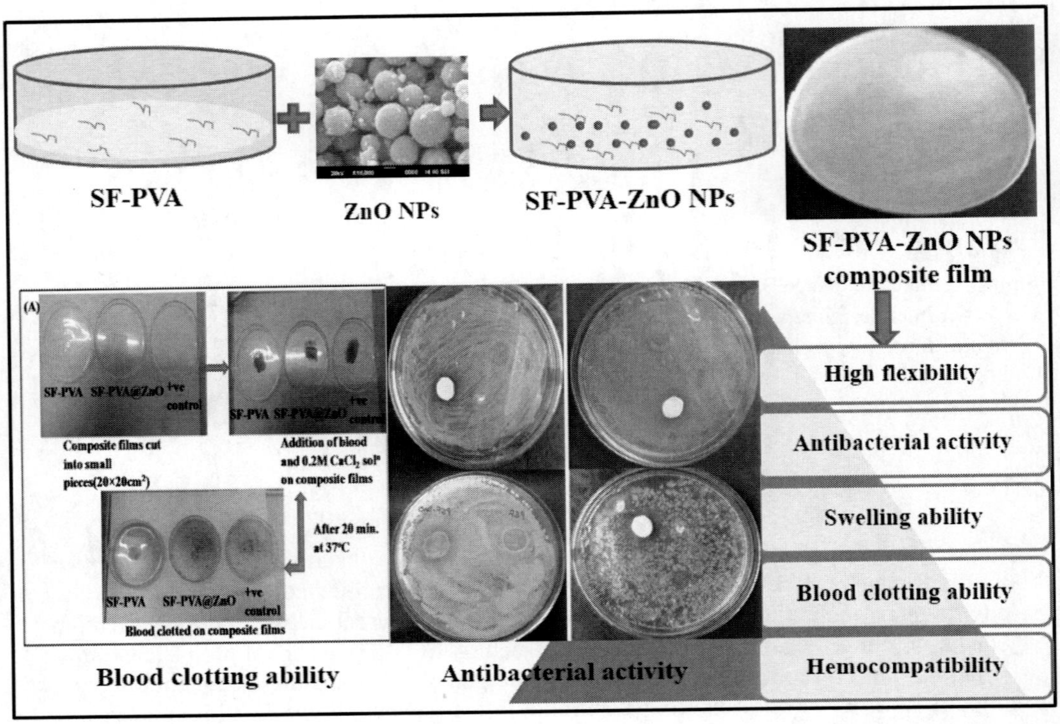

Keywords: ZnO nanoparticles, SF-PVA-ZnO, antibacterial activity, blood clotting ability, wound dressing

1. Introduction

1.1. Wound Dressing and Its Types

A wound dressing is anything that is used in direct contact with a wound to help it heal and prevent future issues or complications. Different wound dressings are used based on the type of the wound, but they all aim to help reduce infection. Wound dressings are also help with the following,

- Stop bleeding and start clotting,
- Absorb excess blood, plasma or other fluid
- Wound debridement [1]

Wound dressings are identified as either primary or secondary. Primary dressings are applied directly to the wound bed, whereas secondary dressings are used to contain or anchor a primary dressing [2]. Custom dressings, which may contain more than one type of primary or secondary dressing, can also be used. When applying dressings, the therapist should always follow universal precautions and use a clean technique to prevent cross-contamination. When choosing a dressing for a wound, four factors related to the wound itself should be considered: (1) the color of the wound (tissue composition), (2) the amount of drainage, (3) the wound depth, and (4) the periwound. Other significant factors include the patient's and caregiver's cognitive and physical ability to apply the dressing and the accessibility of physical therapy or nursing services to the patient. Cost is also a consideration [3].

A large number of wound care products are available, and it is virtually impossible to be aware of the purpose and application instructions of every available dressing. It is the responsibility of the therapist to read the manufacturer's instructions for application, to know the purpose of the dressing, and to make educated decisions when choosing the appropriate dressing for a given wound at each stage of the healing process. Although it is not possible to identify every dressing, it is possible to catalogue most dressings into some basic categories. Wound dressings are of two basic types are traditional dressing and modern dressing. Cloths and cotton gauze included in traditional dressing while hydrogels, hydrocolloids, alginate, foams, sponges, nanofibrous matrices, scaffolds and composite films based on antibacterial agents, biocompatible and biodegradable materials which are depicted in Figure 1 [4].

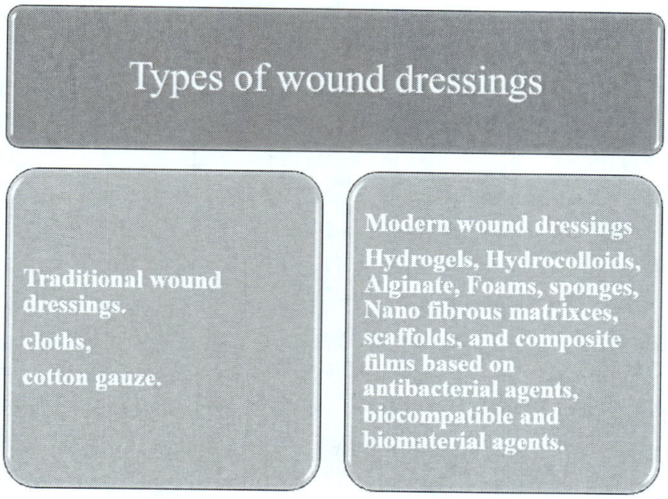

Figure 1. Types of wound dressings.

1.2. SF Structure and Properties

Fibroin and sericin are the two proteins found in silk from the Bombyxmori silkworm: The former is found in the inside of the thread and accounts for roughly 70% of the silk's weight, while the latter is located in the thin layer that surrounds the interior thread and accounts for the remaining 30% and also possesses fat/wax (0.8-1%) and colour/ash (1-1.4%) [5]. Fibroin is produced by the silkworm's posterior gland, while sericin is released by the silkworm's middle and anterior glands, Silk fibroin is made from mulberry silk after the outermost silk sericin is removed, which has the ability to activate an immune response when combined with fibroin [6]. While spinning, the larvae releases two very thin (10 μm diameter) fibroin double strands through the spinnerets, simultaneously bonding them together with sericin from the two exocrine silk glands (aligned on both sides of the body). The major component of silk is fibroin, which serves as the inner core and gives mechanical strength, whereas sericin serves as the exterior glue-like coating. Two silk fibroin filaments are covered with sericin in each silk fibre [7]. It has already been reported that silk fibroin filaments are made up of 3.5 nm diameter nanofibrils, which are the building elements of silk. These nanofibrils intertwine and have a strong interaction with one another and form into bigger fibril units, known as microfibrils that are 20-200 nm in diameter [8]. Parallel to the silk fibroin filaments, microfibrils and nanofibrils are arranged. The major structural protein of silk is silk fibroin, which includes polypeptide chains with molecular weights ranging from 200 to 350 kDa. Silk fibroin's primary structure consists of an H-L complex made up of a light- (L-) chain polypeptide and a heavy- (H-) chain polypeptide joined together by a single disulphide bond at the Cterminus of the H-chain. Furthermore, glycoprotein P25 which is non-covalently connected to the H-L-chains is present in the structure of silk fibroin and contributes to its overall integrity. The H-chain is primarily responsible for the silk fibres fibrous properties. The most common amino acids present in this chain are glycine (46 percent), serine (12 percent), and alanine (30 percent) [9]. Isoleucine, leucine, valine, and other acidic amino acids are mostly found in the nonfibrous L-chains. Because of its biocompatibility, gradual disintegration, low immunogenicity, adaptability, and outstanding mechanical qualities, silk fibroin is increasingly being studied for biomedical applications [10].

Compatibility of Silk Fibroin as a Wound Healing Biomaterial. The human body's skin serves as a barrier against the infectious agents, dehydration, and environment. Wound healing is a complex process including interactions with many cells and matrices, as well as multiple overlapping phases like inflammation, new tissue creation, and tissue remodeling happening at the same time [11]. The initial phase of wound healing is inflammation, which occurs shortly after an injury and can persist up to two days. To prevent further fluid and blood losses, remove dying and dead tissues, and stop infection, the coagulation cascade, inflammatory pathways, and immune system must all be activated. Inflammatory cells like macrophages and neutrophils play a number of important roles in the repair process, including phagocytosis and the production of many cytokines and growth factors [12]. The second phase of wound healing is new tissue development that is correlated with deposition of collagen/matrix, re-epithelialization, angiogenesis, and contraction of wounds [13]. The remodeling or maturation phase of the wound healing process, which can last for a year or longer depending on the degree of the injury, is combined with the rebuilding of the epidermis and extracellular matrix [12]. Skin regeneration is still the target of skin repair in order to minimize the long-term effects of scarring on an individual. Wound healing in mammals is a reparative process that culminates in scar formation rather than a regenerative process that replaces normal tissue architecture. Scar tissue is inferior to normal tissue in terms of function and appearance. It is caused by an excess of extracellular matrix (ECM) being produced following an injury [13]. Collagen I, an important protein influencing the ECM architecture during wound healing, is synthesized majorly by fibroblasts, and its synthesis is majorly regulated by cytokine transforming growth factor β1 (TGFβ)1. (TGFβ)1 which is secreted in inactive form is converted to active form enzymatically in order to execute its effect. (TGFβ)1 changes the morphology of many cells like fibroblasts by stimulating polymerization of the globular form of active cytoskeleton to filamentous form [14]. There really are numerous wound dressing biomaterials on the market today, to mention a few are chitosan, hydrocolloids, polyurethane, alginates, collagen, and hyaluronic acid [15]. The following characteristics should be present in an ideal wound dressing which are shown in Figure 2. It shows (a) Biocompatibility, Biodegradability nontoxicity, (b) maintain high humidity, gaseous and water vapour transport rate, (c) blood clotting ability, (d) mechanical strength and oxygen barrier capacity (e) controlled exudates and photo induced cleaning [16]. The creation of optimal wound dressings, on the other hand, is still ongoing.

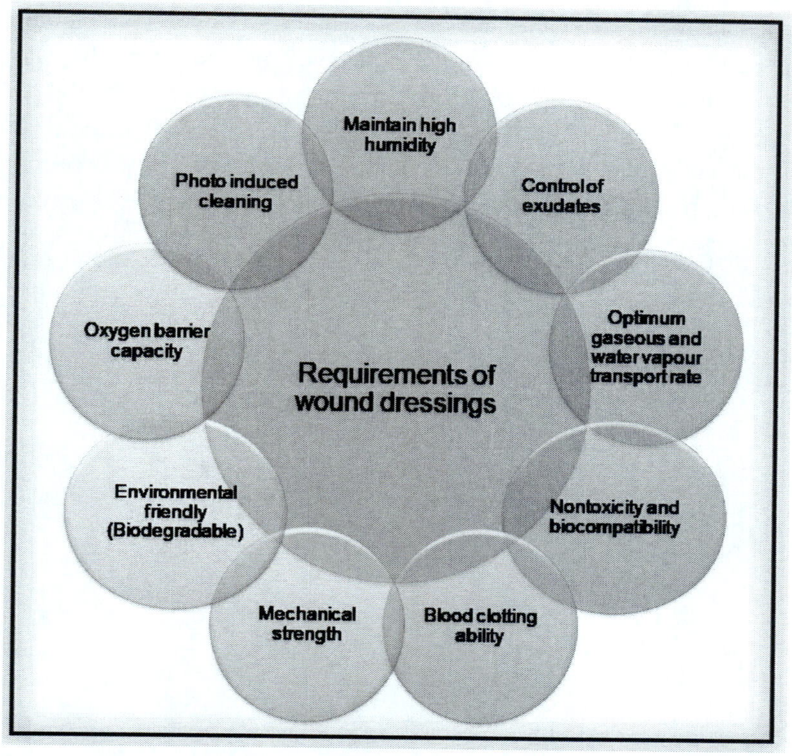

Figure 2. Requirements of wound dressings.

1.3. Polyvinyl Alcohol (PVA)

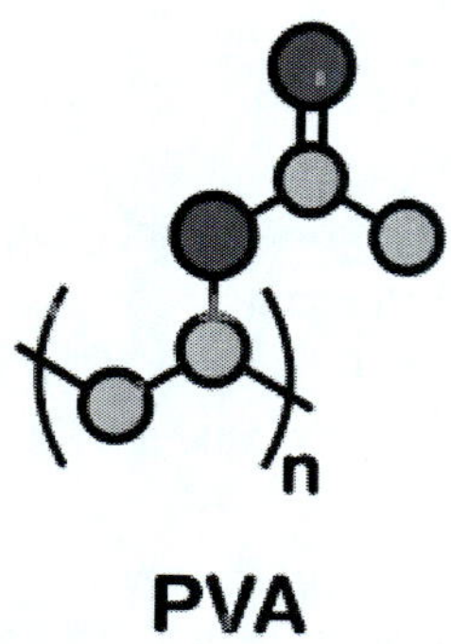

Figure 3. Structure of PVA.

PVA has various useful properties including non-toxicity, biocompatibility, high hydrophobicity, fiber/film forming ability and the chemical and mechanical resistance. It has been widely commercialized and studied in the chemical and medical industries for the productions of fibers, films, coatings, cosmetics, pharmaceuticals, and so on [17]. PVA hydrogels produced by using the freezing–thawing technique form a matrix of physically crosslinked polymeric chains containing uncrosslinked polymer and water. PVA has excellent and easy film-forming properties and has been previously blended with synthetic and natural polymers, due to its good water-soluble, biodegradable, non- carcinogenic, and biocompatible characters. Blended materials either natural or synthetic polymers assemble the desirable properties of each material on its own, while blended polymer materials are always mixed with PVA for improving mechanical and physicochemical properties of obtained polymeric materials [18]. In this sense, the final properties of the blended material depend on the properties of the embedded materials (i.e., natural or synthetic polymers), and thus PVA properties change after blending. Natural and synthetic polymers which have been blended previously with PVA to form film or hydrogel membranes for wound dressing applications [19]. Figure 3 shows structure of PVA.

1.4. ZnO NPs

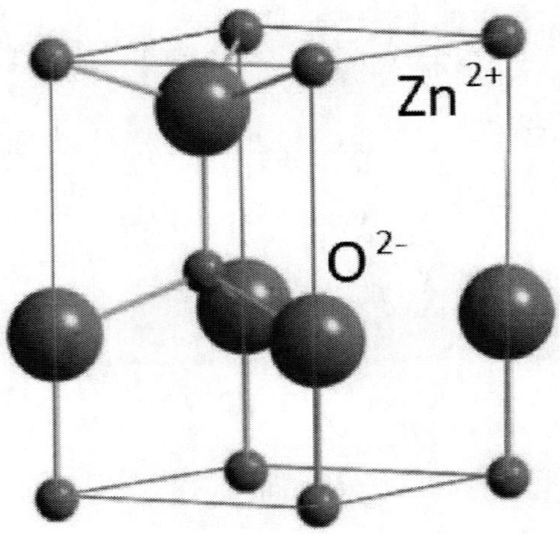

Figure 4. Wurtzite structure of ZnO NPs.

Recently NPs are used for creation of novel dressing materials due to its unique structure, interesting catalytic effects, high surface area, strong mechanical stability, biocompatibility and antibacterial activity [65]. Several NPs shows antibacterial activity, among theseNPs, ZnO NPs are of special interest. ZnO NPs belongs to inorganic, metal oxide NPs are increasingly important as one of the newgroup of NPs. Due to its antibacterial, biocompatible, and tissueadhesive properties, ZnO NPs are capable and best filler in composite films [20]. ZnO NPs crystals form wurtzite structure shows in Figure 4.

Rapid wound healing is the basic aim of the requirement of acute or chronic wounds today. Chronic wounds were impaired in healing prevail because of infections. The infection comes from indigenous microflora or the environments which grow immediately in wounded area and causesa delay in healing process [21]. This can be prevented by protecting the wound with proper dressing material. The role of a wound dressing is to provide the optimum conditions for wound healing, whilst protecting the wound from further trauma and invasion by pathogenic microorganisms.Conventional approaches of wound dressings are based on hydrocolloids andhydrogels that absorb excessive exudates and provide a moist environment to the wound to optimize repair and protect against infections [22]. Moreover, some dressings are based on silver ions to prevent infections or comprise biodegradable polymers, such as hyaluronic acid to promote healing and chitosan to stop bleeding. To sum up the main drawback of the current dressing material is a lack of efficient antibacterial activity along with biocompatibility and biodegradability [23].Thus there is urgent need to design innovative dressings based on biodegradable and biocompatible materials possessing antibacterial activity and targeting specific events in the healing process to accelerate wound closure and restore the functionality of the healed tissue [24].

The present investigation deals with synthesis of blend of SF-PVA and ZnO NPs. The SF-PVA film was prepared by casting method at room temperature. The physical chemical characterizations of synthesized ZnO NPs and prepared composite films were studied. In addition, the composite films were evaluated for mechanical properties, morphological study, antibacterial activity, blood clotting and hemocompatibility. *In vitro* cytotoxicity study by using L929 mouse fibroblasts cell lines were also investigated. The objective of this work was to develop a wound dressing made of a SF-PVA-ZnONPs to be applied on wounds with low level of exudate production, with the ability to release antiseptics for infection control. The most promising systems were evaluated concerning their biocompatibility, cytotoxicity and the antibacterial activity.

2. Preparation of SF-PVA-ZNO Nanocomposite Films

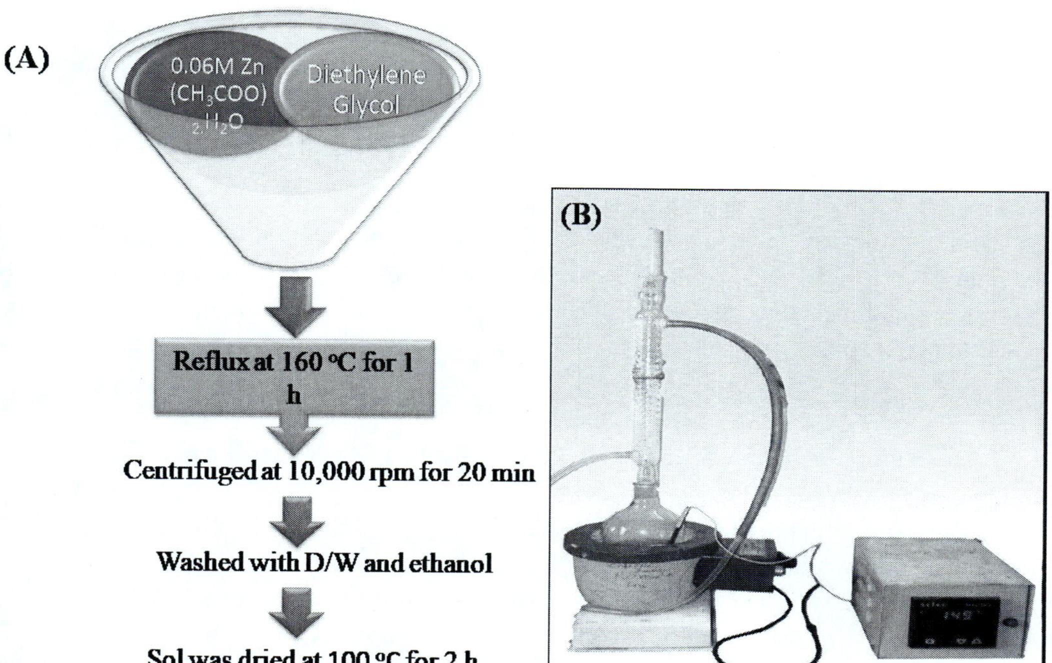

Figure 5. (A) Synthesis of ZnO NPs, (B) Reflux instrument.

2.1. Synthesis of ZnO NPs

ZnO NPs were synthesized by our earlier reported method [25]. Briefly, 0.06M Zn $(CH_2COO)2H_2O$ was added to DEG in a flat bottom flask and reflux at 160°C for 1 h. Figure 5 (A) shows synthesis of ZnO NPs and (B) shows Reflux instrument which was used for synthesis process. The solution was centrifuged at 10,000 rpm for 20 min and the supernatant was removed and dried at 100°C for 2 h.

2.1.1. Mechanism of Synthesis of ZnO NPs

In this study, the reflux method was used for the synthesis of ZnO NPs as mentioned in above. The reaction mechanism for the formation of ZnO NPs is expressed below [26]:

$$nZn\ (CH_3COO_2).\ 2H_2On + nDEG \rightarrow [Zn\ (OH)_2\text{–}\ DEG]\ n \quad (1)$$

$$[Zn\ (OH)_2\text{-}DEG]n \rightarrow [ZnO\text{-}DEG]n + nH_2O \rightarrow ZnO \quad (2)$$

In the above reaction, Zn $(CH_3COO_2).2H_2O$ is dissolved in DEG solvent forming [ZnO $(OH)_2$–DEG] n as shown in equation 1. The [ZnO $(OH)_2$–DEG] n formed is a complex with DEG molecules. The hydroxyl group present on ZnO surface interacts with other molecules. This might be due to the electrostatic interaction between the hydroxyl groups present on the surface of NPs. The complex compound of [ZnO $(OH)_2$- DEG] n is then broken at a given temperature to release water molecules and then forming ZnO NPs as described in equation 2. The prepared solution is washed using double distilled water and ethanol resulting in the formation of pure ZnO NPs.

2.2. Preparation of SF Solution

Cocoons of *Bombyxmori* silkworm were degummed by using 0.5% (w/v) Na_2CO_3 solution at 98°C for 30 min and then rinsed with distilled water. The solution was dissolved in LiBr solution (9.5 Mol/L) for 2 h, followed by a dialysis. The SF was further purified by centrifugation. Final concentration of SF solution was 5% (w/v). All solutions were preserved at 4°C to avoid premature precipitation [27].

2.3. Preparation of SF-PVA and SF-PVA-ZnO NPs Composite Films

SF-PVA-ZnO composite film was prepared by casting method at room temperature. SF-PVA composite film was prepared for control. First SF, PVA stock solutions were prepared at a concentration of 5 wt% and ZnO NPs stock solution was prepared at a concentration of 1wt%. Blending films were performed by mixing the stock solutions of SF, PVA and ZnO NPs with the same initial concentration but different volumes in a beaker. For example SF: PVA: ZnO NPs wt. ratio 1:4:1 and SF: PVA wt. ratio 1:4.After mixing, the solutions were stirred at 150 rpm for 2 h at room temperature, adjusting a neutral pH and then transferred to Petri dishes. All the Petri dishes were placed at a room temperature to dry overnight. The dried composite films were then peeled off and stored in a sealed container at room temperature before use Figure 6 shows synthesis process of SF-PVA and SF-PVA-ZnO NPs composite films.

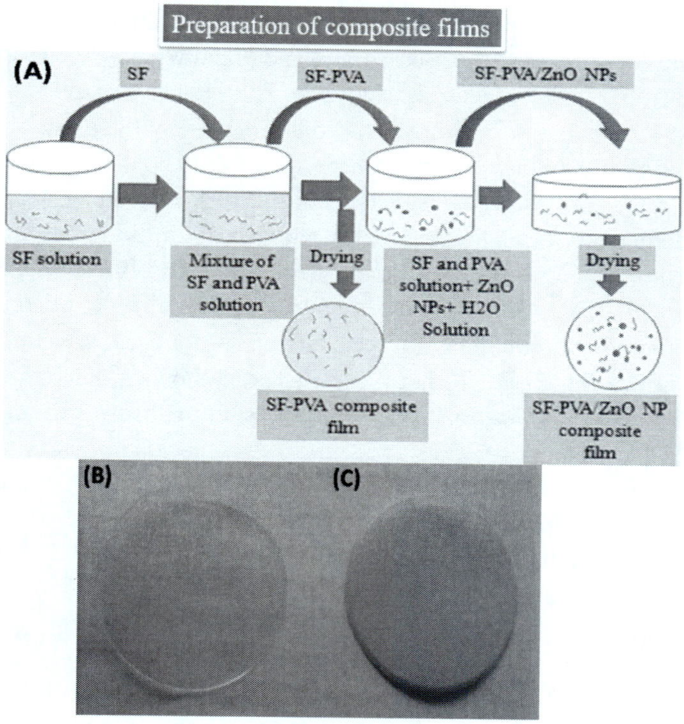

Figure 6. Preparation of SF-PVA and SF-PVA-ZNO NPs composite films.

3. Physical-Chemical and Mechanical Study of SF-PVA-ZnO NPs Composite Film

The prepared SF-PVA and SF-PVA-ZnO composite films were characterized using X-ray diffraction (XRD), Rigaku 600 Miniflex X-ray diffraction instrument in the range 10°-60°. The UV-visible absorption spectra of ZnO NPs and prepared composite films were recorded using Shimadzu UV 1600 spectrophotometer in the range of 300 nm-600 nm. The functional groups presented in the synthesized ZnO NPs and prepared composite films were examined using Fourier transform infrared (FT-IR) spectroscopy. All spectra were recorded in the wave number range from 4000 cm^{-1} to 600 cm^{-1}.

The mechanical properties of as-prepared SF-PVA and SF-PVA-ZnO composite films were characterized using Universal Testing Machine (UTM) with Winsoft tensile and compression testing machine (Shimadzu AG) 100 KN load cell at 25°C. Composite films were cut into 40 mm long and 10 mm wide. The gauge length was 20 mm between clamps. These were examined at a crosshead speed of 5 mm/min. The tensile strength per cross-sectional area and the ratio of the relative elongation to the initial film length at break (%) were determined from the stress and strain curves.

XRD analysis was carried out to study the crystalline nature of composite films. XRD patterns of prepared composite films have been shown in Figure 7 (A). Both the composite films of SF-PVA and SF-PVA-ZnO showed the prominent peaks at 2θ = 19.98, 23.26, 28.56 and 40.96. Diffraction peaks of 2θ = 19.98 and 40.96 attributed to the semi-crystalline nature of PVA. The diffraction peaks at 2θ = 23.26 and 28.56 exhibited to silk II, β- sheet crystalline structure of SF [28]. XRD spectrum of SF-PVA-ZnO composite film depicts different diffraction peaks at 2θ = 31.76,° 34.6, 36.1, 47.66 and 56.60 which were ascribed to the (100), (002), (101), (102) and (110) planes respectively. The intensity of diffraction peaks which attributed to SF-PVA was very high than that of SF-PVA-ZnO composite film because the concentration of PVA was higher compared with that of ZnO NPs.

Figure 7 (B) represents the UV-visible spectra of SF-PVA and SF-PVA-ZnO composite films. SF-PVA composite film showed an absorption peak in the region 250-300 nm with a wide peak at 290 nm, which was the confirmation of small traces of tyrosine present in SF. The absorption peak of SF-PVA-0ZnO composite film showed a minor increase in the overall absorption in the region 350-400 nm with a peak at 365 nm for ZnO NPs. Pure ZnO NPs shows absorption peak in UV region at 370 nm [29]. The peak shows a blue shift in SF-PVA-ZnO composite film compared to pure ZnO NPs. The shift in the absorbance due to the ZnO NPs embedded into SF-PVA composite film.

Figure 7 (C) shows the FT-IR spectra of composite films. The FT-IR spectrum of the SF-PVA composite film showed the characteristic peak at about 3267.93 cm^{-1} was assigned to the stretching vibration of a hydroxyl group (OH) and peaks about 2920.53 cm^{-1}, 2851.30 cm^{-1} were assigned to the band of asymmetric CH_2 stretching.

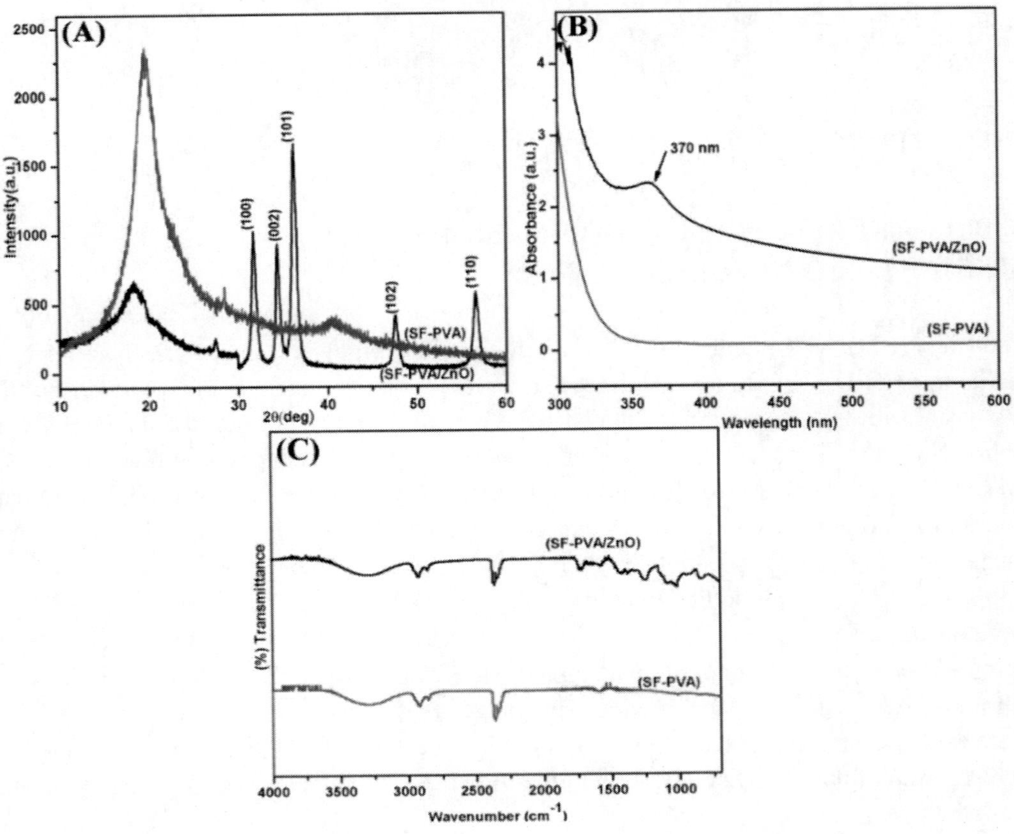

Figure 7. Characterization of prepared SF-PVA and SF-PVA-ZnO composite films of (A) XRD analysis, (B) UV-visible spectroscopy, (C) FT-IR spectrum.

The peak position at about 1018.15 cm^{-1} was assigned to the (-C-O) resonance. The absorption bands at 1631.05 cm^{-1} indicated random coil structure of amide I (C-O stretching), 1545 cm^{-1} was amide II (NH deformation and N-stretching). The FT-IR spectrum of the prepared SF-PVA/ZnO composite film showed the characteristic absorption peaks of PVA at about 3267.93 cm^{-1} was assigned to the stretching vibration of the hydroxyl group (O-H) and at about 1018.15 cm^{-1} and 1413.21 cm^{-1} (-C-O) were observed. The two bands observed at 2920.53 cm^{-1} and 2851.30 cm^{-1} were assigned to the band of asymmetric CH_2 stretching. The peak positions at 1631.05 cm^{-1}, indicated random coil structure of amide I (C-O stretching). The peaks at around 1579.46 cm^{-1} showed amide II (N-H deformation and C-N stretching) and those at 1250 cm^{-1} indicated random coil structure of amide III (C-N stretching and N-H deformation). Metal oxides generally give absorption bands in fingerprint region i.e., below 1000 cm^{-1} arising from inter-atomic vibrations. The absorption peak at about 1413.21 cm^{-1} appeared the stretching mode of vibration of C=O. Also, the band at 835.77 cm^{-1} and 668.30 cm^{-1} indicated the stretching vibrations of ZnO NPs [30].

Figure 8 depicts the mechanical property of SF-PVA-ZNO nanocomposite film. It includes tensile strength and % elongation curves of SF-PVA and SF-PVA-ZnO composite films. The result represents the tensile strength of SF-PVA composite film has about 3. 125 MPa and % elongation is about 178%. SF-PVA-ZnO composite film has a tensile strength of about 3. 791 MPa and % elongation is about 207%. However, the tensile strength and % elongation of SF-PVA-ZnO composite film was increased by the addition of ZnO NPs. The obtained tensile strength and % elongation was higher than reported β-chitin hydrogel/nZnO composite bandage [31]. Therefore, it was concluded that the obtained tensile strength and % elongation of prepared SF-PVA-ZnO composite film was sufficient for wound dressing application.

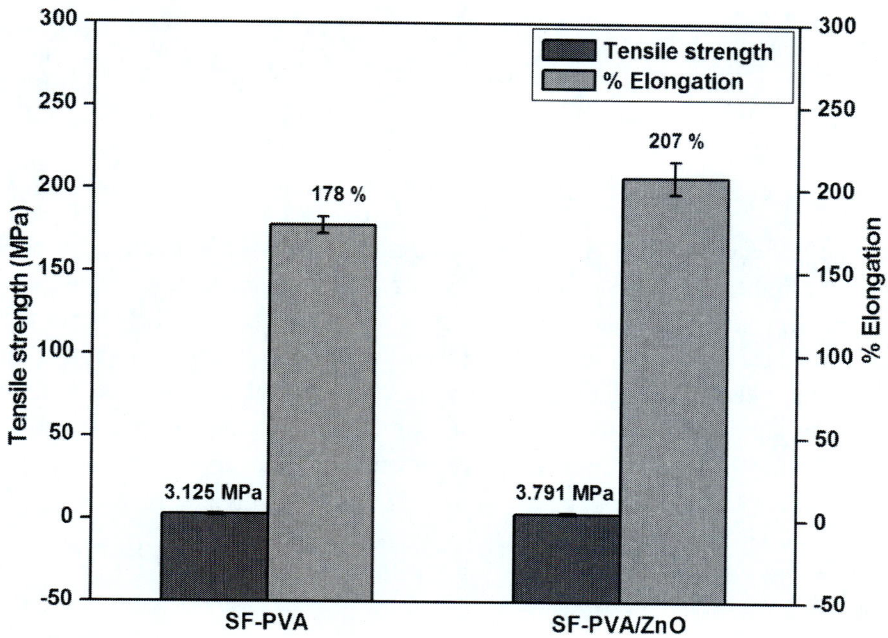

Figure 8. The mechanical property of SF-PVA and SF-PVA-ZnO composite films.

4. Antibacterial Study

Antibacterial study is involving two basic activities such as bactericidal activity and bacteriostatic activity. The bactericidal activity is the use of antibacterial agents or other substances that kill the bacteria, whereas the bacteriostatic activity is defined as the use of antibacterial agents or other substances that kill the bacterial growth. Antibacterial materials are used to fight against infections. Hence, there are different methods employed for investigation of antibacterial activity of different antibacterial agents or other substances. These methods are listed below [32],

- Agar disc diffusion method
- Agar well diffusion method
- Shake flask method

4.1. Agar Disc Diffusion Method

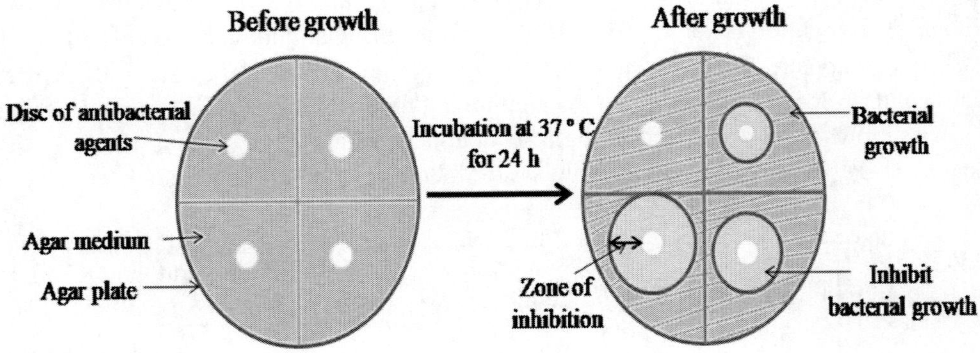

Figure 9. Schematic representation of the agar disc diffusion method for antibacterial activity.

The disc diffusion method is easy, important and most widely method for antibacterial activity study. In the disc diffusion method the first target organism suspension spread on the surface of nutrient agar medium and then the paper disc of antibacterial agents or tested samples deposited on the surface of sterile medium plate. Further, the tested sample discs diffused in medium by plates kept in refrigerator for 10 min. After the plates are incubated at 37ºC for 24 h, the zone of inhibition (ZOI) around each discs are determined by using the scale in millimeter. Figure 9 shows the schematic representation of agar disc diffusion method for antibacterial activity.

4.2. Agar Well Diffusion Method

In agar well diffusion process, the wells (6 mm in diameter) are prepared by applying sterile cork borer and the target organism suspension is spread over the nutrient agar medium. The tested liquid sample of volume 70-80 μl is aseptically dispensed into each well and plates are kept at 4°C for 10 min to facilitate the diffusion of sample in the agar medium. After plates are incubated at 37°C for 24 h and the ZOI is measured around each well. Figure 10 shows the schematic representation of agar well diffusion method for antibacterial activity.

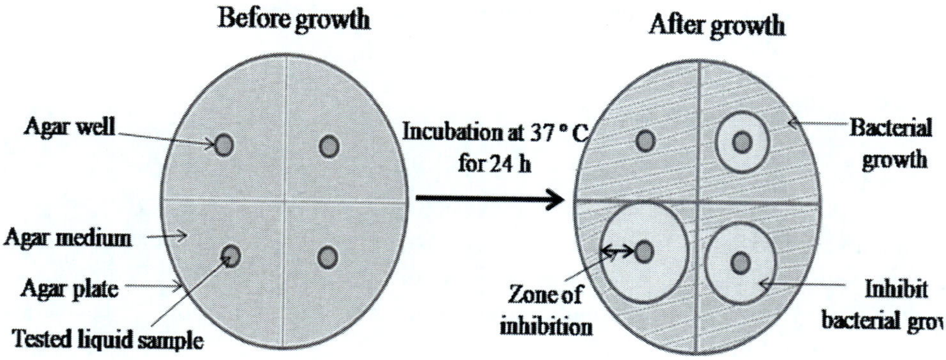

Figure 10. Schematic representation of the agar well diffusion method for antibacterial activity.

4.3. Shake Flask Method

In this method, the nutrient broth are prepared and dispensed in flask aseptically. The tsted samples as well as microbial suspension are aseptically dispensed into the flask and the flasks are kept in shaking incubator at 180 rpm for 24 h at 37 ° C. The growth of organisms is observed in the form of turbidity. The growth of organism is determined by using optical density. Figure 11 shows the schematic representation of shake flask method for antibacterial activity.

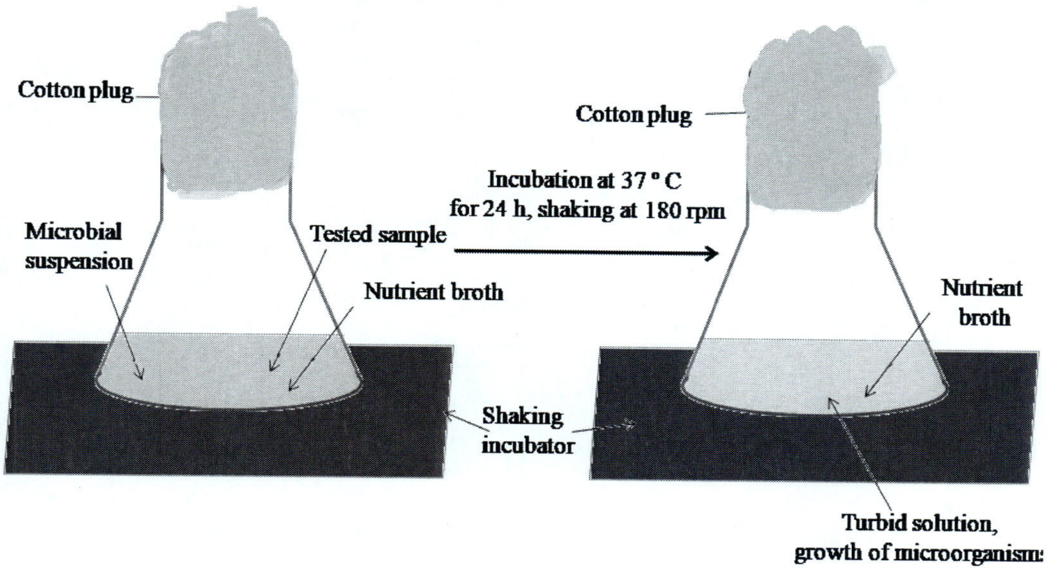

Figure 11. Schematic representation of shake flask method for antibacterial activity.

4.4. Minimum Inhibition Concentration

Minimum inhibition concentration (MIC) refers the lowest concentration of antibacterial agents which restrict the growth of bacteria. The determination of MIC of antibacterial agents are carried out in two different methods, which are listed below [33]:

- Agar dilution method
- Broth dilution method.

In agar dilution method, the defined number of bacterial cell suspension deposited directly onto the sterile medium plates that have integrated different antibacterial agent concentrations. After, incubation period the presence of the bacterial growth on the plates displays the growth of the organisms. This process is generally applicable for antibiotics available in ample amounts. The advantages of this method are to determine large number of bacterial isolates simultaneously under exactly identical condition.

In broth dilution method, the liquid growth medium is used for determination of MIC. In this method, the liquid growth medium incorporated the concentrations of antibacterial agents, which is inoculated with a defined number of bacterial cells. The final volume of the sample is prepared, whether the method is termed microdilution. When using a total volume is 2 ml. The microdilution is achieved in Microtiter plates using < 500 μl per well. The presence of turbidity of samples indicates the growth of organisms, after incubation period. The method of MIC is calculated as the lowest concentration of antibacterial agent, which restricts the growth of bacteria. This method is applicable for testing the different antibiotics as well as other antibacterial agents.

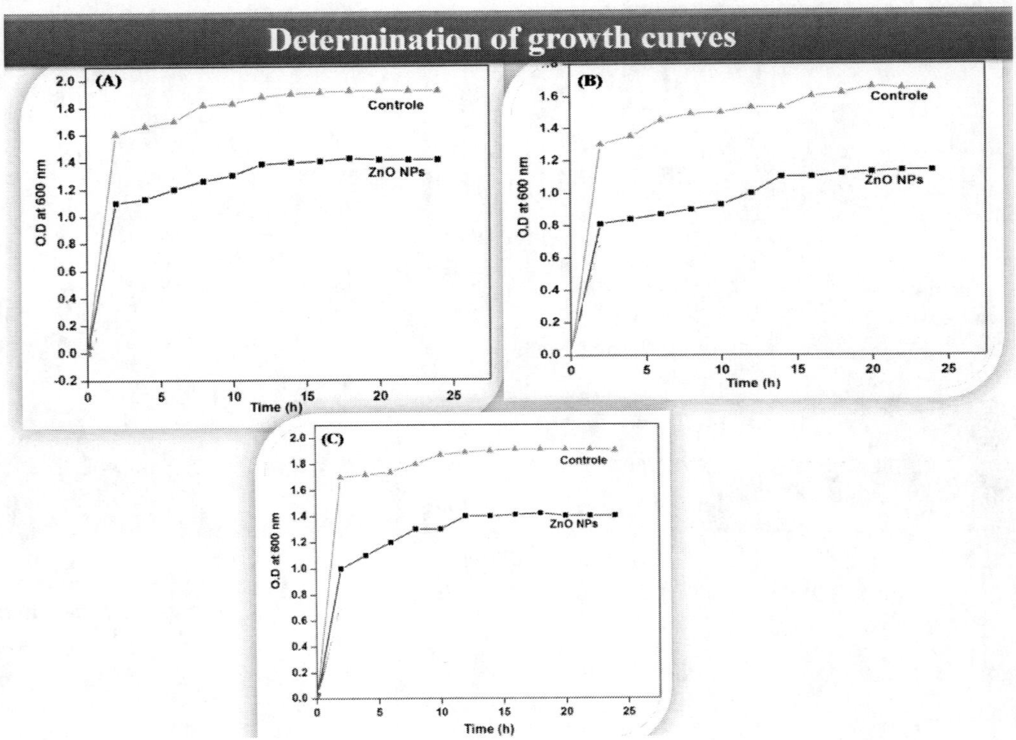

Figure 12. Determination of growth curve of ZnO NPs.

4.5. Determination of Bacterial Growth

Growth curves were examined by bacterial culture exposed to ZnO NPs., nutrient agar broth with MIC concentration of ZnO NPs (*E. coli*: 100 μg/mL, *S. aureus*: 50 μg/mL, P. mirabilis: 100 μg/mL), was used. The bacterial cell concentration was 106 CFU/ml. Bacterial culture flasks were incubated in a shaking incubator at 150 RPM at 37°C. The bacterial cell growth was examined for every 2 h under UV-vis spectrophotometer at 600 nm wavelength.

Figure 12 represents the growth curve of ZnO NPs for *E. coli*, *S. aureus* and *P. mirabilis* bacterial cells, respectively. These results reveal that the effective bacteria killing ability of ZnO NPs against bacteria. The absorbance of the sample increases with enhancing the bacterial cellgrowth. Hence, it is examined that the ZnONPs performs an essential role in antibacterial activity.

4.6. Antibacterial Activity of SF-PVA/ZnO Nanocomposite Films

The antimicrobial activity of prepared composite films of SF-PVA, ZnO NPs embedded SF-PVA composite films on various microbial strains have been studied in detail, which may be used as effective candidate for further application in wound dressings. Metal oxide NPs such as ZnO NPs are increasingly important and considered as reinforcing filler for polymer matrix. ZnO NPs have properties of biocompatible, excellent antibacterial and currently used in cosmetic materials. The decontamination action of ZnO NPs has been widely examined with various pathogenic and nonpathogenic microorganisms. Antibacterial properties of ZnO NPs exhibited owing to enhanced specific surface area and smaller particle size. There are various factors affected the decontamination activity of ZnO NPs. These factors are UV illumination, shape and size of particles, concentration, surface modifications, surface faults, and the minimum inhibitory concentration (MIC) of NPs [34]. There are several reported have been studied for the antimicrobial tests of ZnO NPs. The spherical shaped NPs noticed high bacteria killing ability against the pathogenic bacteria [35]. Thus, ZnO NPs doped materials exhibited considerably high antibacterial property towards *S. aureus* than *E. coli* and *P. aeruginosa* bacteria [36]. ZnO NPs have mostly used in antibacterial coatings for food packaging and wound dressing applications. Bacterial infection generally exposed on the surfaces of wound, this issue controlled the growth of microorganisms by using ZnO NPs based antibacterial surfaces of composite films. In the present work, the antibacterial activity composite films were investigated by using agar disc diffusion procedure. The paper discs of pure composite materials were prepared and used for study. The collected glassware for the experiment was cleaned with distilled water and autoclaved at 121°C for 20 min prior to use. The five types of bacteria like Gram-positive *S. aureus*, Gram-negative *E. coli, P. aeruginosa, P. mirabilis*, and *S. pyogenes* were selected for the antibacterial study. *S. aureus*, *E. coli*, *P. aeruginosa*, *P. mirabilis*, and *S. pyogenes* were collected from the National Collection of Industrial Microorganisms (NCIM), National Chemical Laboratory (NCL) in Pune, India. Nutrient media was supplied from himedia laboratories.

4.7. Bacteria Sample Preparation

The bacterial culture was maintained on the surface of sterile nutrient agar slants at refrigerator. The nutrient agar media composition used for the growth of bacterial cells is mentioned below. Firstly, the nutrient medium was prepared and sterilization process was done in an autoclave at 121°C for 20 min. After, the medium slants were prepared and bacterial culture suspension streaked on the surface of nutrient medium. The cultures of microorganisms were grown at 37°C for 24 h.

4.8. Nutrient Broth Composition

- Peptone (1g),
- Beef extract (1g),
- Sodium chloride (0.5g),
- Double D/W (100 mL),
- pH after sterilization (7.3).

The optical density (O.D.) of this bacterial culture was adjusted to having a concentration of approximately 1 x 107 CFU/mL [37].

4.9. Determination of Antibacterial Activity by Disc Diffusion Assay

The disc diffusion method was used to study the antimicrobial activities of SF-PVA, ZnO NPs embedded SF-PVA composite films. A typical procedure was as follows: Bacterial cultures were grown overnight on nutrient agar medium. 100 μL of cultured bacterial cell suspension was spread over the agar plate using glass spreader. The discs of coated and uncoated composite films were first soaked into double distilled water for 1 h prior to a test. Discs of composite films were placed on the surfaces of agar plates. The plates were refrigerated at 10°C for 10 min to diffuse the sample in the media. Then the plates were incubated at 37°C for 24 h for the growth of microorganisms. After 24 h of incubation, plates were observed for ZOI (Zone of Inhibition). The ZOI was measured for each sample in mm. Determination of antibacterial activity study of SF-PVA and SFPVA/ZnO NPs composite films.

Antibacterial studies of prepared composite films were studied using the disc diffusion method, which showed in Figure 13. The bacterial cultures of *S. aureus*, *E. coli*, *P. aeruginosa*, *P. mirabilis* and *S. pyogenes* were used as model gram-positive and gram-negative organisms respectively. It is very essential to examine the bactericidal effect against these microorganisms. The test specimens are SF-PVA and SF-PVA-ZnO composite films which were sterilized for 2 h using UV light. SF-PVA-ZnO composites produced clear ZOI on nutrient agar plates inoculated with *S. aureus*, *E. coli*, *P. aeruginosa*, *P. mirabilis* and *S. pyogenes* bacteria.The diameter of ZOI around diameter of ZOI of SF-PVA-ZnO shows3 mm in *S. aureus*, 2 mm in *E.coli*, 4 mm in *P. aeruginosa*3 mm in*P. mirabilis* and 3 mm against *S. pyogenes* as calculated. Reported literature study indicated that the ZOI increased with increasing ZnO NPs content of composite films and also observed that the SF-PVA does not show any antibacterial activity against bacteria.

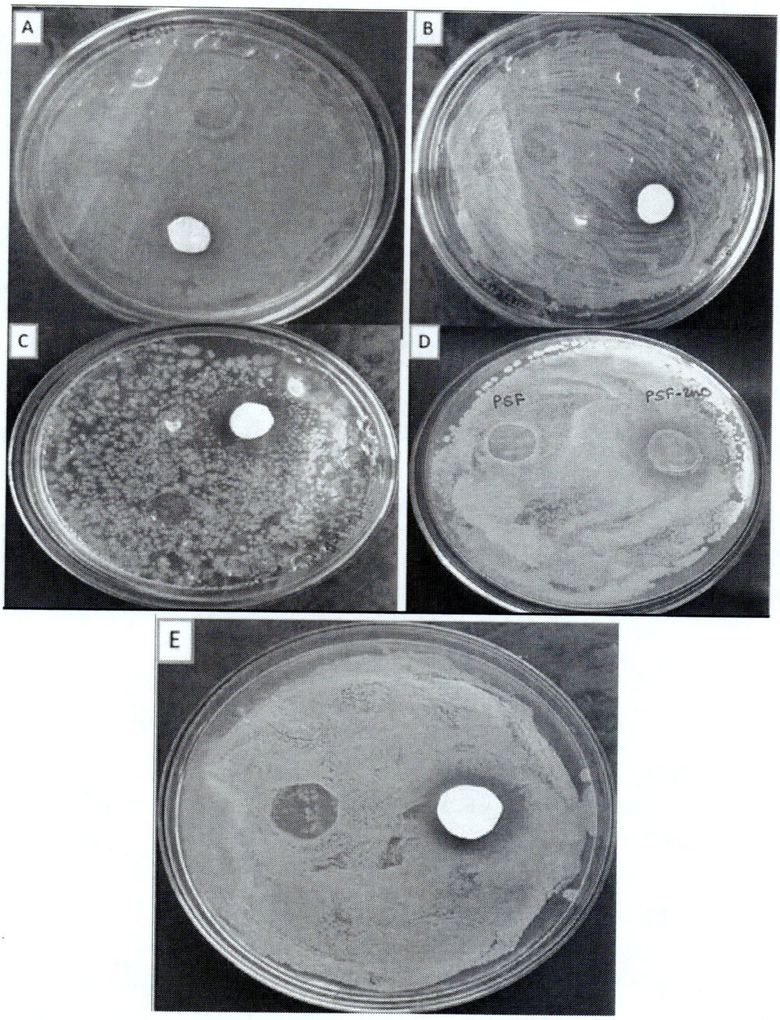

Figure 13. Antimicrobial activity of SF-PVA and SF-PVA/ZnO composite films against (A) *S.aureus* (B) *E.coli,* (C) *P. aeruginosa,* (D) *P. mirabilis* and (E) *S. pyogenes* microorganisms.

4.10. Mechanism of Antibacterial Activity of ZnO NPs

Literature studies revealed that the ROS production as the major contribution responsible forZnO NPs antibacterial activity [38]. Representation of antimicrobial is derived from molecular oxygen. It is term for different reactive molecules and free radicals. ROS includes superoxide anion (O hydroxyl radicals (OH−) and singlet oxygen (ZnO NPs are enhanced by enhancing the production of ROS from ZnO NPis damage the cellular biomolecules the apoptosis pathway and generation of ROS by ZnO NPs h interaction of NPs with cellular organelles i.e., mitochondria, disrupting the electron transport chain and interaction ofNPs with redox active protein i.e., NADPH oxidases etc. Figure 14 shows schematic representation of mechanism of a ZnONPs.

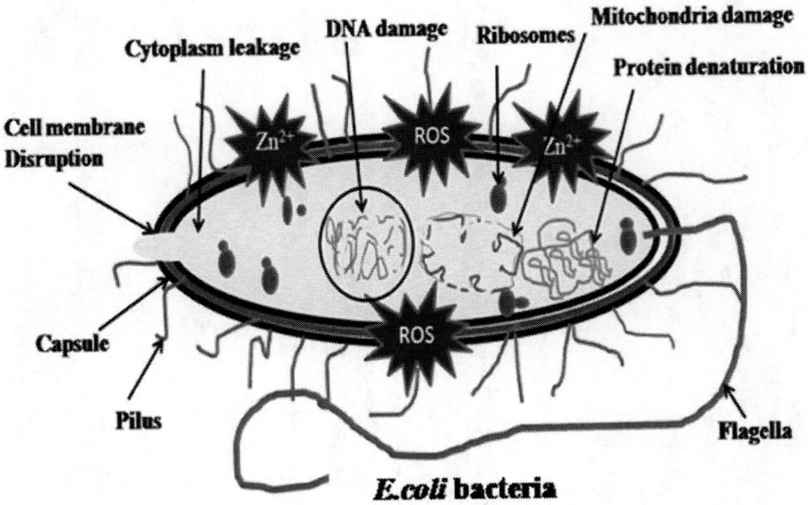

Figure 14. Mechanism of antibacterial activity of ZnO NPs.

4.11. Minimum Inhibitory Concentration of ZnO NPs

Determination of minimum inhibition concentration (MIC) of ZnONPs is the ability of NPs for the inhibition of microorganism's growth at lowest concentration. The most frequently used methods for determination of MIC of ZnONPs are agar dilution or tube dilution and broth microdilution methods. In these methods, the serial dilutions are prepared by using the ZnONPs in bacterial growth medium. Thereafter, test organisms are mixed to the test tubes containing the dilutions of the ZnONPs and tubes incubated for growth. Further, calculate the MIC value. This method is a standard assay procedure for determination of MIC of ZnONPs. The MIC is essential in diagnostic laboratories for confirmation of resistance of bacteria to an antibacterial agent. Moreover, it is used to detect the activity of newer antimicrobial agents [39].

MIC of synthesized ZnO NPs by agar dilution method by using bacteria was evaluated. Table 1 shows the MIC of ZnO NPs against *E. coli*, *S. aureus* and *P. mirabilis*. Inhibit the bacterial growth at minimum concentration of 50 µg/mL against S. aureus, 100 µg/mL against *E. coli*and 100 µg/mL against *P. mirabilis* bacteria.

Table 1. The MIC of ZnO NPs against *E. coli*, *S. aureus* and *P. mirabilis*

No.	Samples	*S. aureus*	*E. coli*	*P. mirabilis*
1.	ZnO NPs	50 µg/mL	100 µg/mL	100 µg/mL

5. Biocompatibility Study of Prepared Composite Film

The biocompatibility study of composite films becomes an important issue in nanotechnology. The term biocompatibility is defined as the capability of biomaterials to perform host response to biomaterials [40]. The capability of biomaterials is to be in contact with a living system without producing adverse effects. Hence, the evaluation of biocompatibility is one of the important parts of the overall safety assessment of nanomaterials. The composite materials include a variety of different materials. To consider the biocompatibility in nanocomposite materials, one should examine all the materials which are coming in biocompatibility of materials on humans may vary based on its application [41]. Therefore, the biocompatibility of nanomaterials is becoming an increasingly important issue regarding nano-bioapplications, especially in wound healing and dressing applications [42]. According to the literature, there is a number of research articles published regarding nanomaterials toxicity and it has published in the subject of dressing bandages. Further, the detailed study indicated that the biocompatibility involves cytotoxicity, sensitization, genotoxicity, implantation, hemo-compatibility, and blood coagulating testing etc. Among these studies, the cytotoxicity, hemocompatibility, and blood coagulation test play an important role for the formation of novel dressing materials [43]. The ISO 10993 device category has an essential role in understanding the biocompatibility study of materials, which is based on the Biological Evaluation of Medical Devices (BEMD) [44]. The standard study describes the test on toxicology, hemocompatibility, and blood clotting ability etc. The in vitro test is used because it can be a biological reaction in the occupation of materials when they are established on or onto tissue cells, experimentally controllable, repeatable, quick, low cost, ethical and legal issues, human genes carrying transgenic cells can be employed and a small amount of materials has required. Nanomaterials toxicity indicates the capability of the materials to adversely affect the general physiology along with directly interrupt the normal structure of organs and tissues [45]. Hemocompatibility is referred to as the materials compatibility with blood. It plays an essential role when designing novel dressing materials. Blood clotting ability of materials is the test of coagulation of blood. This test is important for controlling excessive bleeding in abnormality cases.

A blood sample was obtained from slaughterhouse, and they have no ethical issues for human and animal right in the ethical statement section. The present study does not involve the use of human and animal subjects. All the chemicals were analytical grade (AR grade) and used without purification. Double distilled water was used throughout all the experiments.

5.1. Hemocompatibility Study

Hemocompatibility of composite films was carried out by using hemolysis assay. The hemolysis is a process in which RBCs ruptured and hemoglobin is released. The content of hemoglobin is measured in percentage. Hemolysis percentage represents the limit of RBCs broken by the composite films in contact with blood. The hemolysis percentage value is increased with increase in the number of ruptured red cells. In hemolysis assay, the blood was obtained from slaughter stabilized with dipotassium EDTA (1.5 mg conducted to obtain healthy red blood cells (HRBCs) for hemocompatibility study of composite films. HRBCs were isolated from fresh blood by centrifugation at 2100 rpm for 10 min and purified by successive rinsing with PBS for five times. After that, HRBCs were diluted ten times with the PBS buffer. The diluted HRBCs suspension (0.2 mL) was mixed to 0.8 mL of water (negative control), pure PBS (positive control), and PBS buffer containing composite films (cut into 1 cm × 1 cm). After gentle shaking, the mixtures were kept at 37°C for 2 h and 20 h of incubation time. Then, centrifuge the mixture at 2500 rpm for 5 min and the absorbance of the supernatant hemoglobin was recorded by using a Perkin Elmer Lambda 25 UV–vis spectroscopy at a wavelength of 540 nm. The hemolytic percentages (HP) of the composite films were calculated using the formula equation 3,

$$HP(\%) = \frac{D_t - D_{nc}}{D_{pc} - D_{nc}} \times 100 \quad (3)$$

where,

D_t = the absorbance of the test sample,
D_{pc} = the absorbance of the positive control and
D_{nc} = the absorbance of negative control.

Figure 15 shows the hemolytic behavior of SF-PVA-ZnO and SF-PVA composite films. The obtained results indicate that the hemolytic percentage of SFPVA-ZnO composite film is higher as compared to SF-PVA composite film. The induced hemolysis percentage of the SF-PVA composite film is 2.1% and 2% for 2 h and 20 h of incubation times, respectively. However, the hemolysis percentage for the SF-PVA-ZnO composite film is found to be 3.4% for 2 h and 2.7% for 20 h of incubation. The hemolysis percentages of both the films show below 5%, indicating that both films are slightly hemolytic materials. The obtained results show that the hemolysis percentage decreased with the increasing incubation period of up to 20 h. ZnO NPs cause hemolysis when in contact with blood in a concentration-dependent but not in size or shape.

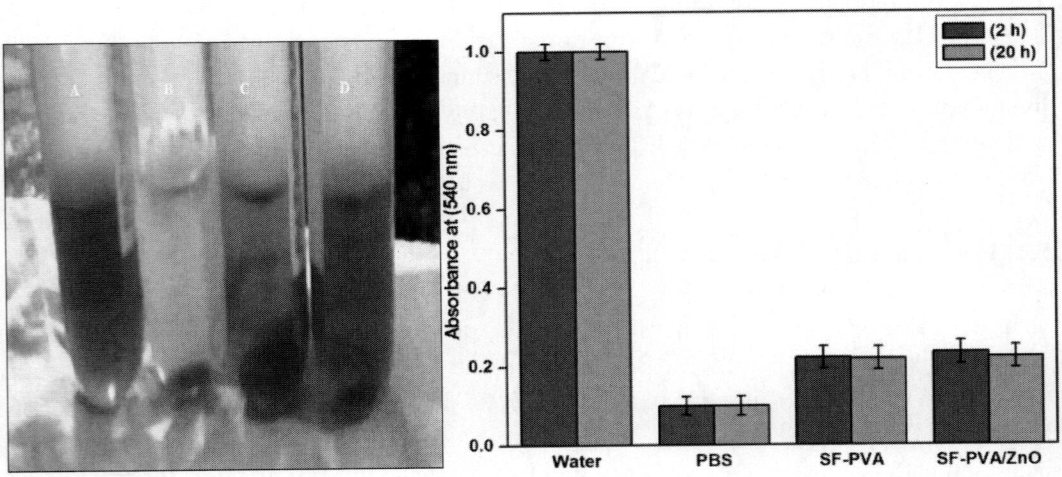

Figure 15. Hemocompatibility of SF-PVA and SF-PVA-ZnO composite films.

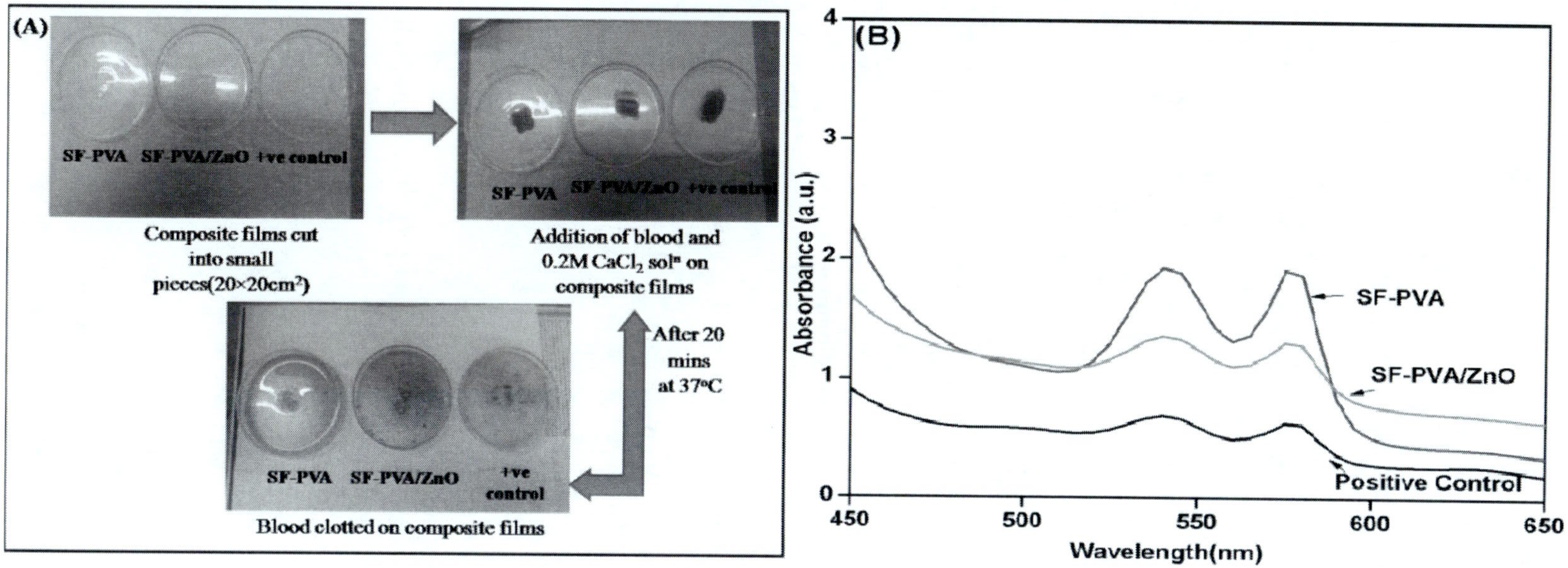

Figure 16. (A) Blood clotting abilities SF-PVA and SF-PVA/ZnO composite films, (B) absorbance at wavelength 540 nm PVA/ZnO composite films.

5.2. Blood Clotting Ability

A wound dressing material should have good capacity to stop bleeding faster. Hence, the determination of blood clotting ability of prepared composite films has important and interesting property. The blood clotting abilities of prepared composite films were determined by blood clotting method [46]. It is also called as anticoagulant assay. In blood clotting assay method, the composite films were cut into small pieces (2 cm × 2 cm). The anticoagulant blood (100 μL, collected from the slaughterhouse) was poured onto the surfaces of composite films, and 10 μL of 0.2 M CaC12 was added onto them to initiate the clotting. The films were kept for incubation at 37°C for 20 min on a shaking incubator. After the incubation period, 5 mL of double distilled water was added dropwise onto the mixture, without disturbing the clot. Further, the solutions were taken from the test tubes and centrifuged at 1000 rpm for 1 min. The supernatant of each sample was collected and kept at 37°C for 1 h. Blood clotting was determined by measuring the optical density at wavelength of 540 nm using UV-visible spectroscopy.

Figure 16 (A) it can be observed that the blood clotting ability of the ZnO NPs embedded composite film is higher than that of the SF-PVA composite film. Figure 16 (B) shows lower optical density of SF-PVA-ZnO composite film; this means that it shows more blood-clotting capability. ZnO NPs is hemostatic, as ZnO NPs can form a certain complex with fibrin, which is the most abundant protein in blood plasma, thus decreasing fluorescence in the presence of ZnO NPs [47]. Pure SF shows poor blood clotting ability when it is used as a blood-contact [48]. Blood clotting property of SF is improved with the help of heparin and ferulic acid [49]. Our results indicate that the blood clotting ability of the SF-PVA-ZnO composite film is enhanced by the integration of ZnO NPs because ZnONPs denatured anticoagulant proteins and affected the intrinsic pathway of blood coagulation by producing a short clotting time [50]. SF-PVA composite film shows poor blood clotting ability when it is used as a blood-contact.

5.3. Cell Viability Study of Composite Films

5.3.1. Selection of Cell Type

The toxicity of nanomaterial's is based on the organ and the type of cell lines, which is owing to the difference in cell physiology (e.g., epithelial or lymphoid), the proliferation of the cell, membrane characteristics and phagocyte characteristics among different cell types [51]. The important factor in cytotoxicity assay is the selection of suitable cell lines. In addition, it is also needed to give priority to the importance of optimizing the numbers and conditions of cells. In order to the exact effects of nanomaterials on the cells of interest, the cytotoxicity assay must include cells that represent the exposure route and organs targeted by NPs. Among the commonly used cell lines, fibroblast cells are conducive for the trial. In addition, cancer cells are also a better choice than normal cells owing enhanced rate of growth, metabolic activity and easy to use [52]. In the present study, L929 mouse fibroblast cell lines have selected for confirmation of biocompatibility study of the prepared SF-PVA, ZnO NPs embedded SF-PVA composite films. The cell lines are commonly used to study the biocompatibility of materials.

5.3.2. Selection of Cytotoxicity Assay

There are different assays can be employed to study the toxicity of nanomaterials on cell cultures. There are lactate dehydrogenase (LDH), leakage, 3-(4, 5-dimethylthiazol-2-yl) 2, 5-diphen-yltetrazolium bromide (MTT) assay, Dye exclusion assay (Trypan blue) including recognition of cytokine/chemokine production etc [53]. Choice of accurate toxicity assay is a key to the accurate assessment of nanomaterials toxicity. The NPs can absorb dyes, be redox active and it is significant that the cytotoxicity study is suitable. Among different assays, MTT assay is commonly used and easy to use. Therefore, in the present investigation, the cytotoxicity has been evaluated by means of MTT assay to minimize the errors and to obtain the quality data. The assay is simple cheap and easy to evaluate. Figure shows MTT assay of SF-PVA and SF-PVA-ZnO NPs composite films. Figure 17 Shows Cytotoxicity of SF-PVA and SF-PVA/ZnO composite films. The perfect wound dressing should nontoxic and biocompatible. The cytotoxicity of prepared SF-PVA and SF-PVA-ZnO composite films under in-vitro conditions in L929 cell lines are examined in terms of their viabilities using the MTT assay.SF-PVA composite films are taken as the control. It is commented that the cell viability of SF-PVA/ZnO composite films slightly increased after 48 h and after 72 h than 24 h of incubation time. SF-PVA composite films is 100% for 24 h and 105% for 48 h and 107% for 72 h of incubation time. However, the cell viability of SF-PVA-ZnO composite films is 72% for 24 h and 76% for 48 h and 81% for 72 h of incubation time. From these results, it can be concluded that the prepared composite films are nontoxic in nature. Such as Zn+ ions. These ions form the acidic condition, which damage the cytoskeleton of the cell. Reactive oxygen species include superoxide anion (O_2-), hydrogen peroxide (H_2O_2), hydrogen radical (HO•) and singlet oxygen ($1O_2$), these are more reactive compared to oxygen (O_2) itself. This ROS production causes big cellular damage through induction of genotoxicity, inflammation, and disruption of biomolecules like DNA, proteins.

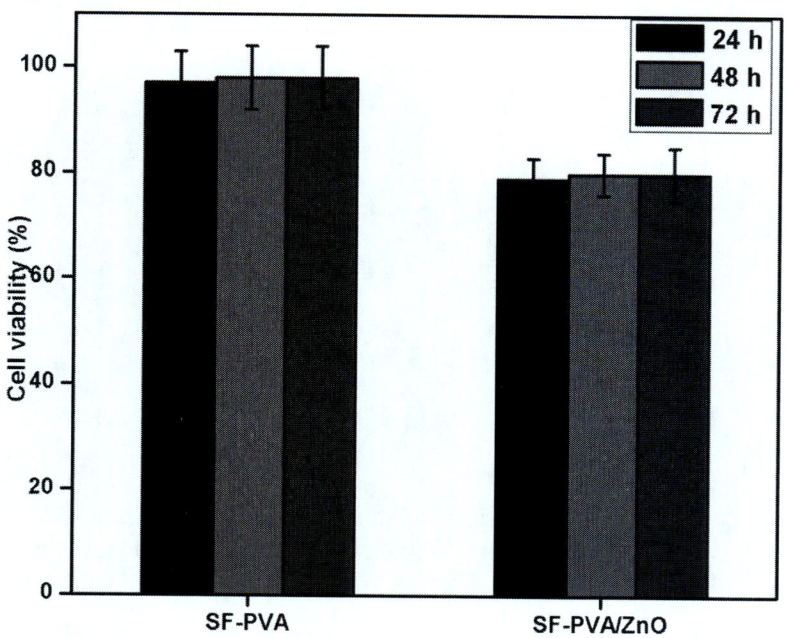

Figure 17. Cytotoxicity of SF-PVA and SF-PVA/ZnO composite films.

This process causes the apoptosis and damages the cells. Physicochemical characteristics of the NPs involve the toxicity of NPs. It includes the size, shape and surface charge of NPs. The size of the NPs decreases with an increase in the surface area and permits a greater potential of its atoms or molecules to be displayed either on the surface or the interior of the materials [56]. Surfaces of NPs should also be created to minimize protein adsorption leading to a well-controlled and reproducible protein corona of certain composition and thickness. The size of NPs has also involved the toxicity of NPs. The spherical shaped NPs are suitable for endocytosis than the rod-shaped or fiber-like NPs. The surface charge of NPs promotes the adsorption of ions and biomolecules, which may change organism or cellular responses toward particles. As compared to anionic particles, cationic particles have been stable and exert the greatest cytotoxic effects [57]. Toxic effects of NPs can be controlled by mixing the ZnO NPs with other biocompatible and biodegradable materials [58].

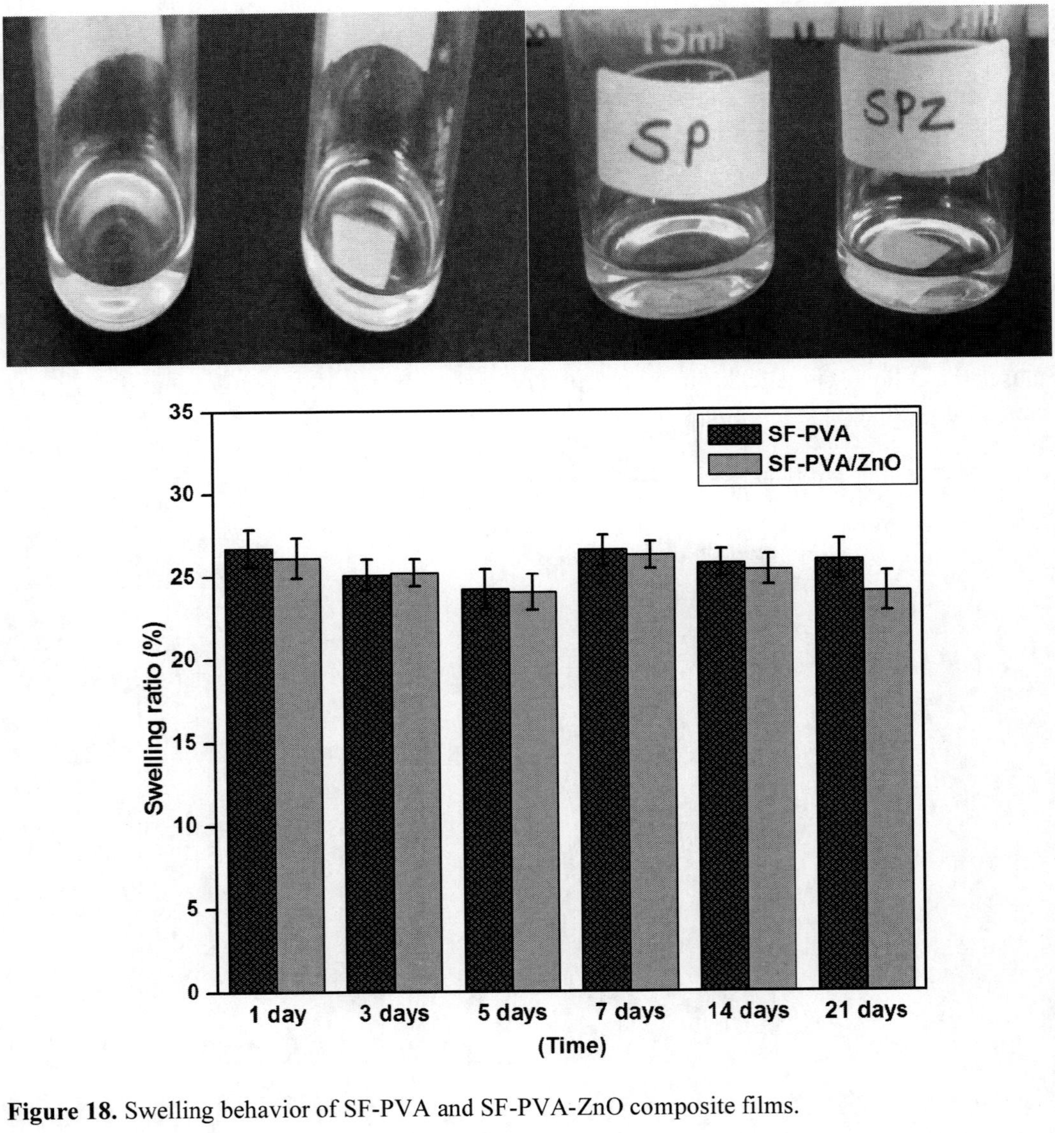

Figure 18. Swelling behavior of SF-PVA and SF-PVA-ZnO composite films.

5.3.3. Mechanism of Cytotoxicity of Composite Films

The properties of metal oxide materials especially Nano size make them desirable for nano-bio applications, but it also leads them to have some disadvantages like Nano size related toxicity [54]. Nanocomposite films receive the nanosized particles and the use of these NPs is very important to investigate their potential toxic effects. The toxicity of NPs on human and environment is a topic that has received increasing interest. There are different mechanisms can cause NPs toxicity. Nanotoxicity of ZnO NPs related to the following mechanisms [55], NPs release of the corresponding toxic ions. NPs produced excess reactive oxygen species. NPs toxicity will base on the physical-chemical characteristics of the NPs corresponding to size, shape, surface charge, chemical stability of the core legends etc. ZnO NPs released toxic ions of dressing up to 1 week.

5.4. Swelling Behavior of Composite Films

Figure 18 represents the swelling ratios of SF-PVA and SF-PVA-ZnO composite films. The swelling ratio of SF-PVA composite film shows slightly higher as compared to SF-PVA-ZnO composite film. The presence of ZnO NPs slightly reduced the swelling ratio due to the composite film contains ZnO NPs have lower porosity compared to SF-PVA composite film. SF and PVA both materials are hydrophilic in nature. Water uptake and retention capacity of SF and PVA are higher compared to ZnO NPs due to slightly reduce the swelling ratio [59]. Additionally, this might be the interaction of SF, PVA between ZnO NPs. The swelling ratios of composite films were increased after 7 days of incubation than 1 day. Water uptake capacity of the composite films was decreased after 14 days and started degradation of composite films. Therefore, the swelling behavior data concluded that the composite film was capable of absorbing a large volume of liquid in which it was immersed and there was no change.

Conclusion

In this study, SF-PVA-ZnO composite film was successfully prepared by casting method at room temperature for effective antibacterial activity, mechanical activity, and enhanced blood clotting ability. The physical-chemical characterizations of ZnO NPs and prepared composite films were done by XRD, FT-IR and UV-visible spectroscopy.SF-PVA-ZnOcomposite film showed higher tensile strength and % elongation, appropriate swelling behavior which was helpful for absorbing large volumes of wound exudates. Biocompatibility is one of the most important issues that have to be addressed before the materials can be used for in vivo applications, especially for materials required to contact with blood. In hemocompatibility the prcentage of hemolysis less than 2% is considered as nonhemolytic, 2-5% is slightly hemolytic and more than 5% is hemolytic as per ASTMF-756-08. The results indicated that the hemolytic percentage of SF-PVA-ZnOcomposite film was higher as compared to SF-PVA composite film. The induced hemolysis of SF-PVA composite film was 2.1% and 2% for 2 h and 20 h of incubation time respectively. However, hemolysis for SF-PVA-ZnO composite film was found to be 3.4% at 2 h and 2.7% for 20 h of incubation time. The hemolysis percentage of both the films showed below 5%, indicating that both the films showed slightly hemolytic material. The obtained results observed the hemolysis percentage was decreased with increasing incubation period of up to 20 h. Blood clotting ability of SF-PVA-ZnOcomposite film represent increased blood clotting ability than SF-PVA composite film. Antibacterial activity of SF-PVA-ZnOcomposite film shows higher inhibition zone against *S. aureus, P. aeruginosa, P. mirabilis* and *S. pyogenes* as compared to *E.coli* bacteria.*In vitro* cytotoxicity studies prove the nontoxic nature of prepared SF-PVA-ZnO composite film. All these. data reveals the prepared composite film has potential application for an infected wound.

References

[1] Ong, S.-Y., Wu, J., Moochhala, S. M., Tan, M.-H., & Lu, J. "Development of a chitosan-based wound dressing with improved hemostatic and antimicrobial properties." *Biomaterials* 29.32 (2008): 4323-4332.

[2] Vowden, Kathryn, and Peter Vowden. "Wound dressings: principles and practice." *Surgery (Oxford)* 35.9 (2017): 489-494.

[3] Shah, Dana. "Infection Control and Safety Issues in the Clinic." *Early's Physical Dysfunction Practice Skills for the Occupational Therapy Assistant E-Book* (2021): 27.

[4] Weigelt, M. A., McNamara, S. A., Sanchez, D., Hirt, P. A., & Kirsner, R. S. "Evidence-based review of antibiofilm agents for wound care." *Advances in Wound Care* 10.1 (2021): 13-23.

[5] Lehmann, T., Vaughn, A. E., Seal, S., Liechty, K. W., & Zgheib, C. "Silk Fibroin-Based Therapeutics for Impaired Wound Healing." *Pharmaceutics* 14.3 (2022): 651.

[6] Byram, Prasanna Kumar, Lopamudra Das, Krishna Chaitanya S., Gaurav Kulkarni, Santanu Dhara, Nishant Chakravorty. "Silk Fibroin-Based Biomaterials in Biomedical Applications." *Functional Biomaterials: Drug Delivery and Biomedical Applications* (2022): 203.

[7] Wang, Ke, Qian Ma, Hong-Tao Zhou, Ju-Mei Zhao, Miao Cao, Shu-Dong Wang. "Review on Fabrication and Application of Regenerated Silk Fibroin Materials." *Autex Research Journal* (2022).

[8] Wang, Hai-Yan, Su-Fang Zhou, Meng Zhang, Heng-Da Wang, Yu-Qing Zhang. "The post-processing temperature or humidity can importantly control the secondary structure and characteristics of silk fibroin films." *Journal of Biomedical Materials Research Part A* 110.4 (2022): 827-837.

[9] Cheng, Baochang, Zhouyue Lei, and Peiyi Wu. "Bio-derived crystalline silk nanosheets for versatile macroscopic assemblies." *Nano Research* (2022): 1-7.

[10] Wongkrongsak, S., Pangon, A., Pongsak, N., Piroonpan, T., and Pasanphan, W. "Strengthened Silk-Fibroin/Poly (ethylene oxide) Nonwoven Nanofibers: A Dual Green Process Using Pure Water for Electrospinning and Electron Beam-Assisted Cross-Linking." *ACS Sustainable Chemistry & Engineering* 10.8 (2022): 2653-2672.

[11] Lehmann, T., Vaughn, A. E., Seal, S., Liechty, K. W., and Zgheib, C. "Silk Fibroin-Based Therapeutics for Impaired Wound Healing." *Pharmaceutics* 14.3 (2022): 651.

[12] Ellis, Samantha, Elaine J. Lin, and Danielle Tartar. "Immunology of wound healing." *Current Dermatology Reports* 7.4 (2018): 350-358.

[13] Rousselle, Patricia, Marine Montmasson, and Cécile Garnier. "Extracellular matrix contribution to skin wound re-epithelialization." *Matrix Biology* 75 (2019): 12-26.

[14] Balan, Poornima, Anbuthiruselvan Solaimuthu, Padmaja Murali, Aishwarya Satish, Purna Sai Korrapati. "Bioinspired Nanofibers: State of Art in Tissue Regeneration." *Handbook of Research on Nano-Drug Delivery and Tissue Engineering: Guide to Strengthening Healthcare Systems* (2022): 429-487.

[15] Vidya, M., and SenthilkumarRajagopal. "Silk fibroin: a promising tool for wound healing and skin regeneration." *International Journal of Polymer Science* 2021 (2021).

[16] Patil, Priyanka P., Michaela R. Reagan, and Raghvendra A. Bohara. "Silk fibroin and silk-based biomaterial derivatives for ideal wound dressings." *International Journal of Biological Macromolecules* 164 (2020): 4613-4627.

[17] Sun, Dandan, Jiguang Chen, Yongming Fu, Jie Ma. "A real-time response relative humidity sensor based on a loop microfiber coated with polyvinyl alcohol film." *Measurement* 187 (2022): 110359.

[18] Carlos Alberto Gómez-Aldapa, Gonzalo Velazquez, Miguel C. Gutierrez, Esmeralda Rangel-Vargas, Javier Castro-Rosas, Rocio Yaneli Aguirre-Loredo. "Effect of polyvinyl alcohol on the physicochemical properties of biodegradable starch films." *Materials Chemistry and Physics* 239 (2020): 122027.

[19] Srdjan Vidovic, Jasmina Stojkovska, Milan Stevanovic, Bojana Balanc, Maja Vukasinovic-Sekulic, Aleksandar Marinkovic and Bojana Obradovic. "Effects of poly (vinyl alcohol) blending with Ag/alginate solutions to form nanocompositefibres for potential use as antibacterial wound dressings." *Royal Society Open Science* 9.3 (2022): 211517.

[20] Patil, Priyanka, and Shivaji H. Pawar. "Wound dressing applications of nano-biofilms." *Biopolymer-Based Nano Films*. Elsevier, 2021. 247-268.

[21] Liang, Yongping, Jiahui He, and BaolinGuo. "Functional hydrogels as wound dressing to enhance wound healing." *ACS nano* 15.8 (2021): 12687-12722.

[22] Richard N. Puzzitiello, Mariano E. Menendez, Michael A. Moverman, Nicholas R. Pagani, Matthew J. Salzler, Grant E. Garrigues. "Silver-impregnated occlusive dressings are a cost-effective strategy for preventing infection after total shoulder arthroplasty." *Seminars in Arthroplasty: JSES*. Vol. 31. No. 3. WB Saunders, 2021.

[23] Valentin Brumberg, Tatiana Astrelina, Tatiana Malivanova, Alexander Samoilov. "Modern Wound Dressings: Hydrogel Dressings." *Biomedicines* 9.9 (2021): 1235.

[24] Priyanka P. Patil, Jagruti V. Meshram, Raghvendra A. Bohara, Shivdas G. Nanaware and Shivaji H. Pawar. "ZnO nanoparticle-embedded silk fibroin–polyvinyl alcohol composite film: a potential dressing material for infected wounds." *New Journal of Chemistry* 42.17 (2018): 14620-14629.

[25] Meshram, J. V., V. B. Koli, S. G. Kumbhar, L. C. Borde, M. R. Phadatare and S. H. Pawar. "Structural, spectroscopic and anti-microbial inspection of PEG capped ZnO nanoparticles for biomedical applications." *Materials Research Express* 5.4 (2018): 045016.

[26] Wojnarowicz, J., Chudoba, T., Koltsov, I., Gierlotka, S., Dworakowska, S., and Lojkowski, W. "Size control mechanism of ZnO nanoparticles obtained in microwave solvothermal synthesis." *Nano-technology* 29.6 (2018): 065601.

[27] Giorgio Rizzo, Marco Lo Presti, Cinzia Giannini, Teresa Sibillano, Antonella Milella, Giulia Guidetti, Roberta Musio, Fiorenzo G. Omenetto and Gianluca M. Farinola. "Bombyx mori Silk Fibroin Regeneration in Solution of Lanthanide Ions: A Systematic Investigation." *Frontiers in bioengineering and biotechnology* 9 (2021): 466.

[28] Stefan, Nicolaie, FloraliceMarimonaMiroiu, and Gabriel Socol. "Degradable silk fibroin–poly (sebacic acid) diacetoxy terminated, (SF-PSADT) polymeric composite coatings for biodegradable medical applications deposited by laser technology." *Progress in Organic Coatings* 134 (2019): 11-21.

[29] Chang, Y.-C., Hsu, C.-C., Wu, S.-H., Chuang, K.-W., and Chen, Y.-F. "Fabrication of Cu-doped ZnOnanoneedles on different substrate via wet chemical approach: Structural characterization and photocatalytic performance." *Applied Surface Science* 447 (2018): 213-221.

[30] Xiong, G., Pal, U., Serrano, J. G., Ucer, K. B., & Williams, R. T. "Photoluminesence and FTIR study of ZnO nanoparticles: the impurity and defect perspective." *Physica Status Solidi C* 3.10 (2006): 3577-3581.

[31] Mohandas, A., Deepthi, S., Biswas, R., and Jayakumar, R. "Chitosan based metallic nanocomposite scaffolds as antimicrobial wound dressings." *Bioactive Materials* 3.3 (2018): 267-277.

[32] Díez-Pascual, Ana María. "Antibacterial activity of nanomaterials." *Nanomaterials* 8.6 (2018): 359.

[33] Al-Ouqaili, Mushtak TS. "Biofilm antimicrobial susceptibility pattern for selected antimicrobial agents against planktonic and sessile cells of clinical isolates of staphylococci using MICs, BICs and MBECs." *Asian Journal of Pharmaceutics (AJP): Free full text articles from Asian J Pharm* 12.04 (2018).

[34] Basova, T. V., Vikulova, E. S., Dorovskikh, S. I., Hassan, A., and Morozova, N. B. "The use of noble metal coatings and nanoparticles for the modification of medical implant materials." *Materials & Design* 204 (2021): 109672.

[35] Sacoto-Figueroa, F. K., Bello-Toledo, H. M., González-Rocha, G. E., Luengo Machuca, L., Lima, C. A., Meléndrez-Castro, M., and Sánchez-Sanhueza, G. A. "Molecular characterization and antibacterial

activity of oral antibiotics and copper nanoparticles against endodontic pathogens commonly related to health care-associated infections." *Clinical Oral Investigations* 25.12 (2021): 6729-6741.

[36] Archana, P., B. Janarthanan, S. Bhuvana, P. Rajiv, S. Sharmila. "Concert of Zinc Oxide Nanoparticles Synthesized using Cucumismelo by Green Synthesis and the Antibacterial Activity on Pathogenic Bacteria." *Inorganic Chemistry Communications* (2022): 109255.

[37] Ihssen, Julian, Nina Jovanovic, Teja Sirec and Urs Spitz. "Real-time monitoring of extracellular ATP in bacterial cultures using thermostable luciferase." *Plos one* 16.1 (2021): e0244200.

[38] Hernández-Díaz, José A., Jorge J. O. Garza-García, Adalberto Zamudio-Ojeda, Janet M. León-Morales, Julio C. López-Velázquez and Soledad García-Morales. "Plant-mediated synthesis of nanoparticles and their antimicrobial activity against phytopathogens." *Journal of the Science of Food and Agriculture* 101.4 (2021): 1270-1287.

[39] Shipkova, Maria, and Hedi Jamoussi. "Therapeutic drug monitoring of antibiotic drugs: the role of the clinical laboratory." *Therapeutic Drug Monitoring* 44.1 (2022): 32-49.

[40] Crawford, Lars, Meghan Wyatt, James Bryers, Buddy Ratner. "Biocompatibility Evolves: Phenomenology to Toxicology to Regeneration." *Advanced Healthcare Materials* 10.11 (2021): 2002153.

[41] Zhao, Lianjia, Lili Wang, Yiqiang Zheng, Shufang Zhao, Wei Wei, Dawei Zhang, Xiyao Fu, Kai Jiang, Guozhen Shen, Wei Han. "Highly-stable polymer-crosslinked 2D MXene-based flexible biocompatible electronic skins for in vivo biomonitoring." *Nano Energy* 84 (2021): 105921.

[42] Bagheri, Mitra, Majid Validi, Abolfazl Gholipour, Pooyan Makvandi, Esmaeel Sharifi. "Chitosan nanofiberbiocomposites for potential wound healing applications: Antioxidant activity with synergic antibacterial effect." *Bioengineering & Translational Medicine* 7.1 (2022): e10254.

[43] Feng, Yue, Xiang Luo, Fan Wu, Hongzhong Liu, Enyu Liang, Rong-Rong He, Mingxian Liu. "Systematic studies on blood coagulation mechanisms of halloysite nanotubes-coated PET dressing as superior topical hemostatic agent." *Chemical Engineering Journal* 428 (2022): 132049.

[44] Chandra, Girish, and Ajay Pandey. "Design approaches and challenges for biodegradable bone implants: a review." *Expert Review of Medical Devices* 18.7 (2021): 629-647.

[45] Lu, Rick Xing Ze, and MilicaRadisic. "Organ-on-a-chip platforms for evaluation of environmental nanoparticle toxicity." *Bioactive Materials* 6.9 (2021): 2801-2819.

[46] Chong Cheen Ong, Subash C. B. Gopinath, Leong Wei Xian Rebecca, Veeradasan Perumal, Thangavel Lakshmipriya, Mohamed Shuaib Mohamed Saheed. "Diagnosing human blood clotting deficiency." *International Journal of Biological Macromolecules* 116 (2018): 765-773.

[47] Shefa, Anha Afrin, Mirana Taz, Monir Hossain, Yong Sik Kim, Sun Young Lee, Byong-Taek Lee. "Investigation of efficiency of a novel, zinc oxide loaded TEMPO-oxidized cellulose nanofiber based hemostat for topical bleeding." *International Journal of Biological Macromolecules* 126 (2019): 786-795.

[48] Shi, Pange, Ling Zhang, Wei Tian, Helei Li, Qiongyu Wang, Honggen Yi, Yin Yin, Aiqing Wang, Ping Ning, Fenglin Dong. "Preparation and anticoagulant activity of functionalised silk fibroin." *Chemical Engineering Science* 199 (2019): 240-248.

[49] Heichel, Danielle L., Ngoc Chau H. Vy, Shawn P. Ward, Douglas H. Adamson and Kelly A. Burke. "Controlled radical polymerization of hydrophilic and zwitterionic brush-like polymers from silk fibroin surfaces." *Journal of Materials Chemistry B* 8.45 (2020): 10392-10406.

[50] Vijayakumar, Sekar, Baskaralingam Vaseeharan, Raja Sudhakaran, Jeyaraman Jeyakandan, Palaniappan Ramasamy, Avinash Sonawane, Avinash Padhi, Palaniyandi Velusamy, Periasamy Anbu and Caterina Faggio. "Bioinspired zinc oxide nanoparticles using Lycopersiconesculentum for antimicrobial and anticancer applications." *Journal of Cluster Science* 30.6 (2019): 1465-1479.

[51] Pechiappan, Harini. *Effects of Nano-and Microplastics on Inflammatory Responses in Macrophages in vitro*. MS thesis. NTNU, 2021.

[52] De, Shreemoyee, Akshay Joshi, Dinesh M. Tripathi, Savneet Kaur, Neetu Singh. "Alginate based 3D micro-scaffolds mimicking tumor architecture as in vitro cell culture platform." *Materials Science and Engineering: C* 128 (2021): 112344.

[53] Aslantürk, Özlem Sultan. "In vitro cytotoxicity and cell viability assays: principles, advantages, and disadvantages." *Genotoxicity-A Predictable Risk to Our Actual World* 2 (2018): 64-80.

[54] Lyu, Qinghua, Ling Peng, Xiangqian Hong, Taojian Fan, Jingying Li, Yubo Cui, Han Zhang, Jun Zhao. "Smart nano-micro platforms for ophthalmological applications: The state-of-the-art and future perspectives." *Biomaterials* 270 (2021): 120682.

[55] Sharma, Shubham, Swarna Jaiswal, Brendan Duffy and Amit K. Jaiswal. "Nanostructured materials for food applications: spectroscopy, microscopy and physical properties." *Bioengineering* 6.1 (2019): 26.

[56] Imtiaz, Yusha, Beza Tuga, Christopher W. Smith, Alexander Rabideau, Long Nguyen, Yali Liu, Sabahudin Hrapovic, Karina Ckless, Rajesh Sunasee. "Synthesis and cytotoxicity studies of wood-based cationic cellulose nanocrystals as potential immunomodulators." *Nanomaterials* 10.8 (2020): 1603.

[57] Farooq, Amjad, Mohammed Kayes, Patoary, Meiling Zhang, Hassan Mussana, Mengmeng Li, Muhammad Awais Naeem, Muhammad Mushtaq Aamir Farooq, Lifang Liu. "Cellulose from sources to nanocellulose and an overview of synthesis and properties of nanocellulose/zinc oxide nanocomposite materials." *International Journal of Biological Macromolecules* 154 (2020): 1050-1073.

[58] Farooq, Amjad, Mohammed Kayes Patoary, Meiling Zhang, Hassan Mussana, Mengmeng Li, Muhammad Awais Naeem, Muhammad Mushtaq, Aamir Farooq, Lifang Liu. "Cellulose from sources to nanocellulose and an overview of synthesis and properties of nanocellulose/zinc oxide nanocomposite materials." *International Journal of Biological Macromolecules* 154 (2020): 1050-1073.

[59] Agarwal, Yashi, P. S. Rajinikanth, Shivendu Ranjan, Upama Tiwari, J. Balasubramnaiam, Prashant Pandey, Dilip Kumar Arya, Sneha Anand, Payal Deepak. "Curcumin loaded polycaprolactone-/polyvinyl alcohol-silk fibroin based electrospunnanofibrous mat for rapid healing of diabetic wound: An in-vitro and in-vivo studies." *International Journal of Biological Macromolecules* 176 (2021): 376-386.

Chapter 13

Engineering of Silk Fibroin Bio-Nanocomposites: A Glimpse of Advanced Research on Innovative Formulations for Potential Applications

Shivaji H. Pawar*
Center for Interdisciplinary Research,
D. Y. Patil University, Kolhapur, Maharashtra, India
Centre for Innovative and Applied Research (CIAR),
T. C. College, Baramati, Maharashtra, India

Abstract

Silk Fibroin is widely used as a green biomaterial in various fields of research like textiles, biomedical engineering biotechnology, electronics, photonics and energy research. This is because SF can be reconstituted in numerous forms by physical and chemical processes in numbers of studies have attempted to incorporate addition to its unique functional aspects that can be incorporated into SF while maintaining its beneficial natural characters. This new area of biotechnology with bio-nanocomposites is the result of breakthroughs in nanoscience and nanotechnology. SF bio-nanocomposites and their innovative applications for the use of these SF bio-nanocomposites materials have been developed in recent years and these are documented in previous chapters of this book.

In this chapter, we report on the advanced research of the engineering of silk fibroin bio-nanocomposites suitable for emerging technologies. Though the formulation of silk fibroin is a natural process, carried out with silkworms, it can be modified with the mulberry leaves which are the silkworm feed. Further, silk fibroin can be changed by doping rare earth elements or by incorporating their nanoparticles at different stages of its formulations. Thus, the properties of silk fibroin are engineered suitably to meet the requirements of various devices with different methods reported recently in the literature. Lastly, the hypothetical applications of silk fibroin in protecting healthcare buildings (hospitals) from pathogenic infections specifically with photocatalytic disinfection of pathogens have been reported in this chapter. This innovative emerging potential application of silk fibroin seems to be an attractive solution to control the spread of communicable diseases like COVID-19.

The chapter ends with a report on a recent method based on microwave applications in which formulation of time SF bio-nanocomposites. This modification is reduced synthesis time from 52 hours to 4 hours. This alteration is predicted as a significant step towards commercialization of formulation of SF bio-nanocomposites technologies newly developed in recent years.

Keywords: silk fibroin, bio-nanocomposites, photocatalytic paint, electrospun deposition

---

* Corresponding Author's Email: shpawar1946@gmail .com.

1. Introduction

Silk is found in nature that has been utilized primarily on fashion fabrics and surgical sutures for a pretty long time. It is generated predominantly from Bombyxmori and consists of two macromolecular proteins namely fibroin and sericin. In recent years, silk fibroin (SF) has been reconstituted in numerous forms by physical and chemical processes and it has been widely used as a possible biomaterial in a variety of biomedical and biotechnological disciplines. Furthermore, despite the direct use of SF with intrinsic structure and properties, several studies have attempted to introduce additional innovative functional qualities into SF while maintaining its favourable natural traits. In the last few years, silk fibroin has been used in a wide range of medical diagnostic disciplines modifying SF with diverse ways like genetic manipulation and dye feeding. Drug delivery, monitoring surgical and wound healing processes, electronics, photonics, optoelectronics, energy generation and storage, and cryogenics are just a few sectors where these functionalized silks might be used. The first chapter of this book reports on the natural resources and the fundamental physicochemical properties of Silk Fibroin-based materials, In contrast, this chapter highlights the innovative and advanced techniques of engineering of SF bio-nanocomposites. The shortening of formulation time by modifying the process with incorporation of microwaves is a big step toward commercialization.

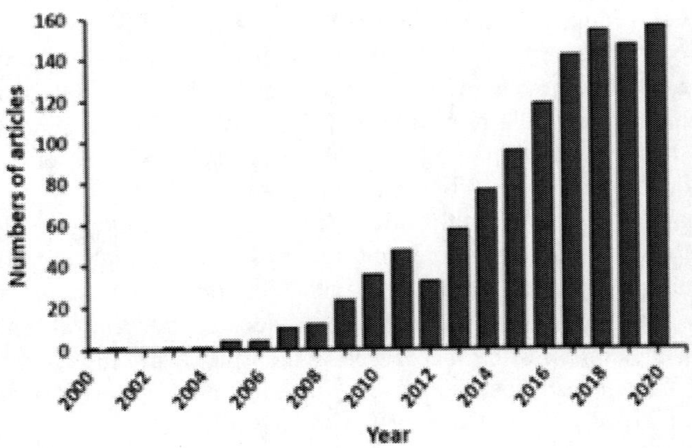

Figure 1. Annual publication on silk fabrion in tissue regeneration from 2000-2020. The search engine was Web of Science. Silk fibroin and tissue regeneration were used for searching [2].

Nowadays switching to more ecologically friendly SF formulation processing procedures have gained appeal in materials science research. This raises awareness of the negative impact of more environmentally harmful approaches. When it comes to the top ten global concerns for the next fifty years, health care is one of the most pressing issues that we have faced in the last two years while we battled the COVID-19 pandemic situation. The author and his colleagues were actively engaged in pathogenic disinfection research and a variety of anti-pandemic medical research activities. Silk Fibroin research was one of the most attractive materials science research in these years. As a result, recently in 2021, Pugno et al. have brought out a special edition titled “Frontiers in Silk Science and Technology” [1]. However, more research is needed to discover the best ways for to exploit SF’s potential in a variety of creative applications outlined in previous chapters as an environmentally benign biomaterial with the most negligible environmental impact. Silk from spiders and silkworms has emerged as useful material for tackling new materials science-based technologies. As a result, in the last two decades, there has been a significant increase in research interest globally. Year after year, a growing number of research organizations have become involved, releasing many research publications [2]. The statistical survey of annual research publications carried out from 2000 is shown in Figure 1 which depicts the rising trend in yearly publications.

Silk fibroin-based materials are found superior to many other polymers, particularly in terms of their mechanical strength, enabling the fabrication of composites with remarkable biological properties. To achieve breakthroughs in silk utility for materials science and technology, researchers must look into raw silks and silk-based natural and artificial composites while controlling and changing processing conditions. From the formulation stage of silk and its derivatives, an interdisciplinary, multiscale approach is required. As a result, the primary goal of this book is to create awareness amongst the researchers by providing literature in collective form through the chapters of a book entitled "Silk Fibroin: Advances in Applications and Research." In support of this approach an update to the knowledge of silk fibroin materials science knowledge with developments in nanoscience and nanotechnology is an essential component. Thus, combination of silk fibroin with nanoparticles leading to bio-nanocomposites has been attempted in recent years. Therefore, this chapter has the following objectives:

1) To describe the creation of repeatable bio-nanocomposite formulation processes that allow for precise control of shape, size, and surface chemistry, as well as the production of long-lasting bio-nanocomposites,
2) To create numerous silk-fibroin-based bio-nanocomposites using a transposable approach,
3) To provide a consistent understanding of the effect of NPs on the structure of silk fibroin, as well as the changes that occur when they are mixed in the same dispersion., and
4) To reveal a proof-of-concept application for various silk-based bio-nanocomposites with their unique properties and a wish to incorporate by incorporating suitable such types of NPs.

2. Bio-Nanocomposites

The information about inorganic NPs has reported in the literature on one hand and, on the other hand, demonstrates the promising potential of combining silk and NPs for the design of bio-nanocomposites with tailored properties and functions on the other hand. As can be seen from the book's contents, chapter 1 provides a detailed description of silk fibroin resources and physicochemical properties of silk fibroin-based materials. The next series of chapters provide information about the variety of potential applications of SF-based materials like bio-nanocomposites. Here, in this chapter, we discuss more concepts and engineering of silk fibroin bio-nanocomposites by different methods of their formulations.

2.1. Concept of Bio-Nanocomposites

Composite materials allow for the combining of qualities from multiple materials as well as the emergence of new ones. Composites are particularly intriguing in the realm of materials because of these gained features. With a smaller amount of the original material, the composites can achieve excellently and outperform bulk material qualities. In many circumstances, it helps to lowers the cost of production. Traditional composites, on the other hand, that rely on the reinforcement of a given material's qualities through the use of others, may result in the loss of critical properties [3]. Nanocomposites are made up of various materials, at least one of which is the nanoscale.

On the other hand, the consequent properties that NPs bring are only advantageous if they are uniformly distributed throughout the final substance. Therefore, it is crucial to take the NPs' surface chemistry, bulk material stabilization and homogeneous dispersion into account while building any type of nanocomposite. As already said, NPs can be quickly stabilized by changing their environment, such as ion concentration. It is unsurprising that NPs aggregate and precipitate inside the substance because combining NPs with another material generates a novel environment. This understanding of nanocomposites also applies to bio-nanocomposites, which have biomaterials as one of their main constituents. Depending on the nature of biomaterials and nanoparticles added therein, a variety bio-nanocomposites has been attempted for their numerous applications in biomedical engineering, tissue engineering electronics, photonics, energy generations and storage of electrical energy. These varieties of bio-nanocomposites are shown in Figure 2.

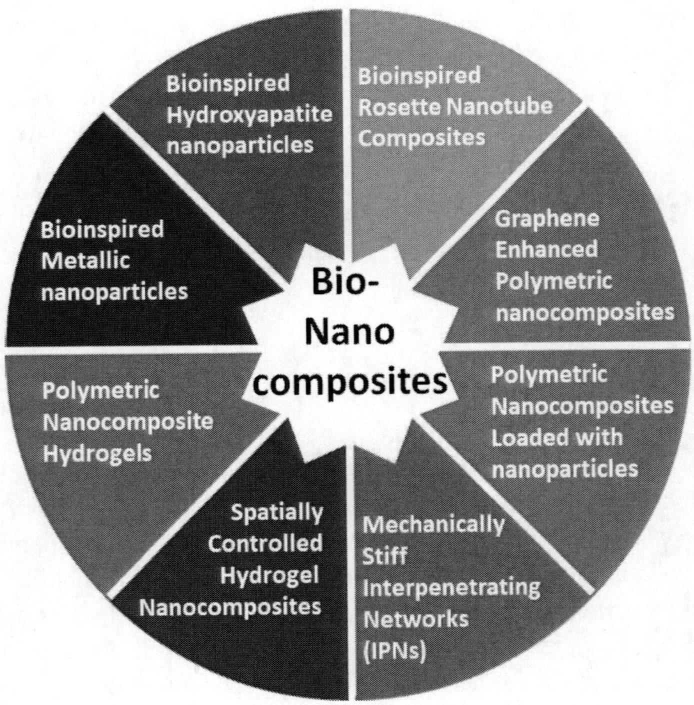

Figure 2. Different types of bio-nanocomposites.

2.2. Properties of Silk Fibroin Bio-Nanocomposites

Amongst the variety of bio-nanocomposites the main focus of our book is related to silk fibroin as the basic prime biomaterial and formulation of its bio-nanocomposites for various applications as mentioned above. It is well known that the properties of bio-nanocomposites are different from that of their original constituents and can be suitably engineered as per the requirements of the applications. Consider a typical example of SF/TiO_2 bio-nanocomposites. The physical structures of these bio-nanocomposites and their constituents are studied with X-ray diffraction (XRD) techniques. Figure 3 shows the XRD patterns of SF, SF/TiO_2, and TiO_2 samples. From the figure, it may be noted that, for SF, no representative peaks were found. SF/TiO_2 samples, on the other hand, show intense (Int/100 > 10) diffraction peaks at 25.3, 37.8, 48.0, 53.9, 55.1, 62.8, which correspond to the 101, 004, 200, 105, 211, and 204 planes, all of which are anatase features (ICDD PDF2 database Set-le 21-1272). The higher peak (Int/100 > 10) in the rutile phase corresponds to 2q of 27.4 (110 planes), a typical rutile peak (ICDD PDF2 database Set-le 21-1276). Although the surface covering appears to be comprehensive, there was a drop in overall intensity for SF/TiO_2 compared to TiO_2, implying reduced crystallinity.

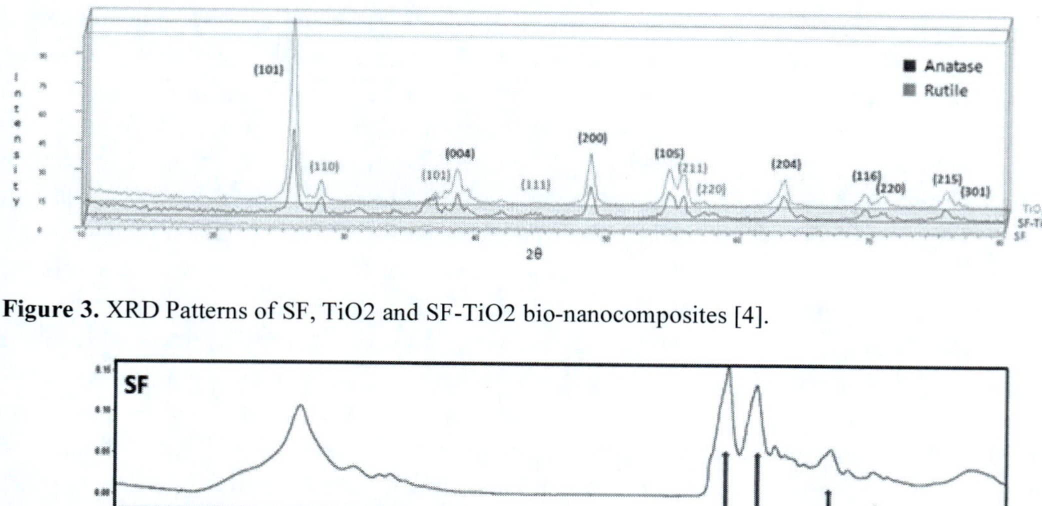

Figure 3. XRD Patterns of SF, TiO2 and SF-TiO2 bio-nanocomposites [4].

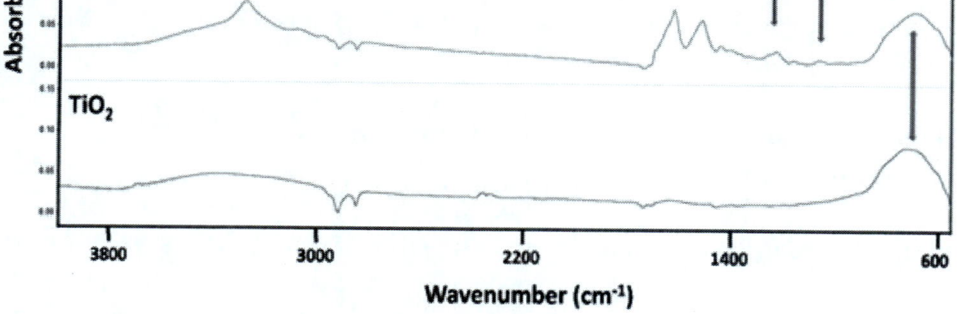

Figure 4. Infrared Spectra of pure SF, TiO2 and SF-TiO2 bio-nanocomposites [4]

Similar to the physical properties, the chemical properties of bio-nanocomposites are also different from that of their original constituents. The chemical properties of nanocomposites are studied by recording their infrared absorption spectra. The infrared absorption spectra recorded for TiO_2 nanoparticles, silk fibroin biomaterial and SF-TiO_2 bio-nanocomposites are recorded in the range from 3800 to 600 wave numbers and these are shown in Figure 4. From the figure it may be noted that the chemical properties of bio-nanocomposites and their constituents are dominated by biomaterial, as a result of this SF than nanoparticles.

3. Engineering of SF-Based Bio-Nanocomposites: A Glimpse of Advanced Research

Bio-Nanocomposites can be made using different variety of methods, consisting of mechanical mixing or in-situ synthesis of NPs without or with cross linking agents [5]. Three alternative uses for silk fibroin materials were discovered in the literature. A variety of reducing agents have been used to study *in situ* synthesis. Some studies suggest that silk can even reduce metal ions on its own [6-8]. Despite the fact that this method requires fewer stages to manufacture the bio-nanocomposite, the size and form of the NPs produces can be quite dispersed. Additionally, the chemistry of these NPs' surfaces is unclear. Overall, the failure to manage these variables could lead to unpredictable properties and toxicity because the features and toxicity of NPs are driven mainly by their properties. Greater control of NP traits can theoretically be achieved by synthesizing NP qualities upstream and inserting them into silk materials later. But in these circumstances, it is essential to manage the surface chemistry of NPs to keep them suspended. A few times, the direct NP inclusion into the dispersion of the newly formed silk can cause the silk to gel. Finally, other investigations have concentrated on directly administering the targeted NPs to silkworms hoping that they may subsequently be found in the silk cocoons. Despite the method's proven practicality, its application is constrained by its ineffectiveness (induced by NP biodistribution within the silk worm) and the necessity of dispersing NPs across the worm's surroundings [9]. Silk bio-nanocomposites is a brand-new and rapidly evolving field of biotechnology is silk bio-nanocomposites. A summary of the silk-based bio-nanocomposites made by embedding nanomaterials using several novel techniques, such as adding Ln ions to cocoon solutions, feeding silk worms leaves bionanofortified with various dopants, feeding silk worms HPMC-dispersed MoS_2 NPs, and electrospinning solutions made from silk worm cocoons are reported in the following sections.

4. Formulation of SF Bio-Nanocomposites by Ln Ion Addition to Cocoon Solution

Lanthanide-doped fibres are frequently used in the field of photonics. The results of numerous investigations have demonstrated that lanthanides can give fibres of fibres photo chromic, luminescent, fluorescent, and up-conversion properties [10, 11]. Fluorescent Pr^{+3} and Dy^{+3} -doped Chalcogenide fibres (Chahal et al., 2016) [12], Er^{+3}/Yb/Tm^{+3} tri-doped tellurite glass microsphere coupled by tapered fibres (Liu et al., 2020) [13] and fluorescent silks obtained from silkworms fed with rare-earth up converting phosphorus (Liu et al., 2020) are examples of fluorescent materials [14]. Additionally, Ln-doped fibres have been used to create smart Dy^{+3} and Eu^{+3} doped stimuli responsive textiles as well as lasers (amplifiers rare-earth doped silica and fluorozirconate fibres, especially erbium-doped fibres) (Digonnet, 2001) [15]. (Shen et al., 2021) [16]. Additionally, lanthanide ion sensitizers like tryptophan and tyrosine can increase emission (da Rocha et al., 2020) [17]. Fluorescent silk fibroins have just recently been used in photonics. Thus, a viable approach for creating new photonic systems could be to combine the mechanical and optical properties of SF with those of lanthanides ions. This is especially true considering SF's high biocompatibility and biodegradability.

The findings of a systematic investigation by G. Rizzo et al. [18] are reported here. Except for Promethium, which was eliminated due to its radioactivity, the behaviour of degummed SF solution in the presence of all lanthanide ions in the Periodic Table, in the form of chloride salts, from Lanthanum to Lutetium, was explored. The primary goals of their paper were to (i) describe a novel process for solubilizing doped fibroin in a fibrous form in the presence of Ln^{+3} ions, and (ii) characterize the structural and morphological characteristics of the resulting fibrous materials. Degummed SF was made using Bombyxmori cocoons purchased from Tajima Shoji (Japan) using previously reported techniques (Rockwood et al., 2011) [19]. To extract sericin, the cocoons were divided into four pieces, shred them, and then boil them for 30 minutes in an aqueous solution of Na_2CO_3 0.02 M. They were then carefully rinsed in double-distilled water to get rid of any sericin that was still present as well as excessive salt. The fibres were dried at room temperature for 24 hours.

The standard Ajisawa's approach (Ajisawa, 1998) [20] was used to dissolve degummed SF using hydrated lanthanide chlorides ($LaCl_3$). The same molar ratios as in Ajisawa's conventional approach were used, but instead of $CaCl_2$, catalytic chaotropic agents such as $7H_2O$, $CeCl_37H_2O$, $PrCl_37H_2O$, $NdCl_36H_2O$, $SmCl_36H_2O$, $EuCl_36H_2O$, $GdCl_36H_2O$, $TbCl_36H_2O$, $DyCl_36H_2O$, $HoCl_36H_2$ (0.528 g of $CaCl_2$, corresponding to 4.76 mol, 0.69 g of H_2O and 0.44 g of EtOH, molar ratio $CaCl_2$: H_2O: EtOH 1:8:2). As a result, for all of the lanthanide chlorides, 0.125 g of degummed SF was added to a solution generated by dissolving 4.76 mol of the salt (see Supplementary Table 1) in 0.69 g of H_2O and 0.44 g of EtOH.

At 60 °C, the mixtures were cooked for four hours. The solutions were dialyzed with distilled water and a regenerated cellulose dialysis membrane to eliminate salt and ethanol (MWCO 12.400 g mol). After 4 hours in the presence of La^{+3}, Ce^{+3}, Pr^{+3}, Er^{+3}, Tm^{+3}, Lu^{+3}, and partially for Dy^{+3}, SF reprecipitated within the dialysis cassette, and the dialysis purification process was stopped after 12 hours with water changes every 4 hours. Before air drying at room temperature, the Ln^{+3}/SF fibres were repeatedly rinsed and washed in double distilled water. The amount of SF recovered was calculated using weighting techniques. After dialysis, clear SF aqueous solutions for Nd^{+3}, Sm^{+3}, Eu^{+3}, Gd^{+3}, Tb^{+3}, and partially for Dy^{+3} were produced. These solutions were stored and maintained at 4 °C. From these solutions, drop-casted films were created for additional analysis. To dissolve degummed SF, hydrated $LnCl_3$ in H_2O/EtOH (in a molar ratio of 1:8:2) was utilized according to Ajisawa's standard procedure (Ajisawa, 1998) [21]. The combinations were cooked for 4 hours at 60 °C. To remove the residual salt and ethanol, the SF solutions were dialyzed against double distilled water. Water was used to rinse the regenerated fibres multiple times before allowing them to air dry. The hydrogel and solutions were kept at 4 °C. In H_2O/EtOH, all lanthanides chlorides were able to dissolve degummed SF. For comparison, calcium chloride standard salt is listed in the last row [22].

Table 1. The details of Lanthanide salts used for SF dissolution/regeneration protocol

Sr. No	Lanthanide	Atomic number (Z)	Salt	Ln^{+3} ionic radius (A^0) (Seaborg 1978)	Dialysis result
1	La	57	$LaCl_3.7H_2O$	1.061	Fibres
2	Ce	58	$CeCl_3.7H_2O$	1.034	Fibres
3	Pr	59	$PrCl_3.7H_2O$	1.013	Fibres
4	Nd	60	$NdCl_3.6H_2O$	0.995	Solution
5	Pm	61	$PmCl_3.6H_2O$	0.979	Solution
6	Sm	62	$SmCl_3.6H_2O$	0.964	Solution
7	Eu	63	$EuCl_3.6H_2O$	0.960	Solution
8	Gd	64	$GdCl_3.6H_2O$	0.938	Solution
9	Tb	65	$TbCl_3.6H_2O$	0.923	Solution
10	Dy	66	$DyCl_3.6H_2O$	0.908	Solution/ Fibres
11	Ho	67	$HoCl_3.6H_2O$	0.894	Gel
12	Er	68	$ErCl_3.6H_2O$	0.881	Fibres
13	Tm	69	$TmCl_3.6H_2O$	0.869	Fibres
14	Yb	70	$YbCl_3.6H_2O$	0.858	Fibres
15	Lu	71	$LuCl_3.6H_2O$	0.848	Fibres
16	Ca	20	$CaCl_2$	1.00	Solution

5. Formulation of Cocoons with Variety of Dopants by Feeding Silk Worms with Bio-Nanofortified Leaves

Mulberry leaves include a variety of colours and nutrients that silkworms ingest and absorb into their bodies. By connecting the yellow carotenoid pigment to the intracellular carotenoid-binding protein (one of the pigments taken into the silkworm body), silkworms create yellow fluids and cocoons (CBP). In the silkworm Bombyxmori, the Y (Yellow Blood) gene controls the uptake of carotenoids into the intestinal mucosa and the silk gland and influences the colour of the cocoon. According to Sakudoh et al. (2007), insufficient CBP mRNA as a result of improper mRNA splicing as a result of a transposon-associated genomic omission led to colourless hemolymph and white cocoons, which resulted in the Y recessive strain. With a modest degree of fibroin absorption, sericin absorbs the colours to a great extent. White SF is produced when sericin is removed using a degumming process. The presence of flavonoids such quercetin derivatives (Tamura et al., 2002; Daimon et al., 2010; [23] Kusurkar et al., 2013) and carotenoid chemicals, however, is strongly suggested to be the cause of the luminosity of silk cocoons (Tamura et al., 2013). (Harizuka, 1953) [24-26]. According to several studies (Mandal and Kundu, 2009; Zhang et al., 2010; Kundu et al., 2012; [27] Hodgkinson et al. [28], 2014; Bhardwaj et al., 2015), SF-derived scaffolds show a fluorescence index in UV, green, and blue lighting [29].

Since SF emits fluorescence at wavelengths identical in colour to the fluorescent dyes themselves, its inherent fluorescence can obstruct the signal from those dyes (Georgakoudi et al., 2007) [30]. To generate an SF scaffold, Amirikia et al. (2018) [31] used the freeze-drying method. Confocal microscopy was then used to assess the SF scaffold's fluorescence before and after cell seeding and fluorescent labelling. The fluorescence index of the SF scaffold was then examined using chemicals. The scaffold itself was fluorescently coloured and cells implanted with fluorophore showed a similar fluorescent appearance when activated by light of a corresponding wavelength. The SF solution displayed absorption maxima at 277 nm and emission maxima at 345 nm in UV-Vis and fluorescence spectroscopy, which is a typical protein-derived signal. They consequently deduced that the SF scaffold's fluorescence caused by intrinsic protein fluorescence rather than by the flavonoid components of the cocoons that the SF protein was produced. Additionally, Kusurkar et al. (2013) [32] used the "water glass" technique to extract fluorophores from silk cocoons.

5.1. Using a Diet Containing Dye as a Method of Feeding

There was an attempt to give colouring agents to silkworms at the beginning of the 20th century (Edwards, 1921) [33]. These studies examined issues such as the toxicity of specific colours to silkworms or the persistence of mulberry leaves in silkworm stomachs (Lombardi, 1920) [34] (Campbell, 1932) [35]. Several studies (Jucci and Ponsevernoi, 1930; [36] Goris and Muehlemann, 1936) revealed that orcein was taken up to silk cocoons from the silkworm diet and those other dyes were absorbed by intestines or even travelled into silkworm eggs [37].

Other dyes, such as gentian violet, berlin blue, safranin, Nile blue, methylene blue, neutral red, alizarin, fuchsin, indigo carmine, and methyl orange, were not absorbed by silk worm (Jucci and Ponsevernoi, 1930) [36]. Due to a lack of appropriate and precise characterization methodologies, the results of the previous research were equivocal in dye incorporation into silk, particularly in SF. The molecular behaviour of these colours or other additions to silk was unrelated. Rhodamine B dye absorption into the SF was initially shown by Tansil et al. They provided dye-containing food to silkworms, and enhanced characterization was used to corroborate it (Tansil et al., 2011) [38]. To create fluorescent silk cocoons and fibres, silkworms were given a mixture of mulberry leaf powder and fluorescent compounds such as rhodamine B, rhodamine 101, and rhodamine 110. It was apparent that SF was competent interact with rhodamine B, rhodamine 101, and rhodamine 110 by molecular recognition (Tansil et al., 2011b) [39]. Recently, genetically altered silk from transgenic silkworms was displayed by Yamao et al. (1999) [40]. Once the silkworm strain has been developed, transgenic silkworms can be easily maintained and propagated. To effectively create an altered silkworm, Tamura et al. (2000) [41] used a customized vector utilizing a Piggy Bac transposon system produced from Trichoplusiani, a Lepidoptera insect. Furthermore, the Piggy Bac transfer vector has already been used in many studies looking into silkworm transformation. A transgenic silkworm that produces recombinant silkworm cocoons expressing human collagen protein was described in 2003 (Tomita et al., 2003) [42]. Different techniques are used to produce recombinant proteins were established using fibroin and sericin. Transgenic fluorescent silkworms have been created using a variety of transformation vectors (Liu et al., 2009). The alteration of the silk protein to attain particular desired properties is made possible by the insertion of a transgene into the genome of the silkworm (Tamura et al., 2000; Iizuka et al., 2013) [43]. Many transformation vectors for silk modification have been made using silk genes (Tomita, 2011) [44].

The fibroin L chain and fibroin H chain genes found in the generated vectors can be used to create modified silks. The vector is used to make a range of available silks that are beneficial for preserving valuable attributes for goods and pharmaceuticals. For instance, the recombinant silkworm/spider silk produced by transgenic silkworms is tougher than regular silk strands and comparable to a natural dragline (Teule et al., 2012) [45]. A chimeric human fibroblast growth factor protein was recently discovered in biologically active silk. The fibroin L chain has been suggested as a possible biomaterial for tissue engineering (Hino et al., 2006) [46]. It has been demonstrated that a silk film containing a chimeric protein comprising the fibroin L chain, a fibronectin sequence, or a portion of a collagen peptide has higher cell adhesion activity than native silk (Yanagisawa et al., 2007) [47]. Designing a cost-effective production process that can yield significant amounts of modified silk is essential to raising the value of transgenic silkworm silks. Iizuka et al. (2013) [43] developed a transgenic silkworm that produces fluorescent silk. They built an isolation procedure to extract sericin at low temperatures to transform it into a useful material for many industrial applications. Using this technology, bright silk robes and illumination were made (Figure 5). By inserting jellyfish EGFP into exon 2 of the silkworm fibroin heavy chain gene, Kim et al. produced a microcarrier. The gene for green fluorescence is handed down to the following generation, and the green fluorescence is kept. Additionally, they developed transgenic silkworms that express the red and blue fluorescent proteins (Kim et al., 2012, 2013, 2014) [48-50].

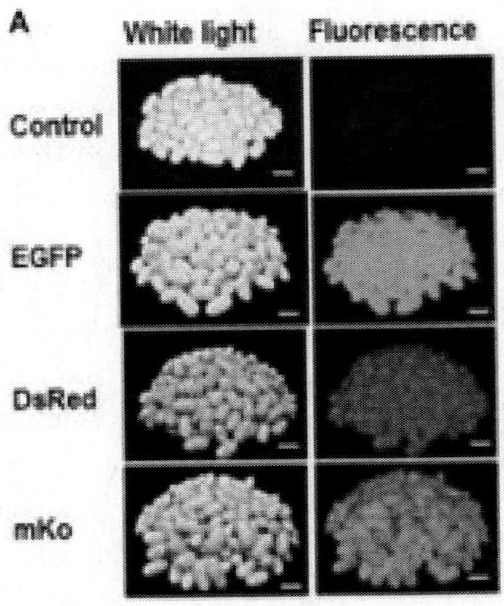

Figure 5. Colored fluorescence cocoons.

6. Feeding Silk Worms with HPMC Dispersal MoS_2 NPs to Yield SF Bio-Nanocomposites for Electronics Applications

In recent years, flexible electronics has been emerged as is an attractive developing technical field that provides platforms for a wide range of wearable gadgets. Silkworm silk, spun by Bombyxmori, has received a lot of attention among the many materials used in the production of flexible electronics. Although recent study has revealed that feeding silkworms with functional nanomaterials-loaded mulberry leaves in vivo enables the synthesis of silkworm silk fibres with better electrical characteristics. However, their performance is restricted by the natural silk's weak conductivity. Carbonized silkworm silk retains high flexibility and conductivity, making it an exciting material for the preparation and creation of flexible electronics.

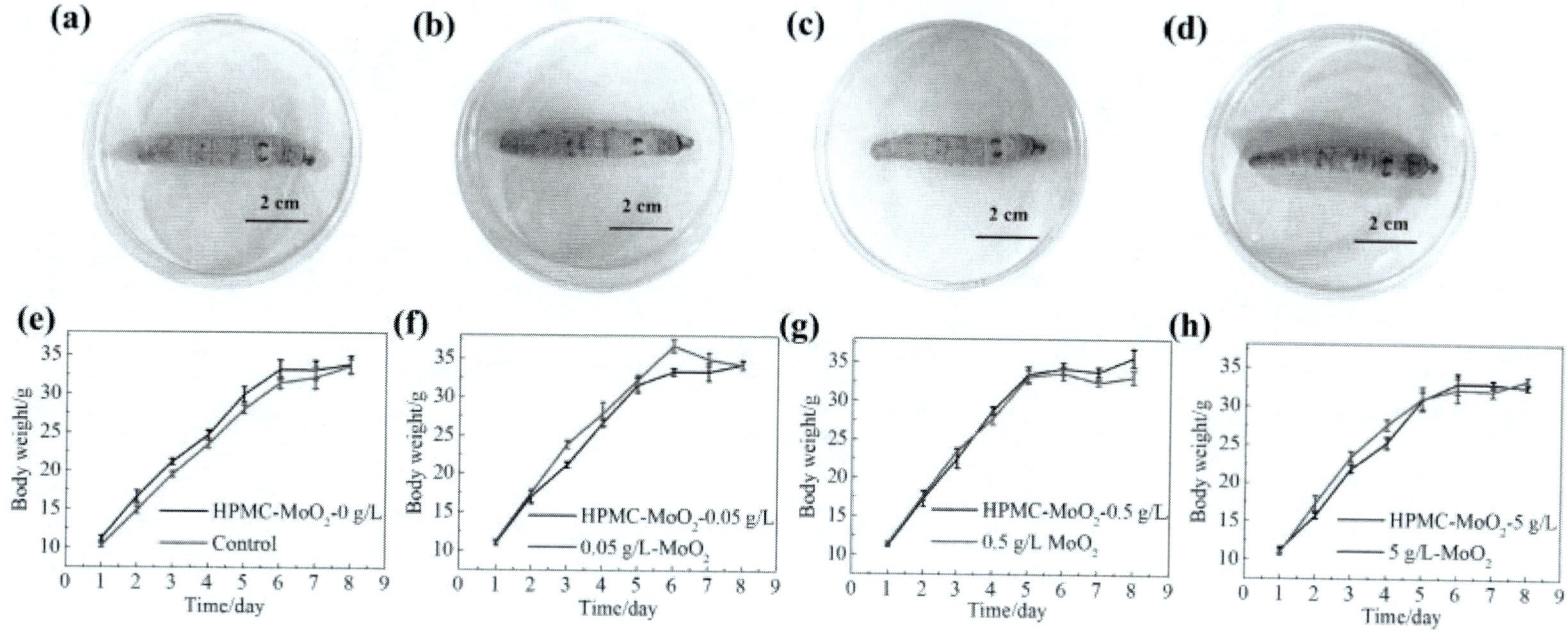

Figure 6. The appearance of silkworms in the following groups: (a) HPMC-MoO_2-0 g/L, (b) HPMC-MoO_2-0.05 g/L, (c) HPMC-MoO_2-0.5 g/L, and (d) HPMC-MoO_2-5 g/L. Every day, the body weight of a silkworm at its fifth in star stage was measured: (e) HPMC-MoO_2-0 g/L and control, (f) HPMC-MoO_2-0.05 g/L and MoO_2-50 mg/L, (g) HPMC-MoO_2-0.5 g/L and MoO_2-500 mg/L, and (h) HPMC-MoO_2-5 g/L and MoO_2-5 g/L [55].

Further, it is revealed that, the weak conductivity of natural silk fibroin is improved with doping of molybledium oxide nanoparticles. Because of this, molybdenum dioxide nanoparticles (MoO_2 NPs) were successfully introduced into spun silk by feeding mulberry leaves coated with MoO_2 NPs to silkworms. The research expanded the possibilities using carbonised silk in energy storage devices.

J. Liang et al. revealed in a recent study that feeding MoO_2 nanoparticles (NPs) to silkworms as a food additive allowed for the incorporation of MoO_2 NPs into the as-spun silk, which can be used to manufacture energy storage electrodes. However, aggregation of MoO_2 NPs was seen at high fractional feeding dosages. The dispersion in this investigation was 0.5 per cent (w/v) hydroxypropyl methyl cellulose (HPMC) solution, which was utilized to disperse the MoO_2 NPs in fibrous silk and increase the specific capacitance of the carbonized silk Hydroxypropyl methyl cellulose is a cellulose derivative (HPMC). It has been approved as a food additive by the Food and Agriculture Organization of the United Nations and the World Health Organization (FAO/WHO) [51]. HPMC solutions are odorless and tasteless. It is commonly used in food and industrial products and systems as dispersion, film forming agent, and emulsifier, to name a few applications [52]. In recent years, HPMC has also been used to increase the water dispersibility of chemicals sprayed on mulberry leaves to make functionalized silk fibres [53, 54]. However, none of these studies investigated whether HPMC keeps nanoparticles separated until they enter the silk gland and are trapped within silk strands. MoO_2 NPs were dispersed using a 0.5 per cent (w/v) HPMC solution, and the mulberry leaves were soaked in this solution to ensure that the MoO_2 NPs were evenly distributed on the mulberry leaves used to feed the silkworms. J. Liang and his colleagues discovered that applying HPMC significantly reduces the aggregation of MoO_2 NPs in as-spun silk. Carbonized silk, on the other hand, preserves its fibrous shape and has a higher specific capacitance. For the first time, J. Liang et al. assessed the effect of HPMC on silkworm growth, nanoparticle feed efficiency, silk yield, and other elements of the results. This is shown schematically in Figure 6.

7. Engineering of Electrospun SF-Based Bio-Nanocomposite Formulations for Photocatalytic Disinfection of Pathogens

Human civilization has contended with disease outbreaks throughout recorded history, from the Antonine Plague of 250 to the 1720s Bubonic Plague and the 2019's COVID pandemic that is still infamous in our collective memory. For the first time, humanity can use nanotechnology meaningfully to combat a signicant infectious disease, namely the COVID-19 pandemic caused by SARSCoV-2. SARS-CoV-2 has spread to millions of people worldwide through a novel mechanism of binding to ACE-2 receptors in the human body using its spike protein. The author of this chapter and his group of researchers have done some pioneering research toward the control of the spread of COVID-19 and reported in recent publication entitled "Nanoparticle Engineered Photocatalytic Paints: A Roadmap to Self-Sterilizing against the Spread of Communicable Diseases" continues here: Engineered nanomaterials for visible light photocatalysis have recently been used to prepare smart paints. In this paper, the authors describe for the first time the potential of these novel paints for the self-sterilization of surfaces in hospitals and public places by merely exposing them to visible light. The purpose of this review is to identify the root cause of this widespread sickness and propose a strategy for halting its spread. We've put up an alternative approach based on the use of SF nanocomposites with strong antipathogenic capabilities in this section. Additionally, electrospun bio-nanocomposites can be uniformly applied to metallic surfaces. This technique is described in detail in the following section.

7.1. Electrospinning Technique

Electrospinning is a simple technique that involves spinning a polymer solution into a non-woven mat made up of nanometer-sized fibres using an electric field. The polymer solution is inserted in a syringe with a conductive needle attached to a high voltage electric field (5–40 kV) during electrospinning. A grounded conductive collector is put in front of the needle at a safe distance. The high voltage electric field produces stretching of the polymer solution as it is extruded through the needle, allowing nanofibers to form. The solvent evaporates at rates determined by its intrinsic properties and experimental settings during the process, and the fibres are deposited on the collector due to the voltage difference (Rockwood et al., 2011) [56]. Figure 7 depicts the electrospun technique's experimental setup.

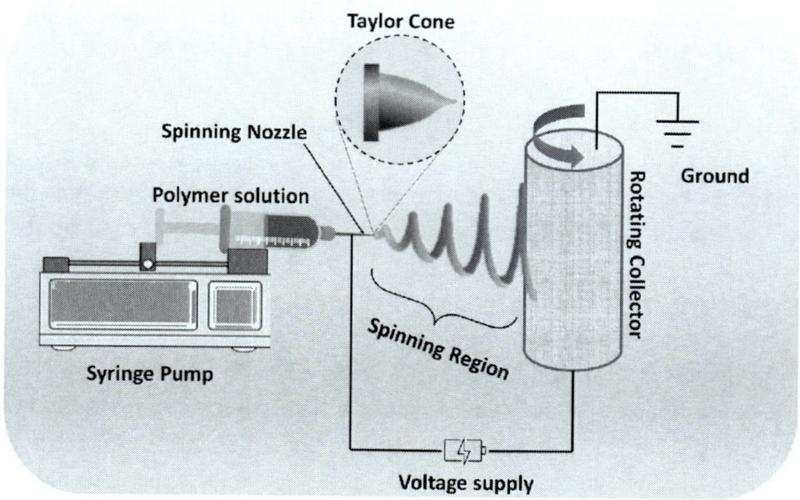

Figure 7. Electrospinning setup.

One can monitor the size of the fibres by changing the polymer concentration, polymer extrusion flow rate, needle-to-collector distance, and applied voltage; you can change the size of the fibres made by electrospinning. Depending on the type of polymer used, the electrospinning process may also be sensitive to other parameters. Electrospinning of silk fibroin also depends a lot on the humidity and temperature. If one of these factors changes, the final material will be different.

The type of collector used influences fibre alignment. If a flat, static collector is used, fibres are deposited at random. The use of a spinning mandrel as a collector causes the fibres of the deposited fiber, to align (Rockwood et al., 2011) [56], which is helpful for some applications such as controlling development direction in brain regeneration (Belanger et al., 2018) [57].

Extrinsic parameters such as polymer concentration, solvent, polymer extrusion flow rate, needle-to-collector distance, and applied voltage can all be changed to alter the diameter of the fibre produced by electrospinning. The electrospinning method could be influenced by inherent properties such as the molecular weight of silk. Humidity and temperature also have an impact on silk fibroin electrospinning. A non-controlled change in one of these characteristics will alter the qualities of the final substance.

7.2. Formulations of Silk Bio-Nanocomposite Formulations

7.2.1. Silk Fibroin Extraction

SF dispersion was successfully achieved by utilizing the Rockwood et al. method [56]. All samples should be boiled at the same time because the boiling time has a considerable impact on the SF molecular weight. Boiling for 30 minutes results in SF with a molecular weight of 4 kDa, according to earlier studies. The molecular weight of SF fragments produced by longer boiling times is lower. There will be 30 minutes of boiling involved. The resultant fibres were successfully dispersed in a LiBr solution due to its capacity to break hydrogen bonds observed in SF sheets. It's easy to spread in water since the hydrogen bonds have been broken. The gelation of the SF dispersion occurs roughly one month after 4 ^{0}C temperature, which is thermodynamically favourable. The SF concentration in the final solution was between 5% and 6%, which can be further reduced.

7.2.2. Electrospinning of SF and SF/NPs Electrospun Mats

Several ways to integrating NPs into electrospun materials have been reported in the literature. NPs were added to the dispersion of the SF electrospinning dispersion and the electrospun material was functionalized in this investigation. To attain a concentration of 10% to 20% SF, the SF dispersion was kept at room temperature until the water had evaporated. As it is, the surface tension of the electrospun SF is not high enough for electrospinning; it needs to be augmented by adding additional polymers, such as polyethene oxide (PEO). The electrospinning dispersion was made by mixing 10% SF with 5% PEO in a 1:4 volume ratio. The electrospinning dispersion was injected into a 5 mL plastic syringe fitted with a 19 G stainless steel needle and connected to a high voltage generator (Gamma High Voltage Research ES-30P Ormond Beach, FL, USA). Aluminium foil wrapped around a flat Plexiglas collector placed 20 cm in front of the needle's tip was used to enhance electrical conductivity. The dispersion was dispensed at a flow rate of 0.015 mLmin^{-1}, while 9-10 kV was applied to ressings, textiles, wearable electrodes, nerve guides, and other devices using an electrospinning dispersion pump from Thermo Scientific (Yukseloglu et al., 2015) [58] as far as the needle's eye. A sample was collected for 50-120 minutes to produce a mat. Relative humidity levels varied from 30% to 50%. At this moment, the substance can be dissolved in water. The samples were vacuumed for at least four hours in the presence of water to accomplish this. PEO was removed from the samples using a 48-hour submersion in 37 °C water. Afterwards the materials were dried and stored at room temperature for further study. Adding NPs to the SF electrospinning dispersion or functionalizing electrospun silk after it was created where the two methods used to functionalize silk electrospun materials.

7.2.3. Addition of NPs to Electrospinning Dispersion

Silk fibres electrospun bio-nanocomposites resulted from the incorporation of NPs into the electrospinning dispersion. The electrospinning process was unaffected by the inclusion of NPs in the dispersion. To create an electrospun material, however, a higher voltage was required due to the presence of IONPs. This behaviour could be explained by the fact of a higher concentration of NPs and a different type of IONP. SEM/EDX examination did not reveal any NPs in silk-NPs electrospun bio-nanocomposites. One-fiber TEM and STEM investigations showed that all NPs were present across all fibres. Because of their aggregated state in suspension, as revealed by DLS, IONPs displayed little aggregation, despite the wide dispersion of Au NPs and Ag NPs within the fibres. In addition, the concentration of NPs in electrospinning dispersions must be restricted by the proportion of silk and PEO dispersions in the process.

Last but not least, although electrospun materials were successfully generated in all situations, there was no reproducibility due to the considerable impact of humidity on the silk electrospinning process and the system's incapacity to manage this parameter. For example, the diameter and fibre form were influenced by these two factors. Cryogels can be made from silk hydrogels by a process called "lyophilization,"

7.2.4. Electrospun Material Functionalization with NPs

When introducing NPs to electrospinning dispersion, the NPs concentration limit can be overcome by post-functionalizing silk electrospun materials. *In situ* synthesis of Au NPs and Ag NPs into electrospun mats was possible due to their one-pot reaction synthesis. Macroscopically, the tissue was consistently coloured with Au NPs that were evenly distributed across the material's surface. It was found that the formation of Ag NPs was heterogeneous, resulting in the construction of significant NPs in several areas of the experiment. There was a yellowish tint to the Ag NPs, as well as grey spots on the surface, even in the electrospun silk mats that had been used in the experiment. Ex situ functionalization with IONPs was accomplished by letting a small volume of NPs suspension dry on top of the silk electrospun mat. It was discovered via MEB imaging that IONPs had coated the whole fibre. However, despite the good outcomes of both *in situ* and ex situ functionalization, consistency still has to be improved. Because NPs might be more likely to enter the environment if these procedures are utilized, further research should be done before the materials are used for other purposes.

7.3. Antibacterial Activity

Silk materials are readily colonized by bacteria and fungi due to their protein composition. Thus, the insertion of antibacterial NPs into silk fibres is essential, since it allows for better preservation of such luxury cloths and inhibits the appearance of undesirable odours. In the biomedical field, such NPs prevent or delay the appearance of infections. Silver NPs are the most commonly utilized NPs for antibacterial applications. Even though the majority of the efforts have been directed toward silk fabrics, additional silk materials have been functionalized with silver NPs. Several authors have pioneered the *in situ* synthesis of silver nanoparticles into silk textiles. UV irradiation was employed to directly synthesize silver NPs in silk dispersions, films, sponges, fibres, fabrics, and electrospun mats [59]. The synthesis of NPs into silk textiles *in situ* was not homogeneous across the silk fibres. This could be due to varying silver ion adsorption based on the chemical groups accessible in the silk. The release of silver ions from electrospun mats was investigated and found to be dependent on the sheet composition. While both materials had antibacterial activity, only the electrospun materials with low-sheet content shown cytotoxicity. This finding is consistent with the known toxicity of silver ions, as stated in section 2.2.2.

The light was used as a reducing agent in another investigation for the *in situ* synthesis of silver NPs into a silk dispersion. The resulting dispersion was combined with Carbopol 934, a gelling agent. Although the antibacterial activity of the silver NPs/silk dispersion was demonstrated, no experiments were conducted on the gel nanocomposite. In animal skin wound models, topical use of this gel resulted in a faster wound closure rate compared to silk, silver NPs, and Carbopol gels, as well as Soframycin gel, a commercially available product. These findings support the notion that silk promotes wound healing. Furthermore, a distinction was seen between silk and silver NPs/silk gels, indicating a synergistic effect of both components in wound healing.

Protein concentration in silk makes it easy for bacteria and fungus to colonise. This is why using antibacterial nanoparticles (NPs) in silk fibres is critical as it improves the longevity of fine textiles while also reducing the likelihood that they may develop offensive odours. When NPs like these is present in the biomedical field, infections can be avoided or delayed. Antibacterial silver NPs, as previously stated, are the most often used NPs. Aside from silk, silver nanoparticles have been employed to improve the performance of various silk materials. Researchers have found ways to incorporate silver nanoparticles into silk fibres in-situ. Silk dispersions, films [60] and sponges, fibres, textiles and electroporation mats [61-65] were irradiated with UV light to synthesize silver NPs directly. NPs were synthesized into silk fabrics *in situ*. However, the process was not uniform across the fibres. Depending on the chemical groups in the silk, the silver ion adsorption may be different. The composition of the electrospun sheet affects the release of silver ions from the mats. However, only electrospun materials with minimal sheet content had cytotoxic effects whereas, both possessed antibacterial capabilities. The light was utilized as a reducing agent in another study to *in situ* synthesizes silver nanoparticles in a silk dispersion, which confirms the previously reported toxicity of silver ions [66]. Carbopol 934, a gelling agent, was added to the resulting dispersion. Though silver NPs/silk dispersion antibacterial activity has been proven, no tests have been done on the gel nanocomposite. Wounds on animals healed more quickly this gel than with silk, silver nanoparticles, Carbopol gels, or the commercially available Soframycin gel.

Findings like this lend credence to the idea that silk speeds up the healing process. As an additional sign that both components function together to heal wounds, there was a difference between silk and silver/silk gels. Bone regeneration has also been achieved using silk hydrogels containing silver nanoparticles (NPs) [64]. *In situ* production of silver nanoparticles (NPs) also used light as a reducing agent. More than 0.5 percent silver nanoparticle content in hydrogels was found to have strong antibacterial effects. Osteoblast cells were used to assess the hydrogels' biocompatibility. There was a correlation between the concentration of silver NPs and cell viability. Natural chemicals, such as honeysuckle extract, have also been employed as reducing agents in silk materials [67–73] *in situ* NP synthesis. Antibacterial capabilities have returned to the polymers, and no cytotoxic effects were seen. Caffeic acid-based products were able to withstand UV exposure. This discovery suggests a new use for silver-containing materials in sun protection gear. The silver nanoparticles in silk sponges and silver-coated films have also been developed [74–75]. The SF dispersion utilized to make these materials was also used to make NPs in this example. They found that silk by itself was capable of converting Ag+ to Ag0 and generating silver nanoparticles (NPs). SF, sericin, and peptides [76–78] have been used by many researchers to reduce the size of their samples. However, these *in situ* synthesis methods resulted in NPs of varying sizes and shapes. NPs' properties are heavily influenced by these two qualities.

Furthermore, under these instances, no information was supplied on the surface chemistry of NPs or the presence of dangerous silver ions. The material produced by this method may not have required properties (antibacterial in this case) or may even have unwanted properties (toxicity) for the final application because of the limited reproducibility of the process. As an alternative, some studies have looked at blending previously synthesized NPs with silk [79, 80]. A two-step silver NP functionalization of silk textiles was studied by Gulrajani et al. Silk fabric was submerged in a suspension of silver nanoparticles after the latter was produced. Scientists found that in acidic conditions, NPs are better absorbed into silk because of the lower pH. The study also found that NPs' absorption of silk is temperature-dependent. Silver NPs were more easily absorbed into silk fibres at 40 °C, but adsorption reduced as temperature rose (up to 80 °C). A recent study devised a two-step strategy for functionalizing silk fibres with silver NPs. A strong association between silk and silver nanoparticles was found even after 30 washings, showing that the final products could still inhibit the growth of E. coli and S. aureus. It's still unclear how silver ions and nanoparticles are released. Some gold NPs, such as [78, 64, 81-85], platinum NPs, and zinc oxide NPs, have been examined for their antibacterial properties [86]. *In situ* synthesis of gold NPs into a silk hydrogel containing HAP was achieved by Ribeiro et al. [64].

After the hydrogel had formed, the fluid was heated to 60 ^{0}C to produce gold NPs. It was surprising that S. aureus, E. coli, and P. aeruginosa all had significant antibacterial activity, but not S. epidermidis, despite the presence of these bacteria in the study. An *in situ* synthesised gold NP was obtained by heating silk fibres with *in situ* synthesised gold NPs to 85 °C, as reported by Tang and colleagues. Textiles could be made fromcaffeic acid, flavonoids, vitamin C, citrate, R. apiculata leaves extract, Streptomyces cell extract irradiation-blocking, and thermally conductive tissue. Silver NPs can be synthesized *in situ*. However, NP size cannot be controlled.

7.4. Photocatalytic Surface Coating/Painting for Self-Sterilizing against Pathogens

Sterilization methods that use liquid chemicals to control the spread of infections must be used regularly. Only small-scale applications are suitable for these. Surface coatings are preferred for large-area applications. However, they are problematic to employ for large-area applications due to their cost, duration, and maintenance requirements. As a result, self-sterilizing, long-lasting, and maintenance-free disinfection methods/techniques are in high demand to combat viral infections like COVID-19. Nanoparticle-designed photocatalytic paints have enabled today's advances and prospects in nanoscience and will serve as a pathway to self-sterilizing against the spread of contagious illnesses. The following sections report and discuss recent advancements in this subject of science and technology.

7.5. Self-Sterilizing of Pathogens with Photocatalytic Disinfection: Principles and Functions

Photocatalysis using semiconducting materials has been studied for a long time, and its principles are well understood in the literature. Light of a suitable wavelength (energy equal to or greater than the semiconductor's band gap) is absorbed by a semiconductor, resulting in the creation of an electron and hole pair. The excited electron–hole pair produced in the semiconducting material subsequently moves toward the surface of the semiconductor. On the surface of the semiconductor, oxidation and reduction reactions occur at the same time. This is depicted in Figure 8 theoretically. The CB produces superoxide and hydroperoxide radicals when electrons combine with oxygen (O2). Reactive oxygen species (ROS) are formed during photocatalysis; additionally, it can cause mineral, genetic material, and protein leaks, resulting in cell death and a reduction in microbial load [87].

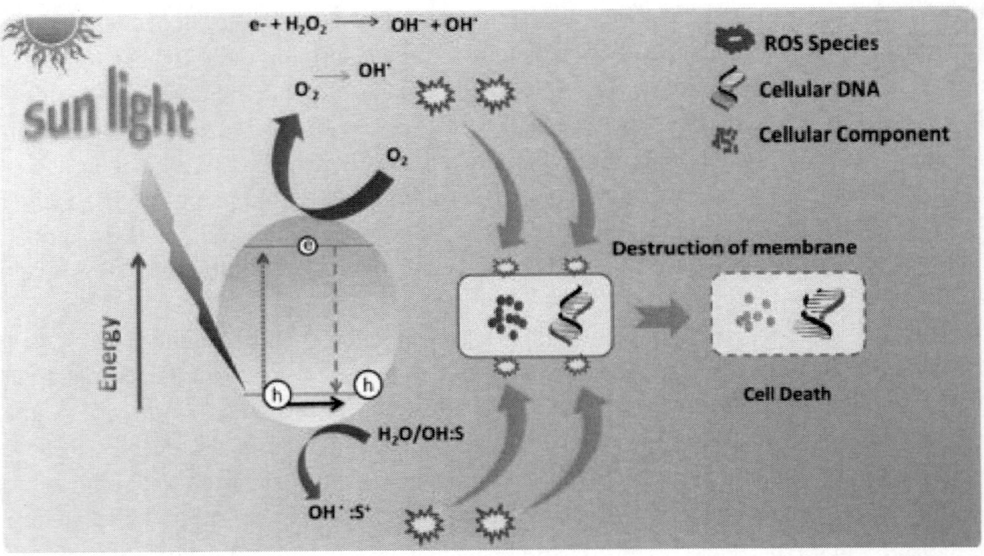

Figure 8. The fundamental mechanism of photocatalysis on a semiconductor particle surface for microbial treatment.

ROS oxidizes viruses' proteins, lipids, bacteria, and enzymatic systems, preventing them from reproducing due to damage to other essential proteins or membranes, such as the spike protein of SARS-CoV-2. The pathogen oxidizes critical proteins and inactivates due to the generation of reactive oxygen species (ROS) when it comes into contact with the photocatalytic surface [88-89]. In general, the photocatalytic material degrades the pollutants (viral proteins, genome, and envelope) in the indoor air. To disinfect air, water, and surfaces, antimicrobial photocatalysis can be used [90-92]. The activation of photocatalytic nanoparticles in the nanocomposite includes adsorption-desorption, electron–hole pair production, electron pair recombination, and chemical reaction. Figure 9 shows the hypothesized mechanism for nanocomposite photocatalytic antibacterial activity. It offers a lot of advantages, including being a low-cost, simple-to-use, low-maintenance approach for disinfecting a variety of pathogens such as bacteria and viruses at the same time.

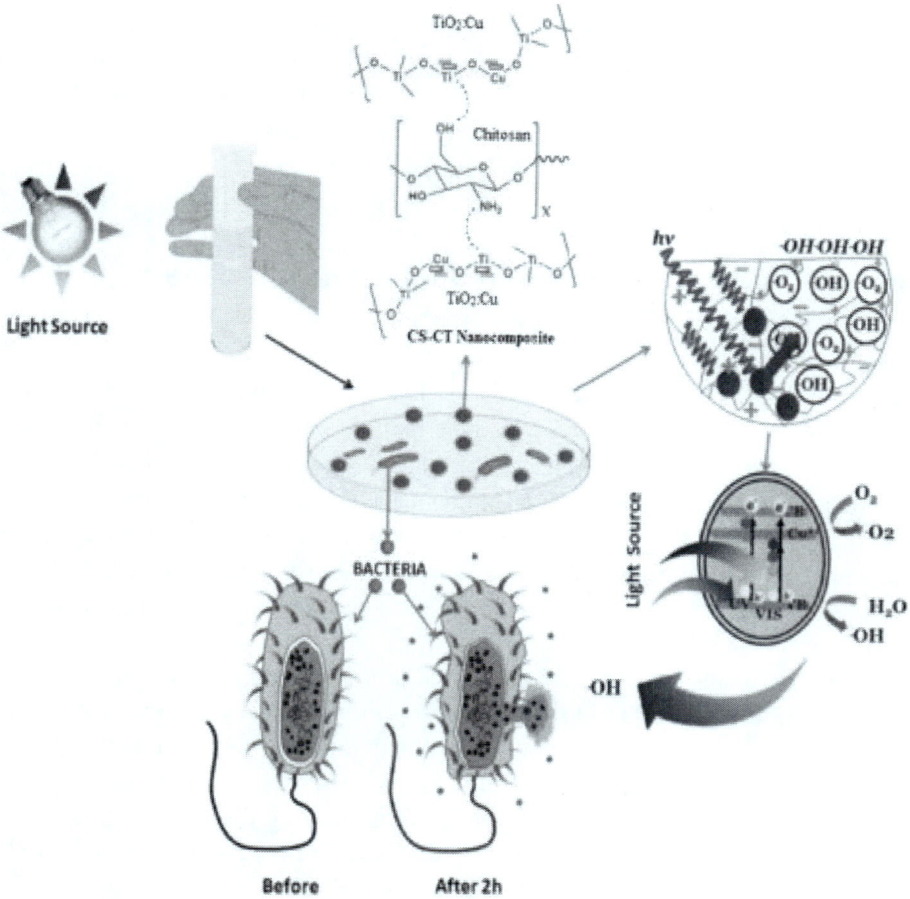

Figure 9. A possible mechanism for photocatalytic antibacterial activity resulting from light stimulation of nanocomposites [93].

7.6. Depollution

On a similar line of photocatalytic disinfection of pathogens, the applications of silk fibroin bio-nanocomposites have been extended for depollution. Besides its mechanical strength, biocompatibility and biodegradability, silk is a good absorber of aromatic dyes. As a result, numerous NPs' catalytic activity was paired with silk's absorbent properties, which are pH- and dye-dependent. Silk electrospun nanofilters and TiO_2 NPs were used to examine the removal of anionic dyes [94]. Unexpectedly, they found that when the concentration of NPs increased, so the materials' absorption capacity, pH was found to have a significant impact on absorption when researchers examined the effect of pH. Materials for anionic dye removal 238 made of silk iron oxide NPs are similar [95]. For dye removal, CuO_2 NPs embedded in silk have also been investigated 239 [96, 97]. A biodegradable substance may be a viable solution for wastewater remediation, according to this research. Some other research has shown that using sodium fluoride and hydroxoapatite for water filtration works well [98]. Dye ions and heavy metal ions may be removed from solutions using this material in a way that was impossible with regular nanofiltration membranes before. To make a fire-resistant and heat-insulating material, silk can be mixed with silica [99]. Using an acid-catalyzed sol-gel synthesis, these materials can be quickly produced. High surface area (311-798 m^2/g), low density (0.11-0.19 g/cm^3), compression flexibility, and fire retardancy are all characteristics of the silk/silica aerogel is the end result of this process.

8. Microwave-Assisted Fast Process to Regenerate SF Solution for Formulations of Bio-Nanocomposites

It is well known that traditionally, silk fibroin has been used for wearable products for a pretty long time. Many more farmers are involved in sericulture industries, and it is the backbone of the silk industry. There are very few countries involved in this type of business. India is one of the major countries in the world in this type of business. The Indian government has established the "Central Silk Board "for the welfare of silk industries, including nine Central research institutes in the country. In recent years there is considerable amount of innovative research on value-added products of silk bio-nanocomposites. However, formulations of SF bio-nanocomposites are complicated and time-consuming process. It takes more than 52 hours and may not be suitable for production of value-added products for the sericulture farmers. There is a need for research in the field of low degradation degumming, dissolving, and reproducible desalting methodology, which would allow the isolation of native-like silk fibroin and might be employed for the ultimate industrial manufacture of regenerated silk fibroin.

The purpose of this section is to report on a recent study by Woltje et al. [100]. Their first goal was to create a softer degumming approach that would allow them to begin the dissolution process with low to non-degraded silk fibroin material. This method is based on Chung et al. observations that the anionic detergent sodium dodecyl sulphate (SDS) improves degumming and the notion that shortening cooking time by employing microwave radiation minimizes degradation of the silk during the degumming process. Another goal of this effort was to reduce the time and temperature required for the complete dissolution of degummed silk fibroin, with the goal to destroy the silk fibroin as little as possible. The method developed is based on results in the field of silk fibre quality control.

So, this work looks into a new way to separate non-degraded regenerated silk fibroin, which cuts the processing time from 52 hours with standard methods to only 4 hours with the new method. When repeated short-term microwave treatments were used instead of the standard degumming protocol, degummed silk fibroin that didn't break down was made. Afterwards a $ZnCl_2$ solution was used to completely dissolve the degummed fibroin in just 1 hour at a temperature of 45 °C. Gel filtration was used to get rid of the salt. Based on these changes, a cyto-compatible aqueous silk fibroin solution could be made from degummed silk in only 4 hours, cutting the total process time by 48 hours without affecting the quality of the isolated silk fibroin solution. Figure 10 shows the standard way to make regenerated silk fibroin and the faster $ZnCl_2$-based method.

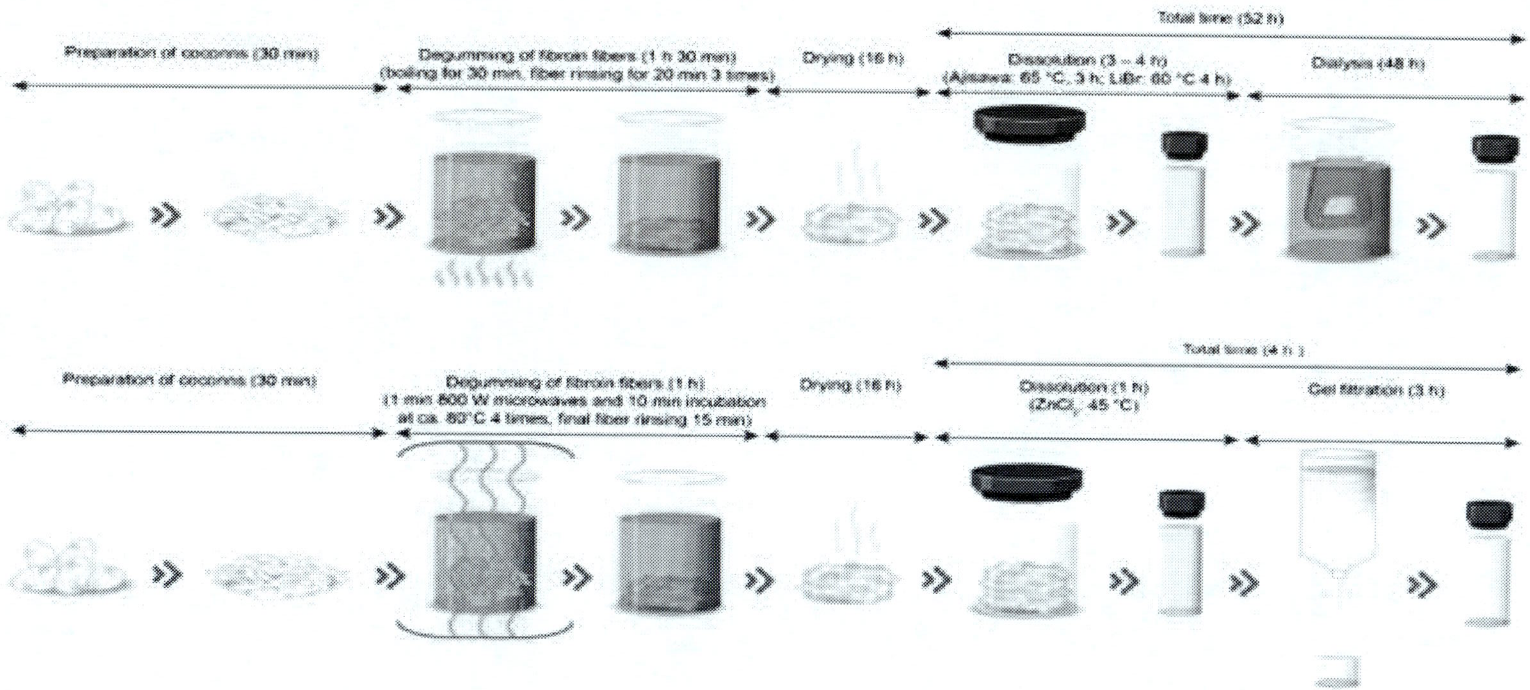

Figure 10. Comparison of a standard process with microwave-assisted SF Solution [100].

Conclusion

For decades, silk worm protein (Bombyx Mori) has been studied as a viable material for several advanced engineering and biological applications. Silkworm silk has recently received a lot of interest in the scientific community, owing to its mechanical qualities. Through bio recognition units, silk has been demonstrated to have individual interactions with tissue cells. Natural silk contains fibroin and sericin, and it has been employed in various human tissues (skin, bone, nerve, and so on). Silk also has anticancer, antityrosinase, anticoagulant, antioxidant, antibacterial, and diabetic effects. Due to advances in nanoscience and nanotechnology, this topic has taken on a new dimension in recent years, opening up new horizons for Silk Fibroin bio-nanocomposites research and applications. This chapter aims to describe the various biomedical characteristics of B. Mori silk as a suitable biomaterial among the different natural and synthetic polymers now available, as such, it is suited for use in regenerative medicine disciplines.

Acknowledgments

The author is grateful to Principal C. V. Murumkar from T.C. College, Baramati and R. K. Mudgal, Vice-Chancellor and Shimpa Shrama, Pro. Vice- Chancellor from D.Y. Patil Education Society, Deemed to Be University, Kolhapur for their constant encouragement and support during this work.

References

[1] Pugno, N.M., Motta, A., Kaplan, D., eds. (2021), *Frontiers in Silk Science and Technology*. Lausanne: Frontiers Media SA . https://doi.org/10.3389/978-2-88971-087-4.

[2] Grabska-Zielińska, S., Sionkowska, A. How to Improve Physico-Chemical Properties of Silk Fibroin Materials for Biomedical Applications?—Blending and Cross Linking of Silk Fibroin—A Review. *Materials* 2021, 14, 1510.

[3] Mishra, R. & Militky, J. Nanocomposites. In *Nanotechnology in Textiles* (ed. The Textile Institute Book Series) 263–310 (Elsevier, 2019). https://doi.org/10.1016/B978-0-08-102609-0.00006-7.

[4] Aznar-Cervantes, S., Aliste, M., Garrido, I., Yanez-Gascon, M.J., Vela,N.,Cenis, J. L., Navarro S. and Fenoll,J. Electrospun silk fibroin/TiO_2 mats. Preparation, characterization and efficiency for the photocatalytic solar treatment of pesticide polluted water. *RSC Adv.*, 2020, 10, 1917.

[5] Aimé, C. & Coradin, T. Nanocomposites from biopolymer hydrogels: Blueprints for white biotechnology and green materials chemistry. *Journal of Polymer Science, Part B: Polymer Physics* vol. 50 669–680 (2012).

[6] Zhang, G., Liu, Y., Gao, X. & Chen, Y. Synthesis of silver nanoparticles and antibacterial property of silk fabrics treated by silver nanoparticles. *Nanoscale Res. Lett.* 9, 1–8 (2014).

[7] Yu, K. et al. *In situ* assembly of Ag nanoparticles (AgNPs) on porous silkworm cocoon based would film: Enhanced antimicrobial and wound healing activity. *Sci. Rep.* 7, 1–13 (2017).

[8] Zhou, L. et al. Minimizing antibiotic dosage through *in situ* formation of gold nanoparticles across antibacterial wound dressings: A facile approach using silk fabric as the base substrate. *J. Clean. Prod.* 243, 118604 (2020).

[9] Xu, Z., Shi, L., Yang, M. & Zhu, L. Preparation and biomedical applications of silk fibroin-nanoparticles composites with enhanced properties - A review. *Materials Science and Engineering* C vol. 95 302–311 (2019).

[10] Rizzo G, Lo Presti M, Giannini C, Sibillano T, Milella A, Guidetti G, Musio R, Omenetto FG and Farinola GM (2021) Bombyxmori Silk Fibroin Regeneration in Solution of Lanthanide Ions: A Systematic Investigation. *Front. Bioeng. Biotechnol.* 9:653033. https://doi.org/10.3389/fbioe.2021.653 033.

[11] Guo, X., Ge, M., and Zhao, J. (2011). Photochromic properties of rare-earth strontium aluminate luminescent fiber. *FibersPolym.* 12, 875–879. https://doi.org/10. 1007/s12221-011-0875-9.

[12] Chahal, R., Starecki, F., Boussard-Pledel, C., Doualan, J.-L., Michel, K., Brillard, L., et al. (2016). Fiber evanescent wave spectroscopy based on IR fluorescent chalcogenide fibers. *Sensors Act.* B 229, 209–216. https://doi.org/10.1016/j.snb.2016.01.09.

[13] Liu, J., Xu, J., Guo, X., Liao, T., and Huang, Y. (2020). The up-conversion luminescence in the Er^{+3} /Yb^{+3} /Tm^{+3} tri-doped tellurite glass microsphere coupled by the tapered fiber. *Opt. Sens. Imag.Technol.* 11567, 1156740– 1156759.

[14] Zheng, X., Zhao, M., Zhang, H., Fan, S., Shao, H., Hu, X., et al. (2018). Intrinsically fluorescent silks from silkworms fed with rare-earth upconverting phosphors. *ACS Biomater. Sci. Eng.* 4, 4021–4027.

[15] Digonnet, M. J. (2001). *Rare-Earth-Doped Fiber Lasers and Amplifiers,* Revised and Expanded. Boca Raton, FL: CRC press.

[16] Shen, X., Hu, Q., and Ge, M. (2021). Fabrication and characterization of multi stimuli-responsive fibers via wet-spinning process. Spectrochim. Acta Part A Mol. *Biomol. Spectrosc.* 250, 119245.

[17] da Rocha, E. G., Pugina, R. S., and Caiut, J. M. A. (2020). Luminescent sensor based on the lanthanide-fibroin composite. *Opt. Mater.* 109, 110236–110244. https://doi.org/10.1016/j.optmat.2020.110236.

[18] Rizzo G, Lo Presti M, Giannini C, Sibillano T, Milella A, Guidetti G, Musio R, Omenetto FG and Farinola GM (2021) Bombyxmori Silk Fibroin Regeneration in Solution of Lanthanide Ions: A Systematic Investigation. *Front. Bioeng. Biotechnol.* 9:653033. https://doi.org/10.3389/fbioe.2021. 653033.

[19] Rockwood, D. N., Preda, R. C., Yücel, T., Wang, X., Lovett, M. L., and Kaplan, D. L. (2011). Materials fabrication from Bombyxmori silk fibroin. *Nat. Protoc.* 6, 1612–1631. https://doi.org/10.1038/nprot. 2011.379.

[20] Ajisawa, A. (1998). Dissolution aqueous of silk fibroin with calcium chloride / ethanol solution. *J. Sericult. Sci.*Japan 67, 91–94.

[21] Altamura, D., Lassandro, R., Vittoria, F. A., De Caro, L., Siliqi, D., Ladisa, M., et al. (2012). X-ray microimaginglaboratory (XMI-LAB). J. Appl. Cryst. 45, 869–873. https://doi.org/10.1107/s0021889 812025733.

[22] Taylor, K. N. R., Darby, M.I. *Physics ofRare Earth Solids,* Chapman and Hall Ltd 11 New Fetter Lane, London EC4, 1972.

[23] Siliqi, D., De Caro, L., Ladisa, M., Scattarella, M., Mazzone, A., Altamura, D., et al. (2016). SUNBIM: a package for X. ray imaging of nano-and biomaterials using SAXS, WAXS, GISAXS, GIWAXS techniques. *J. Appl. Cryst.* 49, 1107–1114.

[24] Harizuka, M. (1953). Physiological genetics of the carotenoids in Bombyxmori, with special reference to the pink cocoon. *Bull. SericExp. Stn. Jpn.* 14, 141–156.

[25] Mandal, B. B., and Kundu, S. C. (2009). Cell proliferation and migration in silk fibroin 3D scaffolds. *Biomaterials* 30, 2956–2965. https://doi.org/10.1016/j.biomaterials.2009.02.006.

[26] Zhang, Y., Fan, W., Ma, Z., Wu, C., Fang, W., Liu, G., et al. (2010). The effects of pore architecture in silk fibroin scaffolds on the growth and differentiation of mesenchymal stem cells expressing BMP7. *ActaBiomater.* 6, 3021–3028. https://doi.org/10.1016/j.actbio.2010.02.030.

[27] Kundu, S. C., Kundu, B., Talukdar, S., Bano, S., Nayak, S., Kundu, J., et al. (2012). Invite dreviewn on mulberry silk biopolymers. *Biopolymers* 97, 455–467. doi: 10.1002/bip.22024.

[28] Hodgkinson, T., Yuan, X. F., and Bayat, A. (2014). Electrospun silk fibroin fiber diameter influences in vitro dermal fibroblast behavior and promotes healing of ex vivo wound models. *J. Tissue Eng.* 5, 2041731414551661. https://doi.org/10.1177/2041731414551661.

[29] Bhardwaj, N., Devi, D., and Mandal, B. B. (2015). Tissue-engineered cartilage: the crossroads of biomaterials, cells and stimulating factors. *Macromol. Biosci.* 15, 153–182. https://doi.org/10.1002/mabi.201400335.

[30] Georgakoudi, I., Tsai, I., Greiner, C., Wong, C., Defelice, J., and Kaplan, D. (2007). Intrinsic fluorescence changes associated with the conformational state of silk fibroin in biomaterial matrices. *Opt. Express* 15, 1043–1053. https://doi.org/10.1364/oe.15.001043.

[31] Amirikia, M., Shariatzadeh, S. M. A., Jorsaraei, S. G. A., and Mehranjani, M. S. (2018). Auto-fluorescence of a silk fibroin-based scaffold and its interference with fluorophores in labeled cells. *Eur. Biophys. J.* 47, 573–581. https://doi.org/10.1007/s00249-018-1279-1.

[32] Kusurkar, T. S., Tandon, I., Sethy, N. K., Bhargava, K., Sarkar, S., Singh, S. K., et al. (2013). Fluorescent silk cocoon creating fluorescent diatom using a "Water glass-fluorophore ferry". *Sci. Rep.* 3:3290. https://doi.org/10.1038/srep03290.

[33] Edwards (1921). Feeding dye stuffs to silkworms. *Textile World* 60, 1111–1113.

[34] Lombardi, P. L. (1920). The duration of the stay of the ingested leaf in the intestine of the silkworm. *Ann. ScuolaAgr.* Portici 2, 1–10.

[35] Campbell, F. L. (1932). Preliminary experiments on the toxicity of certain coaltar dyes for the silkworm. *J. Econ. Entomol.* 25, 905–917. https://doi.org/10.1093/jee/25.4.905.

[36] Jucci, N., and Ponsevernoi. (1930). Migration in to cocoon and eggs of dyes administered to silkworms. *Boll. Soc. Italiana Biol. Speriment.* 5, 1056–1060.

[37] Goris, A., and Muehlemann. (1936). Coloration of silkwormgut. *Bull. Acad. Med.* 116, 268–273.

[38] Tansil, N. C., Li, Y., Koh, L. D., Peng, T. C., Win, K. Y., Liu, X. Y., et al. (2011a). The use of molecular fluorescent markers to monitor absorption and distribution of xenobiotics in a silkworm model. *Biomaterials* 32, 9576–9583. https://doi.org/10.1016/j.biomaterials.2011.08.081.

[39] Tansil, N. C., Li, Y., Teng, C. P., Zhang, S., Win, K. Y., Chen, X., et al. (2011b). Intrinsically colored and luminescent silk. *Adv. Mater.* 23, 1463–1466. https://doi.org/10.1002/adma.201003860.

[40] Yamao, M., Katayama, N., Nakazawa, H., Yamakawa, M., Hayashi, Y., Hara, S., et al. (1999). Gene targeting in the silkworm by use of a baculovirus. *Genes Dev.* 13, 511–516. https://doi.org/10.1101/gad.13.5.511.

[41] Tamura, T., Thibert, C., Royer, C., Kanda, T., Abraham, E., Kamba, M., et al. (2000). Germline transformation of the silkworm Bombyxmori L. using a piggyBac transposon-derived vector. *Nat. Biotechnol.* 18, 81–84. https://doi.org/10.1038/71978.

[42] Tomita, M., Munetsuna, H., Sato, T., Adachi, T., Hino, R., Hayashi, M., et al. (2003). Transgenic silkworms produce recombinant human type III procollagen in cocoons. *Nat. Biotechnol.* 21, 52–56. https://doi.org/10.1038/nbt771.

[43] Iizuka, T., Sezutsu, H., Tatematsu, K. I., Kobayashi, I., Yonemura, N., Uchino, K., et al. (2013). Colored fluorescent silk made by transgenicsilkworms. *Adv. Funct. Mater.* 23, 5232–5239. https://doi.org/10.1002/adfm.201300365.

[44] Tomita, M. (2011). Transgenic silkworms that weave recombinant proteins into silk cocoons. *Biotechnol. Lett.* 33, 645–654. https://doi.org/10.1007/s10529-010-0498-z.

[45] Teule, F., Miao, Y. G., Sohn, B. H., Kim, Y. S., Hull, J. J., Fraser, M. J., et al. (2012). *Silkworms transformed with chimeric silkworm/spider silk genes spin composite silk fibers with improved mechanical properties.* Proc. Natl. Acad. Sci. U.S.A. 109, 923–928. https://doi.org/10.1073/pnas.1109420109.

[46] Hino, R., Tomita, M., and Yoshizato, K. (2006). The generation of germline transgenic silkworms for the production of biologically active recombinant fusion proteins of fibroin and human basic fibroblast growth factor. *Biomaterials* 27, 5715–5724. https://doi.org/10.1016/j.biomaterials.2006.07.028.

[47] Yanagisawa, S., Zhu, Z., Kobayashi, I., Uchino, K., Tamada, Y., Tamura, T., et al. (2007). Improving cell-adhesive properties of recombinant Bombyxmori silk by incorporation of collagen or fibronectin derived peptides produced by transgenic silkworms. *Biomacromolecules* 8, 3487–3492. https://doi.org/10.1021/bm700646f.

[48] Kim, S. W., Yun, E. Y., Choi, K.-H., Kim, S. R., Park, S. W., Kang, S. W., et al. (2012). Construction of fluorescent red silk using fibroin H-chain expression system. *J. SericEntomol. Sci.* 50:8792.

[49] Kim, S. W., Yun, E. Y., Choi, K.-H., Kim, S. R., Park, S. W., Kang, S. W., et al. (2013). Production of fluorescent green silk using fibroin H-chain expression system. *J. Seric Entomol. Sci.* 51, 153–158. https://doi.org/10.7852/jses.2013.51.2.153.

[50] Kim, S. R., Park, S. W., Kang, S. W., et al. (2014). Expression of the blue fluorescent protein in fibroin H-chain of transgenic silkworm. *J. SericEntomol. Sci.* 52, 25–32. https://doi.org/10.7852/jses.2014.52.1.25.

[51] Aguilar, F., Crebelli, R., Domenico, D.D., Dusemund, B., Frutos, M.J., Galtier, P., Gott, D., Herman, L. Safety of A Change in specifications for the food additive hydroxypropyl methyl cellulose (E 464), *EFSA J*13 (5) (2016), https://doi.org/10.2903/j.efsa.2015.4088, 4088-4088.

[52] Burdock, G. A. Safety assessment of hydroxypropyl methylcellulose as A food ingredient, *Food Chem. Toxicol* 45 (12) (2007) 2341–2351, https://doi.org/10.1016/j.fct.2007.07.011.

[53] B. Li, R. Hu, Z. Cheng, J. Cheng, Y. Xie, S. Gui, Q. Sun, X. Sang, X. Gong, Y. Cui, W. Shen, F. Hong, Titanium dioxide nanoparticles relieve biochemical dysfunctions of fifth-instar larvae of silkworms following exposure to phoxim insecticide, *Chemosphere* 89 (5) (2012) 609–614, https://doi.org/10.1016/j.chemosphere.2012.05.061.

[54] Xue, B., F. Li, J. Hu, J. Tian, J. Li, X. Cheng, J. Hu, B. Li, Differentially expressed genes in the silk gland of silkworm (Bombyxmori) treated with TiO2 NPs, *Gene* 611 (5) (2017) 21–26, https://doi.org/10.1016/j.gene.2017.02.022.

[55] J.Liang et al. *Composites Science and Technology* 215 (2021) 109025.

[56] Rockwood, D. N., Preda, R. C., Yucel, T., Wang, X., Lovett, M. L., and Kaplan, D. L. (2011). Materials fabrication from Bombyxmori silk fibroin. *Nat. Protoc.* 6, 1612–1631. https://doi.org/10.1038/nprot.2011.379.

[57] Belanger, K., Schlatter, G., Hébraud, A., Marin, F., Testelin, S., Dakpé, S., et al. (2018). A multi-layered nerve guidance conduit design adapted to facilitate surgical implantation. *Heal. Sci. Reports* 1:e86. https://doi.org/10.1002/hsr2.86.

[58] Yukseloglu, S. M., Sokmen, N., and Canoglu, S. (2015). Biomaterial applications of silk fibroin electrospunnanofibres. *Microelectron. Eng.* 146, 43–47. https://doi.org/10.1016/j.mee.2015.04.008.

[59] Jia, M., Chen, Z., Guo, Y., Chen, X. & Zhao, X. Efficacy of silk fibroin–nano silver against staphylococcus aureus biofilms in a rabbit model of sinusitis. *Int. J. Nanomedicine* 12, 2933–2939 (2017).

[60] Zhou, H. et al. *In situ* decoration of Ag@AgCl nanoparticles on polyurethane/silk fibroin composite porous films for photocatalytic and antibacterial applications. *Eur. Polym. J.* 118, 153–162 (2019).

[61] Hu, C. et al. Berberine / Ag nanoparticle embedded biomimetic calcium phosphate scaff olds for enhancing antibacterial function. *Nanotechnology Reviews*, 568–579 (2020).

[62] Raho, R. et al. Photo-assisted green synthesis of silver doped silk fibroin/carboxymethyl cellulose nanocomposite hydrogels for biomedical applications. *Mater. Sci. Eng.* C 107, 110219 (2020).

[63] Lu, Z., Meng, M., Jiang, Y. & Xie, J. UV-assisted *in situ* synthesis of silver nanoparticles on silk fibers for antibacterial applications. Colloids Surfaces A Physicochem. *Eng. Asp.* 447, 1–7 (2014).

[64] Ribeiro, M. et al. Antibacterial silk fibroin/nanohydroxyapatite hydrogels with silver and gold nanoparticles for bone regeneration. Nanomedicine Nanotechnology, *Biol. Med.* 13, 231–239 (2017).

[65] Fei, X. et al. Green synthesis of silk fibroin-silver nanoparticle composites with effective antibacterial and biofilm-disrupting properties. *Biomacromolecules* 14, 4483–4488 (2013).

[66] Patil, S., George, T. &Mahadik, K. Green synthesized nanosilver loaded silk fibroin gel for enhanced wound healing. *J. Drug Deliv. Sci. Technol.* 30, 30–36 (2015).

[67] Shahid, M., Cheng, X. W., Tang, R. C. & Chen, G. Silk functionalization by caffeic acid assisted in-situ generation of silver nanoparticles. *Dye. Pigment.* 137, 277–283 (2017). Chapter I: 74.

[68] Dhas, S. P., Anbarasan, S., Mukherjee, A. &Chandrasekaran, N. Biobased silver nanocolloid coating on silk fibers for prevention of post-surgical wound infections. *Int. J. Nanomedicine* 10, 159–170 (2015).

[69] Zhou, Y. & Tang, R. C. Facile and eco-friendly fabrication of AgNPs coated silk for antibacterial and antioxidant textiles using honeysuckle extract. *J. Photochem. Photobiol. B Biol.* 178, 463–471 (2018).

[70] Zhou, Y. & Tang, R. C. Facile and eco-friendly fabrication of colored and bioactive silk materials using silver nanoparticles synthesized by two flavonoids. *Polymers (Basel).* 10, (2018).

[71] Baygar, T., Sarac, N., Ugur, A. &Karaca, I. R. Antimicrobial characteristics and biocompatibility of the surgical sutures coated with biosynthesized silver nanoparticles. *Bioorg. Chem.* 86, 254–258 (2019).

[72] Meng, M. et al. Controllable *in situ* synthesis of silver nanoparticles on multilayeredfilmcoated silk fibers for antibacterial application. *J. Colloid Interface Sci.* 461, 369–375 (2016).

[73] Gao, A., Chen, H., Hou, A. &Xie, K. Efficient antimicrobial silk composites using synergistic effects of violacein and silver nanoparticles. *Mater. Sci. Eng.* C 103, (2019).

[74] Das, S. & Dhar, B. B. Green synthesis of noble metal nanoparticles using cysteinemodified silk fibroin: Catalysis and antibacterial activity. *RSC Adv.* 4, 46285–46292 (2014).

[75] Yu, W. et al. Preparation of Ag nanoparticle dispersed silk fibroin compact. *Appl. Surf. Sci.* 262, 212–217 (2012).

[76] Zhang, G., Liu, Y., Gao, X. & Chen, Y. Synthesis of silver nanoparticles and antibacterial property of silk fabrics treated by silver nanoparticles. *Nanoscale Res. Lett.* 9, 1–8 (2014).

[77] Yu, K. et al. *In situ* assembly of Ag nanoparticles (AgNPs) on porous silkworm cocoon based would film: Enhanced antimicrobial and wound healing activity. *Sci. Rep.* 7, 1–13 (2017).

[78] Zhou, L. et al. Minimizing antibiotic dosage through *in situ* formation of gold nanoparticles across antibacterial wound dressings: A facile approach using silk fabric as the base substrate. *J. Clean. Prod.* 243, 118604 (2020).

[79] Karthikeyan, K. et al. Fabrication of novel biofibers by coating silk fibroin with chitosan impregnated with silver nanoparticles. *J. Mater. Sci. Mater. Med.* 22, 2721–2726 (2011).

[80] Gulrajani, M. L., Gupta, D., Periyasamy, S. &Muthu, S. G. Preparation and application of silver nanoparticles on silk for imparting antimicrobial properties. *J. Appl. Polym. Sci.* 108, 614–623 (2008).

[81] Tang, B., Sun, L., Kaur, J., Yu, Y. & Wang, X. In-situ synthesis of gold nanoparticles for multifunctionalization of silk fabrics. *Dye. Pigment.* 103, 183–190 (2014).

[82] Zou, F. et al. Functionalization of silk with in-situ synthesized platinum nanoparticles. *Materials* (Basel). 11, 1–13 (2018).

[83] Abbasi, A. R., Noori, N., Azadbakht, A. &Bafarani, M. Dense coating of surface mounted Cu_2O nanoparticles upon silk fibers under ultrasound irradiation with antibacterial activity. *J. Iran. Chem. Soc.* 13, 1273–1281 (2016). Chapter I Chapter II Chapter III: Antibacterial Chapter III: Magnetic properties Chapter III: Depollution State of the art 75.

[84] Patil, P. P., Meshram, J. V, Bohara, R. A., Nanaware, S. G. &Pawar, S. H. ZnO nanoparticle-embedded silk fibroin-polyvinyl alcohol composite film: A potential dressing material for infected wounds. *New J. Chem.* 42, 14620–14629 (2018).

[85] Lu, Z. et al. Fabrication of CeO_2 nanoparticle-modified silk for UV protection and antibacterial applications. *J. Colloid Interface Sci.* 435, 8–14 (2014).

[86] Chung, S., Ercan, B., Roy, A. K. & Webster, T. J. Addition of selenium nanoparticles to electrospun silk scaffold improves the mammalian cell activity while reducing bacterial growth. *Front. Physiol.* 7, 1–6 (2016).

[87] Murugesan, P.; Moses, J.A.; Anandharamakrishnan, C. Photocatalytic disinfection efficiency of 2D structure graphitic carbon nitride-based nanocomposites: A review. *J. Mater. Sci.* 2019, 54, 12206–12235.

[88] Rtimi, S.; Sanjines, R.; Andrzejczuk, M.; Pulgarin, C.; Kulik, A.; Kiwi, J. Innovative transparent non-scattering TiO_2 bactericide thin films inducing increased E. coli cell wall fluidity. *Surf. Coat. Technol.* 2014, 254, 333–343.

[89] Djuriši´c, A.B.; Leung, Y.H.; Ng, A.M.C.; Xu, X.Y.; Lee, P.K.H.; Degger, N.; Wu, R.S.S. Toxicity of metal oxide nanoparticles: Mechanisms, characterization, and avoiding experimental artefacts. *Small* 2015, 11, 26–44.

[90] Kim, J.Y.; Lee, C.; Cho, M.; Yoon, J. Enhanced inactivation of E. coli and MS-2 phage by silver ions combined with UV-A and visible light irradiation. *Water Res.* 2008, 42, 356–362.

[91] Xia, D.; Hu, L.; Wang, Y.; Bohong, X. Immobilization of facet-engineered Ag_3PO_4 on mesoporous Al_2O_3 for efficient industrial waste gas purification with indoor LED illumination. *Appl. Catal. B Environ.* 2019, 256, 117811.

[92] Zhao, H.N.; Guan, X.Y.; Zhang, F. Rational design of a bismuth oxyiodide (Bi/BiO1 - xI) catalyst for synergistic photothermal and photocatalytic inactivation of pathogenic bacteria in water. *J. Mater. Sci.* 2021, 100, 110–119.

[93] Raut, A.; Yadav, H.; Gnanamani, A.; Pushpavanam, S.; Pawar, S. Synthesis and characterization of chitosan-TiO2: Cu nanocomposite and their enhanced antimicrobial activity with visible light. *Colloids Surf. B Biointerfaces* 2016, 148, 566–575.

[94] Aziz, S., Sabzi, M., Fattahi, A. &Arkan, E. Electrospun silk fibroin/PAN double-layer nanofibrous membranes containing polyaniline/TiO2 nanoparticles for anionic dye removal. *J. Polym. Res.* 24, 0–6 (2017).

[95] Liu, H., Wang, Z., Li, H., Wang, H. & Yu, R. Controlled synthesis of silkworm cocoonlikeα-Fe2O3 and its adsorptive properties for organic dyes and Cr (VI). *Mater. Res. Bull.* 100, 302–307 (2018).

[96] Kim, J. W., Ki, C. S., Um, I. C. & Park, Y. H. A facile fabrication method and the boosted adsorption and photodegradation activity of CuO nanoparticles synthesized using a silk fibroin template. *J. Ind. Eng. Chem.* 56, 335–341 (2017).

[97] Fernández-García, L. et al. Safety and tolerability of silk fibroin hydrogels implanted into the mouse brain. *ActaBiomater.* 45, 262–275 (2016).

[98] Ling, S. et al. Design and function of biomimetic multilayer water purification membranes. *Sci. Adv.* 3, 1–12 (2017).

[99] Maleki, H., Montes, S., Hayati-Roodbari, N., Putz, F. &Huesing, N. Compressible, Thermally Insulating, and Fire Retardant Aerogels through Self-Assembling Silk Fibroin Biopolymers Inside a Silica Structure - An Approach towards 3D Printing of Aerogels. *ACS Appl. Mater. Interfaces* 10, 22718–22730 (2018).

[100] Wöltje, M.; Kölbel, A.; Aibibu, D.; Cherif, C. A Fast and Reliable Process to Fabricate Regenerated Silk Fibroin Solution from Degummed Silk in 4 Hours. *Int. J. Mol. Sci.* 2021, 22, 10565. https://doi.org/10.3390/ijms221910565.

Index

C

F

G

H

M

N

Q

R

S

T

U

V

W

X

Y

Z

About the Editor

Prof. Dr. Shivaji H. Pawar

Professor Shivaji H. Pawar was born at Wasud in Solapur District, India in 1946. He received his Ph.D. degree from Shivaji University, Kolhapur in Solid State Physics under the guidance of Dr A.V. Narlikar in 1974. He started his research and academic career in Shivaji University and became Professor & Head of the Physics Department in 1995. He established "School of Energy Science" in 1985 & worked as coordinator of this school until his retirement in 2006. Professor Pawar then moved to D. Y. Patil University, Kolhapur as Vice-Chancellor and worked for two successive terms until 2016. There he established "Centre for Interdisciplinary Research" with a focus on Nanoscience and Nanotechnology and provided a platform for multi-faculty interdisciplinary research that was utilized by scientists, engineers and doctors. Later on, he shifted to the Sinhgad Institute, working as a director for the Center for Research & Technology Development (CRTD). There he initiated the research work on Atmospheric Nanoscience & Cloud Precipitation.

Presently, Professor Pawar is working as Director at the Centre for Innovative and Applied Research (CIAR), Anekant Education Society, in Baramati, India. He has successfully guided 75 research scholars leading to their Ph.D. degrees; 50 in Shivaji University and 25 in D.Y. Patil University, Kolhapur. He has published more than 800 research papers, contributed to 40 reviews, edited 15 books and obtained 12 patents. He has worked as visiting professor at many universities worldwide including Cambridge University in the UK.

By virtue of his scientific achievements, Professor Pawar was rewarded with the position of 'Emeritus Scientist' by CSIR, New Delhi in 2007, 'Distinguished Professor' by D.Y. Patil University, Kolhapur in 2016 and 'Professor of Eminence' by Sinhgad Institute, Solapur in 2018.

His research interest includes solid state physics, materials science, energy science, nanoscience, nanomedicine, atmospheric nanoscience and agricultural nanotechnology.